PARTICLE PHYSICS
at the Year of Light

Faculty of Physics of Moscow State University

INTERREGIONAL CENTRE
FOR ADVANCED STUDIES

Proceedings of the Seventeenth Lomonosov
Conference on Elementary Particle Physics

PARTICLE PHYSICS
at the Year of Light

Moscow, Russia **20 – 26 August 2015**

Editor

Alexander I. Studenikin

Department of Theoretical Physics
Moscow State University, Russia
and
Joint Institute for Nuclear Research (Dubna), Russia

World Scientific

NEW JERSEY • LONDON • SINGAPORE • BEIJING • SHANGHAI • HONG KONG • TAIPEI • CHENNAI

Published by

World Scientific Publishing Co. Pte. Ltd.

5 Toh Tuck Link, Singapore 596224

USA office: 27 Warren Street, Suite 401-402, Hackensack, NJ 07601

UK office: 57 Shelton Street, Covent Garden, London WC2H 9HE

British Library Cataloguing-in-Publication Data
A catalogue record for this book is available from the British Library.

PARTICLE PHYSICS AT THE YEAR OF LIGHT
Proceedings of the Seventeenth Lomonosov Conference on Elementary Particle Physics

ISBN 978-981-3224-55-1

Desk Editor: Christopher Teo

Typeset by Stallion Press
Email: enquiries@stallionpress.com

Printed in Singapore

Moscow State University
Faculty of Physics
Interregional Centre for Advanced Studies

SEVENTEENTH LOMONOSOV CONFERENCE ON ELEMENTARY PARTICLE PHYSICS

Moscow, August 20-26, 2015

Mikhail Lomonosov
1711-1765

Sponsors

Russian Ministry of Education and Science
Russian Foundation for Basic Research
Interregional Centre for Advanced Studies

Supporting Institutions

Faculty of Physics of Moscow State University
Joint Institute for Nuclear Research (Dubna)
Institite for Nuclear Research of Russian Academy of Sciences
Department of Science and Technology of
Ministry of Education and Science of Russia

International Advisory Committee

Organizing Committee

Moscow State University
Interregional Centre for Advanced Studies

ELEVENTH INTERNATIONAL MEETING ON PROBLEMS OF INTELLIGENTSIA

"The INTELLIGENTSIA and CULTURE"

Moscow, August 26, 2015

Presidium of the Meeting

J.Bleimaier (Princeton)
A.A.Fedyanin (MSU)
A.I.Studenikin (MSU, JINR & ICAS)
N.N.Sysoev (MSU)

FOREWORD

The 17th Lomonosov Conference on Elementary Particle Physics was held at the Moscow State University (Moscow, Russia) on August 20–26, 2015 under the patronage of Victor Sadovnichy, the Rector of the university.

The conference was organized by the Faculty of Physics of the Moscow State University in cooperation with the Joint Institute for Nuclear Research (Dubna) and Institute for Nuclear Research of the Russian Academy of Sciences.

The conference was sponsored by the Ministry of Education and Science of Russia and the Russian Foundation for Basic Research. A reasonable support in organizing the event was provided by the Department of Science and Technology of the Ministry of Education and Science of Russia, the Interregional Centre for Advanced Studies, and the Bruno Pontecorvo Neutrino and Astrophysics Laboratory (MSU).

As early as in 1983 the first of the series of conferences (from 1992 called the "Lomonosov Conferences"), was held at the Department of Theoretical Physics of the Moscow State University (June 1983, Moscow). The second conference was held in Kishinev, Republic of Moldavia, USSR (May 1985).

After the four years break this series was resumed on a new conceptual basis for the conference programme focus. During the preparation of the third conference (that was held in Maykop, Russia, 1989) a desire to broaden the programme to include more general issues in particle physics became apparent. During the conference of the year 1992 held in Yaroslavl (Russia) it was proposed by myself and approved by numerous participants that these irregularly held meetings should be transformed into regular events under the title "Lomonosov Conferences on Elementary Particle Physics".

Since 1993 the conference have been organized on a regular basis each odd year under the title the Lomonosov Conference on Elementary Particle Physics.

Mikhail Lomonosov (1711–1765) is a brilliant Russian encyclopaedias of the era of the Russian Empress Catherine the 2nd, was world renowned for his distinguished contributions in the fields of science and art. He also helped establish the high school educational system in Russia. The Moscow State University was founded in 1755 based on his plan and initiative, and the University now bears the name of Lomonosov.

Since 1993 at the subsequent meetings of this series a wide variety of interesting things, both in theory and experiment of particle physics, astrophysics, gravitation and cosmology, were included into the programmes. It was also decided to enlarge the number of institutions that would take part in preparation of the events.

The 18th Lomonosov Conference will be held at the Moscow State University on August 24–30, 2017. The twenty-fifth anniversary since foundation of the Lomonosov conferences will be celebrated during the conference in the year 2017.

The 6th Lomonosov Conference on Elementary Particle Physics (1993) and all of the subsequent conferences of this series were held at the Moscow State University on each of the odd years. There is a tradition to publish the conferences proceedings as a collection of review and research papers prepared by the conference speakers on the presented talks basis. It started in 1994 when publication of the volume "Particle Physics, Gauge Fields and Astrophysics" containing articles written by speakers of the 5th and 6th Lomonosov Conferences was supported by the Accademia Nazionale dei Lincei (Rome, 1994). Proceedings of the 7th and 8th Lomonosov Conference (entitled "Problems of Fundamental Physics" and "Elementary Particle Physics") were published by the Interregional Centre for Advanced Studies (Moscow, 1997 and 1999). Proceedings of the 9th, 10th, 11th, 12th, 13th, 14th, 15th and 16th Lomonosov Conferences (entitled "Particle Physics at the Start of the New Millennium", "Frontiers of Particle Physics", "Particle Physics in Laboratory, Space and Universe", "Particle Physics at the Year of 250th Anniversary of Moscow University", "Particle Physics on the Eve of LHC", "Particle Physics at the Year of Astronomy", "Particle Physics in the Year of Tercentenary of Mikhail Lomonosov", "Particle Physics in the Year of Centenary of Bruno Pontecorvo") were published by World Scientific Publishing Co. (Singapore) in 2001, 2003, 2005, 2006, 2009, 2011, 2013 and 2015, respectively.

The physics programme of the 17th Lomonosov Conference has included review and original talks on a wide range of items such as neutrino and astroparticle physics, electroweak theory, fundamental symmetries, tests of Standard Model and beyond, heavy quark physics, non-perturbative QCD, quantum gravity effects, physics at the future accelerator. The year 2015 was claimed by the United Nations the Year of Light. That year spans several major scientific anniversaries to be celebrated, in particular, of the electromagnetic theory of light propagation proposed by Maxwell (1865), the Einstein's theory of the photoelectric effect (1905) and general relativity (1915), and the discovery of the cosmic microwave background by Penzias and Wilson (1965). The Year of Light was celebrated at the 17th Lomonosov Conference with a special Round Table discussion on "Particle Physics in the Year of Light: from Maxwell's equations to physics beyond the Standard Model" held during the closing day of the conference.

In total there were more than 350 participants with 180 talks including 97 plenary (25–20 min) talks, 83 session (20–10 min) talks, 36 brief (15–10 min) reports and 8 poster presentations. One of the goals of the conference was to bring together scientists, both theoreticians and experimentalists, working in different fields. The conference has produced a high interest in the scientific community and there were much more applications to present a talk at the conference from scientist all over the world. The amount of applications reasonably surpassed the capacity to allocate oral talks within the six working days of the conference, therefor 13 parallel sessions were organized at the conference.

Following the tradition that has started in 1995, each of the Lomonosov Conferences on particle physics has been accompanied by a conference on problems of intellectuals. The 11th International Meeting on Problems of Intelligentsia held during the 17th Lomonosov Conference (August 26, 2015) was dedicated to discussions on the issue "The Intelligentsia and Culture".

The success of the 17th Lomonosov Conference was due in a large part to contributions of the International Advisory Committee and Organizing Committee. On behalf of these Committees I would like to warmly thank the session chairpersons, the speakers and all of the participants of the 17th Lomonosov Conference and the 11th International Meeting on Problems of Intelligentsia.

We are grateful to Victor Sadovnichy, the Rector of the Moscow State University, Victor Matveev, the Director of the Joint Institute for Nuclear Research, Leonid Kravchuk, the Director of Institute for Nuclear Research of Russian Academy of Sciences, Mikhail Itkis, the Vice Director of JINR, and Anatoly Kozar, the Vice Dean of the Faculty of Physics of the Moscow State University, for the support in organizing these two conferences.

We extremely appreciate continuous support of the Lomonosov Conferences from Nikolay Sysoev, the Dean of Faculty of Physics of Moscow State University.

I would like to thank Fabio Anulli, Paolo Bernardini, Riccardo Brugnera, Anatael Cabrera, Dmitry Denisov, Guido Drexlin, Arnaud Ferrari, Nicolao Fornengo, Gianluca Gemme, Simone Giani, Adalberto Giazotto, Carlo Giunti, Gabriela Gonzalez, Maury Goodman, Karl-Heinz Kampert, Herve de Kerret, Takashi Kobayashi, Yury Kudenko, Luca Latronico, Cristina Lazzeroni, Olivier Leroy, Jukka Maalampi, Daniel McKinsey, Benoit Mours, Masayuki Nakahata, Laura Patrizi, Sergey Petrushanko, Piergiorgio Picozza, Leo Piilonen, Gioacchino Ranucci, Stefan Schmitt, Kate Scholberg, Christian Spiering, Chris Sutton, Stefania Vecchi, Ermanno Vercelli, Guy Wilkinson, Matthew Wing, Yuji Yamazaki, Frank Zimmermann, Kai Zuber, for their help in planning of the scientific programme of the meeting and inviting speakers for the topical sessions of the conference.

We are also very thankful to Dr Christine Sutton, the Editor of "CERN Courier", for providing informational support of the event.

Furthermore, I am very pleased to mention Ilya Balantsev, Oleg Kharlanov, Alexey Lokhov, Alexander Kazantsev and Yana Zhezher, the Scientific Secretaries of the conference. I am also very thankful to Valentina Samsonova, Darya Grechukhina and Mikhail Poplavsky, Vikenty Shakhmatov and Mikhail

Skoptsov for their very efficient work in preparing and running the meeting, and also Alexander Grigoriev and Andrey Egorov for their contribution. I would also acknowledge Vladimir Arkhipov who made video records of the scientific sessions of the conference and interviews with the plenary speakers that are on view at the conference web site.

These Proceedings were prepared for publication at the Interregional Centre for Advanced Studies with support by the Russian Foundation for Basic Research and the Russian Ministry of Education and Science.

Alexander Studenikin

CONTENTS

RELATIVITY AND LIGHT

Vladimir Petrov [a]
Institute for High Energy Physics, NRC "Kurchatov Institute", 142281 Protvino, Russia

Abstract. A brief history of the genesis of the relativity theory with account of the role of light in it is discussed.

1 Introduction

The year 2015 is not only the International Year of Light (whatever it would mean) but also the 110th anniversary of the formulation of the relativity theory. In my talk I will deal with both relativity and light. I am fully aware that much of what I'm going to tell is well known to this audience. I, however, hope that some details and accents will be new and interesting.

2 Short history of relativity

One of the most striking features of the famous Maxwell equations (1862) was the mysterious relationship of the electromagnetic properties of the medium - magnetic permeability μ and electric permittivity ε - with the velocity c of the wave propagation in this medium (including the specific medium - vacuum or "ether")

$$c = \frac{1}{\sqrt{\varepsilon\mu}}.$$

Thus the velocity of light became an integral part of electromagnetism. Later it was discovered that Maxwell equations are invariant under transformations relating inertial frames moving uniformly and rectilinearly with respect to each other. This was the starting point to formulate the relativity theory.

However, much earlier, since ancient Greeks, the notion of relativity was already noted by several thinkers. The most known was the proposal clearly formulated by the great Italian Galileo Galilei in his book "Dialogo sopra i due massimi sistemi del mondo" (1632) [1] in which he considered a ship moving uniformly and rectilinearly over a quiet water surface: "***...have the ship proceed with any speed you like, so long as the motion is uniform and not fluctuating this way and that. You will discover not the least change in all the effects named, nor could you tell from any of them whether the ship was moving or standing still***". Under "effects named" he meant actually any physical processes known to him. Quite often historians of physics insist that Galileo formulated this relativity principles for mechanical phehomena only but he evidently didn't impose such a restriction in his writings.

[a]E-mail: `vladimir.petrov@ihep.ru`

Figure 1: From left to right: Maxwell, Fitzgerald, Larmor, Lorentz.

As to the light, already Maxwell mentioned that the movement which was thought as an ”absolute” one through the ether could influence electromagnetic and optical phenomena due to the ”ether wind”. Among early investigators of this problem we should mention a French physicist E. Mascart who studied the Doppler effects in moving media and here is his conclusion(1872) [2]: ”***The general conclusion of this memoir would be that the translational movement of the earth has no appreciable influence on optical phenomena produced from a terrestrial source or solar light, that these phenomena do not give any means of measuring the absolute motion of a body, and that relative movements are the only we can measure***.” This conclusion was confirmed later (though without reference) by an American experimentalist in the field of optics, A. Michelson. In his two dedicated experiments (the second in collaboration with E. Morley) both measurements have shown the negative result. No influence of the Earth's movement on the light propagation was observed. The second half of the XIXth century was marked by active studies of questions related to the ether and its properties w.r.t electromagnetic phenomena. An Irish professor of physics from Dublin G.F. Fitzgerald published in 1889 an extremely short and without a single formula note [3] where he argued, evidently being inspired by some early results by O. Heavyside [4]on electromagnetic fields of moving charges, that all bodies in movement should be shortened in the direction of movement exactly in such a proportion which enables to explain the negative output of the Michelson-Morley experiment. So if the length of the body in rest is L_0 then the visible length of the same body when moving with velocity v reads $L = L_0\sqrt{1-(v/c)^2}$. Here a special role of the light velocity is seen: it is a limiting velocity for any physical body.To the extent we are interested in purely mechanical properties this can seem strange if not to dwell upon the details of the body inner structure.

So much about the spatial coordinates, but what about time? Since the Newton famous ”Principia” time was thought to be a universal entity, ”absolute time” which conforms well with our everyday experience to mark the flow of events. In 1887 W. Voigt, a German physicist, tried to find what we could call now an invariance group of the wave equation with respect to change of inertial frame. Here are his transformations [5]:

$$x' = x - vt, y' = y\sqrt{1-(v/c)^2}, z' = z\sqrt{1-(v/c)^2}, t' = t - vx/c^2.$$

These transformations do not leave the d'Alembert operator invariant and do not form a group, though the very wave equation remains the same. Sure we immediately recognize the half-stuff of what later will be called Lorentz transformations. We have, nonetheless, to pay a tribute to Prof. W. Voigt because he was the first in the history of physics as analytic science who dared to shake the absoluteness of time, though unconsciously. 10 year ago, in 1897, an Irish physicist J. Larmor has found (though in quite an indirect form) the correct transformations which will be unfairly called in 1905 "Lorentz transformations". Moreover, it was Larmor who first noticed the time dilation :"***...individual electrons describe corresponding parts of their orbits in times shorter for the (rest)system***" [6].

In the last quarter of the XIXth century - the beginning of the XXth, a Hollander Hendrik Antoon Lorentz, who was already reputed as the author of a universal electromagnetic theory of matter , the "electron theory", was maybe the most active researcher in the domain of problems related to the electrodynamics of moving bodies. One of his main concerns was to reconcile the existence of the "world ether" with impossibility to detect its influence via electromagnetic phenomena. In his seminal paper(1904)"Electromagnetic phenomena in a system moving with any velocity smaller than that of light" [7] he almost achieved the symmetry transformations for fields and matter arguing that forces of Nature are so designed that any effect of the ether is compensated completely. During his laborious work Lorentz had a vivid correspondence with a famous French physicist, mathematician and philosopher of science Henri Poincaré who, in parallel with fantastic breakthroughs in mathematics, paid a lot of interest to physics. Having carefully and encouragingly monitoring the progress in the Lorentz research, he managed not only to correct mistakes in the Lorentz papers but came himself to a new and much more general vision of the problem which he summarized as the "relativity principle" [8]: **" *The laws of physical phenomena must be the same for a fixed observer and for an observer in rectilinear and uniform motion so that we have no possibility of perceiving whether or not we are dragged in such a motion.*"** He did this in 1904. In June 1905 he recorded his presentation at the French Academy of Sciences in the form of a short note [9] where all formal basis of the relativity theory was given. By a strange coincidence, three weeks ago a young researcher Albert Einstein from Bern (near 200 km from Paris) sent to "Annalen der Physik" a paper [10] where a new framework for physical theories was proposed which was based on two postulates: **" *1. The laws by which the states of physical systems undergo change are not affected, whether these changes of state be referred to the one or the other of***

Figure 2: From left to right: Poincaré, Einstein, Minkowski.

two systems of co-ordinates in uniform translatory motion. 2.Any ray of light moves in the stationary system of co-ordinates with the determined velocity* c *, whether the ray be emitted by a stationary or by a moving body."

Though I present the history of relativity rather cursorily, it would be unfair not to mention a prominent German mathematician Hermann Minkowski. His celebrated citation (1909) is [11]: "...***The views of space and time which I wish to lay before you have sprung from the soil of experimental physics, and therein lies their strength. They are radical. Henceforth space by itself, and time by itself, are doomed to fade away into mere shadows, and only a kind of union of the two will preserve an independent reality***."

The first who mentioned that relativity principle implies that space and time actually form a unique geometrical 4D manifold was Poincaré. However, in the most complete form this thesis was developed in works of Minkowski who introduced such basic notions as " Minkowski space", "4-vector", "space-like" and "time-like", "light cone", relativistic "interval between events" and gave the first application of the tensor calculus to electrodynamics, the achievement we use up to now, especially in the theory of gravitation.

I wouldn't like to distribute priorities in creation of the relativity theory. If somebody is interested, I could only advise to read attentively the original papers and to draw his own conclusions. Instead I prefer to cite from H. Poincaré's book "The Value of Science"(1912):

" ***Science...is a collective creative work, and it cannot be anything else; it is like a monumental construction that has to be constructed for centuries, and where everybody must bring a stone to, and this stone can cost him a whole life. Hence, it gives us a feeling of a necessary cooperation, solidarity of our labour with the labour of our contemporaries, our predecessors and our followers***".

3 Relativity: Russian connection

If we compare the basis of the relativity theory in Poincaré's and Einstein's expositions we immediately observe that Poincaré refers to " the laws of physical phenomena" in general while Einstein heavily relies on the properties of light related to some of the natural forces, i.e. electromagnetism. It seemingly puts the latter in a privileged position with respect to other forces. This doesn't look to fit very well into the notion of relativity as a universal property of Nature. Probably, such thoughts could come to the mind of a Russian physicist, Vladimir Sergeyevitch Ignatowski, who worked, after having graduated from the University of Giessen, in the Berlin Hochtechnische Schule as professor and studied there diffraction phenomena till the WWI. Here are his motivations:

" ***I've asked myself the question, at which relations or transformation equations one arrives when only the relativity principle is placed at the top of the investigation, and whether the Lorentzian transformation equations are the only ones at all, that satisfy the relativity principle***".

In his paper published in "Physikalische Zeitschrift" [12] and then in a series of more detailed papers (see, e.g. [13]) Ignatowski derived the most general transformations which leave physical phenomena invariant being based on the relativity principle proper, isotropy and homogeneity of space and the natural condition that transformations in question form a group. His reasonings resulted in the following transformations:

$$x' = (x - vt)/\sqrt{1 - v^2 n}, y' = y, z' = z, t' = t - vnx/\sqrt{1 - v^2 n}$$

where n is some a priori unknown - both in magnitude and sign - "world constant" with trivial limits $-\infty < n < \frac{1}{v^2}$. The sign of n can be fixed naturally if to take the expression for the energy of a free point mass m: $E = m/n\sqrt{1 - v^2 n}$. From the positivity of the energy of a free material point follows the positivity of the constant n. From the real-valuedness of energy follows that the quantity $c \doteq \sqrt{\frac{1}{n}}$ is a *universal upper limit for any physical velocity.* Thus from the very general premises Ignatowski *derived* that no physical process can proceed with a velocity more than some universal maximum value.The fact that the constant $\sqrt{\frac{1}{n}}$ appeared to coincide with the velocity of light appeared a pure chance only and is incited by the fact -as we now understand - that the quanta of electromagnetic field are massless. Just imagine that Maxwell equations would correspond to electromagnetic fields of a finite range as it takes place for $W^{\pm}$ and Z bosons. In this case the second Einstein's postulate would be absolutely impossible. So the Ignatowski formulation of the relativity theory has shown that Einstein's second postulate is a consequence of the relativity principle (which is easily seen from the relativistic addition law for velocities) and so is superfluous. It is regretful that an extremely important contribution made

Figure 3: Ignatowski.

by V. S. Ignatowsky to the very foundations of the relativity theory is almost unknown in Russia. In this respect we should celebrate 2015 not only as the International Year of Light and the 110th anniversary of the relativity theory but also as the 140th anniversary of outstanding Russian physicist Vladimir Sergeyevitch Ignatowski, a man of exclusive scientific talent and, unfortunately, of a tragic fate: he was executed in 1942 in besieged Leningrad on false charges of spying for Germany. Rehabilitated posthumously in 1955 . I can only cherish a hope that I was able at least to minimally pay him the due tribute.

4 A little more about the speed of light and the theory of relativity

Sometimes in literature one can stumble some misunderstanding about inertial and non-inertial frames and limiting velocity c. Sometimes one can meet the statement that the relativity theory doesn't work in non-inertial frames where we need to include the gravitational forces for consistency. I believe it would be not without interest and benefit to add some considerations on this issue.In what follows I will keep close to A. A. Logunov's reasonings [14].

In fact, as is pretty long known from the few *good* textbooks on relativity theory, the relativity theory can be fairly well formulated in arbitrary coordinate systems. In Galilean coordinates X^μ the interval(squared) has the form:

$$ds^2 = c^2dt^2 - dx^2 - dy^2 - dz^2 = \eta_{\mu\nu}dX^\mu dX^\nu$$

where $\eta_{\mu\nu} = diag(1, -1, -1, -1)$ is the Minkowski metric tensor. Let us choose arbitrary coordinates $x^\alpha(X^\mu)$ such that inverse transformation $X^\alpha(x^\mu)$ exists. In these new coordinates the interval acquires the form

$$ds^2 = \gamma_{\mu\nu}(x)dx^\mu dx^\nu$$

where $\gamma_{\mu\nu}(x) = \frac{\partial X^\alpha}{\partial x^\mu}\frac{\partial X^\beta}{\partial x^\nu}\eta_{\alpha\beta}$. Light propagation corresponds to the isotropic interval $ds^2 = 0$ being the property of isotropy is invariant under general transformations introduced above. The isotropic interval can be cast into the following form:

$$c^2[\sqrt{\gamma_{00}} + \gamma_{0i}dx^i/c\sqrt{\gamma}_{00}]^2 - [-\gamma_{ik} + \gamma_{0i}\gamma_{0k}/\gamma_{00}]dx^i dx^k = 0.$$

The quantity

$$d\tau \doteq \sqrt{\gamma_{00}} + \gamma_{0i}dx^i/c\sqrt{\gamma_{00}} = \gamma_{0\lambda}dx^\lambda/c\sqrt{\gamma_{00}}$$

is to be considered as (infinitesimal) physical time which doesn't depend on the choice of the time coordinates. In non-inertial frames the quantity $d\tau$ is not a total differential. The second term in the above expression for the interval is exactly what should be considered as the distance squared between infinitesimally close points of 3D space which doesn't depend on the choice of coordinates in this space: $dl^2 = \chi_{ik}dx^i dx^k$. Here $\chi_{ik} = -\gamma_{ik} + \gamma_{0i}\gamma_{0k}/\gamma_{00}$ is the 3D metric tensor. Now we have

$$c = \frac{dl}{d\tau}.$$

Quantities dl and $d\tau$ have local character. So in non-inertial frames the very notion of simultaneity for events at different places makes no sense because the synchronization procedure with help of a light signal is impossible as it depends on the path of synchronization. But physical velocity of light remains constant in distinction with the "coordinate velocity of light"

$$\mathbf{c}_{\text{coordinate}} = \frac{d\boldsymbol{x}}{dt} = c_{\text{coordinate}}\mathbf{e}$$

which is related to the physical velocity as follows:

$$c_{\text{coordinate}} = c[\gamma_{00}dt/(\sqrt{\gamma_{00}} - \gamma_{0i}e^i)]$$

and which can acquire any value between 0 and ∞ dependent on the choice of coordinates. Here we see that it would be too restrictive to base the relativity theory on the postulate of the constancy of the velocity of light.

From the very form of the interval follows that the existence of the limiting velocity for any kind of the physical signal transmission is a direct consequence of the pseudo-Euclidean geometry of Minkowski space. Moreover, the very essence of the relativity theory is that we live in Minkowski space with pseudo-Euclidean geometry. This makes the relativity theory universal and valid for any kind of interaction. Nonetheless we have to recognize and properly appreciate the paramount heuristic role of arguments referring to light signals in the genesis of the relativity theory.

Acknowledgments

I would like to express my deep gratitude to Prof. A. I. Studenikin and his colleagues for a kind invitation to take part in the XVIIth Lomonosov Conference.

References

[1] Galileo Galilei, *"Dialogues Concerning Two New Sciences "* (Dover Books on Physics, 1954).
[2] E.Mascart, Ann.Sc.Ec.Norm.Sup. **1**, 157 (1872).
[3] G. F. Fitzgerald, Science **13**, 390 (1889).
[4] O. Heavyside, Electrician **IV**, 458 (1889).

[5] W. Voigt, Göttinger Nachrichten **8**, 177 (1887).
[6] J. Larmor, Phil. Trans. Roy. Soc.**190**, 205 (1897).
[7] H.A. Lorentz, Proceedings of the Royal Netherlands Academy of Arts and Sciences**VI**, 809 (1904).
[8] H.Poincaré, Bulletin des sciences mathématiques **28**, 302 (1904).
[9] H.Poincaré, Comptes rendus hebdomadaires des séances de l'Académie des sciences.**140**,1504 (1905).
[10] A. Einstein, Annalen der Physik **17 (10)**, 891 (1905).
[11] H. Minkowski, Jahresbericht der Deutschen Mathematiker-Vereinigung **1909**, 75 (1909).
[12] W.von Ignatowski, Physikalische Zeitschrift **11**, 972(1910).
[13] W.von Ignatowski, Physikalische Zeitschrift **12**, 779(1911).
[14] A. A. Logunov, *"Henri Poincaré and Relativity Theory"* (Nauka Publishers, Moscow, 2004 (in Russian); English translation: arXiv:physics/0408077).

Neutrino Physics

A REVIEW ON PRESENT STATUS OF NEUTRINO MIXINGS AND OSCILLATIONS

Carlo Giunti [a]
INFN, Sezione di Torino, Via P. Giuria 1, I–10125 Torino, Italy

Abstract. We review the status of standard three-neutrino mixing and the results of a global analysis of short-baseline neutrino oscillation data in the extended framework of 3+1 neutrino mixing with a sterile neutrino at the eV scale.

1 Introduction

Neutrino oscillations have been measured with high accuracy in solar, atmospheric and long-baseline neutrino oscillation experiments. Hence, we know without doubt that neutrinos are massive and mixed particles (see Ref. [1]). In this short review we discuss the status of the standard three-neutrino mixing paradigm (Section 2) and the indications in favor of the existence of an additional sterile neutrino given by anomalies found in some short-baseline neutrino oscillation experiments (Section 3).

2 Three-Neutrino Mixing

Solar neutrino experiments (Homestake, GALLEX/GNO, SAGE, Super-Kamiokande, SNO, Borexino) measured $\nu_e \to \nu_\mu, \nu_\tau$ oscillations generated by the solar squared-mass difference $\Delta m^2_{\mathrm{SOL}} \simeq 7.4 \times 10^{-5}\,\mathrm{eV}^2$ and a mixing angle $\sin^2\vartheta_{\mathrm{SOL}} \simeq 0.3$. The KamLAND experiment confirmed these oscillations by observing the disappearance of reactor $\bar{\nu}_e$ with average energy $\langle E \rangle \simeq 4\,\mathrm{MeV}$ at the average distance $\langle L \rangle \simeq 180\,\mathrm{km}$.

Atmospheric neutrino experiments (Kamiokande, IMB, Super-Kamiokande, Soudan-2, MACRO, MINOS) measured ν_μ and $\bar{\nu}_\mu$ disappearance through oscillations generated by the atmospheric squared-mass difference $\Delta m^2_{\mathrm{ATM}} \simeq 2.5 \times 10^{-3}\,\mathrm{eV}^2$ and a mixing angle $\sin^2\vartheta_{\mathrm{ATM}} \simeq 0.5$. The K2K and MINOS long-baseline experiments confirmed these oscillations by observing the disappearance of accelerator ν_μ with $\langle E \rangle \simeq 1.3\,\mathrm{GeV}$ and $3\,\mathrm{GeV}$ at distances $L \simeq 250\,\mathrm{km}$ and $730\,\mathrm{km}$, respectively.

The Super-Kamiokande atmospheric neutrino data indicate that the disappearance of ν_μ is likely due to $\nu_\mu \to \nu_\tau$ transitions, in agreement with the observation of $\nu_\mu \to \nu_\tau$ events in the OPERA long-baseline accelerator experiment [2].

The two independent solar and atmospheric Δm^2's are nicely accommodated in the standard framework of three-neutrino mixing in which the left-handed components of the three active flavor neutrino fields ν_e, ν_μ, ν_τ are superpositions of three massive neutrino fields ν_1, ν_2, ν_3 with masses m_1, m_2, m_3:

[a] E-mail: giunti@to.infn.it

$\nu_{\alpha L} = \sum_{k=1}^{3} U_{\alpha k}\nu_{kL}$, for $\alpha = e, \mu, \tau$. The unitary mixing matrix can be written in the standard parameterization in terms of three mixing angles ϑ_{12}, ϑ_{23}, ϑ_{13} and a CP-violating phase[b] δ:

$$U = \begin{pmatrix} c_{12}c_{13} & s_{12}c_{13} & s_{13}e^{-i\delta} \\ -s_{12}c_{23} - c_{12}s_{23}s_{13}e^{i\delta} & c_{12}c_{23} - s_{12}s_{23}s_{13}e^{i\delta} & s_{23}c_{13} \\ s_{12}s_{23} - c_{12}c_{23}s_{13}e^{i\delta} & -c_{12}s_{23} - s_{12}c_{23}s_{13}e^{i\delta} & c_{23}c_{13} \end{pmatrix}, \quad (1)$$

where $c_{ab} \equiv \cos\vartheta_{ab}$ and $s_{ab} \equiv \sin\vartheta_{ab}$. It is convenient to choose the numbers labeling the massive neutrinos in order to have

$$\Delta m^2_{\text{SOL}} = \Delta m^2_{21} \ll \Delta m^2_{\text{ATM}} = \frac{1}{2}\left|\Delta m^2_{31} + \Delta m^2_{32}\right|, \quad (2)$$

with $\Delta m^2_{kj} = m^2_k - m^2_j$. Then, there are two possible orderings the neutrino masses: the normal ordering (NO) with $m_1 < m_2 < m_3$ and the inverted ordering (IO) with $m_3 < m_1 < m_2$.

With the conventions in Eqs. (1) and (2), we have $\vartheta_{\text{SOL}} = \vartheta_{12}$ and $\vartheta_{\text{ATM}} = \vartheta_{23}$. The mixing angle ϑ_{13} generates ν_e disappearance and $\nu_\mu \to \nu_e$ transitions driven by Δm^2_{ATM}, which can be observed in long-baseline neutrino oscillation experiments.

In 2011 the T2K experiment reported the first indication of long-baseline $\nu_\mu \to \nu_e$ transitions [3], followed by the MINOS experiment [4]. In 2013, the T2K Collaboration reported a convincing 7.5σ observation of $\nu_\mu \to \nu_e$ transitions through the measurement of 28 ν_e events with an expected background of 4.92 ± 0.55 events [5]. These transitions are confirmed by the first results of the NOνA experiment in which 6 events have been observed in the Far Detector, with a background expectation of 0.99 ± 0.11 events based on the Near Detector measurement [6].

The first strong evidence of a nonzero value of ϑ_{13} was found in 2012 in the Daya Bay experiment by measuring the long-baseline disappearance of reactor $\bar{\nu}_e$'s with a significance of 5.2 standard deviations [7]. This result is compatible with the previous indication found in the Double Chooz reactor experiment [8] and has been confirmed by the data of the RENO reactor experiment [9]. The current most precise measurement of the value of ϑ_{13} is that obtained in the Daya Bay experiment: $\sin^2 2\vartheta_{13} = 0.082 \pm 0.004$ [10]. It is compatible with the latest results of the RENO ($\sin^2 2\vartheta_{13} = 0.082 \pm 0.011$) [11] and Double Chooz ($\sin^2 2\vartheta_{13} = 0.088 \pm 0.033$) [12] experiments. Hence, we have a robust evidence of a non-zero value of ϑ_{13}. It is very important, because it opens promising perspectives for the observation of CP violation in the lepton sector and matter

[b]For simplicity, we do not consider the two Majorana CP-violating phases which contribute to neutrino mixing if massive neutrinos are Majorana particles, because they do not affect neutrino oscillations (see Ref. [1]).

effects in long-baseline oscillation experiments, which could allow to distinguish the normal and inverted neutrino mass hierarchies.

The three-neutrino mixing parameters can be determined with good precision with a global fit of neutrino oscillation data [13–15]. The largest uncertainty is that of ϑ_{23}, which is known to be close to maximal ($\pi/4$), but it is not known if it is smaller or larger than $\pi/4$. For the Dirac CP-violating phase δ, there is an indication in favor of $\delta \approx 3\pi/2$ [6,16], which would give maximal CP violation, but at 3σ all the values of δ are allowed, including the CP-conserving values $\delta = 0, \pi$.

3 Beyond Three-Neutrino Mixing: Sterile Neutrinos

The completeness of the three-neutrino mixing paradigm has been challenged by the following indications in favor of short-baseline neutrino oscillations, which require the existence of at least one additional squared-mass difference $\Delta m^2_{\rm SBL} \gg \Delta m^2_{\rm ATM}$:

1. The LSND experiment, in which a signal of short-baseline $\bar{\nu}_\mu \to \bar{\nu}_e$ oscillations has been observed with a statistical significance of about 3.8σ [17,18].

2. The Gallium neutrino anomaly [19–23], consisting in a short-baseline disappearance of ν_e measured in the Gallium radioactive source experiments GALLEX [24] and SAGE [25] with a statistical significance of about 2.9σ.

3. The reactor antineutrino anomaly [26], which is an about 2.8σ deficit of the rate of $\bar{\nu}_e$ observed in several short-baseline reactor neutrino experiments in comparison with that expected from the calculation of the reactor neutrino fluxes [27,28].

In this review, we consider the 3+1 neutrino mixing scheme in which there is an additional massive neutrinos at the eV scale and the masses of the three standard massive neutrinos are much smaller. Since from the LEP measurement of the invisible width of the Z boson we know that there are only three active neutrinos (see Ref. [1]), in the flavor basis the additional massive neutrino corresponds to a sterile neutrinos, which does not have standard weak interactions.

In the 3+1 scheme, the effective probability of $\nu_\alpha \to \nu_\beta$ transitions in short-baseline experiments has the two-neutrino-like form [29]

$$P_{\nu_\alpha \to \nu_\beta} = \delta_{\alpha\beta} - 4|U_{\alpha 4}|^2 \left(\delta_{\alpha\beta} - |U_{\beta 4}|^2\right) \sin^2\left(\frac{\Delta m^2_{41} L}{4E}\right) , \tag{3}$$

where U is the mixing matrix, L is the source-detector distance, E is the neutrino energy and $\Delta m^2_{41} = m^2_4 - m^2_1 = \Delta m^2_{\rm SBL} \sim 1\,{\rm eV}^2$. The electron and muon neutrino and antineutrino appearance and disappearance in short-baseline experiments depend on $|U_{e4}|^2$ and $|U_{\mu4}|^2$, which determine the amplitude $\sin^2 2\vartheta_{e\mu} = 4|U_{e4}|^2|U_{\mu4}|^2$ of $\nu_\mu \to \nu_e$ transitions, the amplitude $\sin^2 2\vartheta_{ee} = 4|U_{e4}|^2 \left(1 - |U_{e4}|^2\right)$ of ν_e disappearance, and the amplitude $\sin^2 2\vartheta_{\mu\mu} = 4|U_{\mu4}|^2 \left(1 - |U_{\mu4}|^2\right)$ of ν_μ disappearance.

Since the oscillation probabilities of neutrinos and antineutrinos are related by a complex conjugation of the elements of the mixing matrix (see Ref. [1]), the effective probabilities of short-baseline $\nu_\alpha \to \nu_\beta$ and $\bar{\nu}_\alpha \to \bar{\nu}_\beta$ transitions are equal.

Global fits of short-baseline neutrino oscillation data have been presented recently in Refs. [30–33]. These analyses take into account the final results of the MiniBooNE experiment [34], which was made in order to check the LSND signal with about one order of magnitude larger distance (L) and energy (E), but the same order of magnitude for the ratio L/E from which neutrino oscillations depend. Unfortunately, the data of the MiniBooNE experiment do not allow us to confirm or reject the LSND $\bar{\nu}_\mu \to \bar{\nu}_e$ signal because the experiment had too large backgrounds. Moreover, the ν_e and $\bar{\nu}_e$ MiniBooNE data show an excess in the low-energy bins which is widely considered to be anomalous because it is at odds with neutrino oscillations.

In the following we summarize the results of the 3+1 analysis of short-baseline data presented in Ref. [32] of the following three groups of experiments:

(A) The $\overset{(-)}{\nu}_\mu \to \overset{(-)}{\nu}_e$ appearance data of the LSND [18], MiniBooNE [34], BNL-E776 [35], KARMEN [36], NOMAD [37], ICARUS [38] and OPERA [39] experiments [c].

(B) The following $\overset{(-)}{\nu}_e$ disappearance data: 1) the data of the Bugey-4 [41], ROVNO91 [42], Bugey-3 [43], Gosgen [44], ILL [45], Krasnoyarsk [46], Rovno88 [47], SRP [48], Chooz [49], Palo Verde [50], Double Chooz [51], and Daya Bay [52] reactor antineutrino experiments with the new theoretical fluxes [26–28,53]; 2) the data of the GALLEX [24] and SAGE [25] Gallium radioactive source experiments with the statistical method discussed in Ref. [22], considering the recent $^{71}\mathrm{Ga}(^{3}\mathrm{He}, ^{3}\mathrm{H})^{71}\mathrm{Ge}$ cross section measurement in Ref. [54]; 3) the solar neutrino constraint on $\sin^2 2\vartheta_{ee}$ [23,55–58]; 4) the KARMEN [59] and LSND [60] $\nu_e + {}^{12}\mathrm{C} \to {}^{12}\mathrm{N}_{\mathrm{g.s.}} + e^-$ scattering data [61], with the method discussed in Ref. [62].

(C) The constraints on $\overset{(-)}{\nu}_\mu$ disappearance obtained from the data of the CDHSW experiment [63], from the analysis [64] of the data of atmospheric neutrino

[c] The correct but more complicated analysis of the ICARUS and OPERA data presented in Ref. [40] have not been considered because it would not change significantly the results of the global fits.

	TotGLO	PrGLO	noMB	noLSND
$\chi^2_{\rm min}$	306.0	276.3	251.2	291.3
NDF	268	262	230	264
GoF	5%	26%	16%	12%
$(\chi^2_{\rm min})_{\rm APP}$	98.9	77.0	50.9	91.8
$(\chi^2_{\rm min})_{\rm DIS}$	194.4	194.4	194.4	194.4
$\Delta\chi^2_{\rm PG}$	13.0	5.3	6.2	5.3
$\rm NDF_{PG}$	2	2	2	2
$\rm GoF_{PG}$	0.1%	7%	5%	7%
$\Delta\chi^2_{\rm NO}$	49.2	47.7	48.1	11.4
$\rm NDF_{NO}$	3	3	3	3
$n\sigma_{\rm NO}$	6.4σ	6.3σ	6.4σ	2.6σ

Table 1: Results of the 3+1 fit of short-baseline data [32] taking into account all MiniBooNE data (TotGLO), only the MiniBooNE data above 475 MeV (PrGLO), without MiniBooNE data (noMB) and without LSND data (noLSND). The first three lines give the minimum χ^2 ($\chi^2_{\rm min}$), the number of degrees of freedom (NDF) and the goodness-of-fit (GoF). The following five lines give the quantities relevant for the appearance-disappearance (APP-DIS) parameter goodness-of-fit (PG). The last three lines give the difference between the χ^2 without short-baseline oscillations and $\chi^2_{\rm min}$ ($\Delta\chi^2_{\rm NO}$), the corresponding difference of number of degrees of freedom ($\rm NDF_{NO}$) and the resulting number of σ's ($n\sigma_{\rm NO}$) for which the absence of oscillations is disfavored.

oscillation experiments, from the analysis [65,66] of the MINOS neutral-current data [67] and from the analysis of the SciBooNE-MiniBooNE data neutrino [68] and antineutrino [69] data.

Table 1 summarizes the statistical results. In the TotGLO fit all the MiniBooNE data are considered, including the anomalous low-energy bins, which are omitted in the PrGLO fit following the pragmatic approach proposed in Ref. [31]. There is also a noMB fit without MiniBooNE data and a noLSND fit without LSND data.

From Tab. 1, one can see that in all fits which include the LSND data the absence of short-baseline oscillations is disfavored by about 6σ, because the improvement of the χ^2 with short-baseline oscillations is much larger than the number of oscillation parameters.

The goodness-of-fit in the TotGLO analysis is significantly worse than that in the PrGLO analysis and the appearance-disappearance parameter goodness-of-fit is much worse. This result confirms the fact that the MiniBooNE low-energy anomaly is incompatible with neutrino oscillations, because it would require a small value of Δm^2_{41} and a large value of $\sin^2 2\vartheta_{e\mu}$, which are excluded by the data of other experiments (see Ref. [32] for further details). Therefore, we think that it is very likely that the MiniBooNE low-energy anomaly has an explanation which is different from neutrino oscillations and the PrGLO

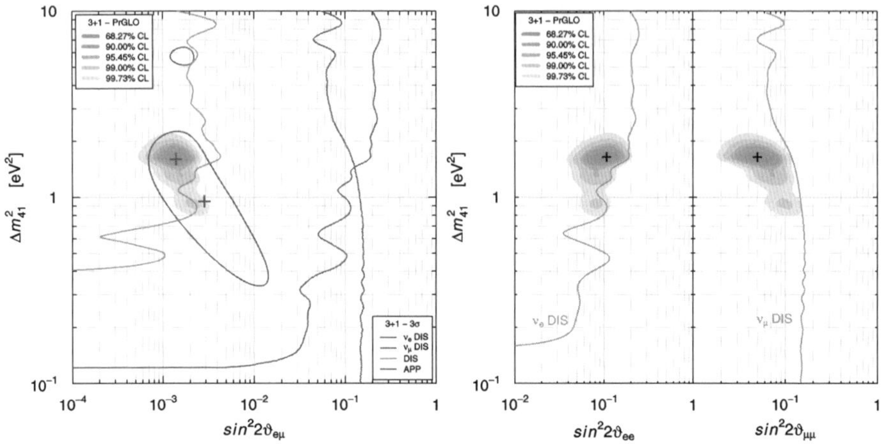

Figure 1: Allowed regions in the $\sin^2 2\vartheta_{e\mu}$–Δm^2_{41}, $\sin^2 2\vartheta_{ee}$–Δm^2_{41} and $\sin^2 2\vartheta_{\mu\mu}$–$\Delta m^2_{41}$ planes obtained in the PrGLO fit [32] of short-baseline neutrino oscillation data compared with the 3σ allowed regions obtained from $\nu_\mu \to \nu_e$ short-baseline appearance data (APP) and the 3σ constraints obtained from ν_e short-baseline disappearance data (ν_e DIS), ν_μ short-baseline disappearance data (ν_μ DIS) and the combined short-baseline disappearance data (DIS). The best-fit points of the GLO and APP fits are indicated by crosses.

fit is more reliable than the TotGLO fit. The cause of the MiniBooNE low-energy excess of ν_e-like events is going to be investigated in the MicroBooNE experiment at Fermilab [70].

Figure 1 shows the allowed regions in the $\sin^2 2\vartheta_{e\mu}$–Δm^2_{41}, $\sin^2 2\vartheta_{ee}$–Δm^2_{41} and $\sin^2 2\vartheta_{\mu\mu}$–$\Delta m^2_{41}$ planes obtained in the PrGLO fit of Ref. [32]. These regions are relevant, respectively, for $\nu_\mu \to \nu_e$ appearance, ν_e disappearance and ν_μ disappearance searches. Figure 1 shows also the region allowed by $\nu_\mu \to \nu_e$ appearance data and the constraints from ν_e disappearance and ν_μ disappearance data. One can see that the combined disappearance constraint in the $\sin^2 2\vartheta_{e\mu}$–Δm^2_{41} plane excludes a large part of the region allowed by $\nu_\mu \to \nu_e$ appearance data, leading to the well-known appearance-disappearance tension quantified by the parameter goodness-of-fit in Tab. 1.

It is interesting to investigate what is the impact of the MiniBooNE experiment on the global analysis of short-baseline neutrino oscillation data. With this aim, the authors of Ref. [32] performed two additional 3+1 fits: a noMB fit without MiniBooNE data and a noLSND fit without LSND data. From Tab. 1 one can see that the results of the noMB fit are similar to those of the PrGLO fit and the exclusion of the case of no-oscillations remains at the level of 6σ. On the other hand, in the noLSND fit, without LSND data, the exclusion of the case of no-oscillations drops dramatically to 2.6σ. In fact, in this case the main indication in favor of short-baseline oscillations is given by the reactor and Gallium anomalies which have a similar statistical significance. Therefore, it is clear that the LSND experiment is still crucial for the indication in favor of short-baseline $\bar{\nu}_\mu \to \bar{\nu}_e$ transitions and the MiniBooNE experiment has been rather inconclusive.

In conclusion, the results of the global fit of short-baseline neutrino oscillation data presented in Ref. [32] show that the data can be explained by 3+1 neutrino mixing if the low-energy MiniBooNE anomaly is omitted following the pragmatic approach proposed in Ref. [31]. Moreover, the crucial indication in favor of short-baseline $\bar{\nu}_\mu \to \bar{\nu}_e$ appearance is still given by the old LSND data. It is however reassuring that the indication of short-baseline neutrino oscillations will be checked with robust methods in many new experiments in the next years (see the review in Ref. [71]).

References

[1] C. Giunti and C.W. Kim, Fundamentals of Neutrino Physics and Astrophysics (Oxford University Press, Oxford, UK, 2007).
[2] N. Agafonova et al., Phys. Rev. Lett. 115 (2015) 121802.
[3] K. Abe et al., Phys. Rev. Lett. 107 (2011) 041801.
[4] P. Adamson et al., Phys. Rev. Lett. 107 (2011) 181802.
[5] K. Abe et al., Phys.Rev.Lett. 112 (2014) 061802.
[6] P. Adamson et al., arXiv:1601.05022.
[7] F.P. An et al., Phys. Rev. Lett. 108 (2012) 171803.
[8] Y. Abe et al., Phys. Rev. Lett. 108 (2012) 131801.
[9] S.B. Kim et al., Phys. Rev. Lett. 108 (2012) 191802.
[10] F.P. An et al., arXiv:1603.03549.
[11] J. Choi et al., arXiv:1511.05849.
[12] Y. Abe et al., JHEP 01 (2016) 163.
[13] D.V. Forero, M. Tortola and J.W.F. Valle, Phys.Rev. D90 (2014) 093006.
[14] J. Bergstrom et al., JHEP 09 (2015) 200.
[15] F. Capozzi et al., arXiv:1601.07777.
[16] K. Abe et al., Phys.Rev. D91 (2015) 072010.
[17] C. Athanassopoulos et al., Phys. Rev. Lett. 75 (1995) 2650.
[18] A. Aguilar et al., Phys. Rev. D64 (2001) 112007.
[19] J.N. Abdurashitov et al., Phys. Rev. C73 (2006) 045805.
[20] M. Laveder, Nucl. Phys. Proc. Suppl. 168 (2007) 344.
[21] C. Giunti and M. Laveder, Mod. Phys. Lett. A22 (2007) 2499.
[22] C. Giunti and M. Laveder, Phys. Rev. C83 (2011) 065504.
[23] C. Giunti et al., Phys. Rev. D86 (2012) 113014.
[24] F. Kaether et al., Phys. Lett. B685 (2010) 47.
[25] J.N. Abdurashitov et al., Phys. Rev. C80 (2009) 015807.
[26] G. Mention et al., Phys. Rev. D83 (2011) 073006.
[27] T.A. Mueller et al., Phys. Rev. C83 (2011) 054615.
[28] P. Huber, Phys. Rev. C84 (2011) 024617.

[29] S.M. Bilenky, C. Giunti and W. Grimus, Eur. Phys. J. C1 (1998) 247.
[30] J. Kopp et al., JHEP 1305 (2013) 050.
[31] C. Giunti et al., Phys.Rev. D88 (2013) 073008.
[32] S. Gariazzo et al., J. Phys. G43 (2016) 033001.
[33] G.H. Collin et al., arXiv:1602.00671.
[34] A. Aguilar-Arevalo et al., Phys.Rev.Lett. 110 (2013) 161801.
[35] L. Borodovsky et al., Phys. Rev. Lett. 68 (1992) 274.
[36] B. Armbruster et al., Phys. Rev. D65 (2002) 112001.
[37] P. Astier et al., Phys. Lett. B570 (2003) 19.
[38] M. Antonello et al., Eur.Phys.J. C73 (2013) 2599.
[39] N. Agafonova et al., JHEP 1307 (2013) 004.
[40] A. Palazzo, Phys.Rev. D91 (2015) 091301.
[41] Y. Declais et al., Phys. Lett. B338 (1994) 383.
[42] A. Kuvshinnikov et al., JETP Lett. 54 (1991) 253.
[43] B. Achkar et al., Nucl. Phys. B434 (1995) 503.
[44] G. Zacek et al., Phys. Rev. D34 (1986) 2621.
[45] A. Hoummada et al., Applied Radiation and Isotopes 46 (1995) 449.
[46] G.S. Vidyakin et al., Sov. Phys. JETP 71 (1990) 424.
[47] A.I. Afonin et al., Sov. Phys. JETP 67 (1988) 213.
[48] Z.D. Greenwood et al., Phys. Rev. D53 (1996) 6054.
[49] M. Apollonio et al., Eur. Phys. J. C27 (2003) 331.
[50] F. Boehm et al., Phys. Rev. D64 (2001) 112001.
[51] Y. Abe et al., JHEP 1410 (2014) 86.
[52] F.P. An et al., Phys. Rev. Lett. 116 (2016) 061801.
[53] K.N. Abazajian et al., arXiv:1204.5379.
[54] D. Frekers et al., Phys. Lett. B706 (2011) 134.
[55] C. Giunti and Y. Li, Phys.Rev. D80 (2009) 113007.
[56] A. Palazzo, Phys. Rev. D83 (2011) 113013.
[57] A. Palazzo, Phys. Rev. D85 (2012) 077301.
[58] A. Palazzo, Mod.Phys.Lett. A28 (2013) 1330004.
[59] B. Armbruster et al., Phys. Rev. C57 (1998) 3414.
[60] L.B. Auerbach et al., Phys. Rev. C64 (2001) 065501.
[61] J. Conrad and M. Shaevitz, Phys. Rev. D85 (2012) 013017.
[62] C. Giunti and M. Laveder, Phys. Lett. B706 (2011) 200.
[63] F. Dydak et al., Phys. Lett. B134 (1984) 281.
[64] M. Maltoni and T. Schwetz, Phys. Rev. D76 (2007) 093005.
[65] D. Hernandez and A.Y. Smirnov, Phys. Lett. B706 (2012) 360.
[66] C. Giunti and M. Laveder, Phys.Rev. D84 (2011) 093006.
[67] P. Adamson et al., Phys. Rev. Lett. 107 (2011) 011802.
[68] K.B.M. Mahn et al., Phys. Rev. D85 (2012) 032007.
[69] G. Cheng et al., Phys. Rev. D86 (2012) 052009.
[70] A.M. Szelc, AIP Conf. Proc. 1666 (2015) 180001.
[71] C. Giunti, arXiv:1512.04758.

DEEP UNDERGROUND NEUTRINO EXPERIMENT

Robert J. Wilson (for the DUNE collaboration) [a]
Department of Physics, Colorado State University, Fort Collins, CO 80523, USA

Abstract. The Deep Underground Neutrino Experiment (DUNE) collaboration will perform an experiment centered on accelerator-based long-baseline neutrino studies along with nucleon decay and topics in neutrino astrophysics. It will consist of a modular 40-kt (fiducial) mass liquid argon TPC detector located deep underground at the Sanford Underground Research Facility in South Dakota and a high-resolution near detector at Fermilab in Illinois. This configuration provides a 1300-km baseline in a megawatt-scale neutrino beam provided by the Fermilab-hosted international Long-Baseline Neutrino Facility.

1 Introduction

Neutrinos are the most abundant known matter particles in the Universe, but they are among the least well-understood of the stable particles. The observation that neutrinos produced as one type (flavor) may later interact as a different type provides evidence for the phenomenon of neutrino flavor mixing and for small, but nonzero, neutrino masses. The framework characterizing these observations is similar to that describing corresponding phenomena in the quark sector, but with a very different pattern of mixing angle values. As in the quark case, this framework involves a phase parameter, $\delta_{\rm CP}$, that changes sign under combined charge conjugation and parity (CP) reversal operations and thus would lead to CP symmetry-violating asymmetries between the pattern of oscillations for neutrinos and antineutrinos. While groundbreaking on its own, the observation of such asymmetries would also provide an experimental underpinning for the basic idea of leptogenesis as an explanation for the baryon asymmetry of the universe [1].

Neutrino oscillation data to date tell us about the magnitudes of the differences in the squared masses of the neutrino mass states, and about the sign of the mass-squared difference between the two states that are closest in mass, but not about the difference of that pair with respect to the third, which may be heavier (normal ordering) or lighter (inverted ordering) than the other two. Resolving this neutrino mass hierarchy ambiguity, along with precise measurements of neutrino mixing angles, would have significant theoretical, cosmological and experimental implications. Because the forward scattering of neutrinos in matter alters the oscillation pattern in a hierarchy-dependent way, GeV-scale neutrinos traveling several hundred kilometers through the Earth's crust enables a decisive determination of the hierarchy independent of the value of $\delta_{\rm CP}$. Taken together, such a comprehensive set of measurements will thoroughly test the standard three-neutrino flavor paradigm that guides our current

---

[a]E-mail: `wilson@colostate.edu`

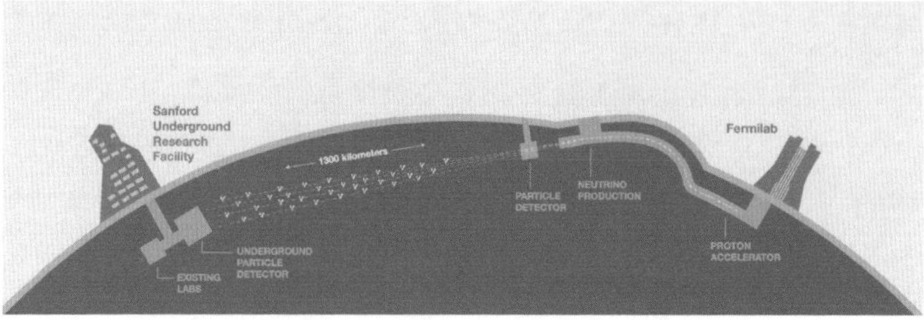

Figure 1: Long-Baseline Neutrino Facility/Deep Underground Neutrino Experiment.

understanding and will provide greatly extended sensitivity to signatures for nonstandard neutrino interactions in matter.

Massive detectors deep underground have broad discovery potential beyond the accelerator neutrino program. The observation of nucleon decay would be a watershed event for the understanding of physics at high energy scales. Neutrinos from supernovae would provide key insights into the physics of gravitational collapse and may also reveal fundamental properties of the neutrino.

The Deep Underground Neutrino Experiment (DUNE) science program and the Long-Baseline Neutrino Facility, as illustrated schematically in Figure 1, are very well described in the Conceptual Design Report (CDR) [2], here I present just a few highlights from that report.

2 Experiment Method and Measurement Sensitivities

The essential method of long-baseline experiments is to produce a (nearly) pure muon-neutrino beam with an energy spectrum matched to the oscillation pattern at the chosen distance (the baseline) then measure the spectrum of muon-neutrinos and electron-neutrinos at the distant detector. DUNE has chosen an optimal beam and distance to have sensitivity to CP violation, CP phase, neutrino mass hierarchy, and other oscillation parameters in the same experiment [3].

Figure 2 shows the disappearance and appearance spectra that would be observed on axis in a liquid argon detector at a distance of 1300 km from a neutrino (antineutrino) source with an exposure of 150 kt.MW.yr; for example, a 40-kt fiducial mass detector exposed to a 1.2 MW, 80 GeV proton beam for 3 years. Spectra for $\delta_{\mathrm{CP}} = 0$ and normal mass hierarchy are shown for the CDR design along with a recent optimized beam design. The full power of the experiment comes from a simultaneous fit to these spectra.

2.1 Mass Hierarchy, Charge-Parity Violation and the Measurement of δ_{CP}

The 1300 km baseline establishes one of DUNE's key strengths: sensitivity to the matter effect. This effect leads to a large discrete asymmetry in the neutrino

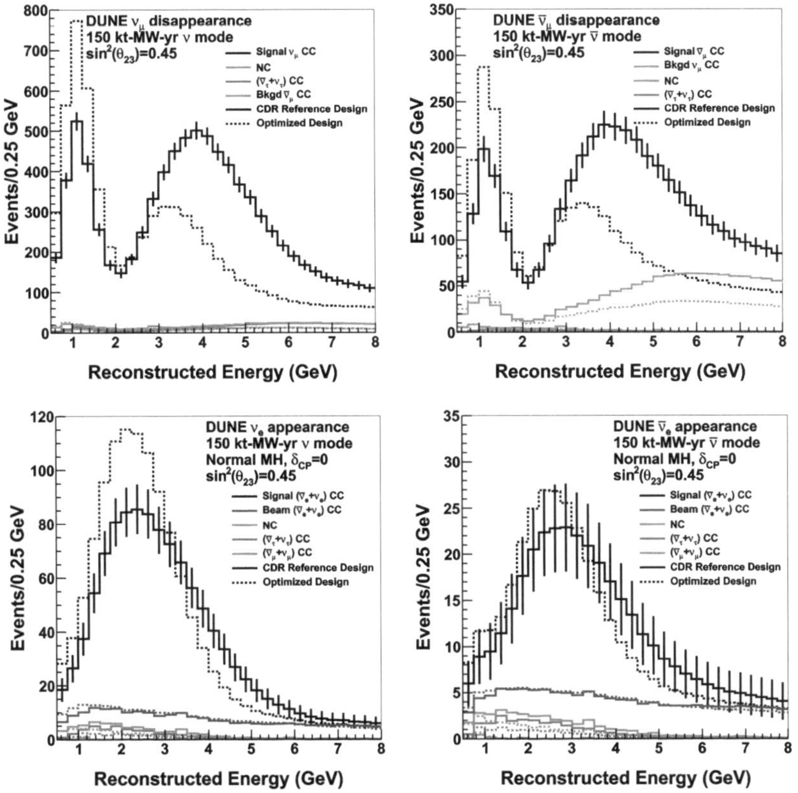

Figure 2: Disappearance (top) and appearance (bottom) spectra for neutrinos (left) and antineutrinos (right).

and antineutrino oscillation probabilities, the sign of which depends on the mass hierarchy (MH). Figure 3[left] shows the mass hierarchy determination significance as a function of the unknown CP phase angle for a total exposure of 300 kt.MW.yr; unambiguous determination is achieved for any value of the phase angle. For some values this discovery arrives much sooner.

The DUNE program has two somewhat distinct objectives with regard to CP symmetry violation in the oscillation channel. First, DUNE aims to make a precise determination of the value of δ_{CP} within the context of the standard three-flavor mixing scenario described by the PMNS matrix. Second, and perhaps more significantly, DUNE aims to observe a signal for leptonic CP violation, independent of the underlying nature of neutrino oscillation phenomenology. Together, the pursuit of these two goals provides a thorough test of the standard three-flavor scenario.

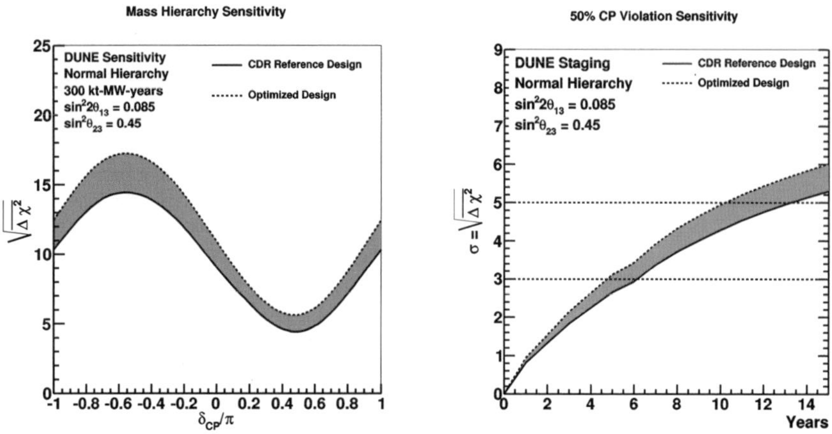

Figure 3: [Left] Mass hierarchy discrimination metric plotted as a function of $\delta_{\rm CP}$ for a 300 kt·MW·year exposure. [Right] Significance with which CP violation can be determined for 50% of $\delta_{\rm CP}$ values as a function of exposure in years using a draft staging plan. The shaded region represents the range in sensitivity due to variations in the beam design.

Figure 3[right] shows the expected sensitivity to non-zero CP phase angle as a function of calendar time for a draft staging scenario outlined in the CDR physics section [2]. Five-sigma determination of CP violation in half the possible phase space is achieved in about 10 years with an optimized horn focusing system under development. The 1σ resolution for $\delta_{\rm CP}$ will be $10° - 18°$ (depending on the value of $\delta_{\rm CP}$) after an exposure of 300 kt.MW.yr and ultimately approaches a precision better than $10°$, which is comparable to the current precision on the CP phase in the CKM matrix in the quark sector. Concurrent analysis of the corresponding atmospheric-neutrino samples will improve the precision with which the mass hierarchy is resolved, which, when combined with the beam data, will increase the CP violation sensitivity at roughly the 1σ level after 10 years operation.

2.2 Deep Underground Science

Among massive detectors designed for neutrino and nucleon decay physics, the LArTPC technology offers unmatched capabilities for position and energy resolution and for high-precision reconstruction of complex interaction topologies over a broad energy range. These features in a detector deep underground will provide a powerful complementarity to the large, underground water Cherenkov or scintillator-based detectors that may be operating at the same time. For example, LArTPC detectors are especially well suited to supersymmetry-favored proton decay modes and uniquely provide detection efficiency and background rejection sufficient to enable a discovery with a single well-reconstructed event.

There is also complementarity with respect to supernova-neutrino detection since liquid argon detectors are primarily sensitive to the electron-neutrino component of the flux, while anti-electron-neutrino interactions dominate for water and scintillator-based detectors. Thus, DUNE will be sensitive to different features of the supernova-neutrino production process. G. Raffelt addresses this topic in more detail in his presentation at this conference. Finally, the LArTPC technology located deep underground opens up an avenue for precision studies of oscillation physics with atmospheric neutrinos, thereby augmenting the results of the beam-based measurements at the core of the experiment.

3 Long-Baseline Neutrino Facility and the DUNE Detectors

The Long-Baseline Neutrino Facility (LBNF) is an international facility hosted in the US – it consists of the neutrino beamline and conventional facilities for both the near and far site detector complexes. The far site is planned for the Sanford Underground Research Facility located at the former Homestake Mine in Lead, South Dakota, well known for the Davis experiment that revolutionized our thinking about neutrinos. Figure 4 shows a schematic of the planned configuration for the DUNE far detector at the 4850 level (4850' below the surface); it consists of four cryostats with dimensions 15(w) x 14(h) x 62(l) m^3, each able to contain a total mass of about 17,000 tonnes of liquified argon.

A new neutrino beamline at Fermilab will provide initial beam power of 1.2 MW, enabled by upgrades to the front end of the accelerator complex carried out within the Proton Improvement Plan-II (PIP-II) program. In anticipation of potential enhancements beyond PIP-II, the beamline is designed to support upgrades to accommodate a beam power of 2.3 MW.

A highly capable near detector will measure the absolute flux and energy scales of all four neutrino species in the LBNF beam, as well as neutrino cross sections on argon, water, and other nuclear targets in the beam's energy range. These measurements are needed to control the systematic uncertainties to a level consistent with the ultimate desired precision for the oscillation parameter measurements. The near detector will also enable a broad range of precision neutrino-interaction measurements, thereby adding a compelling scientific program of its own.

The current baseline near detector design consists of a low-mass straw-tube tracker with radiator foils and pressurized gaseous argon target immersed in a magnetic field and is located about 460 m from the neutrino target. Additional target materials, such as calcium and carbon, can be embedded within the fine-grained tracker. At nominal beam power this detector would collect roughly 10 million neutrino interactions per year, enabling high precision relative and absolute flux measurements. Other options, such as a LArTPC (possibly magnetized) and a high-pressure argon gas TPC are also under consideration. The

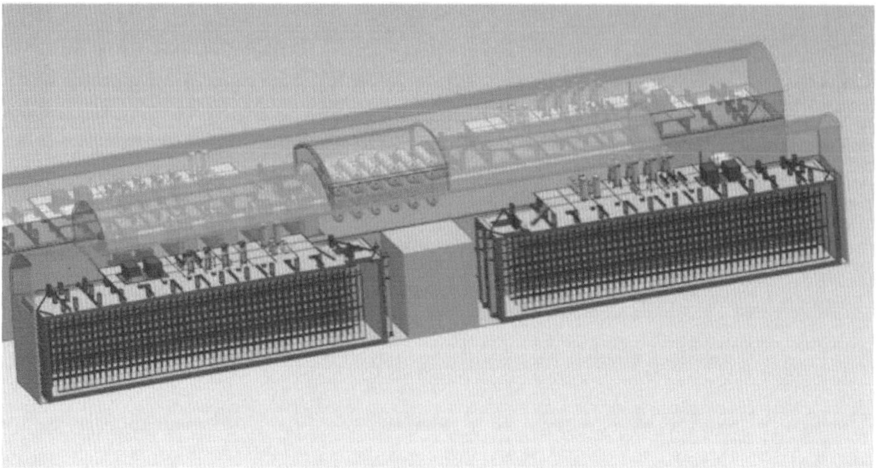

Figure 4: Far site 40-kt fiducial mass detector configuration consisting of four 10-kt liquid argon Time Projection Chambers.

near neutrino detector will be complemented by a system of tertiary beam muon detectors located downstream of the LBNF beamline absorber.

The 10-kt DUNE reference design TPC [4] builds on the pioneering work by the ICARUS collaboration (see [5] and A. Guglielmi this conference) and the former LBNE collaboration. The active volume of 12 m high, 14.5 m wide and 58 m long, is instrumented with anode plane assemblies (APAs), which are 6 m high and 2.3 m wide, and cathode plane assemblies (CPAs), 3 m high by 2.3 m wide. Vertical stacks of two APAs and four CPAs instrument the 12 m height of the active volume. APAs and CPAs are separated horizontally by 3.6 m. The CPAs are held at 180 kV, such that ionization electrons drift in an electric field of 500 $V.cm^{-1}$, which leads to a maximum drift time of 2.3 ms. A wire spacing of 5 mm requires 384,000 electronic readout channels. Planar photon detectors sensitive to the UV scintillation light, for event interaction time determination for underground physics, are mounted within the anode planes.

A dual-phase implementation of the DUNE far detector (which features extraction of track ionization from the liquid into the gas phase for amplification), developed by the former LBNO collaboration, has several potential benefits and is presented as an alternative design in the CDR. If the constructability, cost and performance are confirmed, it could form the basis of the second or subsequent 10-kt far detector modules.

A 35-tonne single-phase prototype has been constructed at Fermilab and a 1m x 1m x 3m dual-phase prototype is under construction at the CERN. Two 8m x 8m x 8m cryostats are planned at CERN to accommodate full-scale single- and double-phase LArTPCs and allow operation in a charged particle beam.

4 Summary and Conclusions

In July 2015 LBNF/DUNE underwent a successful Conceptual Design review by the US Department of Energy, followed in December by a construction readiness review of the Far Site conventional facilities, which may lead to approval to start underground excavation at SURF in 2017. The goal is to begin installation of the first far detector module in 2021, leading to operation with the first 10 kt for non-accelerator physics in 2023/4. It is anticipated that additional far detector modules will be commissioned about every 2 years. The neutrino beam, with a power of at least 1 MW, and the near detector would be operational by 2026.

The International Long-Baseline Neutrino Facility hosted in the US along with the DUNE detectors will perform far-reaching measurements of CP violation, mass hierarchy, non-standard interactions, proton decay and supernova burst neutrinos from intra-galactic distances. DUNE is the only planned experiment able to provide $> 5\sigma$ determination of the mass ordering and for a favorable CP phase this could be achieved in a few years–intriguingly, current experiments seem to mildly favor this scenario. Progress to further internationalize the design, funding, construction and operation of the facility and the experimental program has proceeded very quickly. It is a challenging, long-timescale project but the science goals demand it!

Acknowledgments

The author would like to acknowledge the dedication and contributions of the members of the former LBNE Science Collaboration and Project staff who laid the groundwork for DUNE/LBNF. The work was supported by the US Department of Energy Office of High Energy Physics.

References

[1] S. Pascoli, S. Petcov, and A. Riotto, *"Leptogenesis and Low Energy CP Violation in Neutrino Physics,"* Nucl.Phys. B774 (2007) 1-52.

[2] R. Acciarri et al., *"Long-Baseline Neutrino Facility (LBNF) and Deep Underground Neutrino Experiment (DUNE) Conceptual Design Report Volume 2: The Physics Program for DUNE at LBNF,"* arXiv:1512.06148 [physics.ins-det], 2015 and see http://www.dunescience.org/.

[3] M. Bass et al., *"Baseline optimization for the measurement of CP violation and mass hierarchy in a long-baseline neutrino oscillation experiment,"* Phys. Rev. D 91, 052015.

[4] DUNE Collaboration, *"DUNE/LBNF CDR Volume 4: The DUNE Detectors at LBNF,"* arXiv:1601.02984 [physics.ins-det] 2016.
[5] S. Amerio et al., ICARUS Collaboration, *"Design, construction and tests of the ICARUS T600 detector,"* Nucl.Inst.Meth. A527 (2004) 329-410.

Jiangmen Underground Neutrino Observatory: Status and Prospectives [a]

Yu-Feng Li [b]

(On behalf of the JUNO collaboration)

Institute of High Energy Physics, Chinese Academy of Sciences, P.O. Box 918, Beijing 100049, China

Abstract. The Jiangmen Underground Neutrino Observatory (JUNO) is a 20 kton liquid scintillator (LS) detector, which is planed to determine the neutrino mass hierarchy and measure the oscillation parameters at the sub-percent level using reactor antineutrino oscillations. As a multipurpose neutrino experiment, JUNO is also capable of measuring supernova burst neutrinos, the diffuse supernova neutrino background, geo-neutrinos, solar neutrinos and atmospheric neutrinos. After a brief introduction to the physics motivation, we discuss the status of the JUNO project, including the conceptual design of the detector systems. Finally the latest civil progress and future prospectives are also highlighted.

1 Introduction

The recent discovery of non-zero θ_{13} in reactor [1–3] and accelerator [4,5] neutrino experiments has opened a new window for neutrino oscillation studies. With relatively large θ_{13}, the unknown neutrino mass hierarchy (i.e., the sign of Δm^2_{31} or Δm^2_{32}) and lepton CP violation are anticipated to be measured in the near future. The determination of the neutrino mass hierarchy (MH) is applicable in the long-baseline accelerator neutrino experiments [6–8], atmospheric neutrino experiments [9,10], and medium baseline reactor antineutrino experiments [12–14].

The Jiangmen Underground Neutrino Observatory (JUNO) is a multipurpose liquid scintillator (LS) neutrino experiment, whose primary goal is to determine the neutrino mass hierarchy and measure the oscillation parameters at the sub-percent level using reactor antineutrino oscillations [12,16]. The layout of the JUNO experiment [13] is shown in Fig. 1, where the experimental site is located at Jiangmen in South China, ~53 km away from the Yangjiang and Taishan reactor complexes, which are planed to have six reactor cores of 2.9 GW_{th} and four reactor cores of 4.6 GW_{th}, respectively. Both reactor complexes are under construction. JUNO will build a LS detector of 20 kton with $3\%/\sqrt{E}$ energy resolution. The detector is planed to be located in an experimental hall with the overburden of larger than 700 meters for the purpose of reducing the muon-induced backgrounds.

---

[a] Talk presented at the 17th Lomonosov Conference on Elementary Particle Physics, 20-26 August 2015, Moscow State University.

[b] E-mail: `liyufeng@ihep.ac.cn`

Figure 1: Layout of the experimental design for JUNO. The Yangjiang and Taishan nuclear reactor complexes are used for reactor antineutrino studies.

2 Physics motivations

The neutrino MH is among the most important issues in the future neutrino oscillation program. It can be resolved in reactor antineutrino experiments at the medium baseline using interference between the two fast oscillation components. Because the oscillation amplitudes of these fast components are different (i.e., $\sin^2\theta_{12} < \cos^2\theta_{12}$) and the relative size of their frequencies is also different for two neutrino MHs ($|\Delta m^2_{31}| > |\Delta m^2_{32}|$ or $|\Delta m^2_{31}| < |\Delta m^2_{32}|$), the interference between the fast oscillation components provides us the discrimination ability of the MH. Additional information of the MH can be obtained by comparing the effective atmospheric mass-squared differences in experiments of the medium baseline reactor antineutrino oscillation and long baseline accelerator muon (anti)neutrino disappearance, respectively.

The future prospective of the MH measurement at JUNO is shown in Fig. 2, where a 3σ sensitivity is achieved using the relative interference effects of reactor antineutrino oscillations [12, 15, 16]. The absolute spectral measurement taking account of both interference effects and prior information of the effective atmospheric mass-squared difference can yield an improved sensitivity of 4σ. In our sensitivity studies, a 20-kton liquid scintillator detector with 3% energy resolution and 6 years of data-taking time are assumed. In the left and right panels of Fig. 2, the uncertainties of the effective atmospheric mass-squared difference are taken as 1.0% and 1.5%, and the corresponding MH sensitivity turns out to be 3.7σ and 4.4σ, respectively.

As the prequisite to achieve the MH determination, the precision antineutrino spectral measurements from reactors provide an excellent opportunity to precisely determine the oscillation parameters. The JUNO experiment will be the first experiment to have an unprecedented precision measurement of $\sin^2\theta_{12}$,

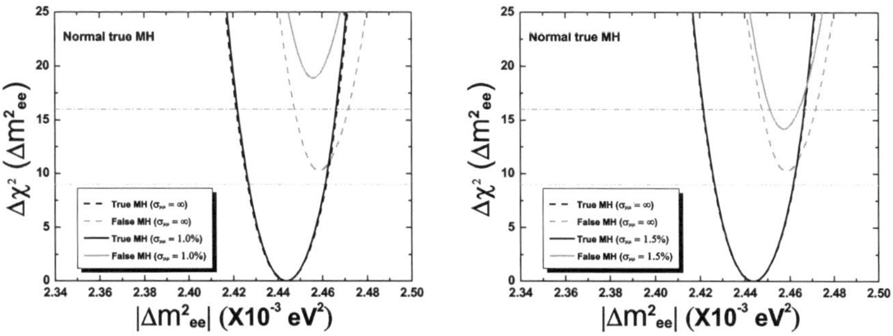

Figure 2: The MH sensitivity of JUNO, where the comparison of black and red lines represents the discrimination power of the MH. The solid and dashed lines are for the interference effects and the combined effects, respectively. The uncertainties of the effective atmospheric mass-squared difference are taken as 1.0% and 1.5% for the left and right panels, respectively.

	Nominal	+ B2B (1%)	+ BG	+ EL (1%)	+ NL (1%)
$\sin^2\theta_{12}$	0.54%	0.60%	0.62%	0.64%	0.67%
Δm^2_{21}	0.24%	0.27%	0.29%	0.44%	0.59%
$\lvert\Delta m^2_{ee}\rvert$	0.27%	0.31%	0.31%	0.35%	0.44%

Table 1: Precision of $\sin^2\theta_{12}$, Δm^2_{21} and $|\Delta m^2_{ee}|$ from the nominal setup to those including additional systematic uncertainties. The systematics are added one by one from left to right.

Δm^2_{21} and $|\Delta m^2_{ee}|$ (defined as the linear combination of Δm^2_{31} and Δm^2_{32} [17]) better than 1% [12, 18]. A systematic analysis of above-mentioned parameter measurements is summarized in Tab. 1, where the nominal setup is defined as the statistical uncertainty plus the reactor spectral uncertainty and detector uncertainty of 1%. Additional systematic uncertainties from left to right include the bin-to-bin (B2B) energy uncorrelated uncertainty, the background (BG), the energy linear scale (EL) and energy non-linear (NL) uncertainties. In summary, for the precision measurements we can achieve the precision level of 0.5%–0.7% for the three oscillation parameters. Moreover, precision measurements of neutrinos are suitable to probe new physics beyond the standard three-neutrino paradigm, including the hypotheses of light sterile neutrinos, nonstandard interactions, unitarity violation, and the Lorentz and CPT invariance violation.

As a large underground observatory, the JUNO experiment is also unique in the astrophysical neutrino measurements. For a typical galactic supernova burst at a distance of 10 kpc, it will register about 5000 events from the inverse beta decay, 2000 events from elastic neutrino-proton scattering, more than 300

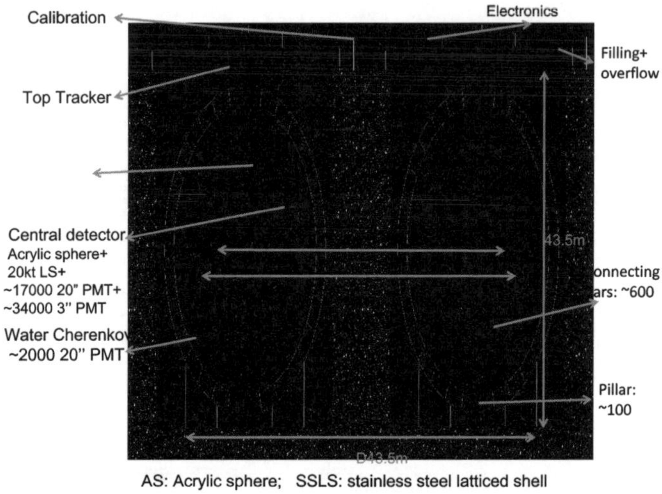

Figure 3: The nominal design of the JUNO detector system (see the context for details).

events from neutrino-electron scattering, as well as the charged current and neutral current interactions on the ^{12}C nuclei. the supernova neutrino observation at JUNO will provide a unique probe for both particle physics and astrophysics, including the absolute neutrino masses and the neutrino mass hierarchy as well as identification of different phases of the supernova burst. Moreover, detection of the diffuse supernova neutrino background is also promising considering the detector size and background levels of JUNO.

There are many other aspects for the JUNO physics potential, including the observation of solar ^{8}B and ^{7}Be neutrinos, detection of geo-neutrinos, studies of atmospheric neutrino oscillations, indirect dark matter search using neutrinos, and observation of the proton decay in the $p \to K^+ + \overline{\nu}$ channel. All the physics studies are thoroughly summarized in Ref. [12].

3 Experimental design

JUNO will deploy a LS detector of 20 kton in the underground experimental hall. The total overburden is $\sim$700 m rock. The baseline design of the JUNO detector system is shown in Fig. 3, which consists of the central detector (CD), the veto detectors and the systems for calibration, electronics, data acquisition and detector control [13].

The CD, which contains 20 kton LS, 17,000 20-inch PMTs and upto 34,000 3-inch PMTs, is designed to reach the energy resolution of $3\%/\sqrt{E_{\rm vis}}$ and the lifetime of over 20 years. The baseline CD option [13] is an acrylic sphere of 35.4 m in diameter supported by stainless-steel structures, where the acrylic sphere will be built from hundreds of acrylic sheets with a thickness of 12 cm using the bulk-polymerization technology. The PMT assembly design including water sealing, implosion protection and installation on the detector is in progress, and a prototype to study the performance of PMTs is being operation. The CD is filled with LS serving as the neutrino detection target. The current LS recipe is linear alkyl benzene (LAB) as the solvent, with PPO as the scintillation fluor and Bis-MSB as the wavelength-shifter. The veto detector system consists of a Water Cherenkov (WC) detector and a top tracker using for the muon detection and muon induced background reduction. The WC detector is a pool filled with purified water and instrumented with PMTs surrounding the CD. The OPERA target tracker will be re-used as the top tracker to reconstruct the direction of cosmic muons. The 20-inch PMTs are mounted on a spherical surface of ~40 m in diameter, covering ~75% of the surface area of the sphere. The total number of inward facing PMTs to detect scintillating photons generated in the CD is ~17,000. In addition, upto 34,000 3-inch PMTs will fill the gaps among the large PMTs in order to improve the energy and vertex resolution.

To achieve the primary goal of the MH determination, an unprecedented energy resolution of $3\%/\sqrt{E_{\rm vis}}$ is a critical parameter in the experimental design, which requires: a) the PMT photocathode coverage $\geq 75\%$, b) the PMT maximal quantum efficiency $\geq 35\%$ [19,20], c) the LS attenuation length ≥ 20 m at 430 nm, which is equivalent to an absorption length of 60 m with a Rayleigh scattering length of 30 m. [c]

4 Project status

The JUNO project was approved by the Chinese Academy of Sciences under the Strategic Priority Research Program in 2013. The international collaboration was formed in July of 2014, which includes over 300 collaborators and over 60 institutions from Asia, Europe and the Americas. The groundbreaking ceremony at the experimental site was hold on January 10th, 2015. It would take around 3 years to finish the civil construction. After the detector component production and the following detector installation and filling, JUNO is planed to start operation in 2020.

5 Conclusion

The primary goal of JUNO is to resolve the neutrino mass hierarchy and measure the oscillation parameters at the sub-percent level using reactor antineutrino oscillations. Meanwhile, as a multipurpose neutrino experiment, it is also

[c] The LAB Rayleigh scattering length is measured to be 28.2 ± 1.0 m at 430 nm [21,22].

capable of measuring supernova burst neutrinos, the diffuse supernova neutrino background, geo-neutrinos, solar neutrinos and atmospheric neutrinos. JUNO has received approval since 2013 and the civil construction has started from 2015. The R&D activities addressing the technical challenges are ongoing within the collaboration. JUNO plans to start data-taking in 2020.

Acknowledgments

The author is grateful to Prof. Alexander I.Studenikin for kind invitation and warm hospitality in Moscow, where this wonderful conference was held. This work was supported by the National Natural Science Foundation of China under grant Nos. 11135009 and 11305193, by the Strategic Priority Research Program of the Chinese Academy of Sciences under Grant No. XDA10010100, and the CAS Center for Excellence in Particle Physics (CCEPP).

References

[1] Daya Bay Collaboration, (F.P. An *et al.*), Phys. Rev. Lett. **108**, 171803 (2012).
[2] Double Chooz Collaboration, (Y. Abe *et al.*), Phys. Rev. Lett. **108**, 131801 (2012).
[3] RENO Collaboration, (J.K. Ahn *et al.*), Phys. Rev. Lett. **108**, 191802 (2012).
[4] T2K Collaboration, (K. Abe *et al.*), Phys. Rev. Lett. **107**, 041801 (2011).
[5] MINOS Collaboration, (P. Adamson *et al.*), Phys. Rev. Lett. **107**, 181802 (2011).
[6] K. Abe *et al.* [Hyper-Kamiokande Proto-Collaboration], PTEP **2015**, 053C02 (2015).
[7] R. Acciarri *et al.* [DUNE Collaboration], arXiv:1512.06148 [physics.ins-det].
[8] J. Cao *et al.*, Phys. Rev. ST Accel. Beams **17**, 090101 (2014).
[9] S. Ahmed *et al.* [ICAL Collaboration], arXiv:1505.07380 [physics.ins-det].
[10] M. G. Aartsen *et al.* [IceCube PINGU Collaboration], arXiv:1401.2046 [physics.ins-det].
[11] S. Adrian-Martinez *et al.* [KM3Net Collaboration], arXiv:1601.07459 [astro-ph.IM].
[12] F. An *et al.* [JUNO Collaboration], J. Phys. G **43**, 030401 (2016)
[13] Z. Djurcic *et al.* [JUNO Collaboration], arXiv:1508.07166 [physics.ins-det].
[14] S. B. Kim, Nucl. Part. Phys. Proc. **265-266**, 93 (2015).
[15] L. Zhan, Y. Wang, J. Cao and L. Wen, Phys. Rev. D **79**, 073007 (2009).
[16] Y. F. Li, J. Cao, Y. Wang and L. Zhan, Phys. Rev. D **88**, 013008 (2013),

[17] H. Nunokawa, S. J. Parke and R. Zukanovich Funchal, Phys. Rev. D **72**, 013009 (2005).
[18] Y. F. Li, Int. J. Mod. Phys. Conf. Ser. **31**, 1460300 (2014).
[19] Y. Wang *et al.*, Nucl. Inst. Meth. A **695**, 113 (2012).
[20] X. C. Lei *et al.*, Chin. Phys. C **40**, 026002 (2016).
[21] Q. Liu *et al.*, Nucl. Instrum. Meth. A **795**, 284 (2015).
[22] X. Zhou *et al.*, Eur. Phys. J. C **75**, 545 (2015).

REVIEW ON θ_{13} MEASUREMENTS FROM REACTOR EXPERIMENTS

Soo-Bong Kim [a]

KNRC, Department of Physics and Astronomy, Seoul National University, Seoul 151-742, Republic of Korea

Abstract. Definitive measurements of the smallest neutrino mixing angle θ_{13} were made by Daya Bay, Double Chooz and RENO experiments in 2012, based on the disappearance of electron antineutrinos emitted from reactors. The new generation reactor experiments have significantly improved a sensitivity for θ_{13} down to the $\sin^2(2\theta_{13}) \sim 0.01$ level. A rather large value of θ_{13} has opened a new window to find the CP violation phase and to determine the neutrino mass hierarchy. The θ_{13} measurements by the three reactor experiments are presented with their future expected sensitivities.

1 Measurement of the Last and Smallest Neutrino Mixing Angle

In the presently accepted framework to describe the neutrino oscillations, there are three mixing angles (θ_{12}, θ_{23} and θ_{13}) and one phase angle in the Pontecorvo-Maki-Nakagawa-Sakata matrix [1,2]. Neutrino oscillation was discovered in the atmospheric neutrino by the Super-Kamiokande experiment in 1998 [3]. It was until 2012 that θ_{13} was the most poorly known as the smallest mixing angle. It took 14 years to measure all of the three mixing angles. Next round of neutrino experiments are expected to determine the CP violation phase and the neutrino mass hierarchy.

Measurements of θ_{13} are possible using reactor neutrinos and accelerator neutrino beams. Reactor measurements can determine the mixing angle without the ambiguities associated matter effects and CP violation. The reactor neutrino detector is not necessarily large, and construction of a neutrino beam is not needed. Past reactor experiments had a single detector located about 1 km from reactors. The new generation reactor experiments, Daya Bay, Double Chooz and RENO, have significantly reduced uncertainties associated with the measurement of θ_{13} using two identically performing detectors at near and far locations from reactors. An accurate value of θ_{13} by the reactor experiment will be able to offer the first glimpse of the CP phase angle, if combined with a result from an accelerator neutrino beam experiment.

Previous attempts of measuring θ_{13} have obtained only upper limits [4, 5]. Indications of a nonzero θ_{13} value were reported by two accelerator appearance experiments, T2K [6] and MINOS [7], and by the Double Chooz reactor disappearance experiment in 2011 [8]. Global analyses of all available neutrino oscillation data have indicated central values of $\sin^2(2\theta_{13})$ that are between 0.05 and 0.1 (see e.g., [9, 10]). In 2012, Daya Bay and RENO reported definitive

[a]E-mail: sbk@snu.ac.kr

measurements of the mixing angle θ_{13} based on the disappearance of reactor electron antineutrinos [11,12]. A combined result of the θ_{13} measurements was reported by the Particle Data Group as $\sin^2(2\theta_{13}) = 0.098 \pm 0.013$ in ref. [13]. Measured values of θ_{13} by the second generation reactor experiments are summarized in Table 1.

Table 1: Summary of θ_{13} measurements by RENO, Daya-Bay and Double Chooz.

Experiments	Measured values of $\sin^2(2\theta_{13})$	Dates
Double Chooz	0.085±0.051	November 2011
Daya Bay	0.092±0.017	March 2012
RENO	0.113±0.023	April 2012
Double Chooz	0.109±0.039	June 2012
Daya Bay	0.089±0.011	October 2012
RENO	0.100±0.018	March 2013
RENO	0.100±0.016	September 2013

2 Overview of Current Reactor Neutrino Experiments

2.1 Principle of θ_{13} measurement

Reactor experiments with a baseline distance of ~1 km can neglect the disappearance of electron antineutrinos driven by θ_{12} and Δm^2_{21} and, thus, unambiguously determine the mixing angle θ_{13} based on the survival probability of electron antineutrinos,

$$P_{survival} \approx 1 - \sin^2 2\theta_{13} \sin^2(1.267\Delta m^2_{31} L/E), \tag{1}$$

where E is the energy of antineutrinos in MeV, and L is the baseline distance in meters between the reactor and detector. A reasonably well measured value of $\Delta m^2_{32} = (2.32^{+0.12}_{-0.08}) \times 10^{-3}$ eV^2 [13] can be substituted for Δm^2_{31} in Eq. (1).

Two identical antineutrino detectors are located at near and far locations from reactors to allow a relative measurement from a comparison of the measured neutrino rates. The measured far-to-near ratio of antineutrino fluxes can considerably reduce systematic errors coming from uncertainties in the reactor neutrino flux, target mass, and detection efficiency.

2.2 Detection method for reactor antineutrinos

A nuclear reactor produces about 10^{20} antineutrinos per GW and per second, mainly coming from the beta decays of fission products of ^{235}U, ^{238}U, ^{239}Pu, and ^{241}Pu. The reactor antineutrino is detected via the inverse beta decay (IBD) reaction, $\bar{\nu}_e + p \to e^+ + n$. Reactor neutrino detectors based on hydrocarbon liquid scintillator (LS) provide free protons as a target. The coincidence of a prompt positron signal and a delayed signal from neutron capture by Gadolinium (Gd) provides the distinctive IBD signature. The three reactor experiments have similar detector designs. The inner most acrylic vessel holds ~0.1% Gd-doped LS as a neutrino target.

2.3 Experimental arrangements

The RENO experiment runs a near detector and a far detector at the Hanbit (Yonggwang) nuclear power plant consisting of six well-aligned and equal-distant reactors, in South Korea. The Daya Bay experiment has two near detectors at two different locations and a far detector, at the Daya Bay nuclear power plant consiting of six reactors, in the southern part of China. The Double Chooz experiment has a far detector only at present and a near detector currently under construction at the Chooz nuclear power plant consisting of two reactors, in France. Three reactor experiments have similar experimental arrangements, but slightly different features in reactor thermal output, detector target mass and overburden, and baselines of near and far detectors. Comparison of their interesting parameters is given in Table 2.

Table 2: Comparison of three reactor experimental parameters

Experiments	Location	Thermal Power (GW)	Flux weighted Baselines Near/Far (m)	Overburden Near/Far (mwe)	Target Mass (tons)
Double Chooz	France	8.5	410/1050	120/300	8.6/8.6
RENO	Korea	16.8	409/1444	120/450	16/16
Daya Bay	China	17.4	470 & 576/1648	250/860	40+40/80

3 Results on Neutrino Mixing Angle θ_{13}

The first generation reactor experiments of Chooz and Paolo Verde could not continue data-taking longer than a half year because of unexpected problems with Gd-doped LS [4,5]. Paolo Verde had problems with precipitation, condensation, and slow deterioration of Gd-doped LS developing in time. In Chooz experiment, Gd-doped LS turned yellow a few months after deloyment. A rapid decay of attenuation length of Gd-doped LS had been observed. It is crucial for an reactor experiment to keep LS chemically stable for several years of experimental duration. The second generation reactor experiments were successful to develop Gd compounds, and have not suffered yet from the troubles that Chooz and Paolo Verde experienced.

RENO was the first reactor experiment to take data with both near and far detectors in operation from August 2011. The RENO collaboration reported a definitive measurement of θ_{13} based on 220 days of data taken through March 2012 as $\sin^2(2\theta_{13}) = 0.113 \pm 0.013(\text{stat.}) \pm 0.019(\text{syst.})$ in April 2012 [12]. Improved results of $\sin^2(2\theta_{13}) = 0.100 \pm 0.010(\text{stat.}) \pm 0.015(\text{syst.})$ and $\sin^2(2\theta_{13}) = 0.100 \pm 0.010(\text{stat.}) \pm 0.012(\text{syst.})$ were reported based on ~400 days of data through October 2012 at 2013 Neutrino Telescope workshop [14] and at 2013 TAUP conference [15], respectively. The improvements came from better understanding of the detector energy scale, more accurate estimation of a cosmic ray induced background uncertainty, and more data. The RENO collaboration has been performing an analysis of energy-dependent neutrino oscillation effects and an effort of reducing the background uncertainty. They are expected to obtain a more precise measurement of the θ_{13} value from about 800 days of data set in early 2014.

Daya Bay began data-taking with a part of near detectors from August 2012, and with 75% near and far detectors in operation from late December 2011. The Daya Bay collaboration reported a measured value of θ_{13} as $\sin^2(2\theta_{13}) = 0.092 \pm 0.016(\text{stat.}) \pm 0.005(\text{syst.})$ in March 2012 [11]. Data-taking was temporarily stopped in August 2012 to complete the rest of detector construction, and resumed in October 2012. An updated result of $\sin^2(2\theta_{13}) = 0.089 \pm 0.010(\text{stat.}) \pm 0.005(\text{syst.})$ was reported based on 140 days of data through May 2012 [16]. The collaboration is expected to have a doubled statistics of data set through July 2013, and to obtain a new result on θ_{13} from both rate and shape analysis.

Double Chooz has been taking data with a far detector only since April 2011, and reported an evidence non-zero value of θ_{13} as $\sin^2(2\theta_{13}) = 0.085 \pm 0.029(\text{stat.}) \pm 0.042(\text{syst.})$ in November 2011 [8], and as $\sin^2(2\theta_{13}) = 0.086 \pm 0.041(\text{stat.}) \pm 0.030(\text{syst.})$ in March 2012 [17]. An updated result of $\sin^2(2\theta_{13}) = 0.109 \pm 0.030(\text{stat.}) \pm 0.025(\text{syst.})$ was reported based on an improved analysis on rate and energy shape with 228 days of data [18]. The collaboration recently reported a θ_{13} measurement by identifying a delayed signal of neutron capture on Hydrogen: $\sin^2(2\theta_{13}) = 0.097 \pm 0.034(\text{stat.}) \pm 0.034(\text{syst.})$ [19]. A near detector is under construction, and will be ready for data-taking in spring 2014.

4 Summary and Prospects

Three reactor experiments have definitively measured the value of θ_{13} by the disappearance of electron antineutrinos. Based on unprecedentedly copious data, Daya Bay and RENO have performed rather precise measurements of the value. Averaging the results of the three reactor experiments with the standard Particle Data Group method, one obtains $\sin^2(2\theta_{13}) = 0.098 \pm 0.013$ [13]. The exciting result provides a comprehensive picture of neutrino transformation among three kinds of neutrinos and opens the possibility of search for CP violation in the leptonic sector. The surprisingly large value of θ_{13} will strongly promote the next round of neutrino experiments to find CP violation effects and determine the neutrino mass hierarchy. The successful measurement of θ_{13} has made the very first step on the long journey to the complete understanding of the fundamental nature and implications of neutrino masses and mixing parameters.

Precise measurements of θ_{13} by the reactor experiments will provide the first glimpse of the CP phase angle if accelerator beam results are combined. Based on total 5 years of data, the reactor experiments are expected to obtain a measured $\sin 2(2\theta_{13})$ value with a precision of 7% at RENO, 4% at Daya Bay, and 10% at Double Chooz. They will also make a direct measurement of Δm^2_{31} from the energy dependent oscillation effects, and measure a precise neutrino spectrum.

References

[1] B. Pontecorvo, Sov. Phys. JETP **7**, 172 (1958).
[2] Z. Maki, M. Nakagawa, and S. Sakata, Prog. Theor. Phys. **28**, 870 (1962).
[3] Y. Fukuda *et al.*, Phys. Rev. Lett. **81**, 1562-1567 (1998).
[4] M. Apollonio *et al.*, Phys. Lett. B **466**, 415 (1999); Eur. Phys. J. C **27**, 331 (2003).
[5] F. Boehm *et al.*, Phys. Rev. Lett. **84**, 3764 (2000); Phys. Rev. D **64**, 112002 (2010).
[6] K. Abe *et al.*, Phys. Rev. Lett. **107**, 041801 (2011).
[7] P. Adamson *et al.*, Phys. Rev. Lett. **107**, 181802 (2011).
[8] See a talk given by Herve de Kerret at 6th Workshop on Low Energy Neutrino Physics, Seoul, Korea, 9-12 November, 2011.
[9] G.L. Fogli *et al.*, Phys. Rev. D **84**, 053007 (2011).
[10] T. Schwetz *et al.*, New J. Phys. **13**, 109401 (2011).
[11] F.P. An *et al.*, Phys. Rev. Lett. **108**, 171803 (2012).
[12] J.K. Ahn *et al.*, Phys. Rev. Lett. **108**, 191802 (2012).
[13] J. Beringer *et al.* (Particle Data Group), Phys. Rev. D **86**, 010001 (2012).
[14] See a talk given by Seon-Hee Seo at 15th Workshop on Neutrino Telescopes, Venice, Italy, 11-15 March, 2013.
[15] See a talk given by Seon-Hee Seo at 13th International Conference on Topics in Astroparticle and Underground Physics, Asilomar, USA, 8-13 September, 2013.

[16] F.P. An *et al.*, Chinese Physics C **37**, 011001 (2013).
[17] Y. Abe *et al.*, Phys. Rev. Lett. **108**, 131801 (2012).
[18] Y. Abe et al., Phys. Rev. D **86**, 052008 (2012).
[19] Y. Abe et al., Phys. Lett. B **723**, 66-70 (2013).

NEUTRINO MASS ORDERING FROM ATMOSPHERIC NEUTRINOS

Alexander Kappes [a] [b]

Erlangen Centre for Astroparticle Physics, Friedrich-Alexander-Universität Erlangen-Nürnberg, D-91058 Erlangen, Germany

Abstract. After an introduction to neutrino oscillation, the article discusses how atmospheric neutrinos in the GeV range can be used to determine the mass ordering of neutrinos and presents detectors which aim at realizing this within the next 10 to 20 years.

1 Introduction

Neutrino physics is currently one of the most interesting sectors in particle physics. After their postulation by Pauli in 1930 [1] and the first observation of electron neutrinos in 1956 by Cowan and Reines [2], neutrinos, of which currently three flavors (electron, muon, tau) are known, were long thought to be massless. Though hints concerning a too-low electron-neutrino flux from the Sun long pointed towards neutrino oscillation, only in 1998 the Super-Kamiokande collaboration was able to establish that atmospheric muon neutrinos oscillate into other neutrino flavors (mainly tau neutrinos) [3] and that therefore neutrinos must have a non-vanishing mass.

1.1 Neutrino oscillations and the mass ordering

Neutrino oscillations occur because the three mass eigenstates ($\nu_i = \nu_1, \nu_2, \nu_3$), which are relevant for the propagation of neutrinos in vacuum, are a non-trivial mixture of the flavor eigenstates ($\nu_\alpha = \nu_e, \nu_\mu, \nu_\tau$) that enter the weak interaction. It is governed by three mixing angles θ_{12}, θ_{23}, θ_{13} and two mass square differences $\Delta m^2_{21} = m^2_2 - m^2_1$, $\Delta m^2_{32} = m^2_3 - m^2_2$ with m_i being the neutrino masses in the mass eigenstates. The flavor eigenstates are connected to the mass eigenstates via the unitary PMNS (Pontecorvo-Maki-Nakagawa-Saka) matrix U which depends on the mixing angles and potential CP violating phases (for a representation of U see e.g. standard text books on neutrino oscillations):

$$|\nu_\alpha\rangle = \sum_i U_{\alpha i}\, |\nu_i\rangle \tag{1}$$

The transition probability between two flavors $P_{\nu_\alpha \to \nu_\beta}$ for propagation in

[a]E-mail: `alexander.kappes@physik.uni-erlangen.de`

[b]now at Institut für Kernphysik, Westfälische Wilhelms-Universität Münster, D-48149 Münster, Germany

Parameter	Best fit	$1\,\sigma$ range
$\delta m^2/10^{-5}\,\text{eV}^2$	7.54	7.32–7.80
$\Delta m^2/10^{-3}\,\text{eV}^2$	2.43	2.37–2.49
θ_{12}/deg	33.7	32.6–34.8
θ_{13}/deg	8.80	8.43–9.17
θ_{23}/deg	41.4	40.0–43.3

Table 1: World best-fit estimates for the oscillation parameters assuming the normal mass ordering. The mass square differences are defined here as: $\Delta m^2 = m_3^2 - (m_1^2 + m_2^2)/2$ and $\delta m^2 = \Delta m_{21}^2$. Taken from [5].

vacuum is then given by

$$P_{\nu_\alpha \to \nu_\beta} = \delta_{\alpha\beta} - 4 \sum_{j>i} U_{\alpha i} U_{\beta i} U_{\alpha j} U_{\beta j} \sin^2 \left(\frac{\Delta m_{ij}^2 L}{4E_\nu} \right) \quad (2)$$

Here, L is the oscillation length and E_ν the neutrino energy (note, that Eq. 2 only contains the real components and neglects CP violation). The quantities θ_{12} and Δm_{21}^2 were measured using solar electron neutrinos, θ_{23} and Δm_{32}^2 were determined with atmospheric neutrinos. In 2012, the Daya Bay collaboration announced that also the third mixing angle θ_{13} has a rather large value [4] which is a prerequisite for the measurability of further oscillation parameters. World best-fit estimates for the oscillation parameters are listed in Table 1. Note that the uncertainty on θ_{23} is still sufficiently large such that its true value could be larger than 45°. Another unknown is the so-called neutrino mass ordering, i.e. the question whether two light mass eigenstates m_1 and m_2 and a heavier eigenstate m_3 (normal mass ordering, NH) or a light eigenstate m_3 and two heavier eigenstates m_1 and m_2 (inverted mass ordering, IH) are realized in nature. In vacuum, a signature of the mass ordering is present in the oscillation of anti-neutrinos from nuclear power reactors as targeted e.g. by JUNO [6]. Another signature of the neutrino mass ordering is contained in neutrinos traversing matter.

1.2 Matter effects

In contrast to muon and tau neutrinos, electron neutrinos and their anti-neutrinos can scatter off atomic electrons in matter via the charged current interaction which modifies the oscillation probabilities and can be described via effective mass-square differences and mixing angles. For the simplified two-flavor oscillation, for example, one obtains the effective parameters Δm_M^2 and

$\sin 2\theta_M$:

$$\Delta m_M^2 = \xi \times \Delta m^2 \ ; \ \sin 2\theta_M = \frac{\sin 2\theta}{\xi} \tag{3}$$

$$\text{with} \quad \xi = \sqrt{\sin^2 2\theta + \left(\cos 2\theta - \frac{A_{CC}}{\Delta m^2}\right)^2} \tag{4}$$

$$\text{and} \quad A_{CC} = \pm 2\sqrt{2} E_\nu G_F N_e \ , \tag{5}$$

where Δm and θ are the vacuum oscillation parameters, E_ν is the neutrino energy, G_F is the Fermi constant and N_e is the electron density in matter. The $\pm$ sign in the expression for A_{CC} is for neutrinos and anti-neutrinos, respectively. As a consequence, under the right environmental conditions (electron density)

$$N_e^R = \frac{\Delta m^2 \cos 2\theta}{2\sqrt{2} E_\nu G_F} \tag{6}$$

the effective mixing becomes maximal and hence a total flavor transition is possible even if the vacuum mixing angle is small. This is the so-called MSW (Mikheyev-Smirnov-Wolfenstein) effect [7,8]. The required densities are reached for example in the Sun which allowed to determine the mass ordering of m_1 and m_2 from the MeV neutrinos produced in the Sun's fusion processes [c] from the energy dependence of the resonance condition.

Though Earth's matter density is much lower than that of the Sun, the much higher energy of neutrinos produced in the interaction of cosmic rays with Earth's atmosphere leads to a measurable effect for neutrinos in the GeV range below $\sim 10\,\text{GeV}$ and propagation lengths longer than $\sim 5000\,\text{km}$ corresponding to zenith angles above 143° (see Fig. 1). Several instruments aim at exploiting this imprint to determine the neutrino mass ordering within in the next 10 to 20 years.

2 Experiments

2.1 ORCA and PINGU

Neutrino telescopes reconstruct the energy and direction of neutrinos via the amplitude and timing of Cherenkov photons originating from charged particles which are produced in the interaction of neutrinos in the ice/water or bedrock below the detectors. The light is detected with optical sensors consisting of photomultipliers. As the targeted energy range below 10–20 GeV is not accessible to current neutrino telescopes like ANTARES [10] or IceCube [11] because of

[c] Note that the sign can either be absorbed into the mass ordering or into the mixing angle θ_{12}, i.e. its octant (see $\cos 2\theta$ term in Eq. 4). The general convention is to define m_1 as being the lighter mass and leave the octant free.

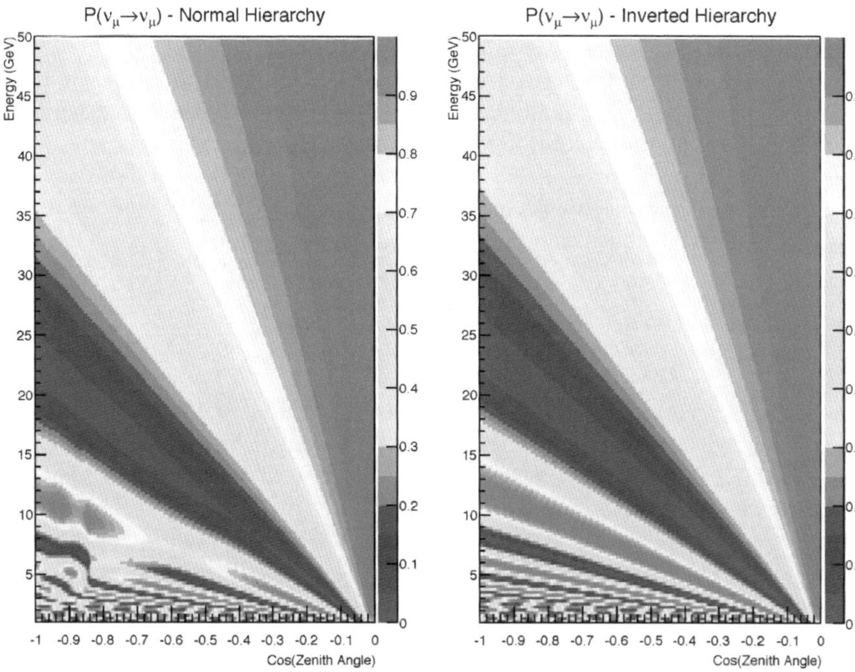

Figure 1: Muon neutrino survival probability for normal and inverted ordering after traveling through the Earth, binned in neutrino energy and cosine of the zenith angle. Taken from [9].

their large sensor spacing, the neutrino telescope community is pursuing plans for the construction of dedicated detectors called ORCA [12] and PINGU [9] in the Mediterranean Sea and the South Pole, respectively, which feature reduced sensor spacing with instrumented volumes of several Mton. For PINGU, the current design foresees 40 additional strings, each equipped with 80 optical sensors, to be placed in the lower center of the current IceCube detector. The horizontal string spacing will be approximately 22 m and the vertical sensor spacing 3 m. The ORCA design features 115 strings to be installed near the ANTARES site in the Mediterranean Sea where each string carries 18 optical sensors. The horizontal string spacing will be about 20 m and the vertical sensor spacing 6 m. Currently, six strings are funded. These will be installed within the next 1–2 years and will allow for vetting the deployed technology and anticipated performance.

In the case of neutrino telescopes, the measurable size of the mass-ordering effect on the oscillation pattern is significantly reduced by the fact that for anti-neutrinos the difference in observed events for NH and IH for a particular zenith angle and energy is almost exactly inverted with respect to neutrinos (in the two-flavor approximation of Eq. 4, a sign change in Δm^2 is equivalent to a sign change of A_{CC} when going from neutrinos to anti-neutrinos). As neutrino telescopes cannot distinguish between neutrinos and anti-neutrinos, the measurement therefore crucially depends on the difference in flux of atmospheric neutrinos and anti-neutrinos and the higher interaction cross section of neutrinos. On the other hand, the large range in observed energies and zenith angles allows neutrino telescopes to utilize extended control regions to mitigate the effect of possible systematic effects on the mass ordering measurement, and they are sensitive to both muon and electron neutrinos.

2.2 Hyper-Kamiokande

The Hyper-Kamiokande design consists [13] of two large water tanks which are subdivided into five compartments to be installed in a mine close to the Super-Kamiokande detector located in Japan. The total water mass amounts to approximately 1 Mton where the detector is structured into a larger inner volume and a smaller outer volume with the latter being used for background suppression. The tank walls are equipped with photomultipliers which observe a fiducial volume in the inner tank of about 560 kT. As neutrino telescopes, Hyper-Kamiokande uses the Cherenkov light emitted by secondary charged particles to reconstruct the energy and direction of neutrinos and, as neutrino telescopes, it is sensitive to both muon and electron neutrinos but cannot distinguish between neutrinos and anti-neutrinos. In addition to being sensitive to the neutrino mass ordering via the observation of atmospheric neutrinos, Hyper-Kamiokande will be located inside the J-PARC neutrino beam, allowing to utilize the high flux of $\sim 600\,\text{MeV}$ off-axis neutrinos and the oscillation baseline of about 300 km to measure the CP phase in the neutrino sector.

2.3 INO

The main detector of the India-based Neutrino Observatory [14] to be built in Bodi West Hills, South India, ICAL, is a magnetized iron calorimeter with a mass of 50 kt. Its target consists of 150 horizontal, 5.6 cm thick iron plates which are interleaved with resistive plate chambers that act as active detector elements. Myon neutrinos undergoing a charged current interaction are detected by reconstructing the track that a muon leaves inside the active elements. The ICAL detector is planned to be magnetized to about 1.5 T within the plane of the iron plates. Together with the ability to sample a track with high resolution, this allows to determine the charge of the produced muon and hence discriminate between muon neutrinos and muon anti-neutrinos. This unique features partially compensates the rather low target mass of 50 kt.

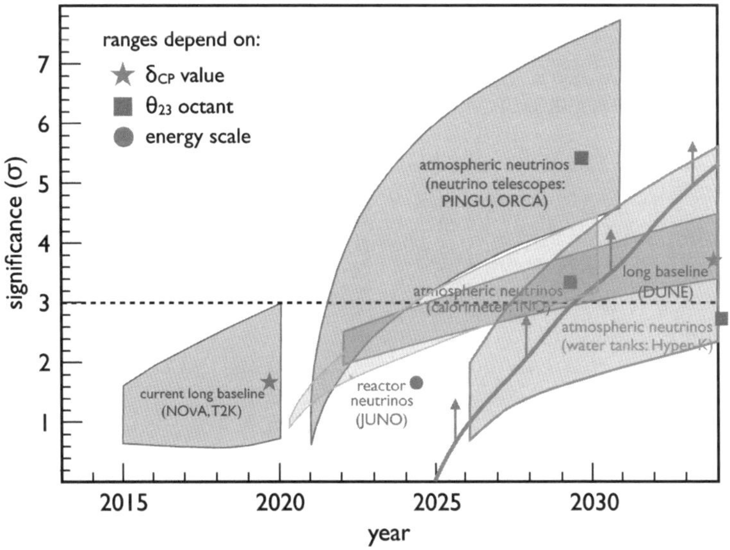

Figure 2: Sensitivity of various experiments to the neutrino mass ordering as a function of time. The width is defined by the respective best/worst case which for example for PINGU and ORCA is determined by the true value of θ_{23} (its octant). Adapted from [9, 15]. The DUNE curve was taken from [16]) and marks the minimum significance (most unfavorable δ_{CP}).

3 Sensitivities to the neutrino mass ordering

Apart from the detectors discussed above, several other running or planned experiments aim at determining the neutrino mass ordering using either reactor neutrinos (JUNO [17]) or long baseline oscillations in neutrino beams (NOνA [18], T2K [19], DUNE [16]). Figure 2 shows an overview over current and planned experiments and their sensitivity as a function of time. It should be stressed that the starting dates of all future experiments (DUNE, Hyper-K, INO, JUNO, ORCA, PINGU) are subject to large uncertainties and should only be taken as indicative for a possible timeline. As can be seen, the sensitivity of most experiments heavily depends on "external" oscillation parameters like the true δ_{CP} or θ_{23} value. JUNO, on the other hand, faces the challenge to reach the required energy resolution. In the end, a single experiments might have difficulties to measure the neutrino mass ordering with sufficient significance, the more so as in all cases systematic effects have to be well under control. Therefore, from today's perspective, consistent results from at least two experiments with sufficiently different systematics are desirable. Though not all of these experiments might in the end be realized, the chances are good that the mass ordering will be known towards the end of the next decade, and experiments utilizing atmospheric neutrinos will play a crucial role in this.

Acknowledgments

The author gratefully acknowledges the support by the German Ministry for Education and Research (BMBF), grant number 05A14WE3.

References

[1] W. Pauli. Liebe radioaktive Damen und Herren. Letter from 1930 printed in Physics Today 31N9 (1978): 72.
[2] C.L. Cowan, F. Reines, F.B. Harrison, H.W. Kruse, and A.D. McGuire. *Science*, 124:103, 1956.
[3] Y. Fukuda et al. (Super-Kamiokande Collaboration). *Phys. Rev. Lett.*, 81:1562, 1998. arXiv:hep-ex/9807003.
[4] F.P. An et al. (DAYA-BAY Collaboration). *Phys.Rev.Lett.*, 108:171803, 2012. 1203.1669.
[5] F. Capozzi, G.L. Fogli, E. Lisi, A. Marrone, D. Montanino, et al. *Phys. Rev.*, D89:093018, 2014. arXiv:1312.2878.
[6] Yu-Feng Li, Jun Cao, Yifang Wang, and Liang Zhan. *Phys.Rev.*, D88:013008, 2013. arXiv:1303.6733.
[7] L. Wolfenstein. *Phys. Rev.*, D17:2369, 1978.
[8] S.P. Mikheyev and A. Yu. Smirnov. *Prog. Part. Nucl. Phys.*, 23:41, 1989.
[9] M.G. Aartsen et al. (IceCube-PINGU Collaboration), Preprint arXiv:1401.2046, 2014.
[10] M. Ageron et al. (ANTARES Collaboration). *Nucl. Inst. Meth.*, A656:11, 2011. arXiv:1104.1607.
[11] A. Achterberg et al. (IceCube Collaboration). *Astropart. Phys.*, 26:155, 2006. arXiv:astro-ph/0604450.
[12] S. Adrian-Martinez et al. (KM3NeT Collaboration). 2016. arXiv:1601.07459.
[13] SISSA. *Proceedings, 16th International Workshop on Neutrino Factories and Future Neutrino Beam Facilities (NUFACT 2014)*, volume NUFACT2014. SISSA, 2015.
[14] Shakeel Ahmed et al. (ICAL Collaboration). 2015. 1505.07380.
[15] Mattias Blennow, Pilar Coloma, Patrick Huber, and Thomas Schwetz. *J. High Energy Phys.*, 03:028, 2014. arXiv:1311.1822.
[16] R. Acciarri et al. (DUNE Collaboration). 2016. arXiv:1601.05471.
[17] Zelimir Djurcic et al. (JUNO Collaboration). 2015. arXiv:1508.07166.
[18] D. S. Ayres et al. (NOvA Collaboration). 2004. arXiv:hep-ex/0503053.
[19] Yuichi Oyama (T2K Collaboration). Current status of the T2K experiment. In *18th International Conference From the Planck Scale to the Electroweak Scale (Planck 2015) Ioannina, Greece, May 25-29, 2015*, 2015. arXiv:1510.07200.

MATTER EFFECTS ON NEUTRINO OSCILLATIONS

Michael B. Smy [a]

Department of Physics and Astronomy, 3117 Frederick Reines Hall, University of California, Irvine, Irvne, California 92697, USA

Abstract. Neutrino flavor oscillations are a consequence of relative phase shifts from the propagation of neutrinos of different mass. Neutrino interactions with other particles produce additional phase shifts and therefore influences these oscillations. This paper reviews the existing experimental data on such "matter effects".

1 Neutrino Oscillations

The propagation of near-massless neutrino of energy E_ν and mass m is described by a plane wave $\nu(x,t) = e^{iE_\nu(x-t)}e^{-i\frac{m^2}{2E_\nu}x}$ where x is the propagation distance and t the propagation time for $E_\nu \gg m$. Neutrinos are produced in a flavor eigenstate α which is a linear combination of different mass eigenstates k: $\nu_\alpha = \sum_k U^{\text{PMNS}}_{\alpha k}\nu_k$ where U^{PMNS} is the Pontocorvo-Maki-Nakagawa-Sakata mixing matrix. After the different mass eigenstates ν_k have propagated a distance L, the relative phase difference of their wave functions are $\Delta\phi_{jk} = \frac{(m_i^2-m_j^2)L}{2E_\nu} = \frac{\Delta m^2_{ij}L}{2E_\nu}$, and therefore the linear combination (and consequently the neutrino flavor composition) is different than at production. The mixing matrix U^{PMNS} is usually written as a product of the "atmospheric mixing", the "reactor mixing", and the "solar mxining":

$$\begin{pmatrix} 1 & 0 & 0 \\ 0 & \cos\theta_{23} & \sin\theta_{23} \\ 0 & -\sin\theta_{23} & \cos\theta_{23} \end{pmatrix} \begin{pmatrix} \cos\theta_{13} & 0 & \sin\theta_{13}e^{-i\delta} \\ 0 & 1 & 0 \\ -\sin\theta_{13}e^{i\delta} & 0 & \cos\theta_{13} \end{pmatrix} \begin{pmatrix} \cos\theta_{12} & \sin\theta_{12} & 0 \\ -\sin\theta_{12} & \cos\theta_{12} & 0 \\ 0 & 0 & 1 \end{pmatrix}$$

using four parameters: the "solar angle" θ_{12}, the "atmospheric angle" θ_{13}, the "reactor angle" θ_{13}, and the CP-violating phase δ (Majorana neutrinos would have two additional CP-violating phases which do not affect neutrino oscillations).

2 Matter Effects and Mass Ordering

In-flight interactions of neutrinos in a medium lead to additional phase shifts, since the electron-flavor eigenstate undergoes charged- and neutral-current interactions with electrons, while the other flavor eigenstates only do neutral-current interactions (interactions with quarks are "flavor-blind"). This is described by an effective Hamiltonian (in the flavor basis) that adds the "potential" $\sqrt{2}G_F\rho_e$ to the "electron-electron component" of the vacuum propagation

[a]E-mail: `msmy@uci.com`

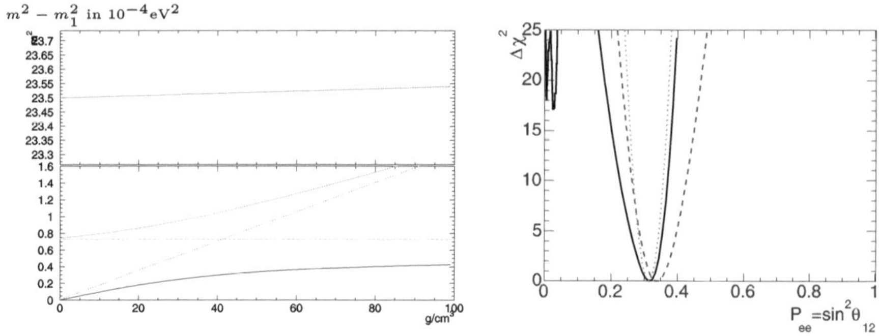

Figure 1: Left: effective mass^2 difference to m_1^2 in the sun for a typical ^{8}B neutrino at 10 MeV. The bottom panel shows the adiabatic conversion of ν_e neutrinos from the solar core to the second mass eigenstate at the solar surface due to a "level crossing" (green dotted lines). There is no level crossing for the $m_3^2 - m_1^2$ (top panel). Right: determination of θ_{12} with solar ^{8}B neutrinos by SNO (solid black line) and SK (dashed blue line). The green dotted line shows the combined fit.

Hamiltonian (which is diagonal in the mass basis with elements $m_k^2/2E_\nu$). If θ is the vacuum two-neutrino mixing angle, the effective mixing in matter is

$$\sin^2 2\theta_{\text{eff}} = \frac{\sin^2 2\theta}{\sin^2 2\theta + \left(\cos 2\theta - 2\sqrt{2}\frac{E_\nu G_F \rho_e}{\Delta m_{ij}^2}\right)^2}.$$

Large effective mixing results, if the oscillation length $l_{\text{vac}} = 2\Delta m_{ij}^2/E_\nu$ of two mass eigenstates i and j matches the "refraction length" $l_{\text{ref}} = \sqrt{2}G_F\rho_e$ ($l_{\text{ref}} = l_{\text{vac}} \cos 2\theta$) This "matter effect" and the possibility of large effective mixing was first explored by Mikheyev, Smirnov and Wolfenstein [1]: in the sun, the electron-density varies adiabatically as the neutrinos pass (compared to oscillation length) from their production location to the solar surface: the effective mass^2 of the solar neutrinos – born as ν_e's in the solar core – varies linearly with electron density, until it approaches the m_2^2 level. This "level crossing" is illustrated by Figure 1 (left panels). For perfect adiabaticity, the neutrinos emerge in a pure m_2 eigenstate. The effect is sensitive to the sign of Δm_{ij}^2, so such matter effects are useful tools to experimentally determine this sign, and therefore the "ordering" of the neutrino masses. In this fashion, ^{8}B solar neutrino data selected the ordering of the first two neutrinos masses, since a sign change in the "solar" Δm_{21}^2 is equivalent to changing θ_{12} to $\pi/2 - \theta_{12}$. The Sudbury Neutrino Observatory [2] (SNO) and Super-Kamiokande [3] (SK) consistently determine θ_{12} below $\sim \pi/2$ (see Figure 1), corresponding to a "normal ordering" of the the first and second mass eigenstate, or $\Delta m_{21}^2 > 0$.

Inside the earth, the oscillations of GeV-scale neutrinos (such as atmospheric neutrinos) are enhanced, if the 'resonance condition' $l_{\text{ref}} = l_{\text{vac}} \cos 2\theta$ is met.

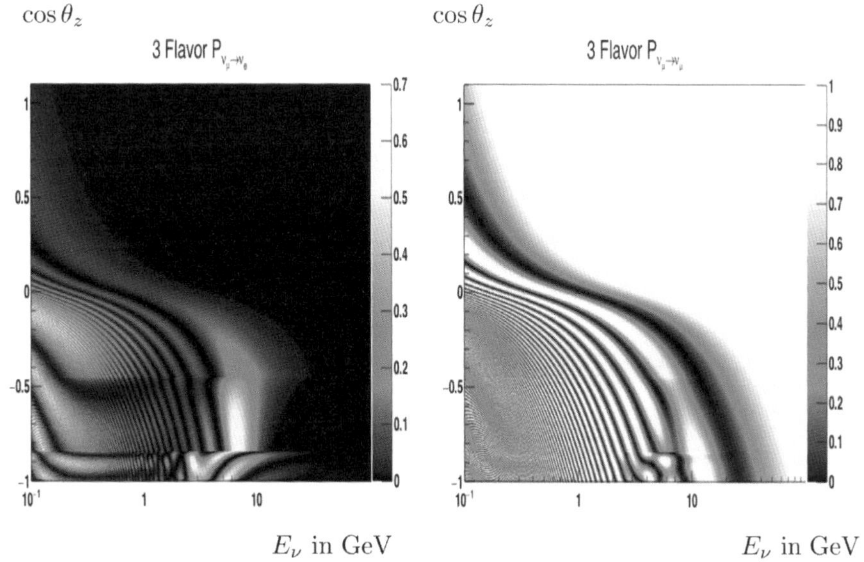

Figure 2: Oscillation probability for atmospheric neutrinos as function of zenith angle θ_z and neutrino energy. The left plot shows the electron-appearance probability, while the right panel is the muon-disappearance probability.

Figure 2 shows the oscillation probabilities inside the earth as a function of zenith angle θ_z and neutrino energy E_ν. The resonant enhancement below $\theta_z = -0.5$ for energies around 7 GeV (as well as the mantle-core boundary at $\theta_z = -0.838$) is clearly visible in the electron-appearance probability. Matter effects in the muon-disappearance channel are smaller, but noticeable. The earth matter effects can be used to determine the yet unknown sign of Δm^2_{31} and thereby decide the neutrino mass ordering: the mass ordering is considered "normal", if $\Delta m^2_{31} > 0$ ($m_1^2 < m_2^2 < m_3^2$) and "inverted", if $\Delta m^2_{31} < 0$ ($m_3^2 < m_1^2 < m_2^2$). Since solar data imply $\Delta m^2_{21} > 0$, and since $|\Delta m^2_{31}| \gg |\Delta m^2_{21}|$ there are only these two possibilities.

Neutrino data is used to measure the "strength" of the matter potential [4] by replacing the matter potential $\sqrt{2} G_F \rho_e$ with $\xi_{\rm MSW} \sqrt{2} G_F \rho_e$ and fit for $\xi_{\rm MSW}$. Figure 3 shows the result for solar *and* earth matter effects (using solar neutrinos and reactor anti-neutrinos) and earth matter effects (using only SK atmospheric neutrinos). For solar neutrinos, $\xi_{\rm MSW} > 0$ with much more than 5σ confidence, but the data also disfavor the standard MSW case $\xi_{\rm MSW} = 1$ at about the 2σ level. The SK atmospheric neutrino data sample favors the presence of matter effects and the normal hierarchy at about the 1 to 1.5σ level.

The earth matter effects can also be probed by long-baseline neutrino beams from accelerators. Figure 4 compares the SK atmospheric neutrino results to

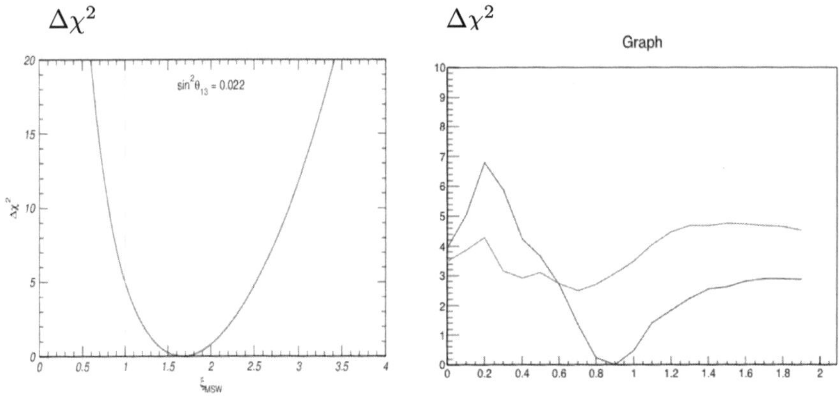

Figure 3: Strength of matter effects obtained by scaling the matter potential $\sqrt{2}G_F\rho_e$ to $\xi_{\rm MSW}\sqrt{2}G_F\rho_e$. Solar neutrino and KamLAND [5] anti-neutrino data [6] (left) imply non-zero solar and earth matter effects. Super-Kamiokande's atmospheric neutrino data (right) favors the presence of earth matter effects. The blue (red) line is for normal (inverted) mass ordering.

the T2K [7] and NOνA [8] (Library Event Matching analysis) long-baseline data. At present, all three data sets prefer normal ordering over inverted ordering. They also favor similar regions of the CP-violating phase δ. While this preference is not yet significant, it illustrates how the earth matter effects can eventually determine the ordering of the neutrino mass eigenestates and impact the search for CP violation in neutrino oscillations.

3 Test of the Standard Matter Effects on Neutrino Oscillations

The presence of matter effects is only statistically significant for solar neutrinos, so only for solar neutrinos, they can be tested in more detail. The adiabatic conversion described in section 1 is energy-dependent; there is very little matter effect for neutrinos below 1 MeV, and neutrinos above ~ 7 MeV are completely converted. The transition between both happens at about 3 MeV. Above ~ 7 MeV, the earth matter effects start to modify the oscillation probability. This modification is mostly a "regeneration" of electron-flavor content. Obviously, the earth matter effects only apply to solar neutrinos passing through the earth: at night time. Earth matter effects can be studied by forming a "day-night asymmetry" $A_{DN} = 2\frac{D-N}{D+N}$, where D (N) stands for the observed neutrino interaction rate during the day (night). Therefore, solar neutrinos allow a direct and straight-forward comparison of neutrinos *with* earth matter effects to neutrinos *without*.

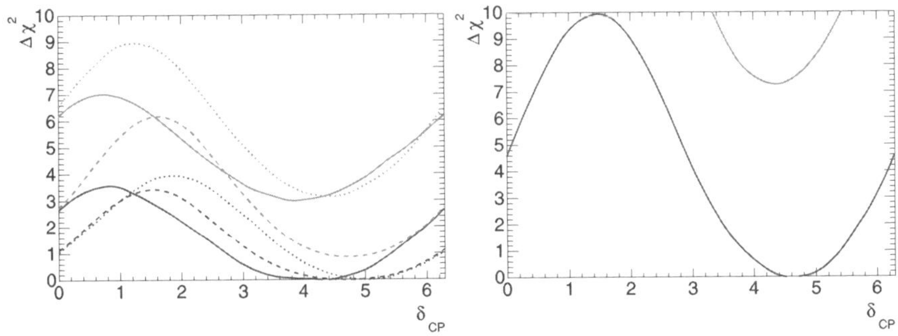

Figure 4: Left: $\Delta\chi^2$ of the SK atmospheric (solid), the T2K long base-line accelerator (dashed) and the NOνA long-baseline accelerator (dotted) analysis. neutrino oscillation analysis as a function of the CP-phase δ for normal (lower, blue) and inverted (upper, red) mass ordering. All three experiments weakly prefer normal mass ordering and a similar value of δ. Right: a naive sum of all three $\Delta\chi^2$ distributions neglecting any correlations (including the other oscillation parameters) shows the consistent favoring of the normal mass hierarchy.

Figure 5 shows all solar neutrino data and the solar neutrino electron-survival probability P_{ee} as a function of neutrino energy: the upper (lower) blue data point is a measurement of the solar *pp* neutrino flux [9] by BOREXINO (all solar neutrino experiments [2, 3, 10–13]. The golden data point is the implied solar *CNO* neutrino flux from the Homestake radiochemical measurement [14] and the water Cherenkov detector data. The red band summarizes the spectral and rate measurements of the ^{8}B neutrino flux by SK [3] and SNO [2]. The other data points are ^{7}Be [11], *pep* [13], and ^{8}B [12] measurements of BOREXINO. The standard MSW resonance describes the solar neutrino data well, but the data uncertainties are large, and measurements rely on the standard solar model predictions for the different neutrino fluxes, except for the ^{8}B data. In particular, the data from SNO and SK *by itself* shows no significant energy dependence in the transition region.

The right panel of the same figure shows the energy dependence of the earth matter effects in SK. Solar neutrinos in SK are detected by elastic scattering with electrons, and the energy shown on the horizontal axis is the kinetic energy of the recoil electrons. The significance of the presence of earth matter effects is about 3σ. The energy dependence is consistent with expectations. The measured day/night asymmetry is $-3.3 \pm 1.1\%$. The equivalent asymmetry for SK implied by SNO data would be $-2.8 \pm 1.8\%$ which is consistent. The day/night asymmetry measured by BOREXINO [15] is $-0.1 \pm 1.4\%$, but it is for neutrinos of a much smaller energy, where little day/night asymmetry is expected.

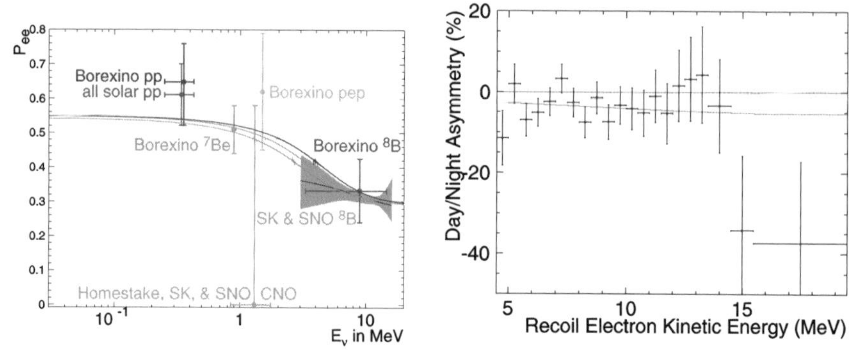

Figure 5: Left: Solar neutrino electron survival probability P_{ee} as a function of energy for three different values of Δm^2_{21}. Right: Observed day/night asymmetry in SK using elastic scattering with electrons. The shown energy-dependence is the kinetic energy of the recoiling electron. The red line indicates the MSW prediction based on the best fit Δm^2_{21} value from all solar data.

4 Measurement of Δm^2_{21} with Matter Effects

The most precise determination of Δm^2_{21} are the reactor anti-neutrino data from KamLAND [5]. It is interesting to compare this value of Δm^2_{21} with one obtained by solar neutrino data, since CPT conservation implies that both are the same. Unfortunately, the uncertainty in Δm^2_{21} from the solar data is considerably larger than that from KamLAND. As shown in Figure 6, there is about 2σ tension between the solar and the KamLAND data. SK data dominate the Δm^2_{21} fit to all solar data, and most of its sensitivity is due to the earth matter effects. The SK day/night measurement contributes about 60% of the tension between solar data and KamLAND. The remaining 40% comes almost exclusively from the energy dependence of the survival probability P_{ee} from solar matter effects. Therefore, this tension is related to the $\sim 2\sigma$ excess in the measured matter potential discussed in section 2.

5 Conclusion

Matter effects on neutrino oscillations don't just enhance neutrino mixing, but also offer interesting tools to understand the parameters governing neutrino oscillations such as mass ordering and the value of Δm^2_{21}. Current experimental data have reached the necessary precision to start making use of these tools: the sign of Δm^2_{21} is already determined from solar matter effects, the sign of Δm^2_{31} may soon follow (using GeV scale neutrinos). Earth (and solar) matter effects provide an interesting independent measurement of Δm^2_{21} using neutrinos to be compared to the KamLAND anti-neutrino measurement of that parameter.

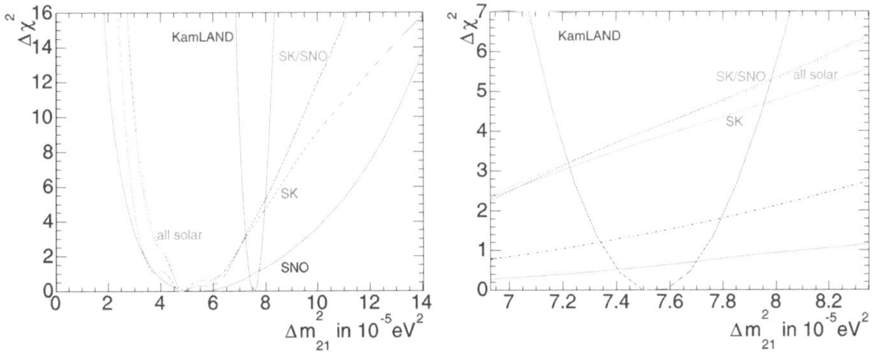

Figure 6: Determinations of Δm^2_{21}. The blue lines use KamLAND reactor anti-neutrino data, the green (black) dotted line in the left panel is a fit to all solar neutrino (SNO) data, the red lines show SK solar neutrino data. There is some tension between the KamLAND best fit and the solar neutrino data. The right panel zooms in and shows in addition the fit to all solar neutrino data without SK (lower black dotted line), and the fit without SK day/night data (upper black dashed-dotted line).

References

[1] S. P. Mikheyev and A. Y. Smirnov, Sov. Jour. Nucl. Phys. **42**, 913 (1985); L. Wolfenstein, Phys. Rev. D **17**, 2369 (1978).

[2] B. Aharmin *et al.*, Phys. Rev. C **88**, 025501 (2013).

[3] Y. Koshio, XXVI International Conference on Neutrino Physics and Astrophysics, AIP Conf. Proc. **1666**, 090001 (2015).

[4] G. L. Fogli *et al.*, hep-ph 0309100; H. Minakata and C. Peña-Garay, hep-ph 1009.4869

[5] S. Abe *et al.*, Phys. Rev. Lett. **100**, 221803 (2008);
The KamLAND Collaboration, arxiv:1303.4667v2 (2013). M. Maltoni and A. Y. Smirnov, arXiv:1507.05287 (2015).

[6] M. Maltoni and A. Y. Smirnov, arXiv:1507.05287 (2015).

[7] K. Abe *et al.*, Phys. Rev. D **91** (2015) 7, 072010.

[8] P. Adamson *et al.*, arXiv:16-1.05022 (2016).

[9] G. Bellini *et al.* Nature 512, 383 (2014).

[10] J. N. Abdurashitov *et al.* (SAGE collaboration), Phys. Rev. C **80**, 015807 (2009); M. Altmann *et al.* (GALLEX Collaboration), Phys. Lett. B **616**, 174 (2005).

[11] G. Bellini *et al.* (Borexino Collaboration), Phys. Rev. Lett. **107**, 141302 (2011).

[12] G. Bellini *et al.* Phys. Rev. D **82**, 033006 (2010).

[13] G. Bellini *et al.* Phys. Rev. Lett. **108**, 051302 (2012).

[14] R. Davis *et al.* (Homestake Experiment), Phys. Rev. Lett. **20**, 1205 (1968).

[15] G. Bellini *et al.* ,Physics Letters **B707**, 22 (2012).

NEW RESULTS FROM THE OPERA EXPERIMENT

Tomoko Ariga[a] on behalf of the OPERA Collaboration
Albert Einstein Center for Fundamental Physics, Laboratory for High Energy Physics, University of Bern, Sidlerstrasse 5, 3012 Bern, Switzerland

Abstract. The OPERA experiment was designed to study $\nu_\mu \to \nu_\tau$ oscillations in appearance mode in the CNGS neutrino beam. The observation of $\nu_\mu \to \nu_\tau$ appearance was previously reported with four ν_τ candidate events. A fifth candidate event has been recently found in an increased data sample. Given the number of analyzed events and the low background, we report the discovery of ν_τ appearance with a significance of 5.1σ.

1 Introduction

Neutrino oscillations have been studied by many experiments with solar, atmospheric, reactor and accelerator-based neutrinos [1]. Most of the current knowledge of neutrino flavor transitions came from the observation of the disappearance of neutrinos of the initial flavor. In order to prove three-flavor neutrino oscillations unambiguously, the detection of neutrino oscillations in appearance mode has been considered.

In recent years, the Super-Kamiokande experiment reported the evidence of a ν_τ appearance signal in atmospheric neutrino data [2], with a signal-to-noise ratio of about one tenth. The T2K experiment has observed $\nu_\mu \to \nu_e$ oscillations in appearance mode [3].

The OPERA experiment was designed to search for $\nu_\mu \to \nu_\tau$ oscillations in appearance mode through the detection of the τ leptons produced in ν_τ charged current interactions. It has operated in low background conditions and with a signal-to-noise ratio of about ten. In 2010, a first ν_τ candidate event was observed [4]. Since then, three more candidate events [5–7] have been detected and allowed us to claim the observation of ν_τ appearance with a significance of 4.2σ [7]. In this paper, new results of the ν_τ analysis based on an enlarged data sample are described.

2 Detector and neutrino beam

In order to meet the experimental requirements of a large target mass and a very high position resolution to detect short-lived τ lepton decays, a target made of lead plates interspaced with emulsion films has been employed. Emulsion is a particle tracking device with an intrinsic resolution of 50 nm and a key technology to detect short-lived particle decays.

The OPERA detector [8] is composed of two identical parts, each consisting of a target section followed by a spectrometer. The target has an average mass

[a]E-mail: tomoko.ariga@lhep.unibe.ch

of about 1.2 kt, consisting of an average number of 141431 target units called bricks. A brick consists of 57 emulsion films [9], interleaved with 1 mm thick lead plates. Bricks are arranged in walls and each of these walls are followed by scintillator strip planes (Target Tracker or TT). Magnetic spectrometers, consisting of iron core magnets instrumented with Resistive Plate Chambers (RPC) and drift tubes, are used for the measurement of the muon charge and momentum. Removable pairs of emulsion films called Changeable Sheet (CS) doublets [10] are used as interface trackers between the TT and the bricks.

The OPERA detector was exposed to the CERN Neutrinos to Gran Sasso (CNGS) beam [11, 12] from 2008 to 2012. The CNGS beam, produced by the SPS 400 GeV protons, had a mean energy of 17 GeV. A sample of 19505 contained neutrino interactions corresponding to 17.97×10^{19} protons on target (pot) has been registered by the detector. A three-dimensional track is tagged as a muon if the product of its length and the density along its path is larger than 660 g/cm^2. An event is classified as 1μ if either it contains at least one track tagged as a muon or the total number of fired TT or RPC planes is larger than 19. The complementary sample is defined as 0μ. A muon momentum cut at 15 GeV/c was introduced to accelerate the analysis since most of signal events are expected in low muon momentum.

3 Event analysis

Bricks predicted to contain a neutrino interaction are extracted from the target wall and the CS doublets are analyzed. Once a brick is validated by the analysis, its emulsion films are developed and dispatched to the scanning laboratories of the Collaboration. All tracks measured in the CS doulets are followed upstream in the brick. The stopping points are considered as either a primary or a secondary vertex.

A further analysis called τ decay search procedure [13] is applied to those located vertices to detect possible decay or interaction topologies. The applied selection criteria for the τ decay search is summarized in Table 1. The variable z_{dec} is the longitudinal coordinate of the decay vertex with respect to the downstream face of the lead plate containing the primary vertex, θ_{kink} is the average three-dimensional angle between the parent and its daughters, p_{2ry} is the scalar sum of the daughters' momentum, p_T^{2ry} is the transverse momentum of the daughter w.r.t. the parent direction, p_T^{miss} is the vectorial sum of the transverse momenta of primaries (except the parent) and daughters w.r.t. beam direction, $\Delta\phi_{\tau H}$ is the angle between the parent and the vectorial sum of the other primaries in the transverse plane, m is the invariant mass of the daughters, and m_{min} is the minimal invariant mass. The details of the event analysis procedure are described in [5].

Charmed particles have similar decay topologies as τ leptons. The detection

Variable	$\tau \to 1h$	$\tau \to 3h$	$\tau \to \mu$	$\tau \to e$
z_{dec} (μm)	[44,2600]	< 2600	[44,2600]	< 2600
θ_{kink} (mrad)	> 20	< 500	> 20	> 20
p_{2ry} (GeV/c)	> 2	> 3	> 1 and < 15	> 1 and < 15
p_T^{2ry} (GeV/c)	$> 0.6(0.3^*)$	/	> 0.25	> 0.1
p_T^{miss} (GeV/c)	< 1	< 1	/	/
$\Delta\phi_{\tau H}$ (rad)	$> \pi/2$	$> \pi/2$	/	/
m, m_{min}	/	> 0.5 and < 2	/	/

Table 1: Selection criteria for the τ decay search. Cut marked with * is applied if there is at least one γ-ray originating from the decay vertex.

of charmed particles was used to verify the efficiency of the search for short-lived particle decays. A total of 50 charm events was observed in a sub-sample of data from the 2008-2010 runs, in good agreement with the expectations [13].

The data set used in the present analysis consists of the 0μ events and the 1μ events with a muon momentum smaller than 15 GeV/c. The analysis has been performed on the first and second most probable bricks for all runs. The numbers of analyzed events are summarised in Table 2.

	2008	2009	2010	2011	2012	Total
pot (10^{19})	1.74	3.53	4.09	4.75	3.86	17.97
0μ events	149	253	269	270	204	1144
1μ events ($p_\mu < 15$ GeV/c)	542	1020	968	966	768	4264
Total of events	691	1273	1236	1236	972	5408

Table 2: Number of events used in this analysis.

4 Description of the ν_τ candidate events

Five ν_τ candidate events have been detected in the analyzed samples: three in the $\tau \to 1h$ decay channel, one in $\tau \to 3h$ and one in $\tau \to \mu$. Four of them were reported previously [4–7] and Fig. 1 shows event displays of these events as reconstructed in the brick.

The fifth candidate event in the $\tau \to 1h$ decay channel has been reported recently [14]. As shown in Fig. 2, a primary track exhibits a kink topology with an angle of (90 ± 2) mrad at the flight length of (960 ± 30) μm. The daughter track interacts in the brick and identified as a hadron. The charm background hypothesis was discarded since the other primary particle is identified as a

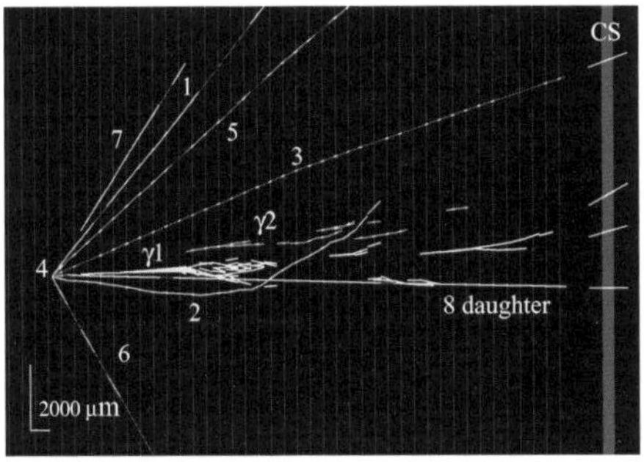

(a) The first candidate [4]

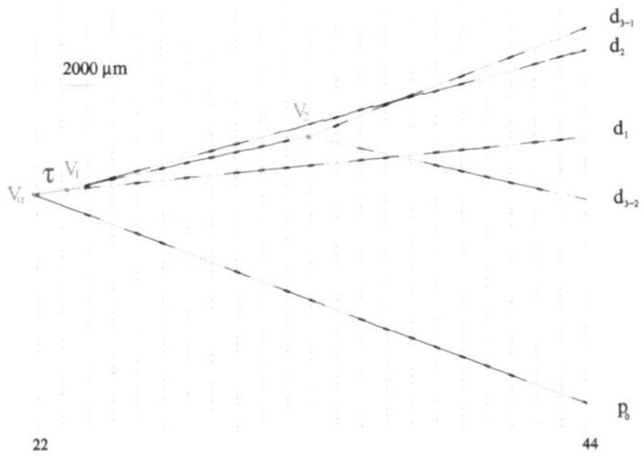

(b) The second candidate [5]

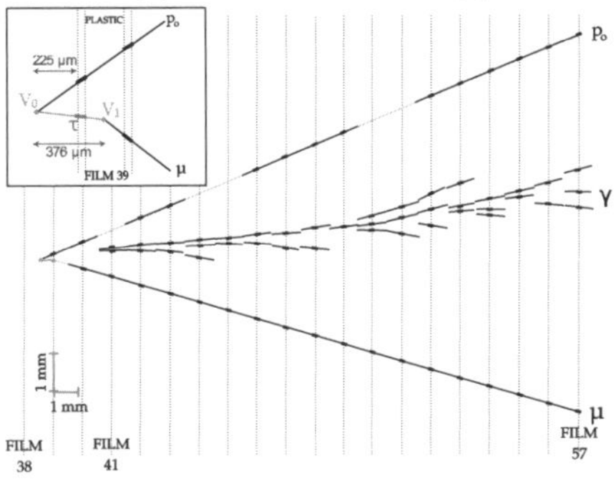

(c) The third candidate [6]

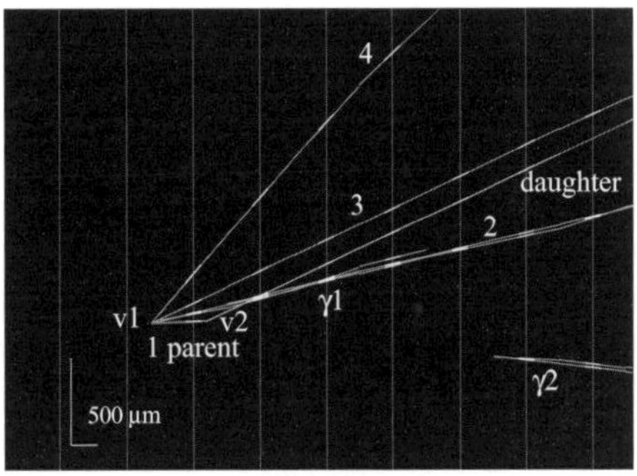

(d) The fourth candidate [7]

Figure 1: Event displays of the four ν_τ candidate events in the projection longitudinal to the neutrino direction.

hadron from its interaction in the downstream brick. The measured values of the kinematical variables of this event satisfy the selection criteria for the $\tau \to 1h$ decay channel.

The scalar sum of the momenta of all particles measured in the emulsion films (p_{sum}) was measured for all ν_τ candidate events and the values are shown in Fig. 3, being compatible with the expectation obtained by the Monte Carlo simulation.

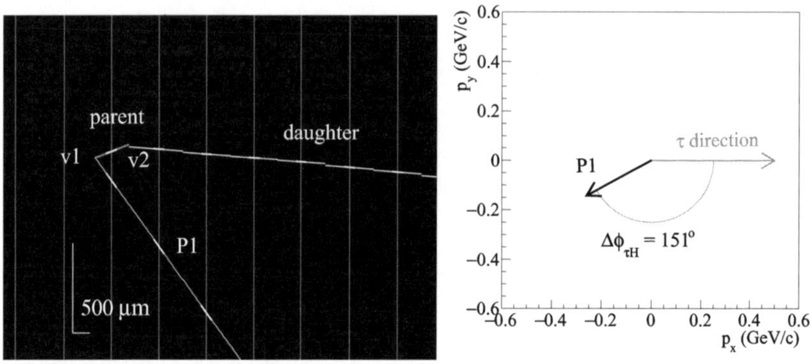

Figure 2: Left: event display of the fifth ν_τ candidate event [14]. Right: the τ direction (red arrow) and the other primary particle (black arrow) in the plane transverse to the beam.

5 Significance of the observation

The estimated signal and background for the observation of ν_τ candidate events in the data sample analyzed in this paper are summarised in Table 3. The details of the estimation are described in [5]. The main background sources are: charmed particles decays, hadronic interactions and large-angle muon scattering (LAS). The estimation of the LAS background has been recently updated [15]. The predicted background events in the $\tau \to \mu$ signal region amounts to $(1.2 \pm 0.1) \times 10^{-7}/\nu_\mu$ CC which is well below the values considered so far.

The significance of the observation of the five ν_τ candidate events is estimated by considering the confidence for excluding the null hypothesis. Two calculations based on the Fisher's method and the likelihood-ratio approach yield a p-value of 1.1×10^{-7}, corresponding to a significance of 5.1σ for the exclusion of the background-only hypothesis.

Given the five observed events and the expected background, the confidence interval of Δm^2_{23} is estimated using three different ways: the profile likelihood

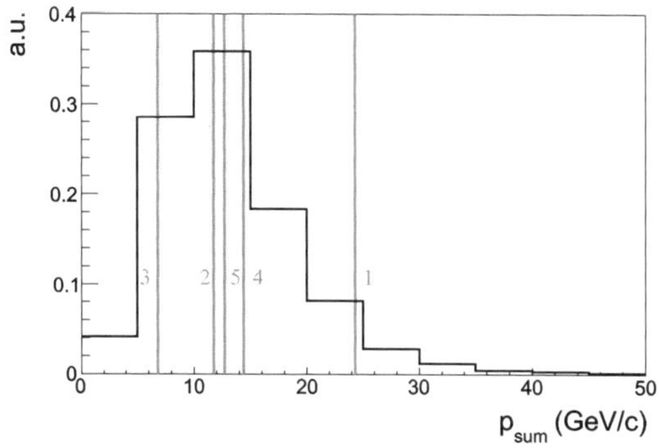

Figure 3: Monte Carlo distribution of the scalar sum of the momenta of all particles measured in the emulsion films, p_{sum}, for ν_τ events passing all the cuts in all decay channels. Red lines show the measured values for the five ν_τ candidate events.

Decay channel	Expected background			Expected signal	Observed
	Charm decays	Hadronic re-interactions	Large-angle muon scattering		
$\tau \to 1h$	0.017 ± 0.003	0.022 ± 0.006	/	0.52 ± 0.10	3
$\tau \to 3h$	0.17 ± 0.03	0.003 ± 0.001	/	0.73 ± 0.14	1
$\tau \to \mu$	0.004 ± 0.001	/	0.0002 ± 0.0001	0.61 ± 0.12	1
$\tau \to e$	0.03 ± 0.01	/	/	0.78 ± 0.16	0
Total	0.22 ± 0.04	0.02 ± 0.01	0.0002 ± 0.0001	2.64 ± 0.53	5

Table 3: Estimated signal and background for the analysed sample and the number of observed events.

ratio, the Feldman-Cousins method and Bayesian statistics. The 90% confidence interval of Δm^2_{23} is $[2.0,\ 5.0] \times 10^{-3}$ eV^2, assuming maximal mixing. The difference among the methods were negligible. More sophisticated analysis using multivariate techniques would be performed in order to improve the statistics of the ν_τ appearance signal and the measurement of Δm^2_{23}.

6 Conclusions

We have observed five ν_τ candidate events in a data sample including the first and second most probable bricks for all runs. The total expected background for the sample is (0.25 ± 0.05) events. A significance of 5.1σ for the exclusion of the null hypothesis is obtained. We therefore report the discovery of ν_τ appearance in the CNGS neutrino beam.

References

[1] K.A. Olive et al. (Particle Data Group), Chin. Phys. **C 38**, 090001 (2014).
[2] K. Abe et al. (Super-Kamiokande Collaboration), Phys. Rev. Lett. **110**, 181802 (2013).
[3] K. Abe et al. (T2K Collaboration), Phys. Rev. Lett. **112**, 061802 (2014).
[4] N. Agafonova et al. (OPERA Collaboration), Phys. Lett. **B 691**, 138-145 (2010).
[5] N. Agafonova et al. (OPERA Collaboration), JHEP **11**, 036 (2013).
[6] N. Agafonova et al. (OPERA Collaboration), Phys. Rev. **D 89**, 051102(R) (2014).
[7] N. Agafonova et al. (OPERA Collaboration), PTEP 2014 **10**, 101C01 (2014).
[8] R Acquafredda et al. (OPERA Collaboration), JINST **4**, P04018 (2009).
[9] T. Nakamura et al., Nucl. Instrum. Meth. **A 556**, 80-86 (2006).
[10] A. Anokhina et al. (OPERA Collaboration), JINST **3**, P07005 (2008).
[11] G. Acquistapace et al., CERN-98-02 (1998).
[12] R. Baldy et al., CERN-SL-99-034-DI (1999).
[13] N. Agafonova et al. (OPERA Collaboration), Eur. Phys. J. **C 74**:2986 (2014).
[14] N. Agafonova et al. (OPERA Collaboration), Phys. Rev. Lett. **115** 12, 121802 (2015).
[15] A. Longhin, A. Paoloni, and F. Pupilli, IEEE Trans. Nucl. Sci. **62**, 5 (2015).

THE MAJORANA DOUBLE BETA DECAY EXPERIMENT: PRESENT STATUS

A.S. Barabash[a], N. Abgrall[b], I.J. Arnquist[c], F.T. Avignone III[d,e], F.E. Bertrand[e], A.W. Bradley[b], V. Brudanin[f], M. Busch[g,h], M. Buuck[i], A.S. Caldwell[j], Y-D. Chan[b], C.D. Christofferson[j], P.-H. Chu[k], C. Cuesta[i], J.A. Detwiler[i], C. Dunagan[j], Yu. Efremenko[l], H. Ejiri[m], S.R. Elliott[k], A. Galindo-Uribarri[e], T. Gilliss[n,h], G.K. Giovanetti[n,h], J. Goett[k], M.P. Green[e], J. Gruszko[i], I.S. Guinn[i], V.E. Guiseppe[d], R. Henning[n,h], E.W. Hoppe[c], S. Howard[j], M.A. Howe[n,h], B.R. Jasinski[o], K.J. Keeter[p], M.F. Kidd[q], S.I. Konovalov[a], R.T. Kouzes[c], B.D. LaFerriere[c], J. Leon[i], M. Lopez[l] J. MacMullin[n,h], R.D. Martin [r], R. Massarczyk[k], S.J. Meijer[n,h], S. Mertens[b], J.L. Orrell[c], C. O'Shaughnessy[n,h], A.W.P. Poon[b], D.C. Radford[e], J. Rager[n,h], K. Rielage[k], R.G.H. Robertson[i], E. Romero-Romero[l,e], B. Shanks[n,h], M. Shirchenko[f], A.M. Suriano[j], D. Tedeschi[d], J.E. Trimble[n,h], R.L. Varner[e], S. Vasilyev[f], K. Vetter[s,b], K. Vorren[n,h], B.R. White[e], J.F. Wilkerson[n,h,e], C. Wiseman[d], W. Xu[k], E. Yakushev[f], C.-H. Yu[e], V. Yumatov[a], I. Zhitnikov[f]

[a] *National Research Center "Kurchatov Institute", Institute for Theoretical and Experimental Physics, Moscow, Russia*
[b] *Nuclear Science Division, Lawrence Berkeley National Laboratory, Berkeley, CA, USA*
[c] *Pacific Northwest National Laboratory, Richland, WA, USA*
[d] *Department of Physics and Astronomy, University of South Carolina, Columbia, SC, USA*
[e] *Oak Ridge National Laboratory, Oak Ridge, TN, USA*
[f] *Joint Institute for Nuclear Research, Dubna, Russia*
[g] *Department of Physics, Duke University, Durham, NC, USA*
[h] *Triangle Universities Nuclear Laboratory, Durham, NC, USA*
[i] *Center for Experimental Nuclear Physics and Astrophysics, and Department of Physics, University of Washington, Seattle, WA, USA*
[j] *South Dakota School of Mines and Technology, Rapid City, SD, USA*
[k] *Los Alamos National Laboratory, Los Alamos, NM, USA*
[l] *Department of Physics and Astronomy, University of Tennessee, Knoxville, TN, USA*
[m] *Research Center for Nuclear Physics and Department of Physics, Osaka University, Ibaraki, Osaka, Japan*
[n] *Department of Physics and Astronomy, University of North Carolina, Chapel Hill, NC, USA*
[o] *Department of Physics, University of South Dakota, Vermillion, SD, USA*
[p] *Department of Physics, Black Hills State University, Spearfish, SD, USA*
[q] *Tennessee Tech University, Cookeville, TN, USA*
[r] *Department of Physics, Engineering Physics and Astronomy, Queen's University, Kingston, ON, Canada*
[s] *Alternate address: Department of Nuclear Engineering, University of California, Berkeley, CA, USA*

Abstract. The MAJORANA Collaboration is constructing the MAJORANA DEMONSTRATOR, an ultra-low background, 44-kg (29 kg ^{76}Ge and 15 kg ^{nat}Ge) modular high purity Ge detector array to search for neutrinoless double-beta decay in ^{76}Ge. In view of the next generation of tonne-scale Ge-based neutrinoless double-beta decay searches that will probe the neutrino mass scale in the inverted hierarchy region, a major goal of the DEMONSTRATOR is to demonstrate a path forward to achieving a background rate at or below 1 count/tonne/year in the 4 keV region of interest around the Q-value at 2039 keV. The current status of the DEMONSTRATOR is discussed, as are plans for its completion.

1 Introduction

Interest in $0\nu\beta\beta$ decay has seen a significant renewal in recent 10-15 years after evidence for neutrino oscillations was obtained from the results of atmospheric, solar, reactor, and accelerator neutrino experiments (see recent reviews [1,2]) . These results are impressive proof that neutrinos have a nonzero mass. The detection and study of $0\nu\beta\beta$ decay may clarify the following problems of neutrino physics (see recent review [3], for example): (i) lepton number non-conservation, (ii) neutrino nature: whether the neutrino is a Dirac or a Majorana particle, (iii) absolute neutrino mass scale, (iv) the type of neutrino mass hierarchy (normal, inverted, or quasidegenerate), (v) CP violation in the lepton sector (measurement of the Majorana CP-violating phases), vi) fundamental weak interections associated with right handed currents, SUSYs and others. The best present limits on effective Majorana neutrino mass $\langle m_\nu \rangle$ are on the level $\sim$ (0.14-0.6) eV and were obtained in experiments with ^{136}Xe [4], ^{76}Ge [5], ^{130}Te [6] and ^{100}Mo [7] (see also recent review [8]). The main goal of future experiments is to reach sensitivity to $\langle m_\nu \rangle \sim$ (0.02-0.05) eV (probe the neutrino mass scale in the inverted hierarchy region). The MAJORANA Collaboration will build an array of isotopically enriched, high purity Germanium (HPGe) detectors in order to search for neutrinoless double-beta decay in ^{76}Ge. Achieving a sensitivity to the $0\nu\beta\beta$ half-life of $\sim 6{\cdot}10^{27}$ years ($\langle m_\nu \rangle \sim$ 0.02-0.04 eV, depending on the nuclear matrix element to be used)) will require a tonne-scale array of detectors with a background rate below 1 count/ROI-t-y in the 4 keV region of interest (ROI) around the 2039 keV Q-value for $\beta\beta$ decay in ^{76}Ge. The MAJORANA DEMONSTRATOR is currently being constructed in order to demonstrate the feasibility of constructing such an array.

2 Overview of the MAJORANA DEMONSTRATOR

The MAJORANA DEMONSTRATOR [9] is an array of enriched and natural germanium detectors that will search for the $0\nu\beta\beta$-decay of ^{76}Ge. The specific goals of the MAJORANA DEMONSTRATOR are several: to demonstrate a path forward to achieving a background rate at or below 1 count/(ROI-t-y) in the 4 keV region of interest around the 2039 keV $Q_{\beta\beta}$ of the ^{76}Ge $0\nu\beta\beta$-decay, when scaled up to a tonne scale experiment; show technical and engineering scalability toward a tonne-scale instrument; prove or disprove the Klapdor-Kleingrothaus claim of the neutrino mass and perform searches for other physics beyond the Standard Model, such as dark matter and axions.

The experiment is composed of 44 kg of high-purity Ge (HPGe) detectors which also act as the source of ^{76}Ge $0\nu\beta\beta$ decay. The benefits of HPGe detectors are that Ge is an intrinsically low-background source material, with understood enrichment chemistry, excellent energy resolution, and event reconstruction capabilities. P-type point contact (PPC) detectors [10, 11] were chosen after extensive R&D by the collaboration for their powerful background rejection capabilities. Twenty nine kg of the detectors are built from Ge material that is enriched to $> 87\%$ in ^{76}Ge and 15 kg are fabricated from natural Ge (7.8% ^{76}Ge). The average mass of the enriched detectors is $\sim$ 850 g and energy resolution lies in the range of 1.6-2.2 keV (FWHM) at 1.33 MeV.

A modular instrument composed of two cryostats built from ultra-pure electroformed copper is being constructed. Each module hosts 7 strings of 3-5 detectors. The modules are operated in a passive shield that is surrounded by a 4π active muon veto. To mitigate the effect of cosmic rays and prevent cosmogenic activation of detectors and materials, the experiment is being deployed at 4850 ft depth (4260 m.w.e. overburden) at the Sanford Underground Research Facility (SURF) in Lead, South Dakota [12]. Passive shield consist of high density polyethylene (30 cm), lead (45 cm), and copper (10 cm). Active background rejection approaches include a muon-veto system, rejection of multi-site events, anti-coincidence cut and rejection of time correlated events. Most γ-ray backgrounds will appear as multi-site events due to Compton scattering, and the low electric fields provided by PPC detectors produce multiple rises in these events that can be found by various analysis techniques. Furthermore, all components inside of the shielding will consist of radio-pure materials. In particular, the copper that forms the structure of the array will be ultra-pure electroformed copper (EFCu), produced and machined on the 4850 ft level of SURF. The amount of material placed inside of the shielding other than the HPGe detectors and EFCu is minimized. In order to verify the purity of materials used for the MAJORANA DEMONSTRATOR, an extensive radio-assay and background-modeling campaign has been carried out. A schematic drawing of the MAJORANA DEMONSTRATOR is shown in Figure 1.

The main technical challenge of the MAJORANA DEMONSTRATOR is to reach a background rate of 3 counts/(ROI-t-y) after analysis cuts, which projects to a background level of 1 count/(ROI-t-y) in a large scale experiment after accounting for additional improvements from thicker shielding, better self-shielding, and if necessary, increased depth. This background level represents a substantial improvement over previous generation experiments [5]. To achieve this goal, background sources must be reduced and offline background rejection

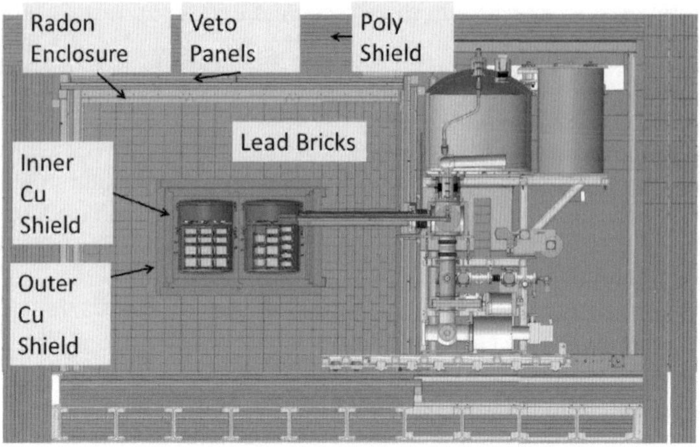

Figure 1: Schematic drawing of the MAJORANA DEMONSTRATOR shown with both modules installed.

must be improved. The estimated ROI contributions based on achieved assays of materials sum to <3.5 counts/(ROI-t-y) in the MAJORANA DEMONSTRATOR, and work is in progress to get the final estimate.

3 The MAJORANA DEMONSTRATOR Implementation

The MAJORANA DEMONSTRATOR follows a modular implementation to be easily scalable to the next generation experiment. The modular approach allows the assembly and commissioning of each module independently, providing a fast deployment and minimum interference with already-operational detectors.

As a first step, the prototype module, an initial prototype cryostat fabricated from commercially produced copper, was loaded with three strings of natural-abundance germanium detectors and placed into the shielding. It has taken data from June 2014 through July 2015. It has served as a test bench for mechanical designs, fabrication methods, and assembly procedures to be used for the construction of the electroformed-copper Modules 1 & 2. The data analysis is in progress.

The second step of deployment is Module 1 which has already been assembled. Module 1 houses 17 kg of enriched germanium detectors and 6 kg of natural germanium detectors. The strings were assembled and characterized in dedicated String Test Cryostats. Then, Module 1 was moved into the shield; it is taking commissioning data at the moment. Figure 2 shows Module 1 detectors and Module 1 being moved to the shield. Module 1 data taking, searching for $0\nu\beta\beta$ decay and including data blindness is foreseen to start soon. Finally, the last step is Module 2 which is composed of 12 kg of enriched and 9 kg of natural Ge detectors. The assembly of Module 2 has already started and its commissioning will take place by the summer of 2016.

Figure 2: Module 1 detectors (left) and Module 1 being moved into the shield (right).

Acknowledgments

This material is based upon work supported by the U.S. Department of Energy, Office of Science, Office of Nuclear Physics. We acknowledge support from the Particle Astrophysics Program of the National Science Foundation. This research uses these US DOE Office of Science User Facilities: the National Energy Research Scientific Computing Center and the Oak Ridge Leadership Computing Facility. We acknowledge support from the Russian Foundation for Basic Research. We thank our hosts and colleagues at the Sanford Underground Research Facility for their support.

References

[1] S.M. Bilenky, hep-ph/1408.2864.
[2] M.C. Gonzalez-Garcia, M. Maltoni, T. Schwetz, hep-ph/1512.06856.
[3] S.M. Bilenky and C. Giunti, Int. J. Mod. Phys. **A 30**, 1530001 (2015).
[4] A. Gando et al., Phys. Rev. Let. **110**, 062502 (2013).
[5] M. Agostini et al., Phys. Rev. Let. **111**, 122503 (2013).
[6] E. Andreotti E. et al., Astropart. Phys. **34**, 822 (2011).
[7] R. Arnold et al., Phys. Rev. **D 92**, 072011 (2015).
[8] A.S. Barabash, AIP Conf. Proc. **1686**, 020003 (2015).
[9] N. Abgrall et al., Adv. High Energy Phys. **2014**, 365432 (2014).
[10] P. S. Barbeau et al., JCAP **9**, 009 (2007).
[11] P. N. Luke et al., IEEE Transactions on Nuclear Science **36**, 926 (1989).
[12] J. Heise, J. Phys.: Conf. Ser. **606**, 012015 (2015).

LATEST RESULTS FROM T2K

Marcela Batkiewicz [a], for the T2K collaboration

Institute of Nuclear Physics Polish Academy of Sciences, Cracow, Poland

Abstract. The T2K (Tokai to Kamioka) experiment is a long-baseline accelerator neutrino oscillation experiment. Intense muon neutrino and antineutrino beams are produced at the J-PARC accelerator complex situated in Tokai. After 280 metres the beam goes through the set of two near detectors: INGRID and ND280, where the beam parameters before oscillations are measured and neutrino cross-sections are studied. After another 295 km, the distance corresponding to the first oscillation maximum, the neutrino beam passes through the Super-Kamiokande far detector. Comparison of the beam characteristics in the near and far detectors makes it possible to determine the oscillation parameters.

Based on muon neutrino disappearance, T2K has delivered the world's leading measurement of the θ_{23} mixing angle. It has been also the first experiment to observe electron neutrino appearance, with a significance of 7.3σ. These results made it possible to determine the θ_{13} mixing angle and to provide the first hint of a non-zero value of the δ_{CP} phase. The T2K experiment has also delivered several neutrino cross-section measurements at neutrino energies around 1 GeV. Currently, T2K is collecting data with a muon antineutrino beam, for the δ_{CP} and antineutrino cross-section measurements.

A summary of the most recent oscillation measurements, including the new electron antineutrino appearance results, is presented.

1 Neutrino oscillations

Neutrino oscillations is a quantum-mechanical phenomenon in which neutrinos change their flavour from one to another while propagating in space. This results from the fact that the neutrino flavour eigenstates $(\nu_e, \nu_\mu, \nu_\tau)$ are linear combinations of mass eigenstates (ν_1, ν_2, ν_3) and their masses (m_1, m_2, m_3) are different from each other, which means that at least two of the masses are non-zero.

The relation between the neutrino flavour and mass eigenstates is described by the Pontecorvo-Maki-Nakagawa-Sakata (PMNS) unitary mixing matrix U:

$$\begin{pmatrix} \nu_e \\ \nu_\mu \\ \nu_\tau \end{pmatrix} = U \begin{pmatrix} \nu_1 \\ \nu_2 \\ \nu_3 \end{pmatrix} \tag{1}$$

The U matrix can be represented using three mixing angles $(\theta_{12}, \theta_{13}, \theta_{23})$ and the imaginary phase responsible for CP symmetry violation δ_{CP}:

$$U = \begin{pmatrix} 1 & 0 & 0 \\ 0 & c_{23} & s_{23} \\ 0 & -s_{23} & c_{23} \end{pmatrix} \begin{pmatrix} c_{13} & 0 & s_{13}e^{-i\delta} \\ 0 & 1 & 0 \\ -s_{13}e^{i\delta} & 0 & c_{13} \end{pmatrix} \begin{pmatrix} c_{12} & s_{12} & 0 \\ -s_{12} & c_{12} & 0 \\ 0 & 0 & 1 \end{pmatrix} \tag{2}$$

[a]E-mail: Marcela.Batkiewicz@ifj.edu.pl

where $c_{ij} = \cos\theta_{ij}, s_{ij} = \sin\theta_{ij}$ and $\delta = \delta_{CP}$. The last two independent parameters describing the neutrino oscillations are the neutrino squared mass differences: $\Delta m^2_{21} = m^2_2 - m^2_1$ and $\Delta m^2_{32} = m^2_3 - m^2_2$.

The current values of the known oscillation parameters are as follows [1]:

$$\begin{aligned} \sin^2 2\theta_{12} &= 0.857 \pm 0.024 \\ \sin^2 2\theta_{23} &= > 0.95 \\ \sin^2 2\theta_{13} &= 0.095 \pm 0.010 \\ \Delta m^2_{21} &= (7.50 \pm 0.20) \times 10^{-5}\ eV^2 \\ |\Delta m^2_{32}| &= (2.32^{+0.12}_{-0.08}) \times 10^{-3}\ eV^2 \end{aligned} \tag{3}$$

The outstanding questions in neutrino oscillations that have not yet been solved are:

- the value of δ_{CP},
- the sign of Δm^2_{32} which defines the neutrino mass hierarchy: $m_1 < m_2 < m_3$ for normal hierarchy (NH) or $m_3 < m_1 < m_2$ for inverted hierarchy (IH),
- the octant of the θ_{23} angle ($\theta_{23} < 45°$ or $\theta_{23} > 45°$).

Neutrino oscillations provide evidence for non-zero neutrino masses, something which is often not accounted for in the Standard Model. This discovery was honoured by the last year's Nobel Prize in Physics and this year's Breakthrough Prize in Fundamental Physics.

2 The T2K experiment

T2K (Tokai to Kamioka) is a long baseline neutrino oscillation experiment, performed in Japan by an international collaboration of around 500 people from 59 institutions in 11 countries [2]. An overview of the T2K experiment is shown in Figure 1. The muon neutrino and the muon antineutrino beams are produced in the Japan Proton Accelerator Research Complex (J-PARC) in the Tokai village on the east coast of Japan. After 280 m the beam of unoscillated neutrinos goes through the set of two near detectors: INGRID and ND280. After another 295 km, the distance corresponding to the first oscillation maximum, the beam reaches the far detector Super-Kamiokande (SK). The INGRID near detector is placed on the beam axis, while the ND280 near detector and the SK detector are located 2.5° away from the neutrino beam axis. The off-axis beam has a narrower energy spectrum and the peak energy to be selected in order to maximise the oscillation probability.

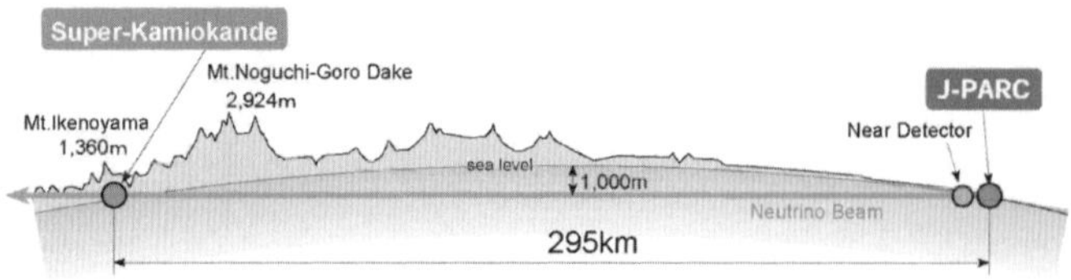

Figure 1: Overview of the T2K experiment.

The main goals of the T2K experiment are the measurements of the probability of muon neutrino disappearance, electron neutrino appearance, the investigation of CP symmetry violation/conservation in the neutrino sector and neutrino-nucleus cross-section measurements at T2K beam energies.

2.1 Neutrino beam

The T2K neutrino beam is produced in the J-PARC accelerator complex, in which protons are accelerated to 30 GeV and impinge on a graphite target, producing mainly pions ($\sim 90\%$) and kaons ($\sim 10\%$). Then, a set of three magnetic horns focuses positive particles in order to obtain the ν_μ beam (neutrino mode) or negative particles for $\bar{\nu}_\mu$ (antineutrino mode). The positive (negative) pions directed to the decay tunnel produce antimuons (muons) and muon neutrinos (muon antineutrinos). The muons and other particles are stopped by the beam dump and in the ground, and the unimpeded neutrino beam travels farther towards the near detectors and the Super-Kamiokande far detector.

2.2 Near detectors

The set of two near detectors (Fig. 2) is located at the J-PARC site 280 m from the graphite target.

The INGRID detector is located at the centre of the neutrino beam and is composed of iron and plastic scintillator layers. Its main goal is the day-to-day monitoring of beam direction and intensity. It also serves to measure the neutrino cross-sections at the energies higher than that of the off-axis beam.

The ND280 detector, positioned 2.5° away from the beam axis, is a magnetised detector consisting of a number of specialised subdetectors, i.e. Time Projection Chambers and subdetectors containing plastic scintillator layers sandwiched with iron, lead, water layers or other material. ND280 is adjusted to measure the neutrino cross-sections on different nuclear targets, as well as to determine the off-axis neutrino flux and flavour composition at the 2.5° angle. These measurements significantly reduce systematic errors in the oscillation analysis.

2.3 Far detector

The Super-Kamiokande detector [3] (Fig. 2) is a 50 kton water Cherenkov detector instrumented with approximately 13000 photomultipliers to register charged particles from the neutrino and antineutrino interactions with water. The particle identification is based on the shape of the Cherenkov light rings: sharp for muons and more fuzzy for electrons. The ring analysis can also provide the information about particle directions and energies. The detector has no magnetic field, therefore it cannot distinguish particles from antiparticles.

Figure 2: Set of the T2K near detectors (left) and the T2K SK far detector (right).

3 Oscillation analysis

The results presented in this paper are based on the data collected by the T2K experiment up to June 2015 which is 7.0×10^{20} POT for the neutrino mode data and 4.0×10^{20} POT for antineutrino mode data.

For both samples the probabilities of the muon neutrino disappearance and electron neutrino appearance were measured. First, the expected number and type of neutrino events in the far detector in the case of no oscillations are predicted. These predictions are based on the neutrino flux simulation and model of neutrino interactions constrained by the measurements of the ND280 detector. The ND280 constraint results in a reduction in the systematic error for the neutrino beam mode analysis from over 20% to around 7% [4]. Then, based on the ratio between measured and predicted number of SK events as a function of neutrino energy, the oscillation parameters are fitted. T2K uses fixed solar oscillation parameters: $\sin^2\theta_{12} = 0.306$ and $\Delta m^2_{21} = 7.5 \times 10^{-5}$ eV^2/c^4 [4,5].

3.1 Neutrino oscillation results

The difference between the muon neutrino spectra measured and expected without oscillations, is shown in Fig. 3 [4,6]. The expected number of the ν_μ

events without oscillation is equal to 446.0 ± 22.5, while the observed number is 120. T2K has provided the world's leading measurement of the θ_{23} angle. The best fit values for muon neutrino disappearance are: $\sin^2\theta_{23} = 0.514$ and $\Delta m^2_{32} = 2.51 \times 10^{-3}$ eV^2/c^4 ($\sin^2\theta_{23} = 0.511$ and $\Delta m^2_{13} = 2.48 \times 10^{-3}$ eV^2/c^4) for NH (IH).

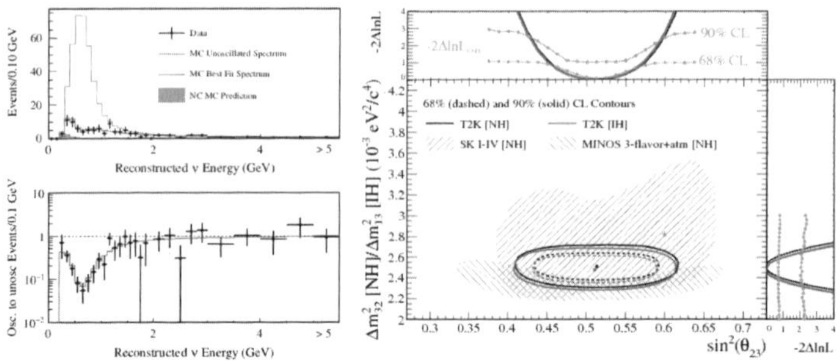

Figure 3: The reconstructed neutrino energy spectrum for the data, best-fit prediction, and prediction for no oscillation (top left) and the ratio to the prediction for no oscillation (bottom left) [4], as well as the 68% and 90% C.L. confidence regions for $\sin^2\theta_{23}$ and Δm^2_{32} (NH) or Δm^2_{13} (IH) (right) [6].

For the ν_e appearance analysis [4,7] 4.92 ± 0.55 events were expected in the case of no oscillation, and the measured number of events was 28. This means that the electron neutrino appearance was confirmed at the level of 7.3σ. The best fitted value, assuming $\delta_{CP} = 0$, is $\sin^2 2\theta_{13} = 0.14$ ($\sin^2 2\theta_{13} = 0.17$) for NH (IH). The allowed regions for $\sin^2 2\theta_{13}$ as a function of δ_{CP} are shown in Fig. 4. The yellow band shows the average θ_{13} value from [5]. Based on the T2K and reactor experiments data [5,8–10] the region of δ_{CP} excluded at 90% C.L. for NH is $(0.19, 0.80)\pi$ and for IH: $(-1.00, -0.97)\pi$ $(-0.04, 1.00)\pi$.

3.2 First antineutrino oscillation results

In antineutrino mode the beam contamination from neutrinos is significantly higher than the antineutrino contamination in the neutrino beam. As previously stated, the SK detector cannot distinguish particles from antiparticles and, therefore, the selected events contain both neutrinos and antineutrinos ($\nu + \bar{\nu}$). The preliminary results for muon antineutrino disappearance in the antineutrino beam mode data were shown in [11], the number of predicted $\nu_\mu + \bar{\nu}_\mu$ events without oscillation is 104 (Fig. 5). The measured number of such events is 34. The best fit values for the oscillation parameters for antineutrinos are as follows: $\sin^2\theta_{23} = 0.515$ and $|\Delta m^2_{32}| = 2.33 \times 10^{-3}$ eV^2/c^4. With this result, T2K has reached the world's best measurement of θ_{23} angle in antineutrino mode. The parameter values obtained for neutrinos and antineutrinos are in agreement within the errors, as predicted by CPT conservation.

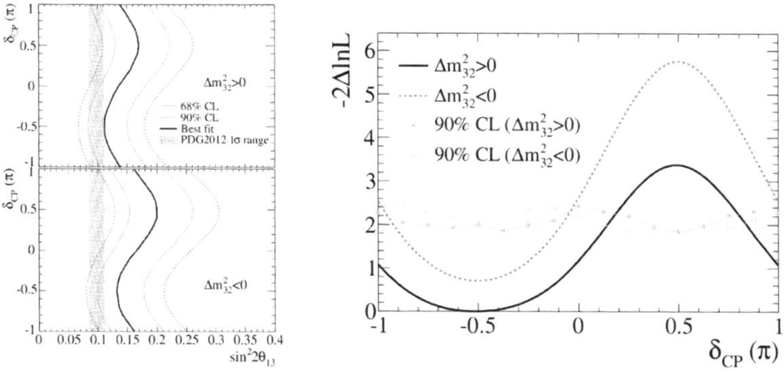

Figure 4: The 68% and 90% C.L. allowed regions for $\sin^2 2\theta_{13}$ as a function of δ_{CP} assuming NH (top left) and IH (bottom left). The δ_{CP} values above 90% C.L. (right) [7].

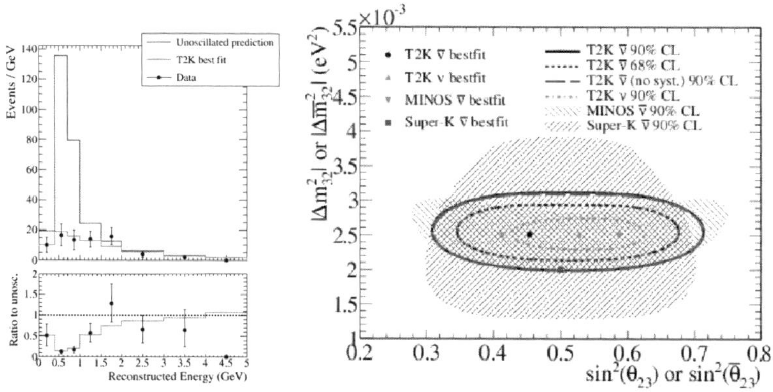

Figure 5: The reconstructed antineutrino energy spectrum for the data, best-fit prediction, and prediction for no oscillation (top left) and their ratio to the prediction for no oscillation (bottom left), as well as the 68% and 90% C.L. confidence regions for $\sin^2\theta_{23}$ and Δm^2_{32} (NH) or Δm^2_{13} (IH) (right) for antineutrino beam mode [11].

In the antineutrino mode $\bar{\nu}_e$ appearance analysis [12], the predicted number of $\nu_e + \bar{\nu}_e$ events for the most probable value of $\delta_{CP} = -\pi/2$ is 3.7 (4.2) for normal (inverted) hierarchy. The observed number of events is 3. At this point the statistical uncertainty is too large to draw any conclusions. Collecting more data is crucial here.

4 Summary and outlook

The T2K experiment is a leading long-baseline neutrino oscillation experiment that has provided one of the world's best measurements of the θ_{23} mixing angle for neutrinos and antineutrinos. Both results are consistent with each other as is expected from the CPT conservation. T2K has observed electron neutrino appearance in a muon neutrino beam at the level of 7.3σ. Using ν_e results combined with the measurements of θ_{13} from the reactor experiments the first hint of the non-zero value of the δ_{CP} phase was provided.

To obtain statistically significant results for oscillation parameters in antineutrino mode it is planned to increase the antineutrino beam statistics by an order of magnitude and perform a joint neutrino and antineutrino 3 flavour analyses.

Acknowledgements

I thank the T2K collaboration for selecting me to present these results. This work was partially supported by the Polish National Science Centre, project number: UMO-2014/14/M/ST2/00850.

References

[1] K. A. Olive et al. (Particle Data Group Collaboration), *"The Review of Particle Physics"*, Chinese Physics **C 38**, 090001 (2014)

[2] K. Abe et al. (T2K Collaboration), Nucl. Instrum. Meth. **A 659**, 106 (2011)

[3] Y. Fukuda et al. (Super-Kamiokande Collaboration), Nucl. Instrum. Meth. **A 501**, 418 (2003)

[4] K. Abe et al. (T2K Collaboration), Phys. Rev. **D 91**, 072010 (2015)

[5] J. Beringer et al. (Particle Data Group Collaboration), *"Review of particle physics"*, Phys. Rev. **D 86**, 010001 (2012).

[6] K. Abe et al. (T2K Collaboration), Phys. Rev. Lett. **112**, 181801 (2014)

[7] K. Abe et al. (T2K Collaboration), Phys. Rev. Lett.RL **112**, 061802 (2014)

[8] F. An et al. (Daya Bay Collaboration) Phys. Rev. Lett. **108**, 171803 (2012).

[9] J. Ahn et al. (RENO Collaboration), Phys. Rev. Lett. **108**, 191802 (2012).

[10] Y. Abe et al. (Double Chooz Collaboration), Phys. Rev. Lett. **108**, 131801 (2012).

[11] A. Kaboth, seminar at the KEK High Energy Accelerator Research Organization, 2015, https://www2.kek.jp/physics-seminar/files2015/20150518_T2K.pdf

[12] M. Ravonel, presentation at EPS-HEP 2015 Conference, 2015, https://indico.cern.ch/event/356420/session/10/contribution/322

THE DOUBLE CHOOZ EXPERIMENT - JUNE 2015 RESULTS AND PROSPECTS TO THE NEAR DETECTOR ERA

Thiago Junqueira de Castro Bezerra [a]
Research Center for Neutrino Science, Tohoku University, 980-8578 Sendai, Japan

Abstract. The Double Chooz experiment was designed for the precision measurement of the neutrino mixing angle θ_{13}, using two identical detectors, one far (1050 m) and the other near (400 m) of two commercial nuclear reactors (4 GWth each). The far detector has been taking data since 2011, same year that the collaboration reported an indication of non-zero θ_{13} ever, by reactor neutrinos, while the near detector is operational since last year. In this talk I give an overview of the experiment and its latest results on θ_{13} using the n-H channel.

1 Neutrino Oscillation

Neutrino oscillations, along with gravity and dark matter/energy, is a phenomenon that is not accounted by the Standard Model theory of the elementary particles and their interactions. Since the first strong evidence was delivered in the end of last century, our knowledge about neutrino oscillations increased, reaching what we can call now the "precision era", where the parameters are being measured on the percent level uncertainty.

This talk, at the 17th Lomonosov Conference, is to summary the latest efforts of the Double Chooz (DC) collaboration on the precision measurement of the neutrino mixing angle θ_{13}, given in greater details in [1].

2 Double Chooz Detector

The DC detector [2] is regarded as an upgrade of the CHOOZ experiment, having a geometry that became default for the current neutrino reactor experiments. Figure 1 shows the detector and its parts that are known as the Inner Detector (ID), Inner Veto (IV) and Outer Veto (OV). The ID is formed by four concentric cylindrical volumes. From the inner most to the outermost, they are: (I) the Neutrino Target (NT), a 10 m^3 acrylic vessel filled with a Gadolinium (Gd) doped liquid scintillator. (II) The Gamma Catcher (GC), which is a 23 m^3 acrylic vessel surrounding the NT, and it is filled with non-doped liquid scintillator. (III) The Buffer, a stainless-still vessel containing 110 m^3 of mineral oil. On the Buffer walls, 390 Photomultiplier-Tubes (PMTs) are attached facing the NT, and they convert the visible photons emitted by the scintillator into an electronic signal [3,4].

Optically separated form the ID, is the IV that is another stainless-steel cylinder. The IV is filled with 90 m^3 of liquid scintillator and contains 78 PMTs. Its purpose is to veto cosmic muons (μ) and external radiation entering

[a]E-mail: `thiago@awa.tohoku.ac.jp`

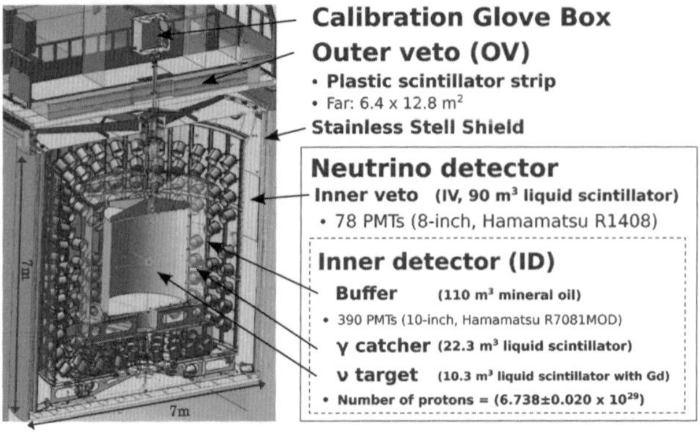

Figure 1: The DC detector and its contents. See the text for more details of each part.

the detector from the surrounding rocks. The OV is an array of plastic scintillator strips (91 m^2) placed above the ID and IV. It also tags cosmic-μs, which come downwards. The ID and IV PMTs signals are read by the Data Acquisition Electronics (DAQ), consisting of custom made Front-End Electronics and trigger boards, and Flash-ADCs [5]. Surrounding the IV and ID the is also a 15 cm thick steel shield, placed in order to reduce further the external radiation entering the detector. Finally, the NT and GC are accessed by a chimney, used for the filling of these volumes, connected to a glove box for radioactive source calibration.

Among the possible weak interactions to detected a neutrino, the Double Chooz collaboration uses the Inverse Beta Decay (IBD), given by

$$\bar{\nu}_e + p \to n + e^+, \tag{1}$$

which is the same process used by Cowan and Reines on their historical first observation of the neutrino in 1956. The positron (e^+) will deposit its energy in the ID and annihilate with a surrounding electron, while the neutron (n) thermalizes and will be captured by the Gd. These two consecutive events are called *prompt* and *delayed*, respectively. The addition of the Gd was envisaged to reduce accidental coincidences that mimics a true IBD signal, because Gd has a large thermal neutron capture cross-section, reducing the time between the prompt and delayed event, and it emits gammas with total energy of about 8 MeV, well above the high rate environmental radioactivity. However, n can also be captured by a Hydrogen nucleus (H), of the liquid scintillator, emitting a 2.2 MeV gamma. While with a lower n capture cross-section and emission of a gamma energy on the environmental high rate energy range, therefore regarded as not ideal for precision experiments at shallow depths, this paper shows the improvements developed by DC that brings the uncertainty of the H channel on the same order as for the Gd, for θ_{13} measurement.

3 IBD Selection and Backgrounds

An IBD candidate is searched in the single event triggers by looking for two time and spatial coincident events, that follows other series of cuts and vetoes to maximise the signal over background ratio. The first cut is on the reconstructed visible energy (E_{vis}), where the prompt candidate is required to have an energy between 1.0 and 20.0 MeV, well covering the reactor neutrino spectrum, while the delayed event needs to have its energy between 1.2 and 3.0 MeV, due to the n-H emitted gamma. As for the correlations, both prompt and delayed need to have a time difference on the (0.5, 800) μs window, and have a reconstructed vertices separated by no more than 120 cm. Finally, an isolation cut requires that there is no other triggers rather than the candidate pair on a time window of (-800, 900) μs, centred in the prompt event. All these cuts were enlarged significantly, when compared with the previous analysis [6], to improve the selection inefficiency.

The backgrounds that can mimic a true IBD signal, passing all the cuts described above, are divided in three categories: (I) Accidentals, when two uncorrelated signal happens with similar energy, time and space correlation as an IBD; (II) Cosmogenic, when an high energy cosmic muon interacts with nuclei inside de detector production long-lived $\beta - n$ emitters radionuclide, such as ^{9}Li; (III) Correlated, when n produced by spallation processes scatters a proton (*prompt*) before being captured by a H nucleus (*delayed*), or when an untagged μ decays inside the NT.

In the n-H analysis, the background that contributes the most is the accidentals, due to the low energy of the gamma released by the neutron capture, which is in the same range as the high rate environmental events. In order to overcome it, the DC collaboration developed an analysis based on multi-variable, instead of the single-cut approach of [6]. A method based on Artificial Neural Network (ANN) is used, taking advantage of the fact that the distributions of the delayed energy, time and space correlations are different when comparing an accidental sample and the MC IBD signal. Figure 2 shows a 3D plot of these three variables for both cases, as illustration. A second method for accidental reduction developed, is the use of the IV to tag high energy gammas produced mainly in the surrounding rocks, that enters the detector and mimicking the delayed event. Such tagging reduced by 25% the accidentals, after the ANN veto is already applied.

Regarding correlated background, a novel analysis using the PMTs signal pulse recorded by the DAQ system was developed that reduced its contribution to the signal. The Multiplicity Pulse Shape (MPS) veto tags fast-n events by

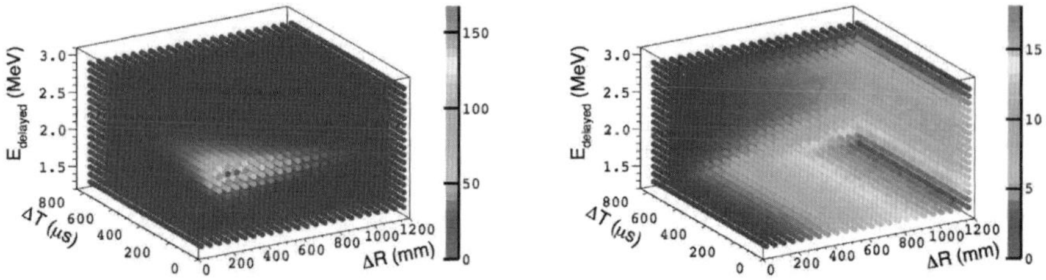

Figure 2: Three variables histogram (delayed energy, time and space correlation) of the MC IBD sample, on the left, and measured accidental sample, on the right, which is used for the training of the ANN.

identifying small energy deposits in the ID, which can be due to proton recoils before the main signal in the FADC window. Other vetoes and analysis used in this new n-H analysis were applied in the similar way as in [7]. Table 1 shows the contribution of each background component to the data daily rates in comparison with the previous n-H data release, while Fig. 3 shows a stacked set

Table 1: Comparison of the prediction, backgrounds and data rate uncertainty of the H-II, H-III and Gd-III analysis.

	H-II (day^{-1})	H-III (day^{-1})
IBD Prediction	73.68 ± 2.28	66.0 ± 1.3
Accidental	73.45 ± 0.16	4.33 ± 0.01
Cosmogenic ^{9}Li/^{8}He	2.8 ± 1.2	$0.95^{+0.57}_{-0.33}$
Fast-n + Stopping μ	3.17 ± 0.54	1.55 ± 0.15
Data	151.12	69.9
S/B	0.9	9.2

of histograms of the prompt energy and each background contribution. In these table and figure, it is possible to see the great reduction of the backgrounds contamination and that the final prompt spectrum is dominated by IBD rather than by accidental as on the previous analysis.

4 Neutrino Oscillation Analysis

DC uses two fits procedure for the neutrino oscillation analysis, one based on the event rate and another one based on the spectral energy shape and rate information. The first method is called the Reactor Rate Modulation (RRM) fit [8], and it is based on the data and prediction of the selected event rates on

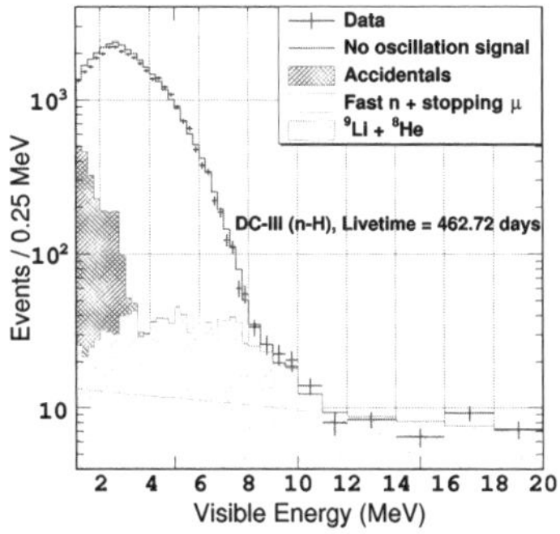

Figure 3: Stacked prompt visible energy spectrum, where the data is compared with the IBD MC signal and background models.

different reactor power scenarios, as expressed by

$$R^{obs} = B + \left(1 - \sin^2 2\theta_{13} \left\langle \sin^2 \frac{1.27\Delta m^2 L}{E_\nu} \right\rangle\right) R^{exp,\, no\, osc}, \tag{2}$$

where R^{obs} is the observed rate, B is the background rate, $R^{exp,\, no\, osc}$ is the predicted event rate without oscillation and the term in parenthesis is the two-flavour survival oscillation probability. This linear dependency is shown on the left of Fig. 4, where the reactor off data [9] is also used to constrain the background rates together with its model, for both n-H and n-Gd selection methods. The right plot of the same figure shows the best-fit values and the confidence level contours for the background rate and θ_{13}. As for the last, the central value and 1σ uncertainty is:

$$\sin^2 2\theta_{13} = 0.088 \pm 0.033. \tag{3}$$

The shape and rate fit, uses as input the prompt spectral shape and rate of the data, prediction and backgrounds to measure the value of $\sin^2 2\theta_{13}$. It is a regular χ^2 method where covariances matrices for the systematic shape uncertainties and pull terms for the known rates and variables. Figure 5 shows the ratio between the best fit and the non-oscillated spectra, given as an output of the fit method. In this figure a "bump" is also present on the 4 to 6 MeV range for the n-H analysis, which was already reported before [7] for the n-Gd analysis. So far, there is no known background model that could accommodate this difference. Comparing data with the MC model of the energy response at this range, also using n-C captures and ^{12}B decay, did not show any energy reconstruction misbehaviour as well. However, a correlation between the excess and the reactor power was observed, giving an evidence that the current models of reactor neutrino flux might be insufficient. Since for DC the oscillation

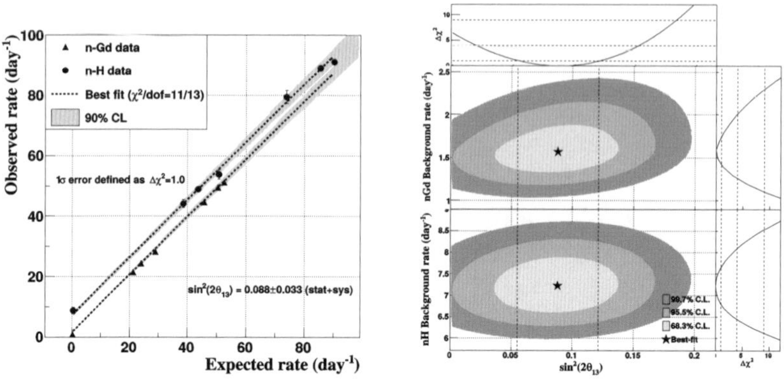

Figure 4: Combined n-Gd and n-H RRM fit on the left, where the reactor off data and background expectation are used, assuming uncorrelated background uncertainty and fully correlated flux uncertainty between the data sets. On the right, the parameter space of the two fitted parameters, $\sin^2 2\theta_{13}$ and total background rate. The curves represents the 68.3%, 95.5% and 99.7% C.L., and the one-dimension χ^2 curves are displayed in the upper and lateral insets.

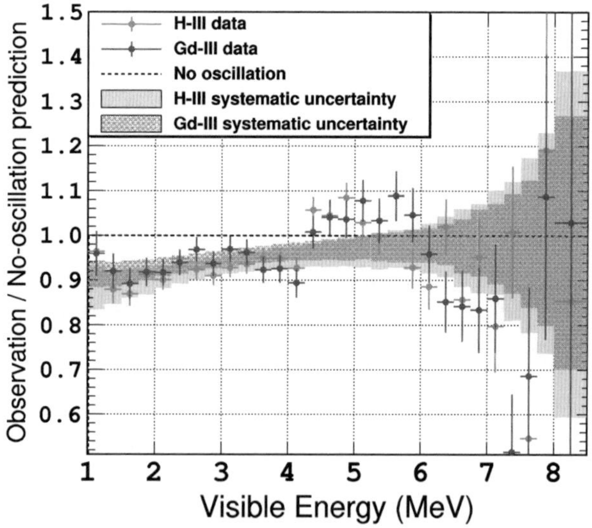

Figure 5: Ratio between the best-fit and non-oscillated background reduced prompt spectra for both n-Gd and n-H analysis. A structure ("bump") can be seen between 4 and 6 MeV for both data-sets.

happens mainly below this energy range, the $\sin^2 2\theta_{13} = 0.124^{+0.030}_{-0.039}$ given by the fit is not affected much by the "bump", and agrees with the value measured by the RRM fit.

5 Summary

Following a regular yearly data-release, in 2015 the DC collaboration released an update of the neutron capture on Hydrogen nuclei. Comparing with the previous analysis, the current one successfully reduced the accidental background, which is the main component, by a factor 17 using two additional vetoes: the ANN and IV tagging methods. Additionally, 25% of the correlated background was further reduced by the new MPS method. Combining this n-H sample with the n-Gd in the RRM fit gives $\sin^2 2\theta_{13} = 0.088 \pm 0.033$. A spectral shape and rate fit gives a similar result and presents a "bump" on the 4 to 6 MeV range in agreement as reported in 2014 by the n-Gd analysis. The near detector is operational since the end of 2014, and a further improvement on the measurement is expected soon.

References

[1] Double Chooz Collaboration, arXiv:1510.08937 [hep-ex] (2015) (accepted for publication on JHEP)

[2] C. Anatael for DC Collaboration, Nucl. Instrum. Meth. A **617** (2010) 473-477; F. Suekane for DC Collaboration, Nucl. Instrum. Meth. A **623** (2010) 440-441; D. Greiner *et al.* Nucl. Instrum. Meth. A **581** (2007) 139-142; A. Tonazzo *et al.* for DC Collbaoration, Nucl. Phys. Proc. Suppl. **172** (2007) 41-44; J.V. Dawson for DC Collaboration, Nuovo Cim. C035N1 (2012) 127-132

[3] C. Buck *et al.* Nucl. Phys. Proc. Suppl. **221** (2011) 372; C. Aberle *et al.* JINST **6** (2011) P11006; C. Aberle *et al.* Nucl. Phys. Proc. Suppl. **448** (2012) 229-232

[4] T. Matsubara *et al.*, Nucl. Instrum. Meth. A **661** (2012) 16-25; C. Bauer *et al.*, JINST **6** (2011) P06008; E. Calvo *et al.*, Nucl. Instrum. Meth. A **621** (2010) 222-230; J. Haser *et al.*, JINST **8** (2013) P04029; F. Sato *et al.*, Phys. Procedia **37** (2012) 1164-1170

[5] Y. Abe *et al.*, JINST **8** (2013) P08015; F. Beissel *et al.* JINST **8** (2013) T01003; T. Akiri for DC Collaboration, Nucl.Phys.Proc.Suppl. **458** (2012) 229-232; J. Maeda for DC Collaboration, J.Phys.Conf.Ser. **331** (2011) 022018; T. Konno *et al.*, DOI: 10.1109/NSSMIC.2009.5401999

[6] Double Chooz Collaboration, Phys. Lett. B **723** (2013) 66-70

[7] Double Chooz Collaboration, JHEP **723** (2014) 086; JHEP **1502** (2015) 074
[8] Double Chooz Collaboration, Phys. Lett. B **735** (2014) 51-56
[9] Double Chooz Collaboration, Phys. Rev. D **87** (2013) 011102

RESULTS FROM BOREXINO AT LNGS

Stefano Davini [a], M. Agostini, I. Drachnev, S. Marcocci
Gran Sasso Science Institute (INFN), 67100 L'Aquila, Italy
G. Bellini, B. Caccianiga, D. D'Angelo, M. Giammarchi, P. Lombardi, L. Ludhova, E. Meroni, L. Miramonti, G. Ranucci, A. Re
Dipartimento di Fisica, Università degli Studi e INFN, Milano 20133, Italy
G. Bonfini, P. Cavalcante, F. Gabriele, A. Goretti, Aldo Ianni, Andrea Ianni, G. Korga, A. Razeto, R. Roncin, N. Rossi, R. Tartaglia
INFN Laboratori Nazionali del Gran Sasso, Assergi 67010, Italy
A. Caminata, L. Di Noto, C. Ghiano, M. Pallavicini, G. Testera, S. Zavatarelli
Dipartimento di Fisica, Università e INFN, Genova 16146, Italy
S. Appel, L. Papp, S. Schönert, F. von Feilitzsch
Physik Department, Technische Universität München, Garching 85747, Germany
D. Bick, C. Hagner, M. Kaiser, M. Meyer
Institut für Experimentalphysik, Universität Hamburg, Germany
D. Bravo, S. Manecki, R. B. Vogelaar
Physics Department, Virginia Polytechnic Institute and State University, Blacksburg, VA 24061, USA
F. Calaprice, C. Galbiati, P. Mosteiro
Physics Department, Princeton University, Princeton, NJ 08544, USA
A. Chepurnov, M. Gromov
Institute of Nuclear Physics, Lomonosov Moscow State University, 119899, Moscow, Russia
A. Derbin, V. Muratova, D. Semenov, E. Unzhakov
St. Petersburg Nuclear Physics Institute, Gatchina 188350, Russia
A. Etenko, E. Litvinovich, G. Lukyanchenko, I. Machulin, M. Skorokhvatov
NRC Kurchatov Institute, Moscow 123182, Russia
K. Fomenko, D. Korablev, O. Smirnov, A. Sotnikov, A. Vishneva, O. Zaimidoroga
Joint Institute for Nuclear Research, Dubna 141980, Russia
A. Empl, E. Hungerford
Department of Physics, University of Houston, Houston, TX 77204, USA
D. Franco, D. Kryn, M. Obolensky
Laboratoire AstroParticule et Cosmologie, 75231 Paris cedex 13, France
A. Pocar, K. Otis
Physics Department, University of Massachusetts, Amherst MA 01003, USA
K. Jedrzejczak, M. Misiaszek, M. Wojcik, G. Zuzel
M. Smoluchowski Institute of Physics, Jagellonian University, Krakow, 30059, Poland
(The Borexino Collaboration)

Abstract. The Borexino experiment is an ultra-pure liquid scintillator detector, running at Laboratori Nazionali del Gran Sasso (Italy). Borexino has completed the real time spectroscopy of the solar neutrinos generated in the proton-proton chain in the core of the Sun. This proceedings reviews the Borexino experiment and the first direct measurement of pp solar neutrinos.

[a] E-mail: `stefano.davini@gssi.infn.it`

1 The Borexino detector at LNGS

Borexino is a real-time solar neutrino detector that is designed to detect low energy solar neutrinos. The motivating goals of low energy solar neutrino detection experiments are to directly probe the nuclear reaction processes in the Sun, and explore neutrino oscillations over a broader range of energies than has been done to date.

The Borexino detector is described in detail in [1]. One of the unique features of Borexino is its very low radioactive background (see table 1). The active detector is 278 tons of ultra-pure liquid scintillator composed of pseudocumene (PC) and 2,5-diphenyloxazole (PPO), a wavelength shifter. The scintillator is contained in a thin nylon vessel, shielded by two PC buffers separated by a second nylon vessel. The scintillator and buffers are contained within a 13.7 m stainless steel sphere that is housed in a 16.9 m domed water tank for additional shielding and muon veto [2].

Solar neutrinos are detected by their elastic scattering on electrons in the liquid scintillator. The scintillation light is detected with an array of 2200 photomultiplier tubes mounted on the inside surface of the stainless steel sphere.

2 Experimental phases of the Borexino experiment

Borexino is in data taking since May 2007. Borexino Phase-I covers the period from May 2007 to May 2010. After the purification of the scintillator performed between May 2010 and August 2011, in November 2011 the Phase-II of Borexino started. Borexino Phase-II is expected to last until the end of 2016, before the begin of SOX [3]. The direct measurement of the solar neutrinos from the fundamental *pp* reaction [4] was achieved in Borexino Phase-II.

Borexino Phase-I solar neutrino results, described in detail in [5], include a high-precision measurement of ^{7}Be solar neutrino interaction rate [6], the measurement of the absence of day-night asymmetry of ^{7}Be neutrinos [7], a measurement of ^{8}B solar neutrino rate with a recoil electron energy threshold of 3 MeV [8], and the first time detection of *pep* solar neutrinos and the strongest constraint up to date on CNO solar neutrino flux [9]. Other results include the study of solar and other unknown anti-neutrino fluxes [10], observation of geo-neutrinos [11, 12], measurement of neutrino velocity [13], and searches for solar axions [14].

The radioactive background levels in Borexino Phase-II are further reduced, thanks to a set of scintillator purifications performed after Phase-I. The residual background contaminations are summarised in table 1. The contamination of ^{210}Bi, which is the dominant internal radioactive background in the region of interest for *pep* and CNO solar neutrino detection, is reduced of a factor $\sim$3. The contamination of ^{85}Kr, the main background for ^{7}Be solar neutrinos, is also strongly reduced.

Isotope	Specs for LS	BX Phase-I	BX Phase-II
^{238}U	$\leq 10^{-16}$ g/g	$(5.3 \pm 0.5)\times10^{-18}$ g/g	$\leq 8 \times10^{-20}$ g/g
^{232}Th	$\leq 10^{-16}$ g/g	$(3.8 \pm 0.8)\times10^{-18}$ g/g	$\leq 9 \times10^{-19}$ g/g
^{14}C/^{12}C	$\leq 10^{-18}$	$(2.69 \pm 0.06)\times10^{-18}$	unchanged
^{40}K	$\leq 10^{-18}$ g/g	$\leq 0.4 \times10^{-18}$ g/g	unchanged
^{85}Kr	≤ 1 cpd/100t	(30 ± 5) cpd/100t	≤ 7 cpd/100t
^{39}Ar	≤ 1 cpd/100t	$\ll$ ^{85}Kr	$\ll$ ^{85}Kr
^{210}Po	not specified	$\sim$ 700 cpd/100t (decaying)	$\sim$ 80 cpd/100t (decaying)
^{210}Bi	not specified	$\sim$ 20 – 70 cpd/100 t	(25 ± 2) cpd/100t

Table 1: Residual radioactive contamination of the Borexino liquid scintillator before and after the purification performed in 2010-2011. ^{210}Po rate is a factor 100 less than at the begin of data taking. ^{210}Bi is a factor 2 less than in Phase-I.

3 First direct detection of *pp* solar neutrinos

The ultra-low radioactive backgrounds of Borexino Phase-II allowed the direct measurement of the solar neutrinos from the fundamental *pp* reaction [4]. A test of the electric charge conservation leading to a new limit on the stability of the electron for decay into a neutrino and a single mono-energetic photon was also obtained with Borexino Phase-II data [15].

The analysis procedures are described in detail in [4]. The main backgrounds in the *pp* region of interest are due to ^{14}C β-decay, whose endpoint is 156 keV, and ^{14}C pileup.

The ^{14}C rate has been estimated independently by the main analysis by looking at at samples of data where the event causing the trigger is followed by a second event within the same data acquisition window. These second events are not subject to the trigger threshold, so scintillation events with energy lower than the trigger threshold can be recorded in this sample. This allows to fit these events against the ^{14}C spectral shape [4].

The contribution due to ^{14}C - ^{14}C pileup, whose endpoint is very close to the endpoint of the *pp* induced recoil spectrum, was estimated in an independent data-driven method. In this method we artificially overlap real triggered events with random data obtained from the ends of real trigger windows, thus uncorrelated with the triggering event. These synthetic events are then reconstructed using the same software used for real events, and selected using the same analysis cuts. With this method, we obtained the true rate and spectral shape of the pileup in Borexino [4].

Once the ^{14}C and pileup rates are tightly constrained, we obtained the *pp* neutrino interaction rate by fitting the data spectrum against the known spectral shapes. In addition to *pp* recoils, ^{14}C, and pileup, the other species included in the fit are ^{7}Be, *pep* and CNO solar neutrino recoils, ^{85}Kr, ^{210}Bi, ^{210}Po and

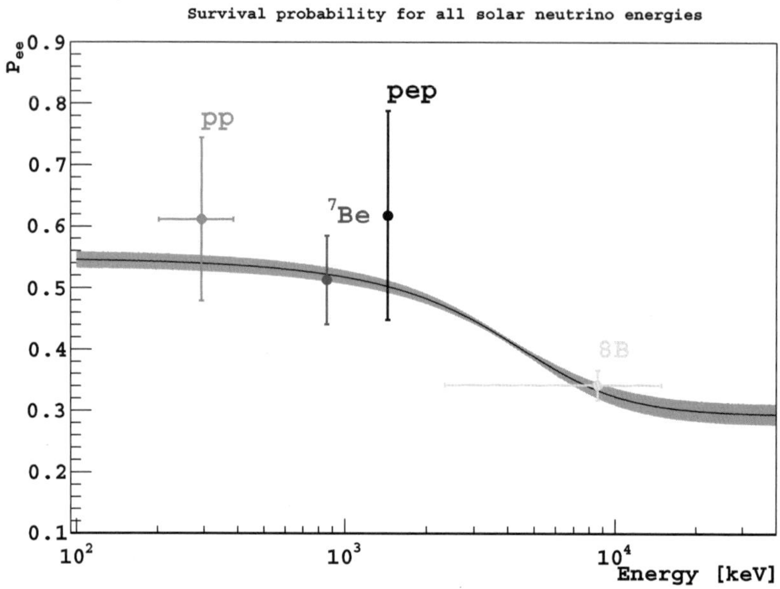

Figure 1: Electron neutrino survival probability (P_{ee}) for solar neutrinos. All the points in this plot are measurements from the Borexino experiment. The shaded gray area is the prediction of the MSW-LMA oscillation model. This figure has been taken from [4].

^{214}Pb. The evaluation of the systematic uncertainty in the fit and methods is described in detail in [4].

The measured *pp* neutrino interaction rate in Borexino is

$$R_{pp} = 144 \pm 13\,(\text{stat}) \pm 10\,(\text{syst})\;\;(\text{days} \cdot 100\text{ ton})^{-1} \tag{1}$$

The electron neutrino survival probability P_{ee} is the probability for a electron neutrino produced in the Sun to be detected in the Earth as electron neutrino. We can measure P_{ee} from our measurement (1), knowing the neutrino interaction cross sections and the electron number density in Borexino and using the prediction of the Solar Standard Model as input. The survival probability for *pp* solar neutrinos is $P_{ee} = 0.64 \pm 0.12$, consistent with the expectation from MSW-LMA oscillation models. The contribution of Borexino to the experimental knowledge of the solar neutrino survival probability is displayed in figure 1.

Using the neutrino oscillation parameters as input, we derived the *pp* solar neutrino flux: $\phi_{pp} = (6.6 \pm 0.7) \times 10^{10}\,\text{cm}^{-2}\,\text{s}^{-1}$. The Solar Standard Model prediction is consistent with our measurement.

4 Conclusion and outlook

Borexino performed the first direct spectral measurement of the neutrinos from the keystone proton-proton fusion. This observation is a crucial step towards the completion of the spectroscopy of the neutrinos emitted in the *pp* chain, as well as further validation of the LMA-MSW model of neutrino oscillation. The measurement strongly confirms our understanding of the Sun.

The main goals for Borexino Phase-II and beyond include the reduction of the uncertainty of the *pep* neutrino flux, the detection of CNO neutrinos, and the reduction of the uncertainty of the ^{7}Be neutrino flux.

One of the aims of Borexino Phase-II is also the search for a fourth neutrino family (sterile neutrinos) with masses in the eV regime and the oscillation length of the order of 1 m. The search will be performed with an artificial neutrino generator in the frame of the dedicated SOX project [3].

Acknowledgments

Borexino was made possible by funding from INFN (Italy), NSF (USA), BMBF, DFG, and MPG (Germany), NRC Kurchatov Institute (Russia), MNiSW (Poland, Polish National Science Center (grant DEC-2012/06/M/ST2/00426)), Russian Foundation for Basic Research (Grant 13-02-92440 ASPERA, the NSFC-RFBR joint research program) and RSCF research program (Russia). We acknowledge the generous support of the Gran Sasso National Laboratories (LNGS). SOX is funded by the European Research Council with grant ERC-AdG-2012 N. 320873.

References

[1] G. Alimonti *et al. (Borexino Collaboration)* Nucl. Instr. Meth. **A600**, 568 (2009).
[2] G. Bellini *et al. (Borexino Collaboration)* JINST **6**, P05005 (2011).
[3] G. Bellini *et al. (Borexino Collaboration)* JHEP **1308**, 038 (2013).
[4] G. Bellini *et al. (Borexino Collaboration)* Nature **512**, 383 (2014).
[5] G. Bellini *et al. (Borexino Collaboration)* Phys. Rev. **D 89**, 112007 (2014).
[6] G. Bellini *et al. (Borexino Collaboration)* Phys. Rev. Lett. **107**, 141302 (2011).
[7] G. Bellini *et al. (Borexino Collaboration)* Phys. Lett. **B 707**, 22 (2012).
[8] G. Bellini *et al. (Borexino Collaboration)* Phys. Rev. **D 82**, 033006 (2010).
[9] G. Bellini *et al. (Borexino Collaboration)* Phys. Rev. Lett **108**, 051302 (2012).
[10] G. Bellini *et al. (Borexino Collaboration)* Phys. Lett. **B 696**, 191 (2011).
[11] G. Bellini *et al. (Borexino Collaboration)* Phys. Lett. **B 687**, 299 (2010).

[12] M. Agostini *et al.* *(Borexino Collaboration)* Phys. Rev. **D 92**, 031101 (2015).

[13] P. Alvarez Sanchez *et al.* *(Borexino Collaboration)* Phys. Lett. **B 716**, 401-405 (2012).

[14] G. Bellini *et al.* *(Borexino Collaboration)* Phys. Rev. **D 85**, 092003 (2012).

[15] M. Agostini *et al.* *(Borexino Collaboration)* Phys. Rev. Lett **115**, 231802 (2015).

NEUTRINOS AND COLLIDER PHYSICS

Frank F. Deppisch [a]
Department of Physics and Astronomy, University College London, London WC1E 6BT, United Kingdom

Abstract. Models of neutrino mass generation constitute well motivated scenarios of Beyond-the-Standard-Model physics. The interplay between high energy collider physics and low energy searches provides us with an effective approach to rule out, constrain and pinpoint such models. In this report, we give a brief overview of examples where collider searches at the LHC can help determine the mechanism of light neutrino mass generation and potentially falsify baryogenesis mechanisms.

1 Introduction

Following the discovery of the Higgs at the LHC, we are close to verifying the mechanism of charged fermion mass generation. What will remain missing though is an understanding of the light neutrino masses. The observation of neutrino oscillations shows that neutrinos have finite masses and that individual lepton flavour is violated. Neutrinos are also usually considered to be Majorana particles, an assumption that facilitates an understanding of their small masses. It is natural to expect that the violation of the individual lepton flavours and, in the case of Majorana neutrinos, the total lepton number will show up in other contexts as well. This for example includes rare lepton flavour violating (LFV) decays of muons/taus and the total lepton number violating (LNV) neutrinoless double beta ($0\nu\beta\beta$) decay.

Quite generally, the possible violation of lepton flavour/number should be searched for at all energies that are experimentally accessible. This is because the observation of such processes would equally allow us a direct insight into the mechanism of neutrino mass generation. The most popular example is the so called seesaw mechanism (of type I) in which heavy right-handed Majorana neutrinos N with masses $\gtrsim 10^{11}$ GeV are added to the Standard Model (SM). Their Yukawa coupling with the left-handed neutrinos induces the light Majorana masses of light neutrinos after electroweak (EW) symmetry breaking. This motivates the lightness of neutrinos through the breaking of lepton number symmetry at a very high scale [1].

Despite its popularity, the generic seesaw mechanism has major phenomenological issues: (i) In the regime $m_N \gtrsim 10^{11}$ GeV, heavy neutrinos are far too heavy to be probed experimentally; (ii) Heavy neutrinos are sterile and they only interact through a small mixing with light neutrinos. In this short proceedings report, we will briefly review two scenarios that instead include TeV scale and potentially non-singlet neutrinos and which can be probed at the LHC. In addition, we will comment on the general impact of the experimental observation of LNV on baryogenesis models.

[a] E-mail: `f.deppisch@ucl.ac.uk`

2 Low Scale Seesaw

In the standard type-I seesaw model with the (one generation) mass matrix for the left- and right-handed neutrino, the mass of the light neutrino ν and its mixing θ with the heavy neutrino N is given by $m_\nu = -m_D^2/m_N$ and $\theta = m_D/m_N = \sqrt{m_\nu/m_N}$, respectively. Here, m_D is the neutrino Dirac mass, expected to be of the order of the EW scale, and m_N is an LNV Majorana mass of the right-handed neutrino. For an observed light neutrino mass scale $m_\nu \approx 0.1$ eV this yields $\theta \approx 10^{-5}\sqrt{\mathrm{GeV}/m_N}$. For a GeV to TeV scale for the heavy neutrino mass, the mixing is rather small. This will be very different in the inverse seesaw scenario [2] described by the mass matrix

$$\begin{pmatrix} 0 & m_D & 0 \\ m_D & \mu_R & m_N \\ 0 & m_N & \mu_S \end{pmatrix}, \tag{1}$$

similarly for the left-handed neutrino, the right-handed neutrino and an additional SM gauge singlet state S. Due to the presence of the small lepton number violating mass parameters μ_R and μ_S, light neutrino masses are achievable for any $\theta = m_D/m_N$ [3]; in the simplest inverse scenario with $\mu_R = 0$ one has $\theta \approx 10^{-2}\sqrt{\mathrm{keV}/\mu_S}$. The reason for this suppression can be understood as the two heavy neutrino states formed by N and S have opposite CP parities and they combine to form quasi-Dirac neutrinos with a fractional mass splitting of order μ_S/m_N. All lepton number violating observables, such as the light neutrino mass, will be suppressed by this small mass splitting.

In order to see the transition between standard and inverse seesaw, we choose $\mu_S = \mu_R$ in eq. (1). Fixing the other terms as $m_N = 1$ TeV and $m_D = 10$ GeV, Fig. 1 (left) shows that successful light neutrino mass generation occurs for $\mu \approx 10^{-6}$ GeV corresponding to the inverse seesaw and $\mu \approx 10^{12}$ GeV corresponding to the normal high-scale seesaw. The inverse case contains two heavy Majorana neutrinos with masses $m_N \pm \mu_S$ constituting a quasi-Dirac state, whereas for large $\mu \gg m_N$ GeV two heavy quasi-degenerate Majorana neutrinos with masses $\mu \pm m_N$ are formed.

3 Heavy Sterile Neutrinos

A large number of laboratory searches put constraints on the mixing between sterile and active neutrinos: For $m_N \ll 1$ MeV, sterile neutrinos are being probed in neutrino oscillation experiments. For pure Majorana sterile neutrinos, $0\nu\beta\beta$ searches provide stringent constraints on the mixing with electron neutrinos [5, 6], but these limits are considerably weakened for quasi-Dirac neutrinos such as found in the inverse seesaw mechanism discussed above. For $1\ \mathrm{MeV} \lesssim m_N \lesssim 1$ GeV, the active-sterile mixing is constrained by peak searches

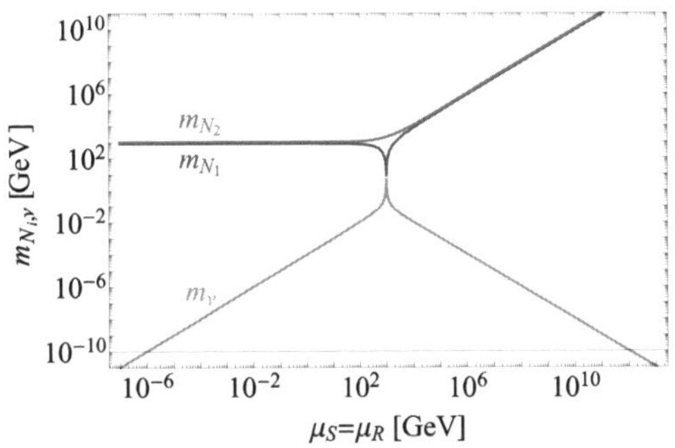

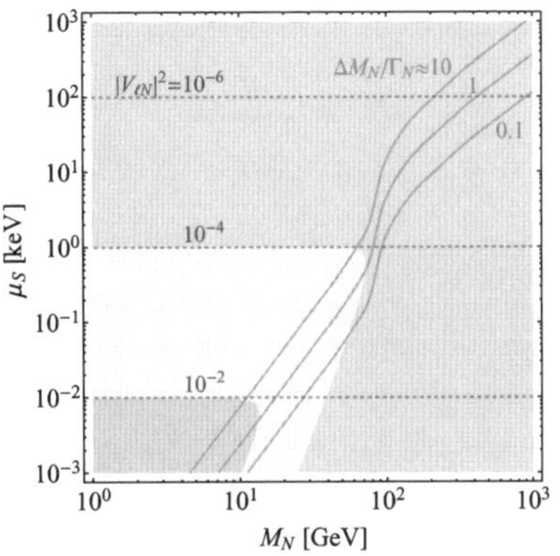

Figure 1: Left: Neutrino mass spectrum as function of $\mu_S = \mu_R$ ($m_N = 1$ TeV, $m_D = 10$ GeV). Right: Ratio of heavy neutrino mass splitting to decay width in the inverse seesaw model (solid red contours). The shaded regions take into account the effect of both the mass and coupling suppression with an LNV signal too small to be observed in the red region.

in leptonic decays of pions and kaons and in beam dump experiments. See [4] for a more complete overview of experimental searches.

Regarding LNV at high energy colliders, a general observation can be made in scenarios with approximately conserved lepton number like the above inverse seesaw mechanism: Like any LNV observable, the rate of an LNV process will be suppressed by the small mass splitting, but for on-shell resonant production of a heavy neutrino, the suppression is with respect to the neutrino width, $\Delta m_N/\Gamma_N$, rather than the absolute mass or the energy scale of the process. For $\Delta m_N \approx \Gamma_N$ it can be resonantly enhanced [7]. The effect of the suppression is shown in Fig. 1 (right) giving contours of $\Delta m_N/\Gamma_N$ as a function of the inverse seesaw parameters m_N and μ_S. Within the red shaded region the suppression by either the mass splitting or the active-sterile mixing $|V_{lN}|^2$ would be too severe to expect an LNV observation in the near future.

In the specific context of the LHC, a Majorana heavy neutrino leads to a LNV signature with two same-sign leptons plus jets and no missing energy: $pp \to W^{(*)} \to N\ell^\pm \to \ell^\pm\ell^\pm jj$ [8]. The CMS and ATLAS collaborations have performed direct searches for the production of heavy neutrinos limiting the mixing to active neutrinos $|V_{e(\mu)N}|^2 \lesssim 10^{-2} - 10^{-1}$ for $m_N \lesssim 500$ GeV at $\sqrt{s} = 8$ TeV [9]. During the ongoing run II of the LHC, the limits could be improved to apply to about a TeV. In addition to the basic s-channel production, it is also very worthwhile to consider other production modes and decay scenarios: Electroweak t-channel processes of the form $pp \to W^*\gamma^* \to N\ell^\pm jj$ can for example give a better sensitivity for higher m_N values.

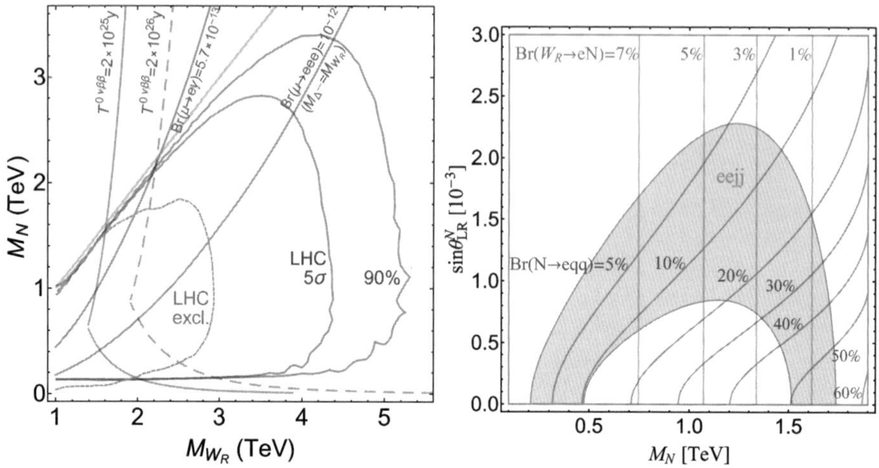

Figure 2: Left: LHC sensitivity of heavy neutrino production via right-handed currents. Solid blue contours give the signal significance at 14 TeV and $\mathcal{L} = 300$ fb^{-1}. Green and red contours show the sensitivity of $0\nu\beta\beta$ and LFV searches. Right: Fitting the CMS $eejj$ excess (red band) in the heavy neutrino mass M_N – light-heavy mixing $\sin\theta^N_{\rm LR}$ plane. The other LRSM parameters are chosen as described in the text. Taken from [16].

4 Left-Right Symmetric Models

One of the simplest options to extend the above sterile neutrino scenario is an additional, broken $U(1)'$ gauge symmetry under which the heavy neutrinos are charged. Under favourable parameter conditions, heavy neutrinos can then be pair-produced abundantly and be probed even for very small mixing with the active neutrinos [10]. Another popular option are left-right symmetric models (LRSMs); the minimal LRSM extends the SM gauge symmetry to $SU(2)_L \times SU(2)_R \times U(1)_{B-L}$ [11]. Leptons are assigned to doublets $L = (\nu, \ell)_L$ and $R = (N, \ell)_R$ under $SU(2)_L$ and $SU(2)_R$, respectively. The Higgs sector of the minimal LRSM consists of a bidoublet and two triplets $\Delta_{L,R}$. The VEV v_R of the neutral component of Δ_R breaks the gauge symmetry $SU(2)_R \times U(1)_{B-L}$ to $U(1)_Y$ and gives masses to the RH gauge bosons W_R, Z_R and the right-handed neutrinos N. LRSMs provide a simple ultraviolet complete seesaw mechanism with the key properties built in: The presence of right-handed neutrinos is a necessary ingredient and the LNV seesaw scale can be identified with the breaking scale of the $SU(2)_R$ symmetry.

With regard to LHC searches, the right-handed current interactions in the LRSM can lead to a significant enhancement of the LNV signal. Even for negligible left-right neutrino mixing, heavy neutrinos can be directly produced via s-channel W_R exchange [12, 13]. Fig. 2 (left) compares the sensitivity of

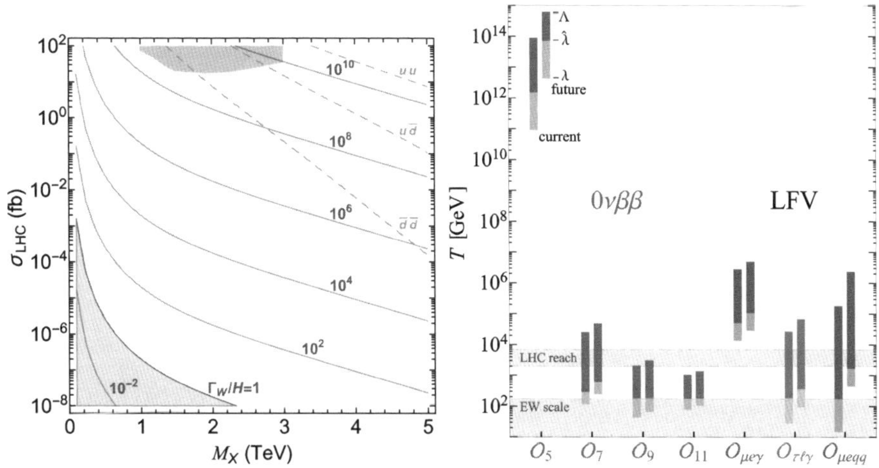

Figure 3: Left: Lepton number washout rate as function of the corresponding LHC cross section and resonance mass (solid contours). Taken from [19]. Right: Temperature intervals where the given LNV and LFV operators are in equilibrium assuming that the corresponding process is observed at the current or future experimental sensitivity. Taken from [20].

such searches with the sensitivity of $0\nu\beta\beta$ and low energy LFV experiments assuming equality of the $SU(2)_L$ and $SU(2)_R$ gauge couplings, $g_R = g_L$.

Both ATLAS and CMS have reported excesses in searches for dibosons, dijets and $e^{\pm}e^{\mp}jj$ using the LHC run I data, situated at invariant masses near 2 TeV. While far from being statistically significant and not supported by the first LHC run II data, their coincidence at a resonant mass of 2 TeV is intriguing. Among several interpretations put forward in the literature, the excesses could be understood as a hint for W_R production with $m_{W_R} \approx 2$ TeV, an $SU(2)_R$ gauge coupling $g_R \approx 0.6 g_L$ and a $W - W_R$ mixing of $\sin\theta^W_{LR} \approx 1.5 \times 10^{-3}$ [14]. Following through in this scenario, the CMS excess in $e^{\pm}e^{\mp}jj$ [15] can be tentatively interpreted as the production of a heavy, quasi-Dirac neutrino as discussed above. This would allow to connect the LHC searches with neutrino physics, specifically the heavy neutrino mass and the strength of its mixing with the light neutrinos [16], cf. Fig. 2 (right).

5 Falsifying High-Scale Baryogenesis

The observed matter-antimatter asymmetry of the universe cannot be understood with SM physics. A large number of possible mechanisms to generate the observed asymmetry have been proposed. An interesting scenario is leptogenesis [17]. In its original formulation, the out-of-equilibrium and CP violating decay of the heavy Majorana neutrinos in the type-I seesaw mechanism create a lepton asymmetry which is then converted into a baryon asymmetry through $(B+L)$ violating EW sphaleron processes [18].

The presence of LNV is a crucial ingredient in leptogenesis. Furthermore, the observation of LNV would have important consequences on the viability of baryogenesis models in general; specifically, it is possible to falsify a large class of high-scale baryogenesis scenarios if LNV was observed at the LHC [19]. For example if a resonant LNV process with the signature $pp \to l^\pm l^\pm jj$ is observed, its LHC cross section $\sigma_{\rm LHC}$ is related to the induced lepton asymmetry washout rate Γ_W/H (relative to the expansion of the universe) [19],

$$\log_{10}(\Gamma_W/H) \gtrsim 6.9 + 0.6(M_X/\text{TeV} - 1) + \log_{10}(\sigma_{\rm LHC}/\text{fb}). \qquad (2)$$

Here M_X is the mass of the hypothetically observed resonance. If $\Gamma_W/H \gg 1$, the dilution of a primordial net lepton number density, understood to be produced in a baryogenesis mechanism at a higher scale, is highly effective and the lepton asymmetry would be washed out before sphaleron processes take effect. This is illustrated in Fig. 3 (left). Observation of LNV at the LHC would therefore strongly constrain baryogenesis scenarios above the scale M_X.

A similar argument can be applied to non-standard mechanisms mediating $0\nu\beta\beta$ decay and low energy LFV processes [20]: if observed, the corresponding processes would be in equilibrium in certain temperature ranges. This is shown in Fig. 3 (right) where the coloured bars denote the efficient equilibration temperatures assuming the relevant observable is seen at the current (left bar) or expected future (right bar) sensitivity. In the case of the 7,9,11-dimensional effective operators $\mathcal{O}_{7,9,11}$ mediating $0\nu\beta\beta$ decay, an electron lepton asymmetry present at higher energies would be washed out. Observation of LFV via 6-dimensional LFV operators at compatible scales would allow to extend the argument to other flavours than the electron.

6 Conclusions

The synergy between low energy/high intensity searches and high energy LHC searches are an effective approach to rule out, constrain and ideally pinpoint neutrino mass models. In this report we have briefly reviewed a few phenomenological scenarios where the LHC can help to determine whether the light neutrino masses, and maybe the matter-antimatter asymmetry as well, are generated in a mechanism close to the electroweak scale or beyond.

Acknowledgements

The author would like to thank the organizers of the 17th Lomonosov Conference on Elementary Particle Physics for the opportunity to contribute to the conference and the proceedings.

References

[1] P. Minkowski, Phys.Lett. **B67**, 421 (1977); M. Gell-Mann, P. Ramond, and R. Slansky, (1979), Print-80-0576 (CERN); R. N. Mohapatra and G. Senjanović, Phys. Rev. Lett. **44**, 912 (1980); J. Schechter and J. W. F. Valle, Phys. Rev. **D22**, 2227 (1980); J. Schechter and J. W. F. Valle, Phys. Rev. **D25**, 774 (1982).

[2] R. N. Mohapatra and J. W. F. Valle, Phys. Rev. **D34**, 1642 (1986).

[3] M. C. Gonzalez-Garcia and J. W. F. Valle, Phys. Lett. **B216**, 360 (1989).

[4] F. F. Deppisch, P. S. Bhupal Dev, and A. Pilaftsis, New J. Phys. **17**, 075019 (2015), 1502.06541.

[5] S. Kovalenko, Z. Lu, and I. Schmidt, Phys.Rev. **D80**, 073014 (2009), 0907.2533.

[6] A. Atre, T. Han, S. Pascoli, and B. Zhang, JHEP **0905**, 030 (2009), 0901.3589.

[7] A. Pilaftsis, Nucl.Phys. **B504**, 61 (1997), hep-ph/9702393.

[8] A. Pilaftsis, Z.Phys. **C55**, 275 (1992), hep-ph/9901206.

[9] CMS Collaboration, V. Khachatryan *et al.*, 1501.05566.

[10] F. F. Deppisch, N. Desai, and J. W. F. Valle, Phys.Rev. **D89**, 051302 (2014), 1308.6789.

[11] J. C. Pati and A. Salam, Phys. Rev. **D10**, 275 (1974); R. N. Mohapatra and J. C. Pati, Phys.Rev. **D11**, 2558 (1975); G. Senjanovic and R. N. Mohapatra, Phys.Rev. **D12**, 1502 (1975).

[12] W.-Y. Keung and G. Senjanovic, Phys. Rev. Lett. **50**, 1427 (1983).

[13] M. Nemevsek, F. Nesti, G. Senjanovic, and Y. Zhang, Phys.Rev. **D83**, 115014 (2011), 1103.1627; J. Aguilar-Saavedra, F. Deppisch, O. Kittel, and J. Valle, Phys.Rev. **D85**, 091301 (2012), 1203.5998; S. Das, F. Deppisch, O. Kittel, and J. Valle, Phys.Rev. **D86**, 055006 (2012), 1206.0256.

[14] F. F. Deppisch, T. E. Gonzalo, S. Patra, N. Sahu, and U. Sarkar, Phys.Rev. **D90**, 053014 (2014), 1407.5384; J. Aguilar-Saavedra and F. Joaquim, Phys.Rev. **D90**, 115010 (2014), 1408.2456; B. A. Dobrescu and Z. Liu, arXiv:1506.06736 [hep-ph]; J. Brehmer, J. Hewett, J. Kopp, T. Rizzo, and J. Tattersall, (2015), 1507.00013.

[15] CMS Collaboration, V. Khachatryan *et al.*, Eur.Phys.J. **C74**, 3149 (2014), 1407.3683.

[16] F. F. Deppisch *et al.*, (2015), 1508.05940.

[17] M. Fukugita and T. Yanagida, Phys. Lett. **B174**, 45 (1986).

[18] V. A. Kuzmin, V. A. Rubakov, and M. E. Shaposhnikov, Phys. Lett. **B155**, 36 (1985).

[19] F. F. Deppisch, J. Harz, and M. Hirsch, Phys.Rev.Lett. **112**, 221601 (2014), 1312.4447; F. F. Deppisch and J. Harz, (2014), 1408.5351.

[20] F. F. Deppisch, J. Harz, M. Hirsch, W.-C. Huang, and H. Päs, Phys. Rev. **D92**, 036005 (2015), 1503.04825.

THE KATRIN EXPERIMENT

Florian M. Fraenkle for the KATRIN Collaboration [a]
Institute for Nuclear Physics, Karlsruhe Institute of Technology, 76344 Eggenstein-Leopoldshafen, Germany

Abstract. The KArlsruhe TRItium Neutrino (KATRIN) experiment is a large-scale experiment for the model independent determination of the effective mass of electron anti-neutrinos with a sensitivity of 200 meV/c^2 (at 90 % C.L.). It investigates the kinematics of electrons from tritium β-decay close to the endpoint of the energy spectrum with a high-resolution electrostatic spectrometer (energy resolution 0.93 eV at 18.6 keV). The experiment is currently in in the final phase of construction at the Karlsruhe Institute of Technology, Germany.
The KATRIN measurement setup consists of a high luminosity windowless gaseous molecular tritium source, a differential and cryogenic pumped electron transport and tritium retention section, a tandem spectrometer section (pre-spectrometer and main spectrometer) for energy analysis, followed by a detector system for counting transmitted β-decay electrons.
This proceedings will give an overview of the KATRIN experiment and its current status. Furthermore, results of the recent commissioning measurements of the KATRIN main spectrometer will be presented.

1 Introduction

The absolute neutrino mass scale is one of the big open questions in particle physics, astrophysics and cosmology. Cosmological observations and neutrinoless double β-decay experiments provide an indirect access to the absolute neutrino mass scale, but are model-dependent. A model independent, direct approach to determine the neutrino mass is the precise investigation of weak decays such as β-decay.

In nuclear β-decay a neutron in the atomic nucleus decays into a proton, thereby emitting an electron (e^-) and an electron anti-neutrino ($\overline{\nu}_e$). The energy released in the decay is divided between the e^- and $\overline{\nu}_e$ in a statistical way. The energy spectra of the electron is given by the well known Fermi theory of β-decay [1]:

$$\frac{dN}{dE} \propto p(E + m_e c^2)(E_0 - E)\sqrt{(E_0 - E)^2 - m_{\overline{\nu}_e}^2 c^4} \tag{1}$$

with the electron energy E, the endpoint energy E_0 , the electron mass m_e and the average electron anti-neutrino mass $m_{\overline{\nu}_e}^2 = \sum |U_{ei}|^2\; m(\nu_i)^2$. This is the incoherent sum of neutrino mass eigenstates and therefore not sensitive to phases of the neutrino mixing matrix (in contrast to neutrinoless double β-decay). As one can see in equation 1, it is the square of the neutrino mass $m_{\overline{\nu}_e}^2$ that enters as a parameter. Its effect on the shape of the spectrum is significant only in a very narrow region close to E_0. The current upper limit on

[a]E-mail: `florian.fraenkle@kit.edu`

the neutrino mass of 2 eV/c^2 [2] was determined from investigating the tritium β-spectrum near the endpoint of 18.6 keV by experiments in Mainz [3] and Troitsk [4].

2 The KATRIN experiment

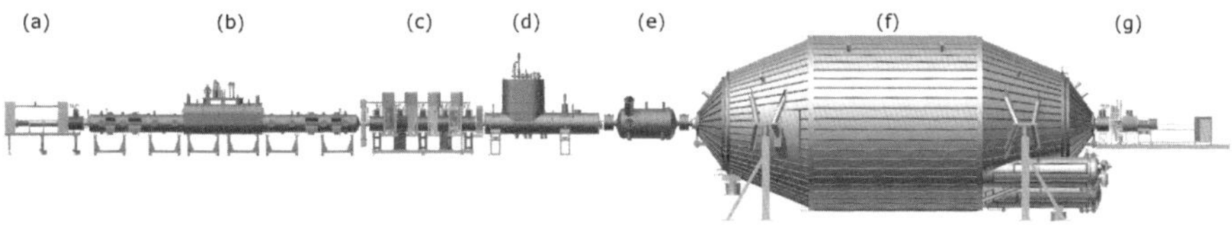

Figure 1: The KATRIN experimental setup with its main components: (a) rear section; (b) windowless gaseous tritium source (WGTS); (c) differential pumping section (DPS); (d) cryogenic pumping section (CPS); (e) pre-spectrometer; (f) main spectrometer; (g) focal plane detector.

The **KA**rlsruhe **TRI**tium **N**eutrino (KATRIN) experiment [5] is a next generation, large scale experiment to determine the mass of the electron antineutrino by investigating the kinematics of tritium β-decay with a sensitivity of 200 meV/c^2 (at 90 % C.L.). The experiment is located at the Karlsruhe Institute of Technology (KIT) in Germany. The measurement setup (see figure 1) has an overall length of $\approx$ 70 m. Molecular tritium is injected into the windowless gaseous tritium source (**a**) where it decays with an activity of 10^{11} Bq, thus providing a sufficient number of β-decay electrons close to the tritium endpoint energy E_0. The tritium is removed from the beamline in the differential pumping section (**b**) and the cryogenic pumping section (**c**) while electrons from the source are magnetically guided towards the spectrometer section. Both pre-spectrometer and main spectrometer are operated as electrostatic retarding high pass filters of MAC-E filter type [6]. The pre-spectrometer (**d**) is operated as a pre-filter in order to reduce the flux of electrons into the main spectrometer (**e**) which performs the energy analysis of the β-decay electrons near the endpoint with an energy resolution of ΔE = 0.93 eV at 18.6 keV. The main spectrometer is equipped with a dual layer wire electrode system for electrostatically shielding secondary electrons from the inner vessel surface [7]. The transmitted β-decay electrons are counted in the detector system (**f**) with a segmented silicon detector [8].

With the arrival of the WGTS on September 10th, 2015, all major components of the experiment are now on site. Commissioning measurements of the complete beamline are expected to start in 2016, with the first tritium runs at the end of the year.

2.1 Windowless gaseous tritium source

The Windowless gaseous tritium source (WGTS) consists of a 10 m long tube with a diameter of 90 mm and is operated at a temperature of about 30 K by circulation of two phase neon. Molecular tritium (T_2) is injected into the center of the source tube and decays with an activity of 10^{11} Bq to provide a sufficient number of electrons close to the tritium endpoint energy E_0. The β-electrons are guided via an axial magnetic field of 3.6 T towards the spectrometer section. The T_2 is collected via turbo-molecular pumps at both ends of the WGTS and is recirculated via an "inner loop" which removes contaminants (particularly ^{3}He) and is capable to process 40 g of T_2 a day. A prototype system to investigate the performance of the temperature stabilization of the beam tube showed that the stringent thermal performance specifications (temperature stability $\pm$30 mK) could be met and a factor of twenty better temperature stability was achieved [9]. The WGTS was delivered to KIT in September 2015 and is currently integrated into the KATRIN beam line.

2.2 Differential and cryogenic pumping section

The task of the Differential Pumping Section (DPS) is to reduce the T_2 partial pressure by a factor of $>10^5$ and to guide β-electrons via a strong magnetic field of 5.6 T. The beam tube has four bends to avoid beaming of T_2 molecules towards the spectrometers. In order to remove tritium ions, the DPS is equipped with electric dipole electrodes. The magnet system was successfully commissioned and the installation of the beam tube is ongoing.

Any remaining T_2 that passes the DPS is trapped in the Cryogenic Pumping Section (CPS) by argon frost frozen on the 4 K cold beam tube, which forms a highly efficient, large-area and radiation-immune surface. The feasibility of this approach was successfully tested in a test experiment called TRAP [10] which achieved a T_2 reduction factor of about 10^7. The CPS was delivered to KIT in July 2015 and is currently integrated into the KATRIN beam line.

2.3 Spectrometer section

The pre-spectrometer, as well as the main spectrometer, are electrostatic retarding potential spectrometers of MAC-E filter type [6] (**M**agnetic **A**diabatic **C**ollimation combined with an **E**lectrostatic Filter).

The pre-spectrometer is intended to be used as a pre-filter on a potential a few hundred volts below E_0. The pre-filtering reduces the flux of β-electrons into the main spectrometer by many orders of magnitude and minimizes β-electron induced background processes in the main spectrometer.

The purpose of the 10 m in diameter and 24 m long main spectrometer is the energy analysis of the β-decay electrons. It has an energy resolution of 0.93 eV at 18.6 keV. In order to reduce the spectrometer background rate, a double layer inner electrode system made of thin wires - mounted with submillimeter precision - is installed. The wire layers are put on a more negative potential with respect to the tank voltage in order to shield secondary electrons produced in the vessel wall. The absolute voltage of -18.6 kV needs to be stable on the 1 ppm level and is monitored with a high precision voltage divider and an independent monitoring and calibration beam line [11]. The vacuum system of the main spectrometer is capable of reaching a pressure of about 10^{-10} mbar with one active non-evaporable getter pump.

2.4 Detector

Electrons that are able to overcome the potential barriers of the spectrometers are detected in a monolithic 148 pixel silicon PIN diode [12]. The energy resolution of the detector system is 1.4 keV (FWHM). Selection of materials, shielding and an active veto are used to keep the intrinsic detector background at a low level of 1.2 mcps/keV.

3 Main spectrometer commissioning measurements

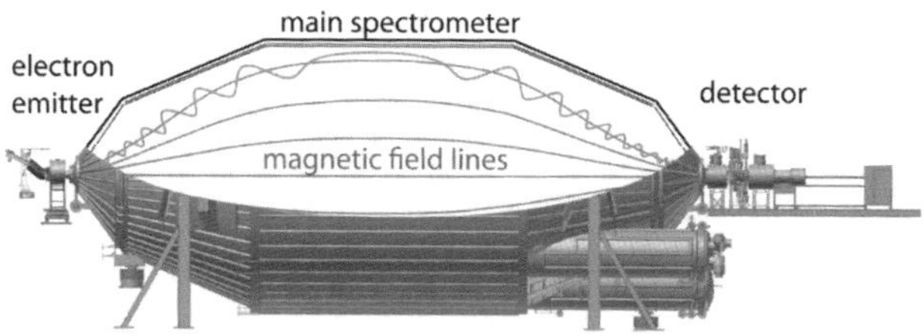

Figure 2: **a)** Measurement configuration during the main spectrometer and detector commissioning phase. A photoelectric electron emitter (e-gun) is mounted on the upstream side of the spectrometer, the detector system on the downstream side.

In order to investigate the performance of the main spectrometer with regard to energy resolution and background performance, a series of commissioning measurements were performed. For this purpose a photoelectric electron emitter (e-gun) was installed at the upstream side of the main spectrometer, which allows the generation of electrons of dedicated energies as well as polar angles, with respect to the guiding magnetic field line. The e-gun was mounted on a manipulator in order to shoot electrons along selectable magnetic field lines through the spectrometer towards the detector system attached at the downstream side of the spectrometer (see Figure 2). For a fixed manipulator position and spectrometer retarding voltage, the e-gun acceleration voltage was

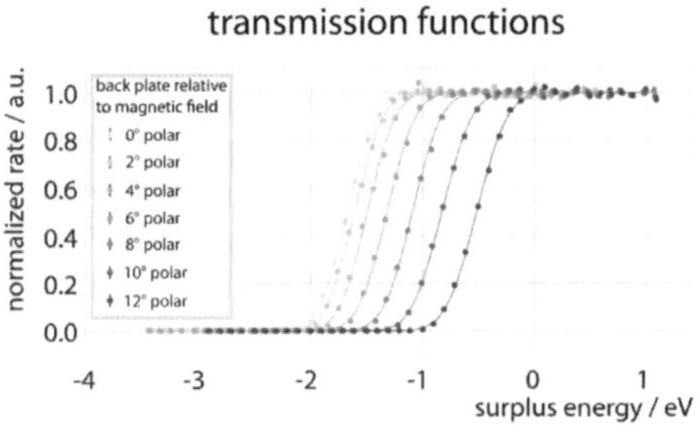

Figure 3: Transmission functions for different electron polar angles at the e-gun and a spectrometer retarding potential of -18.6 kV. The surplus energy is calculated form the difference of e-gun acceleration voltage and spectrometer retarding voltage, multiplied by the elementary charge. The onset of the transmission at negative surplus energies is due to a slightly more positive retarding potential in the analyzing plane, caused by the influence of the beam tube at ground potential on both ends of the spectrometer.

increased in small steps and the rate at the detector was measured. This measurement was repeated for different polar angles (see figure 3). The width of the transmission functions is dominated by systematic effects of the e-gun, in particular energy and angular spread. The shift of the transmission functions with increasing polar angles to larger surplus energies, shows that the magnetic adiabatic collimation is working and that the KATRIN main spectrometer works as a high resolution MAC-E filter.

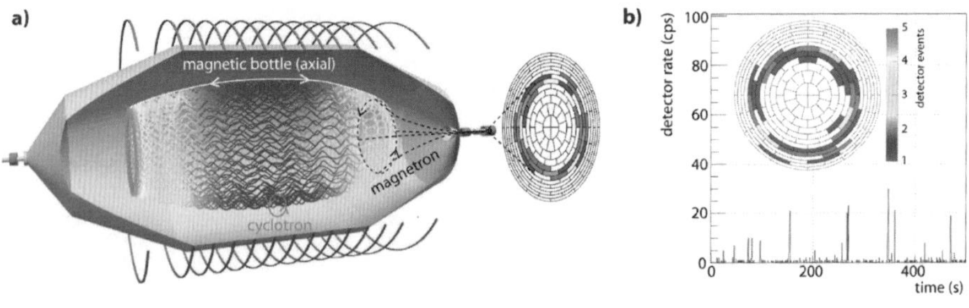

Figure 4: **a)** Simulation of a magnetically stored electron in the volume of the main spectrometer and pixel distribution at the detector of secondary electrons, produced in ionization processes of the trapped electron with residual gas molecules. **b)** Background measurement at elevated pressure (about 10^{-8} mbar). Each spike in the background rate is initially caused by a radon decay in the volume of the spectrometer. As expected from the simulation, the background events during a spike show a ring-shaped pattern on the detector.

Initial measurements of the main spectrometer background rate, with a not yet background optimized system, resulted in a value of about 1 cps. This is a remarkably low initial background rate compared to the commissioning of previous spectrometers such as the KATRIN pre-spectrometer, where Penning discharges caused extreme large backgrounds [13]. A major background source in the KATRIN spectrometers is due to ^{220}Rn and ^{219}Rn [14–16] . Radon emanates from materials inside the vacuum region of the spectrometer, in particular from the non-evaporable getter (NEG) pump and decays in the volume of the spectrometer. Electrons emitted in processes accompanying the Rn α-decay, such as shake-off, internal conversion of excited levels in the Rn daughter atoms, Auger and Coster-Kronig electrons are magnetically trapped inside the spectrometer (see Figure 4a)). Depending on their initial energy, which can be on the order of up to 10 keV, the trapped electrons can produce hundreds of secondary electrons via subsequent ionization of residual gas molecules. These secondary electrons are accelerated by the retarding potential as they leave the spectrometer and have the same energy as signal electrons from tritium β-decay by the time they reach the detector. In a measurement with increased pressure (argon, 10^{-8} mbar), in order to decrease the time between ionizations, the radon induced background is clearly visible as spikes in the background rate with the typical ring-shaped distribution of secondary electrons on the detector (see Figure 4b)). A liquid nitrogen cooled baffle system installed in the pump ports of the main spectrometer, between the NEG pump and the main volume of the spectrometer, could reduce the overall background rate from 1 cps to 0.6 cps with an efficiency of about 97 %. The mechanisms and sources of the remaining background are currently under investigation. Measurements investigating the correlation between variations of the muon flux through the spectrometer surface, and the background rate, exclude cosmic muons as a dominant source of the remaining background.

4 Conclusions

Direct neutrino mass measurements are a model independent way to determine the neutrino mass. A major improvement of the neutrino mass sensitivity by one order of magnitude is expected of the KATRIN experiment, which is currently in the final phase of construction and aims to start taking neutrino mass data with tritium in 2016. Commissioning measurements of the KATRIN main spectrometer showed that it works as a high resolution MAC-E filter. The decay of radon in the volume of the spectrometer is a major source of background but can be efficiently suppressed with a system of LN2-cooled baffles.

References

[1] E. Fermi, *Versuch einer Theorie der β-Strahlen*, *Zeitschrift für Physik* **88** (1934) http://doi.org/10.1007/BF01351864

[2] Particle Data Group, *Review of Particle Physics, Chinese Physics C* **38** (2014) http://doi.org/10.1088/1674-1137/38/9/090001

[3] C. Kraus et al., *Final results from phase II of the Mainz neutrino mass search in tritium β decay, The European Physical Journal C* **40** (2005) http://doi.org/10.1140/epjc/s2005-02139-7

[4] V.N. Aseev et al., *Upper limit on the electron antineutrino mass from the Troitsk experiment, Physical Review D* **84** (2011) http://doi.org/10.1103/PhysRevD.84.112003

[5] KATRIN Collaboration, *KATRIN design report, FZKA scientific report* **7090** (2005) http://bibliothek.fzk.de/zb/berichte/FZKA7090.pdf

[6] G. Beamson et al., *The collimating and magnifying properties of a superconducting field photoelectron spectrometer, Journal of Physics E: Scientific Instruments* **13** (1980) http://doi.org/10.1088/0022-3735/13/1/018

[7] K. Valerius, *The wire electrode system for the KATRIN main spectrometer, Progress in Particle and Nuclear Physics* **64** (2010) http://doi.org/10.1016/j.ppnp.2009.12.032

[8] J.F. Amsbaugh et al., *Focal-plane detector system for the KATRIN experiment, Nuclear Instruments and Methods in Physics Research Section A* **778** (2015) http://dx.doi.org/10.1016/j.nima.2014.12.116

[9] S. Grohmann et al., *The thermal behaviour of the tritium source in KATRIN, Cryogenics* **55-56** (2013) http://dx.doi.org/10.1016/j.cryogenics.2013.01.001

[10] F. Eichelhardt et al., *The Cryogenic Pumping Section of KATRIN and the Test Experiment TRAP, Nuclear Physics B - Proceedings Supplements* **221** (2011) http://dx.doi.org/10.1016/j.nuclphysbps.2011.09.042

[11] M. Erhard et al., *High-voltage monitoring with a solenoid retarding spectrometer at the KATRIN experiment, Journal of Instrumentation* **9** (2014) http://dx.doi.org/10.1088/1748-0221/9/06/P06022

[12] J.F. Amsbaugh et al., *Focal-plane detector system for the KATRIN experiment, Nuclear Instruments and Methods A* **778** (2015) http://dx.doi.org/10.1016/j.nima.2014.12.116

[13] F.M. Fränkle et al., *Penning discharge in the KATRIN pre-spectrometer, Journal of Instrumentation* **9** (2014) http://doi.org/10.1088/1748-0221/9/07/P07028

[14] F.M. Fränkle et al., *Radon induced background processes in the KATRIN pre-spectrometer, Astroparticle Physics* **35** (2011) http://doi.org/10.1016/j.astropartphys.2011.06.009

[15] N. Wandkowsky et al., *Validation of a model for radon-induced background processes in electrostatic spectrometers, Journal of Physics G: Nuclear and Particle Physics* **40** (2013) http://doi.org/10.1088/0954-3899/40/8/085102

[16] S. Mertens et al., *Background due to stored electrons following nuclear decays in the KATRIN spectrometers and its impact on the neutrino mass sensitivity*, *Astroparticle Physics* **41** (2013) http://doi.org/10.1016/j.astropartphys.2012.10.005

FIRST NEUTRINO OSCILLATION RESULTS FROM NOνA

M. J. Frank[a b]
University of Virginia, Charlottesville, VA 22904, USA

Abstract. The NOνA experiment aims to study the mixing behavior of neutrinos and anti-neutrinos using the Fermilab NuMI neutrino beam. The experiment collects data at two detectors, one near the neutrino production target at Fermilab; the other 810 km away. The detectors are 14 mrad off the beam axis, which results in an almost monoenergetic beam of 2 GeV neutrinos. The construction of the 14 kT Far Detector and 300 T Near Detector finished in 2014. Since then, NOνA has collected data and we will discuss the first data analyses using 2.74×10^{20} POT searching for muon neutrino disappearance and electron neutrino appearance in the Far Detector.

1 Introduction

The acronym NOνA stands for NuMI Off-Axis ν_e-Appearance, which summarizes our experiment [1]. The NuMI (Neutrinos at the Main Injector) facility provides the muon neutrino (ν_μ) source [2]. These neutrinos travel north through the earth to the Far Detector (FD) which is located 14 mrad off the beam axis where we expect an almost monoenergetic beam of 2 GeV neutrinos. During their 810 km journey to the FD, the neutrinos oscillate in their flavor states and we measure the disappearance rate of muon neutrinos (ν_μ) and appearance rate of electron neutrinos (ν_e). From this we can measure the transition probabilities $P(\nu_\mu \rightarrow \nu_\mu)$ and $P(\nu_\mu \rightarrow \nu_e)$ which are functions of the oscillation parameters [3]. From the measurement of $P(\nu_\mu \rightarrow \nu_\mu)$ we can extract information about the mixing angle θ_{23} and the neutrino mass difference Δm^2_{32}. From the measurement of $P(\nu_\mu \rightarrow \nu_e)$ we can extract information about the mixing angle θ_{13} and θ_{23}, the CP-violating phase angle $\delta_{\rm CP}$, and the neutrino mass hierarchy.

2 Neutrino Detection

The neutrino interaction cross section is very low, so we require a large detector mass. When an electron neutrino interacts with the detector an electron is produced that will in turn produce an electromagnetic (EM) shower. We therefore designed a "fully active" detector using low Z materials. We use PVC extrusions filled with liquid scintillator, which provides a radiation length of approximately 40 cm and a Molière radius of 11 cm. Each of the extrusions contains one wavelength-shifting fiber that is read out by an avalanche photo-diode (APD). This detection technique is optimized to differentiate EM showers from hadronic showers as will be shown later.

[a]On behalf of the NOνA collaboration.

[b]E-mail: `mjf8u@virginia.edu`

3 The Experiment

The neutrinos are created in the NuMI facility by colliding 120 GeV protons on a graphite target, which results in π^+ particles that are allowed to decay into μ^+ and ν_μ. The μ^+ particles are stopped by muon absorbers and the muon neutrinos travel onward to the Near Detector (ND). The NuMI beam line is operating routinely with a beam power of 420 kW and is being upgraded to increase to a final power of 700 kW.

The ND is located 105 m underground and 14 mrad off the beam axis. It has a mass of approximately 300 tons and a size of 4 m by 4 m by 15 m. The FD is located on the surface a distance of 810 km away and also 14 mrad off the beam axis. Its construction was completed in 2014 and it has a mass of over 14,000 tons and a size of 15 m by 15 m by 60 m.

Figure 1 shows the simulated signatures of different particle interactions in the ND. It shows that we are able to differentiate EM showers (center) from hadronic showers (bottom). With this particle identification we can measure $P(\nu_\mu \to \nu_\mu)$ and $P(\nu_\mu \to \nu_e)$ by looking for a deficit of ν_μ events and an excess of ν_e events, respectively. We can then extract nature's parameters based on the standard oscillation equations [3] as explained above using a baseline of 810 km and a neutrino energy of 2 GeV.

Figure 2 shows what NOνA can achieve by combining three years of neutrino running with three years of anti-neutrino running. If nature's parameters reside at the starred point, we would have optimal sensitivity for the neutrino mass hierarchy and $\delta_{\rm CP}$.

4 Muon Neutrino Disappearance Analysis

The ν_μ disappearance analysis to measure $P(\nu_\mu \to \nu_\mu)$ proceeds by first identifying ν_μ charged-current events together with their energies in both the ND and FD. From the measured ND neutrino energy spectrum, we can extrapolate the expected FD energy spectrum as explained in Figure 3. We can then extract the oscillation parameters ($\sin^2(\theta_{23})$ and Δm^2_{32}) by comparing the extrapolated energy spectrum to the one we measure in the FD. Figure 4 shows the FD ν_μ energy spectrum and Figure 5 shows the extracted oscillation parameters in a two-dimensional contour. For the normal mass hierarchy, we measured the individual oscillation parameters to be:

$$\begin{aligned}\Delta m^2_{32} &= 2.37^{+0.16}_{-0.15} \\ \sin^2\theta_{23} &= 0.51 \pm 0.10\end{aligned}$$

The official NOνA ν_μ disappearance results are described in more detail elsewhere [6].

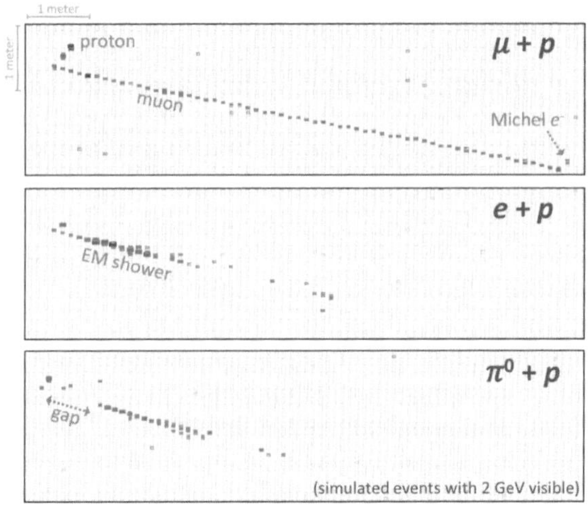

Figure 1: Simulated ND event with a visible energy of 2 GeV. The top event shows a ν_μ charged-current interaction with the characteristically long muon track and short proton track with large energy deposition. The middle event shows a ν_e charged-current interaction that demonstrates the long EM shower in our "fully active" detector. The bottom event shows a hadronic interaction where the majority of the π^0 momentum is carried by one of the two decay photons, which in turn produces a displaced EM shower.

5 Electron Neutrino Analysis

The ν_e appearance analysis to measure $P(\nu_\mu \to \nu_e)$ proceeds by first identifying ν_e charged-current events. We then predict the beam backgrounds using ND events and finally interpret the FD event excess as ν_e appearance. The ν_e signature has been explained in Figure 1, so the task is to identify all such events in the FD. The largest background comes from the cosmic rays passing through the FD at a rate of 10^5 Hz because of the detector's proximity to the earth's surface. We therefore only look at events that are in time with a beam spill onto the NuMI target, which reduces our cosmic ray background events by a factor of 10^5. Additionally, we reduce our background by three additional orders of magnitude by applying event-topology-based requirements on our events. Finally, we use two techniques to identify electron showers. The Likelihood Identification (LID) technique tests event shower $\frac{dE}{dx}$ profiles against probability density functions for different particle hypotheses. The Library Event Matching (LEM) technique tests the entire event topology against a large library of simulated electron neutrino signal and background events. Figure 6 shows the energies of the six identified LID ν_e candidates and eleven LEM ν_e candidates. If we compare this result to the reactor measurement of $\sin^2(2\theta_{13})$ [7–9] (we use the weighted average $\sin^2(2\theta_{13}) = 0.086$), we can

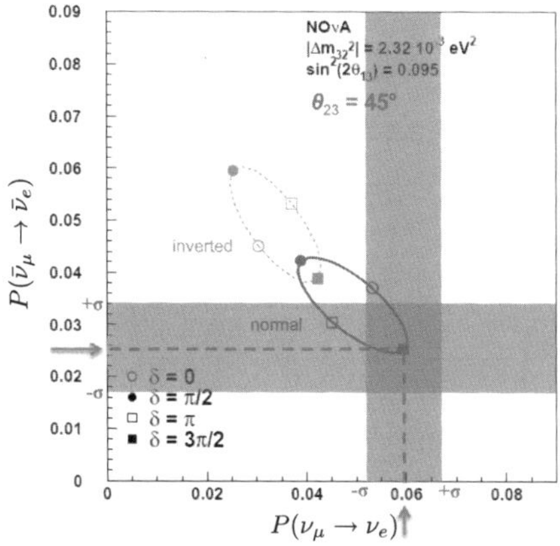

Figure 2: This plot shows the transition probabilities $P(\nu_\mu \to \nu_e)$ and $P(\bar{\nu}_\mu \to \bar{\nu}_e)$ for different values of $\delta_{\rm CP}$ for the normal (solid blue oval) and inverted (dashed red oval) mass hierarchies. If we measure the probabilities indicated by the green arrows we could resolve the mass hierarchy and set limits on $\delta_{\rm CP}$ at different confidence levels. The 1σ range is indicated by the green shaded region.

produce the exclusion significances shown in Figure 7. The official NOνA ν_e appearance results are described in more detail elsewhere [10].

Acknowledgments

I wish to thank the organizers of the 17$^{\rm th}$ Lomonosov Conference on Elementary Particle Physics for a most enlightening and entertaining conference, my NOνA colleagues, and Fermilab. Fermilab is operated by Fermi Research Alliance, LLC under Contract No. De-AC02-07CH11359 with the United States Department of Energy.

References

[1] D. S. Ayres *et al.*, FERMILAB-DESIGN-2007-01 (2007).
[2] J. K. Anderson *et al.*, FERMILAB-DESIGN-1998-01 (1998).
[3] K. A. Olive *et al.* (Particle Data Group), Chin. Phys. C, 38, 090001 (2014).
[4] P. Adamason *et al.*, Phys. Rev. Lett. 112, 191801 (2014).
[5] K. Abe *et al.*, Phys. Rev. D 91, 072010 (2015).
[6] P. Adamson, arXiv:1601.05037 (2016).
[7] F. P. An *et al.*, Phys. Rev. Lett. 115, 111802 (2015).
[8] S. H. Seo, arXiv:1410.7987 (2014).
[9] Y. Abe *et al.*, JHEP 1410 86 (2014).
[10] P. Adamson, arXiv:1601.05022 (2016).

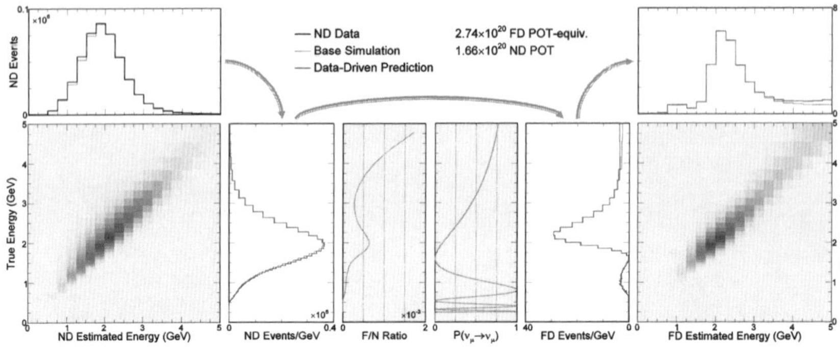

Figure 3: This diagram explains how we predict the number of neutrino events in the FD using a data-driven extrapolation technique. We start with the ND neutrino energy spectrum and estimate the true energy distribution. Next, we multiply by the expected far/near event ratio and oscillation probability. As a final step we convert the extrapolated true energy spectrum to the reconstructed energy spectrum that we measure in the FD.

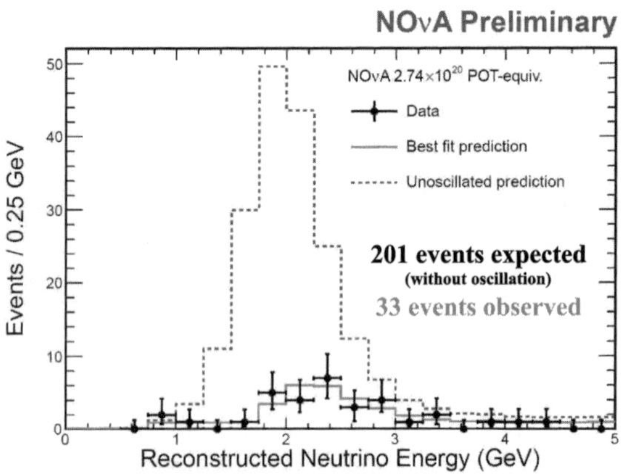

Figure 4: This plot shows the FD ν_μ energy spectrum. The black points show the measured FD data, the green line shows the best fit to the data from which we extract our oscillation parameters. For illustration, the prediction if there were no neutrino oscillations is shown as the blue dotted line. Without neutrino oscillations, we would have expected 201 ν_μ events, but we observe only 33 events, which is unambiguous confirmation of neutrino oscillations.

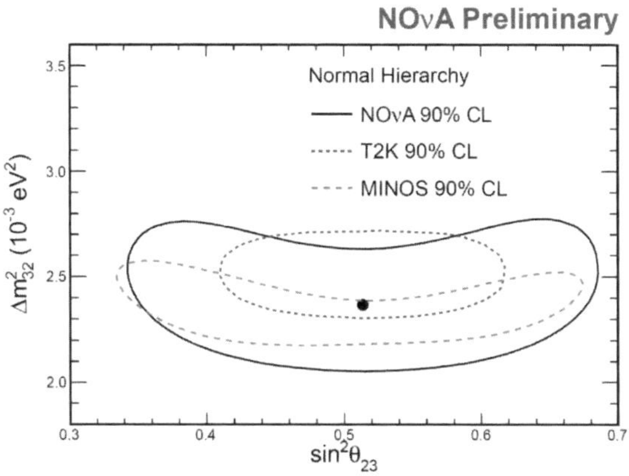

Figure 5: This plot shows the 90% C.L. contours of the normal mass hierarchy oscillation parameters ($\sin^2(\theta_{23})$ and Δm^2_{32}) extracted from the data in Figure 4. For comparison, the MINOS [4] and T2K [5] results are superimposed showing that our first results are in good agreement.

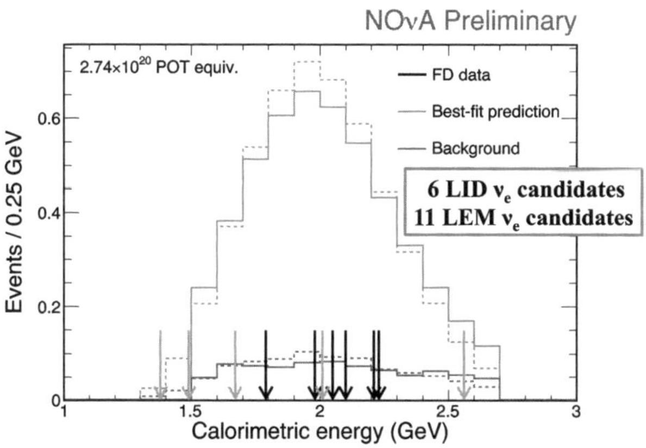

Figure 6: This plot shows the FD ν_e energy spectrum. The black arrows indicate the ν_e candidates identified by both the LID and LEM technique, while the grey arrows indicate the ν_e candidates identified by LEM only. The red lines indicate the best-fit spectrum to the LID (solid) and LEM (dashed) candidate events. The blue lines indicate the estimated background for LID (solid) and LEM (dashed).

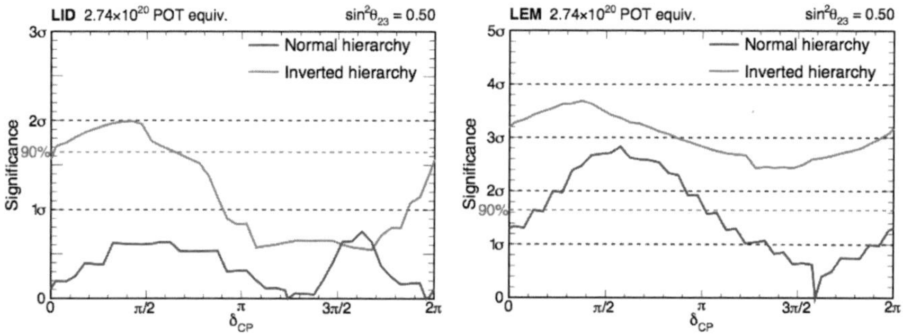

Figure 7: The plots above show the exclusion significance for different values of δ_{CP} for both the normal (blue) and inverted (red) neutrino mass hierarchies. The left plot shows the result using the 6 LID ν_e candidates and the right plot shows the result when we use the 11 LEM ν_e candidates instead.

ICARUS: PRESENT STATUS AND PERSPECTIVE FOR STERILE NEUTRINOS SEARCH AT FERMILAB

Alberto Guglielmi [a]
INFN Sezione di Padova, via Marzolo 8, 35131 Padova, Italy

Abstract. ICARUS T600 liquid argon TPC completed in 2013 a very successful three year run at LNGS underground laboratory exposed to CNGS neutrino and cosmic rays achieving relevant physics and technical results. The T600 will be be exposed to FNAL Booster neutrinos in the framework of the joint ICARUS, SBND and MicroBooNE effort to definitely clarify the presence of sterile neutrinos.

1 Introduction

Several experimental "anomalies" pointing out the possible existence of at least one fourth non standard neutrino state driving neutrino oscillations at a small distance with a mass at $\sim$ 1 eV have been reported at reactors, intense radioactive sources and accelerators [1]. Recently the ICARUS results [2] on LSND-like search with the CNGS beam, confirmed by OPERA [3], allowed to identify a small parameter region $\Delta m^2 \sim 0.5$ eV^2 where all the reported results at accelerators are in agreement.According to Planck measurement and Big Bang cosmology, at most one sterile neutrino, $m < 0.4$ eV is expected [4].

Among the new proposed searches for sterile neutrinos [5], a conclusive clary-fication on such phenomena is expected from the new SBN experiment [6] in preparation at FNAL. The measurement is based on the search for spectral differences in both electron and muon-like channels in three liquid Argon (LAr) TPC detectors, SBND, MicroBooNE and ICARUS T600 at different baselines along the Booster Neutrino Beam (BNB). ICARUS, after the present WA104 refurbishing phase, will operate by 2017 as far detector at 600 m from target.

2 ICARUS T600 LAr-TPC: a powerful detector for neutrino physics

The ICARUS T600 detector consists of two identical adjacent modules filled with 760 t of ultra-pure liquid Argon ($\sim$ 476 t active mass) [7]. Each module is equipped with 2 TPC chambers on the long sides, separated by a central common cathode with 1.5 m drift length. Ionization electrons produced by charged particles along their path are drifted by a 500 V/cm uniform electric field towards the 3 parallel readout planes facing the drift volume with wires (54000 in total, 3 mm pitch and 3 mm plane spacing) oriented at 0^0, $\pm 60^0$ with respect to the horizontal direction. The deposited energy is measured by the charge of the drift electron collected in the last wire plane (Collection view). The LAr-TPCs are complemented by a system of photo-multipliers coated with a TPB wavelength shifter, installed in LAr behind the wire planes to detect

[a]on behalf on the ICARUS Collaboration, e-mail: alberto.guglielmi@pd.infn.it

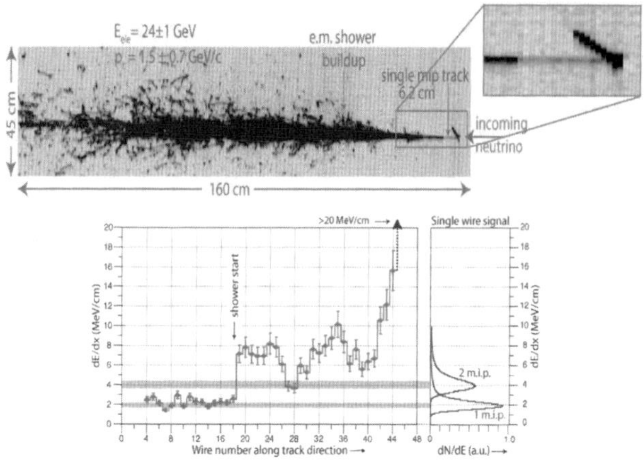

Figure 1: A ν_eCC CNGS event in the 2D Collection view of ICARUS-T600 (top) with the corresponding evolution of ionization density dE/dx in the first wires with the marked shower onset (bottom). The expected dE/dx signals for 1 and 2 m.i.p. are also shown.

the scintillation light emitted at $\sim$ 128 nm by crossing ionizing particles. The relative time of each ionization signal, combined with the electron drift velocity v_D $\sim$1.6 mm/μs, provides the position of the track along the drift coordinate.

Relevant physics and technical results were achieved during the three years CNGS run, demonstrating the excellent performance of T600 as tracking device with $\sim$ 1 mm^3 spatial resolution and homogenous calorimeter measuring the total energy with $\sigma_E/E \simeq 0.03/\sqrt(E(GeV))$ for e.m. contained showers. Momentum of escaping muons has been determined via multiple Coulomb scattering with $\Delta p/p \sim$ 15% in the 0.4-4 GeV/c range.

Exposed to the CNGS neutrino beam in 2010-2012, the T600 collected a total statistics of 8.6×10^{19} p.o.t with a livetime exceeding 93%. The detector demonstrated a remarkable e/γ separation and particle identification exploiting the measurement of dE/dx vs. range. The capabilities to reconstruct the neutrino interaction vertex, to identify and measure e.m. showers generated by primary electrons and to accurately measure invariant mass of photon pairs allowed to reject to unprecedented level NC background in the study of $\nu_\mu \rightarrow \nu_e$ transitions. In total 2650 CNGS neutrino interactions have been identified in 7.9 10^{19} pot statistics, $\sim$ 3.4 neutrino interactions and 12 beam related muons per 10^{17}, consistent within 6% with MC predictions. Globally 7 electron-like events have been observed (Fig.1) to be compared with the 8.5 ± 1.1 expected from the $\sim$ 1% intrinsic beam component and standard 3-flavor oscillations, providing the limit on $\nu_\mu \rightarrow \nu_e$ oscillations $P < 3.86\ 10^{-3}$ at 90 % CL (Fig. 3).

The ICARUS detector collected also cosmic ray data. About 115 atmospheric neutrino events are expected in the full T600 exposure of 0.73 kt year. Event

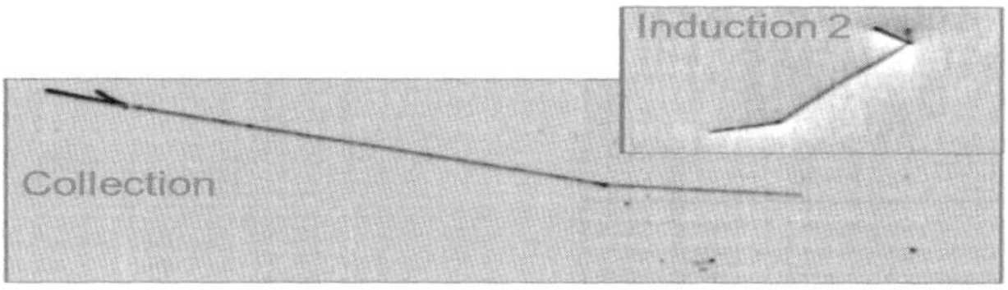

Figure 2: A two prong $E_{dep} \simeq 200$ MeV NC atmospheric neutrino candidate, with a pion (63 cm track lenght) generating 2 protons as detected in Collection and Induction2 views.

candidates with $E_{dep} > 20$ are selected by an automatic procedure to reconstruct 3D tracks and interaction vertex and identify "multi-prong" neutrino event topologies for further visual analysis (Fig. 2).

To ensure long drift path for ionisation electrons without appreciable attenuation, the level of electronegative impurities in LAr was kept exceptionally low (order some ppt) by new industrial LAr purification methods developed to continuously filter and recirculate both in liquid and gas Argon phases. A lifetime in excess of 16 ms (< 20 ppt O_2 equivalent) was obtained since April 2013 [8], thus demonstrating the effectiveness of single-phase LAr-TPC technique, paving the way to huge LAr-TPC detectors (up to 5 m drift).

3 The ICARUS-WA104 refurbishing for the SBN search program

The T600 detector has been transported to CERN for overhauling and refurbishment of the cryogenics and LAr purification system before the installation at FNAL. The TPC chambers, with improved planarity cathodes, will be installed in new cold vessels surronded by a new purely passive thermal insulation. The upgraded "warm" electronics will allow to increase the signal to noise ratio.

While, in the deep underground conditions of the LNGS laboratory, a single prompt trigger has always ensured the unique timing connection of the main image of the event wich determine the track position along the drift coordinate, at the FNAL shallow depth location. The detector will be protected by 3 m overburing from neutrals. Moreover ~ 11 additional cosmic muons are expected in the 1 ms drift time, introducing some difficulties for event track reconstruction. A 4π, segmented cosmic muon tagging system (CTS), composed by plastic scintillation slabs read out by silicon PMTs surrounding the T600 (area ~ 1200 m^2), will allow to unambiguosly identify cosmic rays entering the detector. The inner light detection system will be upgraded for a more precise event timing and localization, exploiting also the BNB bunched beam structure at FNAL (~ 2 ns width batches every 19 ns) to reject out-of-bunch cosmics.

The SBN short-baseline experimental configuration [6] at FNAL will include SBND, MicroBooNE and ICARUS-T600 LAr-TPC's installed on-axis in the BNB neutrino beam at 110 m, 470 m and 600 m distance from target respectively. Multiple LAr-TPC detectors at different baselines will allow a very

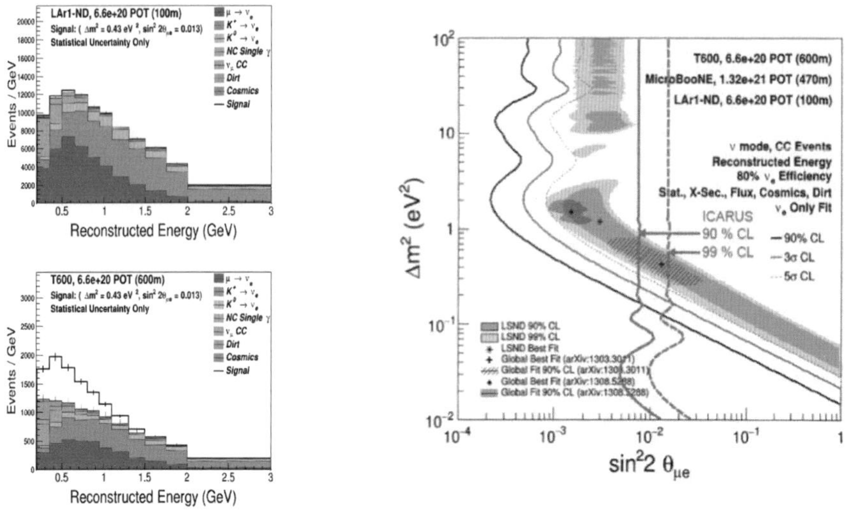

Figure 3: Expected ν_e CC signal distributions for $\Delta m^2 = 0.43$ eV^2, $\sin^2 2\theta = 0.013$ at Near and Far detectors (6.6 10^{20} pot) compared to intrinsic ν_e and background events (Left). Corresponding sensitivity on $\nu_\mu \to \nu_e$ (Right).

sensitive search for high-Δm^2 neutrino oscillations in multiple channels. The expected residual ν_e background in the Near and Far detectors, including intrinsic ν_e beam events, neutral-current and ν_μ CC mis-IDs, out-of-detector beam related backgrounds, and cosmogenic photon induced electromagnetic shower backgrounds is shown in Fig. 3 after some topological cuts. For comparison, an oscillation signal corresponding to the best-fit parameters from the analysis of $\Delta m^2 = 0.43$ eV^2 and $\sin^2(2\theta) = 0.013$ is also shown. The resulting experimetal sensitivity of the SBN program to $\nu_\mu \to \nu_e$ appearance signals in the $(\Delta m^2, \sin^2(2\theta))$ plane is shown in Fig.3 for 6.6 10^{20} p.o.t.,three year data taking (twice for MicroBooNE). The LSND 99% C.L. allowed region is covered at the $\sim 5\sigma$ level everywhere; the region below $\Delta m^2 = 0.1$ eV^2 is already ruled out at more than 5σ by the previous results of ICARUS at Gran Sasso.

Whyle the absolute neutrino flux and cross section uncertainties in any detector are expected larger than 10%, the high correlations between the near detector and the MicroBooNE/ICARUS event samples combined with the excellent statistics at the near site will make the SBN program the most sensitive ν_μ disappearance experiment at $\Delta m^2 \sim 1$ eV^2, more than an order of magnitude beyond the previous combined analysis of Short-Baseline and MiniBooNE.(fig. 4). The disappearance measurement is a critical aspect of the SBN program which is needed to confirm a signal, if seen in ν_e appearance, as oscillations. A genuine $\nu_\mu \to \nu_e$ appearance can also be accompanied by a disappearance of the intrinsic ν_e beam component, since the three oscillation probabilities are

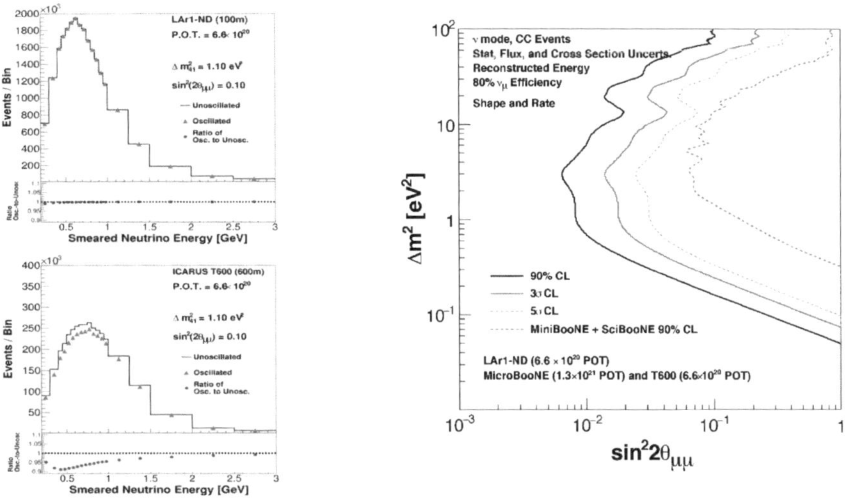

Figure 4: Expected ν_μ CC distributions for $\Delta m^2 = 1.1$ eV^2, $\sin^2 2\theta = 0.10$ at Near and Far detectors (6.6 10^{20} pot) compared to unoscillated events (Left). Corresponding sensitivity on ν_μ disappearance (Right).

related through a common mixing matrix. The ability to perform searches for oscillation signals in multiple channels is a major advantage for the FNAL SBN oscillation physics program. By collecting at the same time the ν_μ and ν_e event samples in the same experiment, correlations between the samples can be well understood and many systematics are common. The simultaneous analysis of ν_e CC and ν_μ CC events will be a very powerful way to explore oscillations and disentangle the effects of $\nu_\mu \to \nu_e$, ν_μ disappearance, and ν_e disappearance, if they exist, in this mass-splitting range.

References

[1] A. Aguilar-Arevalo et al., Phys. Rev. **D 64**, 112007 (2001);
G. Mention et al., Phys. Rev. **D 83**, 073006 (2011);
A. Aguilar-Arevalo et al., Phys. Rev. Lett. **110**, 161801 (2013);
J.N. Abdurashitov et al., Phys. Rev. **C 59**, 2246 (1999);
W. Hampel et. al., Phys. Lett.**B 420**, 114 (1998).

[2] M. Antonello et al., Eur. Phys. J. **C 73** 2599 (2013).

[3] N. Agafona et al., J. High Energy Phys, **1307** 004 (2013).

[4] P.A.R. Ade et al., arXiv:1502.02114, submitted to Astr. and Astroph.

[5] A.S. CuCoanes et al., J. Phys. Conf. Ser. **375** 042063 (2012);
G. Bellini et al., JHEP **1308** 028 (2013);
D. Adey et al., Phys. ReV. **D 89** 071301 (2014).
[6] R. Acciarri et al., Eur. Phys. J. Preprint arXiv:1503.01520 (2015).
[7] C. Rubbia et al., JINST **6** P07011 (2011) and references therein.
[8] M. Antonello et al., JINST **9** P12006 (2014).

MODELS FOR NEUTRINO MASSES WITH IMPLIFICATIONS IN COSMOLOGY

Salah Nasri [a]

Department of Physics, College of Science, United Arab Emirates University, PO Box 15551, Al Ain, United Arab Emirates

Abstract. In this work, I discuss in detail the model of Krauss-Nasri-Trodden (KNT) that can generate small neutrino masses at three-loop level and being consistent with the lepton flavor violating processes. In addition, I argue that the model provide a viable dark matter candidate, and a strongly electroweak phase transition can be achieved. Furthermore, we show that the KNT model belongs to a class of three loop models with interesting implications for the DM.

1 The KNT Model

The KNT [1] model is an extension of the SM with three right handed (RH) neutrinos, N_i, and two charged scalars, $S_1^{\pm}$ and $S_2^{\pm}$, that are singlet under $SU(2)_L$ gauge group. In addition, we impose a discrete Z_2 symmetry on the model, under which $\{S_2, N_i\} \to \{-S_2, -N_i\}$, and all other fields are even. The Lagrangian reads

$$\mathcal{L} = \mathcal{L}_{SM} + \{f_{\alpha\beta} L_\alpha^T C i\tau_2 L_\beta S_1^+ + g_{i\alpha} N_i S_2^+ \ell_{\alpha R} + \frac{1}{2} m_{N_i} N_i^C N_i + h.c\} - V(\Phi, S_1, S_2), \tag{1}$$

where L_α is the left-handed lepton doublet, $f_{\alpha\beta}$ are Yukawa couplings which are antisymmetric in the generation indices α and β, m_{N_i} are the Majorana RH neutrino masses, C is the charge conjugation matrix, and $V(\Phi, S_1, S_2)$ is the tree-level scalar potential which is given by

$$\begin{aligned} V(\Phi, S_{1,2}) = \lambda \left(|\Phi|^2\right)^2 - \mu^2 |\Phi|^2 + m_1^2 S_1^* S_1 + m_2^2 S_2^* S_2 + \lambda_1 S_1^* S_1 |\Phi|^2 + \\ \lambda_2 S_2^* S_2 |\Phi|^2 + \frac{\eta_1}{2} (S_1^* S_1)^2 + \frac{\eta_2}{2} (S_2^* S_2)^2 + \\ \eta_{12} S_1^* S_1 S_2^* S_2 + \{\lambda_s S_1 S_1 S_2^* S_2^* + h.c\}. \end{aligned} \tag{2}$$

Here Φ denotes the SM Higgs doublet.

1.1 Neutrino Mass and Lepton Flavor Violation

The neutrino mass matrix elements arising from Fig. 1, are given by

$$(M_\nu)_{\alpha\beta} = \frac{\lambda_s m_{\ell_i} m_{\ell_k}}{(4\pi^2)^3 m_{S_2}} f_{\alpha i} f_{\beta k} g_{ij} g_{kj} F\left(m_{N_j}^2/m_{S_2}^2, m_{S_1}^2/m_{S_2}^2\right), \tag{3}$$

[a]E-mail: snasri@uaeu.ac.ae

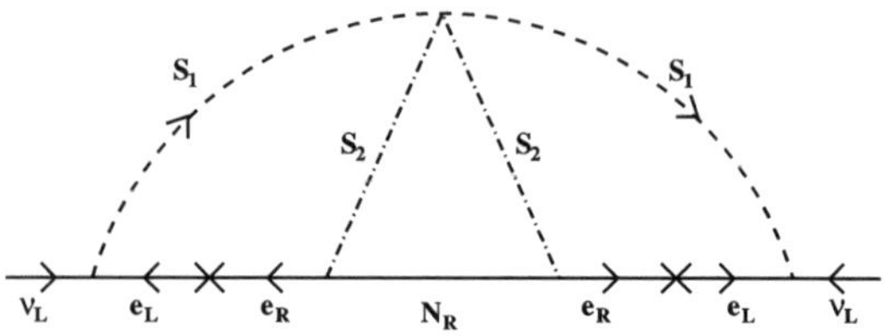

Figure 1: Generation of neutrino mass at three-loops in the KNT model.

where $\rho, \kappa (= e, \mu, \tau)$ are the charged leptons flavor indices, $i = 1, 2, 3$ denotes the three right-handed neutrinos, and the function F is a loop integral which is of order one [2]. It is interesting to note that, unlike the conventional seesaw mechanism, the radiatively generated neutrino masses are directly proportional to the charged leptons and RH neutrino masses.

The Lagrangian (1) induces lepton flavor violation (LFV) processes such as $\ell_\alpha \to \gamma \ell_\beta$ if $m_{\ell_\alpha} > m_{\ell_\beta}$. The branching ratio of such process can be computed following as

$$B(\ell_\alpha \to \gamma \ell_\beta) = \frac{\alpha_{em} v^4}{384\pi} \left\{ \frac{\left| f_{\kappa\alpha} f^*_{\kappa\beta} \right|^2}{m^4_{S_1}} + \frac{36}{m^4_{S_2}} \left| \sum_i g_{i\alpha} g^*_{i\beta} F_2 \left(\frac{m^2_{N_i}}{m^2_{S_2}} \right) \right|^2 \right\}, \tag{4}$$

with $\kappa \neq \alpha, \beta$, α_{em} is the fine structure constant and $F_2(x) = (1 - 6x + 3x^2 + 2x^3 - 6x^2 \ln x)/6(1-x)^4$. For the case of $\ell_\alpha = \ell_\beta = \mu$, this leads to a new contribution to the muon anomalous magnetic moment δa_μ. In our scan of the parameter space of the model we impose the experimental bound $\mathbf{B}\,(\mu \to \mathbf{e}\gamma) < 5.7 \times 10^{-13}$ [3], and use the best fit for neutrino oscillation parameters [4].

1.2 Dark Matter

An immediate implication of the Z_2 symmetry is that that lightest right handed neutrino, N_1, is stable, and hence it could be a candidate for dark matter (DM). The N_1 number density get depleted through the annihilation process $N_1 N_1 \to \ell_\alpha \ell_\beta$ via the t-channel exchange of $S_2^\pm$. As can be seen in Fig. 2, the neutrino oscillation data combined with the the observed DM relic density [5], seems to prefer $m_{S_1} > m_{S_2}$ for large space of parameters. However, the masses of both the DM and $S_2^\pm$ can not exceed $m_{N_1} < 225$ GeV and $m_{S_2} < 245$ GeV, respectively.

1.3 Electroweak Phase Transition

The investigation of the scalar effective potential in the KNT model reveals that, within the allowed parameter space of the model, the strength of the

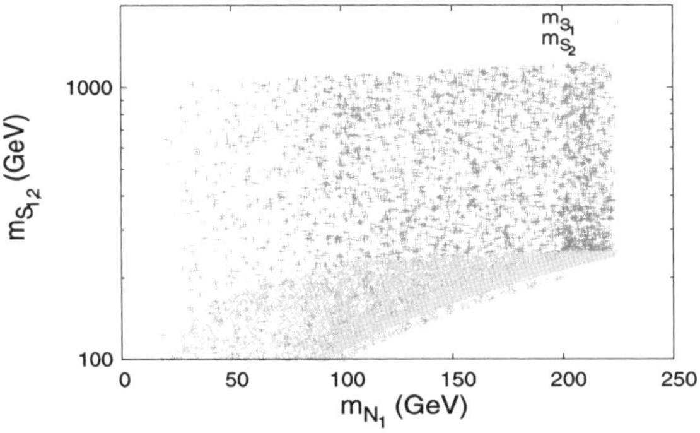

Figure 2: *The charged scalar masses m_{S_1} (red) and m_{S_2} (green) versus the lightest RH neutrino mass, where the constraints from neutrino data, LFV constraints and the DM relic density have been imposed.*

electroweak phase transition (EWPT) can be first order [2] provided that the one-loop corrections to the Higgs mass are sizeable [6]. In Fig. 3, we present the plot for $v(T_c)/T_c$ versus the critical temperature which shows that a strongly first order EWPT can be realized.

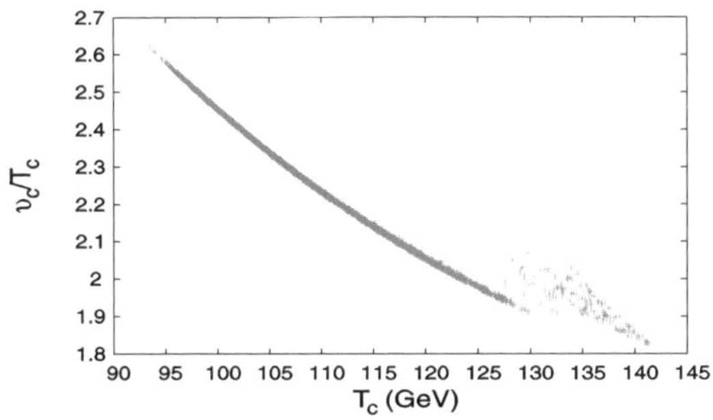

Figure 3: *The critical temperature versus the quantity υ_c/T_c.*

2 A Class of Three-Loop Models with Dark Matter

The KNT model can be generalized where one replaces the charged scalar singlet S_2 and N_R by larger multiplets of the $SU(2)_L$ gauge group. Interesting consequences for the Z_2 symmetry emerge, as we outline now:

- The triplet model: Here $S_2 \sim (1, 3, 2)$ and $N_R \sim (1, 3, 0)$ are taken as $SU(2)_L$ triplets [7]. The DM is the neutral component of the lightest triplet fermion and its mass increases to $M_{DM} \approx 3$ TeV .

- The quintuplet model: This model employs the multiplets $S_2 \sim (1,5,2)$ and $N_R \sim (1,5,0)$ [8]. Unlike the KNT model and the triplet model, the quintuplet variant does not require a Z_2 symmetry to prevent tree-level neutrino mass. It is therefore a viable model of radiative neutrino mass, independent of DM considerations. Due to the gauge interactions, the case with DM requires $M_{DM} \approx 10$ TeV.

- The septuplet model: This case has $S_2 \sim (1,7,2)$ and $N_R \sim (1,7,0)$ [9]. The septuplet model *automatically* possesses an accidental Z_2 symmetry and therefore always contains a stable DM candidate. The DM is relatively heavy, with $M_{DM} \approx 20 - 25$ TeV, due to the Sommerfeld enhancement induced by $SU(2)$ gauge boson exchange [9].

There are additional models of neutrino mass with DM that are related to the KNT model [10]. For example, one can replace the internal SM leptons in Fig 1 with down-type quarks, $e_{L,R} \to d_{L,R}$, and take $S_1 \sim (\bar{3},1,2/3)$ and $S_2 \sim (3,1,-2/3)$ as colored multiplets. Further variant models appears in [10].

3 Conclusion

Three-loop models are particularly interesting as the new physics is generically expected at or around the TeV scale. Furthermore, such models can provide viable DM candidates, thereby solving both the neutrino mass and DM problems. Here we reported recent analyses of the KNT model, which show the model to be a viable theory for neutrino mass and DM that can be tested at collider experiments. We also showed that the KNT model belongs to a larger class of three-loop models that can have interesting observable effects.

Acknowledgments

I would like to thank Alexander I.Studenikin for inviting me to give this talk at the 17th Lomonosov Conference on Elementary Particle Physics in Moscow. I am very thankful to my collaborators A. Ahriche, K. McDonald, and C. Chen for many useful discussions and insights.

References

[1] L. M. Krauss, S. Nasri and M. Trodden, Phys. Rev. D **67**, 085002 (2003).
[2] A. Ahriche and S. Nasri, JCAP **1307**, 035 (2013).
[3] J. Adam et al. [MEG Collaboration], arXiv:1303.0754 [hep-ex].
[4] D.V. Forero, M. Tortola and J.W.F. Valle, Phys. Rev. D 86, 073012 (2012).

[5] P. A. R. Ade *et al.* [Planck Collaboration], arXiv:1502.01589 [astro-ph.CO].
[6] A. Ahriche and S. Nasri, Phys. Rev. D 83, 045032 (2011).
[7] A. Ahriche, C. S. Chen, K. L. McDonald and S. Nasri, Phys. Rev. D **90**, 015024 (2014).
[8] A. Ahriche, K. L. McDonald and S. Nasri, JHEP **1410**, 167 (2014).
[9] A. Ahriche, K. L. McDonald, S. Nasri and T. Toma, Phys. Lett. B **746**, 430 (2015).
[10] C. S. Chen, K. L. McDonald and S. Nasri, Phys. Lett. B **734**, 388 (2014).

LIQUID SCINTILLATOR DEVELOPMENT FOR SOLAR AND SUPERNOVA NEUTRINO DETECTION

Lothar Oberauer [a] and Vincenz Zimmer [b]
Physik-Department, Technical University of Munich, James-Franck-Str. 1, 85748 Garching, Germany

Abstract. We report about new results from elastic neutron scattering experiments on protons contained in samples of a LAB-based liquid scintillator. This measurements have been performed at the MLL accelerator laboratory in Garching, Germany. We measure the quenching factor of the light emitted by recoil-protons as a function of energy and we determine the pulse-shape discrimination capability of LAB-based liquid scintillators in order to probe the feasibility of future low energy neutrino experiments to separate neutrino events from background signals, which are caused by fast neutrons. This studies were performed first in the context of the project LENA (Low Energy Neutrino Astronomy) where a 50 kton liquid scintillator detector in the underground laboratory at Pyhäsalmi, Finland, was proposed. However, the results obtained are also applicable for the upcoming JUNO (Jiangmen Underground Neutrino Observatory) neutrino experiment in China, where a 20 kton neutrino detector with a LAB-based liquid scintillator will be used.

1 Motivation

Large liquid scintillator (LSc) detectors in underground laboratories have the capability to perform high precision measurements on neutrinos from astrophysical sources at low energies. This has been demonstrated successfully by the Borexino experiment at the Italian Gran-Sasso underground laboratory, where solar neutrinos from the most abundant branches of the pp-chains have been measured [1, 2]. Another source of low energy astrophysical sources will be provided in form of a burst by the next galactic supernova of type II. Future LSc-based detectors may have the potential to detect even the very faint signal from the so called diffuse supernova neutrino background (DSNB), which is formed allover the universe by all previous supernovae of type II. In the past, an international collaboration has followed the LENA (Low Energy Neutrino Astronomy) project, proposing a 50 kton large LSc-detector in the underground mine of Pyhäsalmi. In this scope the main characteristics of LAB-based scintillator have been studied over the time-span of several years [3]. Although the realization of LENA seems to be improbable for the moment, the achievement on LAB-based scintillators apply to the upcoming 20 kton large LSc-experiment JUNO (Jiangmen Underground Neutrino Observatory) [4] in China, which aims to start data taking in 2020 and will use a scintillator based on LAB, too. A complete characterization of a liquid scintillator encompasses many aspects and a description of all achievements would be out of scope for this report. Here, we would rather like to focus on the effective light yield of elastic scattering

[a] E-mail: `oberauer@ph.tum.de`
[b] E-mail: `vincenz.zimmer@ph.tum.de`

of fast neutrons on free protons in the scintillator and the possibility to identify this events by pulse-shape discrimination. Both inputs are mandatory for successful background rejection of fast neutrons, which are generated mainly by cosmic muons and can mimic neutrino events. Fast neutron generation via atmospheric neutrino neutral-current (nc) interaction is especially dangerous for DSNB detection, because it may happen inside the fiducial volume without the possibility to veto the event. These neutrons will slow down via multiple scattering off protons mainly in order to get captured on free protons in the scintillator. In this way they resemble the DSNB detection reaction, the inverse beta decay $\bar{\nu}_e + p \rightarrow e^+ + n$, and can be distinguished only via pulse-shape discrimination, which arises from the difference between the scintillation signals from a positron versus a recoil-proton. In addition the light yield due to recoil photons is necessary in order to interpret the reaction rates due to nc-elastic neutrino proton scattering $\nu + p \rightarrow \nu + p$ correctly, a process which will be of interest for measuring the total energy distribution of supernova neutrinos.

In the past, studies on the quenching of recoil-protons on LAB-scintillators have been already performed in the frame of LENA in general [3], and with focus about the potential to measure the DSNB in LENA [5]. Our results reported here are an update of these measurements with increased accuracy. In addition results of proton and alpha-particle quenching measurements in LAB-scintillators have been published in the context of the upcoming SNO+ experiment at the Sudbury underground laboratory in Canada [6, 7] and we will provide a comparison of all these results at the end of this report.

2 Experimental setup

The sample was prepared using LAB (linear alkylbenzene, chemical formula $C_9H_{12}(CH_2)_n$) as a liquid solvent and PPO (2,5-diphenyloxazole) with a concentration of 3 g/l added as a the primary solute, which acts as the first wavelength-shifter. As a second solute we add bisMSB (1,4-bis(2-methylstyryl)-benzene) with a concentration of 20 mg/l in order to match the spectral distribution of the emitted light with the sensitive wavelength region of our photomultiplier, which has a bialkali-photocathode.

The tandem accelerator at the MLL in Garching provides a pulsed (2-3 ns FWHM) ^{11}B-ion beam with 61 MeV kinetic energy which is directed to a target cell filled with gaseous hydrogen at a pressure of 2 bar. Here, neutrons are generated with a maximum kinetic energy of 11.2 MeV in forward direction due to the reaction $^{11}B + p \rightarrow {}^{11}C + n$. As a consequence also gammas with energies in the multi-MeV region are emitted. The neutrons hit a 120 cm^3 large cell, filled with liquid LAB-scintillator, coupled to a 3-inch photomultiplier (PM). The detector position can be selected between about 1° and 35° with respect to the beam axis at distances between about 3.6 m at 1° and 2.1 m at 35°. This allows to select the energy of the incoming neutron between 11.2 MeV and 4.7 MeV. Time information from the pulsed ion beam was used to synchronize the timing of the PM-output. This way a time of flight measurement (TOF) was performed in order to provide an a-priori separation between neutron and gamma induced events.

Detector calibration was performed by the use of several radioactive sources providing gammas with energies between 511 keV to 4.4 MeV. The effect of multiple Compton-scattering was taken into account in a Monte-Carlo based fit to the measured Compton-spectra. This method has been probed by means of coincidence measurements with an additional Ge-detector and so the systematic uncertainties on the energy response have been determined.

In Fig. 1 the time of flight spectrum for forward direction, i.e. with the detector at about 1°, is shown, clearly demonstrating the capability to distinguish between gamma and neutron events. In addition the contributions of accidental background and of neutrons scattered either in the hydrogen cell or in the surrounding materials are shown.

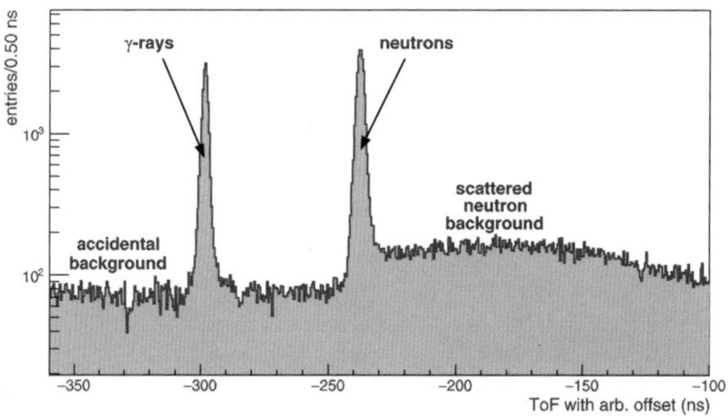

Figure 1: Time of flight spectrum recorded in forward direction. The separation between fast gamma and slow neutron events is evident. Background due to scattered neutrons can be identified, too. The contribution of accidental background is on a $\sim 2.5\%$ level.

3 Results on the proton quenching effect

The energy E_{n} of the incident neutrons at each detector position used was determined by a multi-parameter fit to the respective measured TOF-spectrum. The time resolution of the system is provided by the gamma peak (see fig. 1). In addition a gaussian distribution of E_{n} was assumed. The maximum energy of the recoil protons $E_{\mathrm{rec}}^{\mathrm{max}}$ in the scintillator is approximately equal to the neutron energy, i.e. $E_{\mathrm{rec}}^{\mathrm{max}} \approx E_{\mathrm{n}}$, as the masses of protom and neutron are almost equal. The recoil spectrum measured in forward direction is shown in

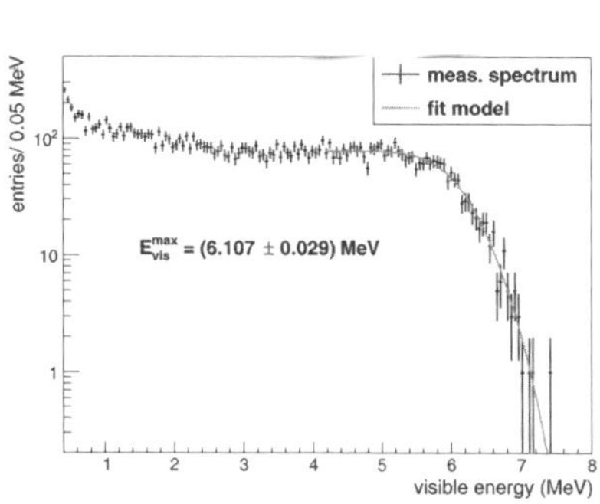

Figure 2: Proton recoil spectrum in forward direction as a function of the visible energy. The maximum visible recoil energy $E_{\rm max}^{\rm vis}$ was determined by a fit with a complementary error function and a constant background.

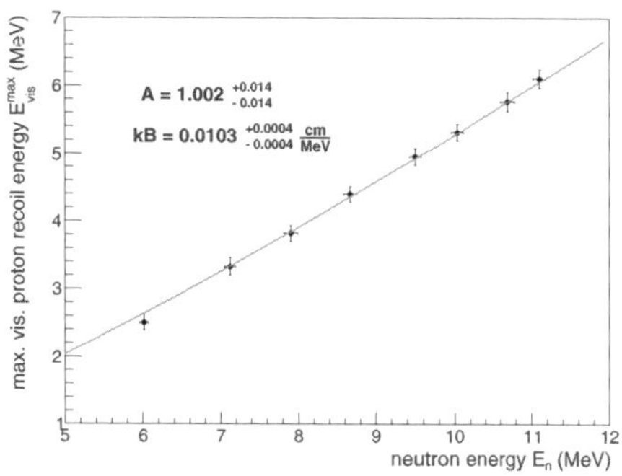

Figure 3: Fit of the Birks model as a function of the neutron energies $E_{\rm n}$ deduced from TOF. The maximum visible proton recoil energies $E_{\rm vis}^{\rm max}$ are determined from the corresponding recoil spectra. For details see text.

Fig. 2. The maximum visible proton recoil energy[c] $E_{\rm vis}^{\rm max}$ was determined from the recoil spectra by a fit of a complementary error function plus a constant background to the region around the edges. The data points comprised of the neutron energies $E_{\rm n}$ and the maximum visible proton recoil energies $E_{\rm vis}^{\rm max}$ were fitted by the semi-empiric Birks quenching model [8]:

$$E_{\rm vis} = \int_0^{E_{\rm dep}} \frac{A}{1 + kB\,\frac{{\rm d}E}{{\rm d}x}(E)}\,{\rm d}E \tag{1}$$

where $E_{\rm dep}$ is the energy deposited by the proton and the Birks factor kB measures the strength of the quenching effect, i.e. the nonlinear energy response of the scintillator to a given particle type. A is the ratio between the detected visible energy and the deposited energy in case of no quenching and hence, $A \approx 1$ is expected as the detector was calibrated. In fig. 3 the fit of the Birks model to the data and the values of the fit parameters obtained for the investigated LAB scintillator are shown. The error bars are dominated by systematic contributions arising from the detector calibration and the models used to extract $E_{\rm n}$ and $E_{\rm vis}^{\rm max}$. The energy dependent energy loss $\frac{{\rm d}E}{{\rm d}x}(E)$ was calculated from values obtained from the PSTAR database [9].

The quenching effect for protons in the LAB scintillator described above was measured in two beam times at the MLL in September 2013 and February 2014. The values obtained for the Birks quenching factor kB are $(0.0103 \pm 0.0005)\,{\rm cm/MeV}$ (Sept. 2013) and $(0.0098 \pm 0.0005)\,{\rm cm/MeV}$ (Feb. 2014) [10], which are in good agreement. The results are also in good agreement with a measurement performed in the scope of the SNO+ experiment of $kB =$

[c]The visible energy denotes the energy scale, which was calibrated using gamma rays.

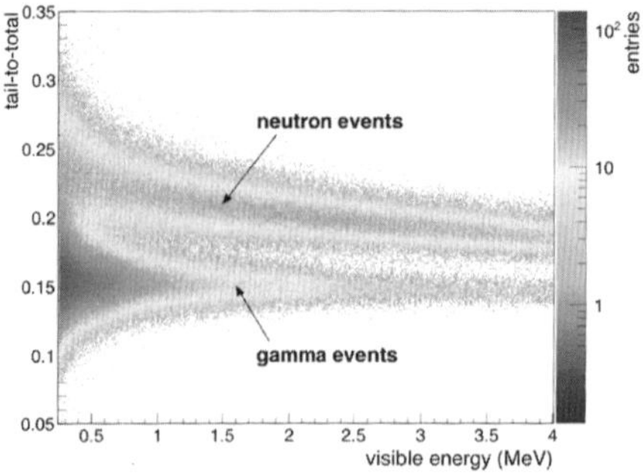

Figure 4: Distribution of the tail to total parameter for all events selected from the gamma and neutron time of flight peaks (see fig. 1). The distribution of events with values for the tail to total parameter of more than about 0.17 are caused by neutron induced events, while those with smaller values are induced by gamma rays.

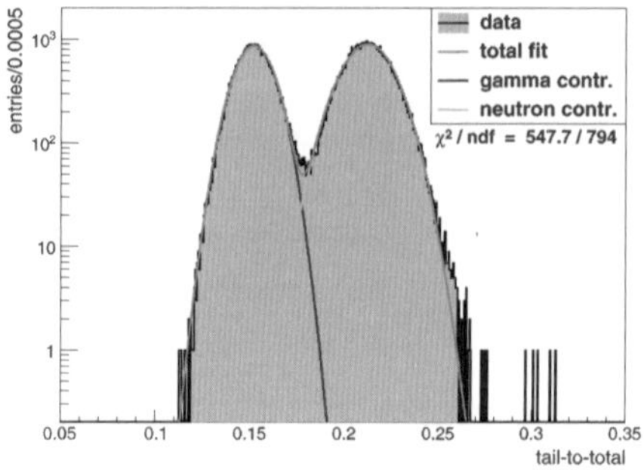

Figure 5: Fit of the sum of two Gaussians (red curve) to the distribution of the tail to total events for energies between 1.0-1.5 MeV. The contributions by gamma (blue, left) and neutron induced events (green, right) are indicated by the result for the corresponding Gaussian.

(0.0098 ± 0.0003) cm/MeV [7] for a slightly different LAB scintillator, featuring 15 mg/l of bisMSB instead of 20 mg/l [d].

4 Results on the pulse-shape discrimination performance

The signals are digitized by a flash-ADC with a resolution of 14 bit and a sampling rate of 2 GHz in order to probe for pulse-shape discrimination (PSD) between recoil protons and gamma events. A 'tail-to-total' (T2T) method was applied by measuring the ratio between the integral over the tail region and the total area of the recorded pulses. The region used for the total integral $[t_{\max} - 50\,\text{ns}, t_{\max} + 350\,\text{ns}]$ and the upper boundary for the tail integral $t_{\max} + 350\,\text{ns}$ were fixed with respect to the maximum position $t_{\max}$ of each pulse. The lower boundary for the tail integral was optimized in order to maximize the PSD performance ($t_{\max} + 28.5\,\text{ns}$ for the LAB scintillator discussed here). In Fig. 4 the distribution of the T2T-parameter as a function of energy for each event is shown, demonstrating the excellent PSD performance of the scintillator by the nicely separated event bands for gamma and neutron induced events. In fig. 5 the distribution of the T2T-parameter for events with energies between 1.0-1.5 MeV are shown. The distribution was fitted by the sum of two Gaussians, and the results are used to quantify the PSD performance. For this we define as a figure of merit $F_M = (\mu_{\text{n}} - \mu_\gamma)/(\sigma_{\text{n}}^2 + \sigma_\gamma^2)^{\frac{1}{2}}$ [11], where μ_{n}, μ_γ are the mean

[d] The concentration of the secondary solute bisMSB was found to have no significant impact on kB [10].

values and σ_n, σ_γ are the widths of the Gaussian distributions, respectively. We obtain $F_M = 3.730 \pm 0.010$ for energies between 1.0-1.5 MeV [e] [10]. This implies a neutron leakage of only (0.175 ± 0.006) % in this energy region, if the gamma selection efficiency is fixed to a value of 99 %. Hence, PSD with LAB-scintillators works even at low energies and the separation power is increasing with energy, which is important e.g. for DSNB detection where the region of interest is at energies above $\sim$ 10 MeV. For instance, in the range 3-4 MeV the neutron leakage is (0.0116 ± 0.0005)% at 99 % gamma detection efficiency. For comparability of the PSD results the photoelectron (pe) yield was determined to be 628±40 pe/MeV [10], using the detector resolution extracted from the energy calibration data. LENA aims for a photoelectron yield of at least 200 pe/MeV [3] and in JUNO goes for an unprecedented yield of more than 1000 pe/MeV [4]. With the shown results we demonstrate clearly, that LAB-scintillators developed for LENA and JUNO have very promising PSD-capabilities in order to distinguish neutrino events from background signals.

Acknowledgments

This research was supported by the DFG cluster of excellence 'Origin and Structure of the Universe' and the MLL laboratory in Garching, Germany.

References

[1] G. Bellini et al., Borexino coll., Phys. Rev. Lett. **108**, 051302 (2012); arXiv:1110.3230
[2] G. Bellini et al., Borexino coll., Nature **512**, 383 (2014)
[3] M. Wurm et al., LENA coll., Astropart. Phys. **35**, 685-732, (2011); arXiv:1104.5620.
[4] F. An et al., JUNO coll., J. Phys. G **43**, 030401 (2016); arXiv:1507.05613.
[5] R. Möllenberg et al., Phys. rev. D **91**, 032005 (2015); arXiv:1409.2240.
[6] B. von Krosigk et al., arXiv:1510.00458 (2015).
[7] B. von Krosigk et al., Eur. Phys. J. C 73 (2013), 2390.
[8] J. B. Birks, *The Theory and Practice of Scintillation Counting*, Pergamon Press, First Edition (1964).
[9] M. J. Berger et al., NIST, http://physics.nist.gov/Star (acc. 13-May-2014).
[10] V. Zimmer, PhD thesis, Technische Universität München (2015)
[11] F. Elisei et al., Nucl. Instr. and Meth. in Phys. Res. A 400 (1997), 53.

[e] Values for the F_M above 1 indicate a good PSD performance [11].

ON QUASIDEGENERACY OF MAJORANA NEUTRINOS AND THE OBSERVED PATTERN OF LEPTONIC MIXING

M. N. Rebelo[a]

Centro de Física Teórica de Partículas - CFTP and Departamento de Física, Instituto Superior Técnico - IST, Universidade de Lisboa, Av. Rovisco Pais, P-1049-001 Lisboa, Portugal

Abstract. We relate the observed pattern of leptonic mixing to the quaidegeneracy of three Majorana neutrinos. We show how lifting the degeneracy may lead to the measured value of $|U_{13}|$ and to sizeable CP violation of Dirac-type. We show some of the correlations obtained among physical observables, starting from some of the most interesting schemes proposed in the literature.

1 Introduction

This talk is based on work done in collaboration with G. C. Branco, J. I. Silva-Marcos and Daniel Wegman [1].

It was already shown, quite some time ago that the limit of exact degeneracy of Majorana neutrinos is non trivial in the sense that it allows for leptonic mixing and CP violation [2]. This result relies on the assumption that neutrinos are Majorana particles. Only fermions that are neutral under all $U(1)$ type symmetries can have Majorana character [3], therefore all other fermions in the Standard Model (SM) are Dirac fermions. Despite great recent experimental progress in the field of Neutrino Physics, the question of whether neutrinos are Dirac or Majorana fermions still remains one of the major open questions in Particle Physics.

The 2015 Nobel prize was awarded to T. Kajita and B. McDonald "for the discovery of neutrino oscillations, which shows that neutrinos have mass". In fact, it is by now established that at least two of the observed neutrinos have non-zero masses. The presently allowed ranges are listed in Table 1. It should be noted that the sign of $\Delta m^2_{31} \equiv m^2_3 - m^2_1$, is not yet known and is at present under intense experimental investigation. Experimentally the mass of the lightest neutrino, either ν_1 or ν_3 is still consistent with zero. However, upper bounds for the neutrino masses allow for quasidegeneracy. It is therefore of great interest to consider the limit of exact degeneracy of neutrino masses and possible ways of lifting this degeneracy. Specially so because the limit of exact degeneracy favours two large mixing angles and leads to one zero mixing angle which upon the lifting of the degeneracy can give rise to the observed small angle. The leptonic mixing angles have been measured experimentally, and are also listed in Table 1. There are two angles that are large when compared to the Cabibbo angle and one small mixing angle, θ_{13}, which was consistent

[a]E-mail: rebelor@tecnico.ulisboa.pt

with zero until the recent measurements performed at reactor [4] and accelerator neutrino experiments [5]. Leptonic mixing is remarkably different from quark mixing. It is not yet known whether or not there is CP violation in the leptonic sector either at low or at high energies [6]. In the absence of a flavour model no relation can be established between these two manifestations [7], [8]. The limit of exact degeneracy with Majorana neutrinos allows for one Majorana-type CP violating phase in the mixing matrix [2]. Lifting of the degeneracy together with the requirement that all three mixing angles should be different from zero allows for Dirac as well as Majorana-type CP violation [1]. In Ref. [1] we studied perturbations of exact degeneracy around some of the very well known mixing textures proposed in the literature, with $\theta_{13} = 0$, [9], [10], [11], [12], [13], [14], [15], [16] leading to quasidegenerate masses and mixing in agreement with all available experimental data. We also showed that in this process it is possible to generate Dirac-type CP violation large enough in strength to be detectable in the next round of neutrino experiments without introducing new sources of CP violation.

2 Present Experimental Knowledge of Neutrino Masses and Leptonic Mixing

Table 1 gives the 2014 update on global fits of neutrino oscillation parameters provided by Forero, Tortola and Valle [17]. The quantities Δm^2_{ij} are defined by $(m_i^2 - m_j^2)$ and the angle θ_{ij} and the phase δ are those of the standard parametrisation [18] Several textures for the leptonic mixing have been studied

Table 1: Neutrino oscillation parameter summary. For Δm^2_{31}, $\sin^2\theta_{23}$, $\sin^2\theta_{13}$, and δ the upper (lower) row corresponds to normal (inverted) neutrino mass hierarchy. [a]There is a local minimum in the first octant, $\sin^2\theta_{23} = 0.467$ with $\Delta\chi^2 = 0.28$ with respect to the global minimum. From [17]

Parameter	Best fit	1σ range
Δm^2_{21} $[10^{-5} eV^2]$	7.60	7.42 – 7.79
$\|\Delta m^2_{31}\|$ $[10^{-3} eV^2](NH)$	2.48	2.41 – 2.53
$\|\Delta m^2_{31}\|$ $[10^{-3} eV^2](IH)$	2.38	2.32 – 2.43
$\sin^2\theta_{12}$	0.323	0.307 – 0.339
$\sin^2\theta_{23}(NH)$	0.567	0.439[a] – 0.599
$\sin^2\theta_{23}(IH)$	0.573	0.530 – 0.598
$\sin^2\theta_{13}$ (NH)	0.0234	0.0214 –0.0254
$\sin^2\theta_{13}$ (IH)	0.0240	0.0221 – 0.0259
δ (NH)	1.34 π	0.96 –1.98 π
δ (IH)	1.48 π	1.16 –1.82 π

in the literature, often in the context of family symmetries [19], [20], [21]. In most of the proposed schemes, the pattern of leptonic mixing is predicted but the spectrum of masses is not constrained by the symmetries. It is therefore consistent to consider these schemes, together with the hypothesis of quasidegeneracy of Majorana neutrinos.

3 The Limit of Exact Degeneracy with Majorana Neutrinos

We work in the framework of three left-handed neutrinos and write the effective Majorana mass term in the weak basis where the charged lepton mass matrix is diagonal, real and positive as:

$$\mathcal{L}_{\text{mass}} = -\,(\nu_{L\alpha})^T\, C^{-1}\, (M_o)_{\alpha\beta}\ \ \nu_{L\beta}\ +\ \text{h.c.} \tag{1}$$

where $\nu_{L\alpha}$ stand for the left-handed weak eigenstates and M_o is a 3×3 symmetric complex mass matrix. Obviously there is no loss of generality in choosing this weak-basis. In general M_o is diagonalized by a unitary matrix U_o through $U_o^T\ M_o\ U_o = diag\ (m_{\nu_1}, m_{\nu_2}, m_{\nu_3})$. In the limit of exact degeneracy, M_o can be written:

$$M_o = \mu\ S_o \tag{2}$$

where μ is the common neutrino mass and $S_o = U_o^* U_o^\dagger$. Therefore, in the limit of exact degeneracy M_o is proportional to a symmetric unitary matrix. Leptonic mixing and even CP violation can occur in this limit provided that neutrinos are Majorana particles [2]. Leptonic mixing can be rotated away if and only if there is CP invariance and all neutrinos have the same CP parity [22]. It was also shown in Ref [2] that in the case of different CP parities the leptonic mixing matrix U_o can be parametrised by two angles and one phase, in the form:

$$U_o\ = O_{23}\,(\phi)\ O_{12}\left(\frac{\theta}{2}\right)\ \begin{pmatrix} 1 & 0 & 0 \\ 0 & i & 0 \\ 0 & 0 & e^{-i\frac{\alpha}{2}} \end{pmatrix} \tag{3}$$

up to an orthogonal rotation of the three degenerate neutrinos and with each orthogonal matrix O_{ij} chosen to be symmetric. Obviously, if U_o diagonalises M_0 so does $U_o O$ with O an arbitrary orthogonal rotation. An important feature is the fact that U_o always has one zero entry which in the above parametrisation appears in the (13) position. Although the location of the zero is not fixed the observed pattern of leptonic mixing suggests that the above choice is a good starting point for a perturbation.

4 Lifting the Degeneracy

In order to lift the degeneracy one may add a small perturbation to S_0:

$$M = \mu\ (S_o + \varepsilon^2\ Q_o) \tag{4}$$

where the matrix Q_0 is fixed in such a way that the correct neutrino masses are obtained. We assume that after lifting the degeneracy the leptonic mixing matrix is given by:

$$U_{PMNS} = U_o \cdot O \tag{5}$$

where O is an orthogonal matrix, parametrized by small angles. The matrix O is denoted by:

$$O = O_{12}O_{13}O_{23} = \begin{pmatrix} c_{\phi_1} & s_{\phi_1} & 0 \\ -s_{\phi_1} & c_{\phi_1} & 0 \\ 0 & 0 & 1 \end{pmatrix} \begin{pmatrix} c_{\phi_3} & 0 & s_{\phi_3} \\ 0 & 1 & 0 \\ -s_{\phi_3} & 0 & c_{\phi_3} \end{pmatrix} \begin{pmatrix} 1 & 0 & 0 \\ 0 & c_{\phi_2} & s_{\phi_2} \\ 0 & -s_{\phi_2} & c_{\phi_2} \end{pmatrix} \tag{6}$$

Notice that U_{PMNS} still diagonalises S_o, thus establishing a strong connection between the degenerate and quasidegenerate case. Different cases were analysed by choosing U_0 to coincide with some of the most interesting cases considered in the literature with a zero in the (13) entry.

One of the examples considered in Ref. [1] consists of perturbing the tribimaximal mixing. In this case, we have: $U_o = U_{TBM} \cdot K$ with:

$$U_{TBM} = \begin{pmatrix} \frac{2}{\sqrt{6}} & \frac{1}{\sqrt{3}} & 0 \\ \frac{1}{\sqrt{6}} & -\frac{1}{\sqrt{3}} & \frac{1}{\sqrt{2}} \\ \frac{1}{\sqrt{6}} & -\frac{1}{\sqrt{3}} & -\frac{1}{\sqrt{2}} \end{pmatrix} \quad \text{and} \quad K = diag(1, i, e^{-i\alpha/2}) \tag{7}$$

In the notation of Eq. (3), this ansatz corresponds to $\phi = 45°$ and $\cos\left(\frac{\theta}{2}\right) = \frac{2}{\sqrt{6}}$ i.e., $\frac{\theta}{2} = 35.26°$. We used data from the global fit of neutrino oscillations provided in 2012 by the authors of Ref. [17], requiring agreement within 1σ range. Although there are deviations in the more recent data from that of 2012 the deviations are slight and therefore the conclusions would not change significantly. In Ref. [1] we concluded that, in this example, complying with the experimental bounds allowed the leptonic strength of Dirac-type CP violation to range from 0 to about 4×10^{-2}, so that it could be within reach of future neutrino experiments.

Here we reproduce one of the figures obtained in the previous reference showing the correlation between I_{CP} and $|U_{13}|^2$. This scenario allows for a particularly simple solution since, one can reach agreement with the experimental data by choosing a matrix O with only one parameter different from zero, namely the angle ϕ_2.

5 Final Remarks

The starting point of our analysis is the limit of exact degeneracy of neutrino masses. In this limit there is no Dirac-type CP violation, only Majorana-type CP violation may occur and one of the leptonic mixing angles is exactly zero. It

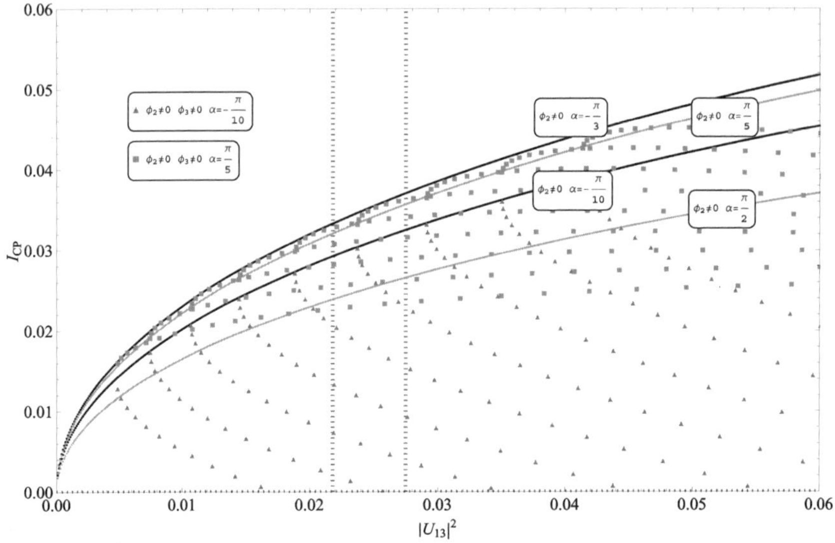

Figure 1: I_{CP} versus $|U_{13}|^2$ obtained by perturbing tribimaximal mixing with $\phi_3 = 0$. Each curve corresponds to a fixed α and to $\phi_1 = 0$, therefore ϕ_2 is the only variable. The points drifting away from each curve were obtained by varying also ϕ_3.

has been shown in Ref. [23] that there is no loss of generality in parametrising a general unitarity matrix in the form given by Eq.(5), with U_0 of the generic form introduced before, however fixing the parameters of U_0, as was done in the previous analysis, by making use of some of the most interesting patterns discussed in the literature with $\theta_{13} = 0$ restricts U_{PMNS}, and gives rise to predictive power while at the same time allowing to relate quasidegeneracy of neutrino masses with the observed pattern of leptonic mixing..

Acknowledgments

The author thanks the Organisers of the 17th Lomonosov Conference on Elementary Particle Physics for the very stimulating scientific environment provided during the Conference as well as the warm hospitality. This work was partially supported by Fundação para a Ciência e a Tecnologia (FCT, Portugal) through the projects PTDC/FIS-NUC/0548/2012, CERN/FIS-NUC/0010/2015, CFTP-FCT Unit 777 (UID/FIS/00777/2013) which are partially funded through POCTI (FEDER), COMPETE, QREN and EU. Part of the work presented here was done at CERN-Theory Division during different periods of time and with partial financial support from CERN.

References

[1] G. C. Branco, M. N. Rebelo, J. I. Silva-Marcos and D. Wegman, Phys. Rev. **D 91**, 013001 (2015) doi:10.1103/PhysRevD.91.013001 [arXiv:1405.5120 [hep-ph]].

[2] G. C. Branco, M. N. Rebelo and J. I. Silva-Marcos, Phys. Rev. Lett. **82**, 683 (1999) doi:10.1103/PhysRevLett.82.683 [hep-ph/9810328].

[3] R. N. Mohapatra and P. B. Pal, *"Massive neutrinos in physics and astrophysics. Second edition"* (World Sci. Lect. Notes Phys. **60**, 1 (1998) [World Sci. Lect. Notes Phys. **72**, 1 (2004)]).

[4] F. P. An *et al.* [DAYA-BAY Collaboration], Phys. Rev. Lett. **108**, 171803 (2012) [arXiv:1203.1669 [hep-ex]]; J. K. Ahn *et al.* [RENO Collaboration], Phys. Rev. Lett. **108**, 191802 (2012) [arXiv:1204.0626 [hep-ex]]; Y. Abe *et al.* [Double Chooz Collaboration], Phys. Rev. **D 86**, 052008 (2012) [arXiv:1207.6632 [hep-ex]]; F. P. An *et al.* [Daya Bay Collaboration], Chin. Phys. **C 37**, 011001 (2013) [arXiv:1210.6327 [hep-ex]].

[5] K. Abe *et al.* [T2K Collaboration], Phys. Rev. **D 88**, 032002 (2013) [arXiv:1304.0841 [hep-ex]].

[6] G. C. Branco, R. G. Felipe and F. R. Joaquim, Rev. Mod. Phys. **84** 515 (2012) [arXiv:1111.5332 [hep-ph]].

[7] G. C. Branco, T. Morozumi, B. M. Nobre and M. N. Rebelo, Nucl. Phys. B **617**, 475 (2001) doi:10.1016/S0550-3213(01)00425-4 [hep-ph/0107164].

[8] M. N. Rebelo, Phys. Rev. D **67**, 013008 (2003) doi:10.1103/PhysRevD.67.013008 [hep-ph/0207236].

[9] P. F. Harrison, D. H. Perkins and W. G. Scott, Phys. Lett. B **530**, 167 (2002) [hep-ph/0202074].

[10] H. Fritzsch and Z. Z. Xing, Phys. Lett. B **372**, 265 (1996) doi:10.1016/0370-2693(96)00107-4 [hep-ph/9509389].

[11] V. D. Barger, S. Pakvasa, T. J. Weiler and K. Whisnant, Phys. Lett. B **437**, 107 (1998) doi:10.1016/S0370-2693(98)00880-6 [hep-ph/9806387].

[12] Y. Kajiyama, M. Raidal and A. Strumia, Phys. Rev. D **76**, 117301 (2007) doi:10.1103/PhysRevD.76.117301 [arXiv:0705.4559 [hep-ph]].

[13] W. Rodejohann, Phys. Lett. B **671**, 267 (2009) doi:10.1016/j.physletb.2008.12.010 [arXiv:0810.5239 [hep-ph]].

[14] C. H. Albright, A. Dueck and W. Rodejohann, Eur. Phys. J. C **70**, 1099 (2010) doi:10.1140/epjc/s10052-010-1492-2 [arXiv:1004.2798 [hep-ph]].

[15] J. E. Kim and M. S. Seo, JHEP **1102**, 097 (2011) doi:10.1007/JHEP02(2011)097 [arXiv:1005.4684 [hep-ph]].

[16] J. E. Kim and M. S. Seo, Int. J. Mod. Phys. A **27**, 1250017 (2012) doi:10.1142/S0217751X12500170 [arXiv:1106.6117 [hep-ph]].

[17] D. V. Forero, M. Tortola and J. W. F. Valle, Phys. Rev. D **90**, 9, 093006 (2014) doi:10.1103/PhysRevD.90.093006 [arXiv:1405.7540 [hep-ph]].

[18] K. A. Olive *et al.* [Particle Data Group Collaboration], Chin. Phys. C **38**, 090001 (2014). doi:10.1088/1674-1137/38/9/090001

[19] For a review and additional references, see: G. Altarelli and F. Feruglio, Rev. Mod. Phys. **82**, 2701 (2010) doi:10.1103/RevModPhys.82.2701 [arXiv:1002.0211 [hep-ph]].

[20] S. F. King, A. Merle, S. Morisi, Y. Shimizu and M. Tanimoto, New J. Phys. **16**, 045018 (2014) doi:10.1088/1367-2630/16/4/045018 [arXiv:1402.4271 [hep-ph]].

[21] R. M. Fonseca and W. Grimus, JHEP **1409**, 033 (2014) doi:10.1007/JHEP09(2014)033 [arXiv:1405.3678 [hep-ph]].

[22] L. Wolfenstein, Phys. Lett. B **107**, 77 (1981).

[23] D. Emmanuel-Costa, N. R. Agostinho, J. I. Silva-Marcos and D. Wegman, Phys. Rev. D **92**, no. 1, 013012 (2015) doi:10.1103/PhysRevD.92.013012 [arXiv:1504.07188 [hep-ph]].

KAMLAND-ZEN

Junpei Shirai [a]
(for the KamLAND-Zen Collaboration)
Research Center for Neutrino Science,
Tohoku University, Sendai, 980-8578, Japan

Abstract. Results of the phase-1 and a part of the phase-2 of the KamLAND-Zen experiment searching for the neutrinoless double beta decay ($0\nu\beta\beta$) of ^{136}Xe nuclei and the current activity for the next phase and future upgrade are presented.

1 Introduction

Fundamental problems of neutrinos are closely related to the mass of neutrinos. Many challenges to the problem have been made on neutrino oscillations for precise measurement of the oscillation parameters (the mass squared differences and the mixing angles), and the next steps are to determine the mass hierarchy and the CP violation phase in the lepton sector.

Another challenge related to the neutrino mass is to find out whether neutrinos and anti-neutrinos are equal or not, that is, they are Majorana particles or Dirac particles. In general, the Lagrangian of the neutral spin 1/2 fermions has not only Dirac mass terms $M_D(\overline{\psi_R}\psi_L + h.c.)$ but also mass terms connecting particle and anti-particle, like $(M_L/2)(\overline{(\psi_L)^c}\psi_L + h.c.) + (M_R/2)(\overline{((\psi)_R)^c}\psi_R + h.c.)$. If neutrinos are Majorana particles, the physical masses (M_ν and M_N) of neutrinos are related to the Dirac mass by an equation $M_\nu = M_D^2/M_N$ through the "see-saw mechanism". This scenario is very attractive and exciting because it would explain extraordinarily small masses of existing neutrinos by suggesting the existence of super-heavy neutrinos providing a key to the matter dominance of the universe through leptogenesis. Therefore, testing whether neutrinos are Majorana particles or not is very important, but there have been no convincing results on this problem to date.

Neutrinoless double beta decay ($0\nu\beta\beta$) of a nucleus has been considered to be a unique and feasible way to test the Majorana nature of neutrinos. Unlike the 2-neutrino double beta decay ($2\nu\beta\beta$) which is the 2nd order weak interaction within the standard model (SM) process, if neutrinos are Majorana particles, then $0\nu\beta\beta$ can happen. This is $\Delta L = 2$ process and not allowed in the SM. Among many possible processes, light-Majorana neutrino exchange is considered to be the dominant process of $0\nu\beta\beta$. Moreover, observation of $0\nu\beta\beta$ implies that neutrinos are Majorana particle irrespective of the processes [1].

If $0\nu\beta\beta$ occurs through the light Majorana neutrino exchange, the half-life ($T_{1/2}^{0\nu}$) is expressed by the following equation, $(T_{1/2}^{0\nu})^{-1}{=}G^{0\nu}|M^{0\nu}|^2\langle m_{\beta\beta}\rangle^2$, where $G^{0\nu}$ is the phase space factor, $M^{0\nu}$ is the nuclear matrix element, and

[a] E-mail: `shirai@awa.tohoku.ac.jp`

$\langle m_{\beta\beta} \rangle$ is the effective mass of neutrinos which contains all the parameters of neutrinos and information of absolute mass scale, mass ordering and CP-phases of neutrinos. Although there was a claim for observation of $0\nu\beta\beta$ of ^{76}Ge nucleus [2], recent highly sensitive experiments with O(100)kg isotopic nuclei are going to exclude the claim and the next step is to explore the inverted mass hierarchy (IH) region by using several hundred kilograms to 1000kg of decaying nuclei.

To increase the $0\nu\beta\beta$ discovery potential large mass of the target nuclei and a clean environment are the most important factors, but background rejection by clever ideas and technologies are also required. So, selection of nuclei is very important. The ^{136}Xe nucleus with a relatively high Q-value (2.548MeV) and long $2\nu\beta\beta$ half-life (10^{21}yr), satisfies many of the requirements; high natural abundance (8.6%), established enrichment and purification technique, chemical stability and easiness to handle. This is largely because Xe is a rare gas. Moreover, Xe is easy to dissolve into organic liquids like liquid scintillator (LS), easily separated from LS and re-purified. Therefore, Xe is very flexible and has excellent scalability. In this point of view Xe+LS combination is excellent strategy for $0\nu\beta\beta$ search. KamLAND-Zen has made full use of these properties of Xe using the existing large volume LS facility of KamLAND (Kamioka Liquid scintillator Anti-Neutrino Detector).

2 KamLAND-Zen detector

KamLAND-Zen (Zero neutrino double beta decay experiment) is realized by minimum modification of the KamLAND detector located deep underground by 1,000m in the Kamioka mine, Gifu prefecture in Japan. The muon flux is 10^5 times smaller than at earth level. The detector is described in [3] and references therein, so a brief description is given here.

In the center of the KamLAND-Zen is an inner balloon (IB) made of clean nylon film of 25μm thickness and 3.08m in diameter. It is filled with Xe dissolved liquid scintillator (Xe-LS) with 320kg of enriched ^{136}Xe (90% enrichment). The IB is suspended by 12 belts made of the same nylon film at the center of the 13m-diameter KamLAND balloon which contains 1,000 ton of ultra-pure liquid scintillator. The KamLAND balloon is surrounded by 2.5m-thick transparent buffer oil and housed in a 18m-diameter stainless-steel spherical tank. Events in the Xe-LS and LS are detected by 1325 17-inch PMTs and 554 20-inch PMTs through the buffer oil. The PMTs are mounted on the inner surface of the spherical tank and provide 34% coverage of 4π and energy resolution, $\sigma_E/E \sim 6.5\%/\sqrt{E_{[MeV]}}$. Outside of the tank is a 3,200 ton purified water Cherenkov detector equipped with 225 20-inch PMTs which serves as the radiation shield from surrounding rocks, the stabilizer of the temperature of the central detector, and allows cosmic-ray muon identification.

The IB was constructed in May 2011 in a class-1 super-clean room in Tohoku

University and installed into the detector in August of 2011. After the calm down of the radon activity data taking was started in October of 2011 and continued up to June of 2012 which we call phase-1 of KamLAND-Zen.

3 Analysis results of phase-1 and phase-2

We published 3 papers for $0\nu\beta\beta$ and $2\nu\beta\beta$ results [3–5] based on the phase-1 data. The 3rd paper [5] which is based on an exposure of 89.5 kg-yr of ^{136}Xe provided a limit on the $0\nu\beta\beta$ half-life ($T^{0\nu}_{1/2}$) to 1.9×10^{25} yr at 90%C.L. and combined results with EXO-200 [6] provided 3.4×10^{25}yr at 90%C.L. By using various calculation of $M^{0\nu}$ the results can be compared to $0\nu\beta\beta$ searches for ^{76}Ge and the combined Xe results is inconsistent with the KK claim [2] for $0\nu\beta\beta$ of ^{76}Ge.

The energy spectrum in phase-1 showed a peak at the visible energy region of 2.2∼3.0 MeV, but it was reproduced in shape and time variation of ^{110m}Ag decays with $T_{1/2}$ of 250 days. Other backgrounds in the energy region are the ^{214}Bi (from the IB surface), ^{10}C produced by cosmic ray muon spallation, and $2\nu\beta\beta$ decay of ^{136}Xe. We stopped data taking (phase-1) and started purification of Xe-LS in June of 2012 as follows.

Xe-LS was extracted from the IB and the collected Xe was purified by distillation and getter. The LS was purified by distillation and transferred to the IB in circulation mode. This was done 3 times for the total IB volume. Finally, purified Xe was dissolved into the LS and pushed into the IB. The Xe amount was increased from 320kg to 380kg. After the purification process the data taking (which we call phase-2) started in November 2013. Unfortunately during the Xe extraction a leakage of a diaphragm pump increased the backgrounds of ^{214}Bi at the bottom region of the IB.

Figure 1 shows comparison of the results of the phase-1 and phase-2 data. The upper panels are cubic R vs visible energies, where R is the radial distance of the reconstructed events from the detector center. The horizontal line at R=1.54m corresponds to the IB film which was contaminated by various radioactive elements; ^{134}Cs, ^{137}Cs, ^{214}Bi and ^{208}Tl. Lower panels show the event rates in the energy region from 2.2 to 3.0 MeV for R<1m. Significant reduction of ^{110m}Ag by a factor of 10 can be seen after the purification.

However, backgrounds are still seen on the IB which is mainly ^{214}Bi, especially in the bottom region of IB. This is caused by the leakage of the diaphragm pump as described above. We conducted a multi-volume analysis dividing the R<2m space into 40 equal-volume bins and made simultaneous fit to the energy spectra with the backgrounds and the $0\nu\beta\beta$ signal.

The results are shown in Figure 2 as an example for the R<1m region. There is no excess over the expected backgrounds. Limits at 90%C.L. on the $0\nu\beta\beta$ signal are obtained as <17.0 events/day/kton-LS which corresponds to $T^{0\nu}_{1/2}$ limit of $> 1.3 \times 10^{25}$yr.

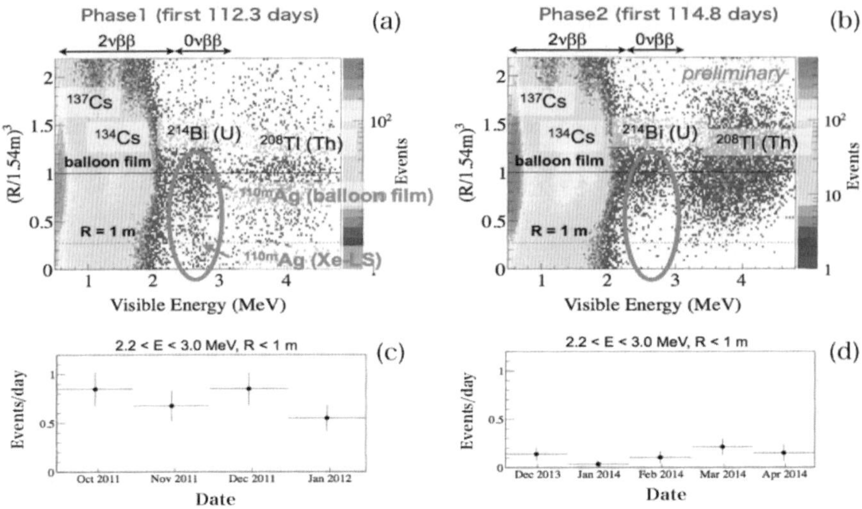

Figure 1: Visible energy vs R^3 plots for (a) phase-1 and (b) phase-2, and event rates in the energy region of 2.2 to 3.0 MeV for (c) phase-1 and (d) phase-2 data.

The study shows that ^{10}C background caused by muon spallation remained. It has a long life ($T_{1/2}$=28s) and the rejection is not trivial. However, in most cases ^{10}C production is associated with neutrons. Therefore ^{10}C background can be rejected by triple coincidence of muon, neutron and ^{10}C events by using a space-time cut. Neutrons are detected by using the dead-time free electronics, Mogura with rejection efficiency of 72±5% with a loss of 7% of the signal.

Combined results of the phase-1 plus the new data of phase-2 give the 90%C.L. limit of $T^{0\nu}_{1/2}$ of 2.6×10^{25}yr. This result gives the most stringent $0\nu\beta\beta$ limit and refutes the KK claim by KamLAND-Zen data alone.

4 Next phase and future plan

As already mentioned the surface of the IB is contaminated by ^{214}Bi background for $0\nu\beta\beta$ search. We decided to construct a new IB under better cleanliness control than the current one. The new IB is 2 times bigger and is now under construction in the class-1 super-clean room in Tohoku University.

We introduced as many methods as possible to protect the film against dust after careful examination of the release and the behavior of the dust in the clean room by a laser-light visualization technique.

The timeline for the next phase is as follows. We will stop data taking of phase-2 in September of 2015, run the calibration, remove the IB, followed by

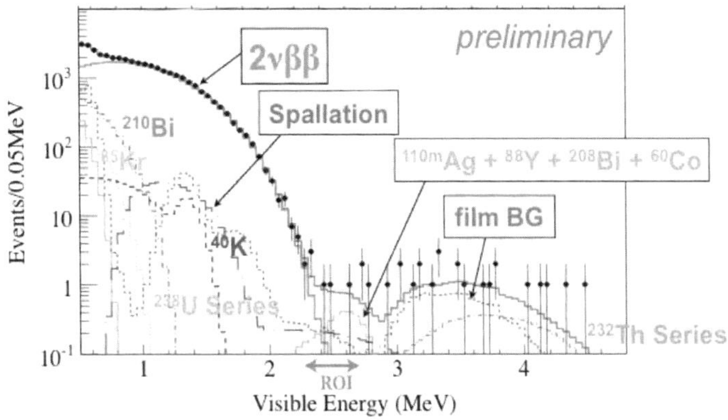

Figure 2: Spectrum for the phase-2 data (dots with error bars) and fitted background components ($R < 1$m).

the outer detector refurbishment, detector tank inspection, purification of the Xe and LS, and deployment of the new IB filled with Xe-LS containing >600kg enriched ^{136}Xe. Data taking will be started next summer.

As a future project we are planning the KamLAND2-Zen using 1 ton enriched ^{136}Xe aiming to achieve the $\langle m_{\beta\beta} \rangle$ sensitivity of 20meV in 5 years. Many R&D efforts are ongoing to increase the energy resolution by introducing new LAB-based LS with high light yield and improved light attenuation length, and light collectors on high quantum efficiency PMTs. Studies are also made of rejection of ^{214}Bi-^{214}Po backgrounds by introducing a scintillating balloon, imaging device for β/γ discrimination, and new purification methods for removing radio-impurities in LS.

Figure 3 shows the expected sensitivity of near-future KamLAND-Zen. We have collected phase-2 data for 1.8 years and by the analysis of the full data set we expect the $T^{0\nu}_{1/2}$ sensitivity of $> 3 \times 10^{25}$yr. Next year with the new larger IB with at least 600kg enriched ^{136}Xe we expect sensitivity, $T^{0\nu}_{1/2} > 10^{26}$yr, which corresponds to $\langle m_{\beta\beta} \rangle$ of ~60meV entering the IH region. In this region the area in $\langle m_{\beta\beta} \rangle$ vs summed neutrino masses plane allowed by oscillation experiments are compared by cosmological observation and theoretical predictions which is very exciting. The ultimate goal of KamLAND2-Zen is to explore the full IH region, $\langle m_{\beta\beta} \rangle$~20meV, and to make clear the Majorana nature of the neutrinos.

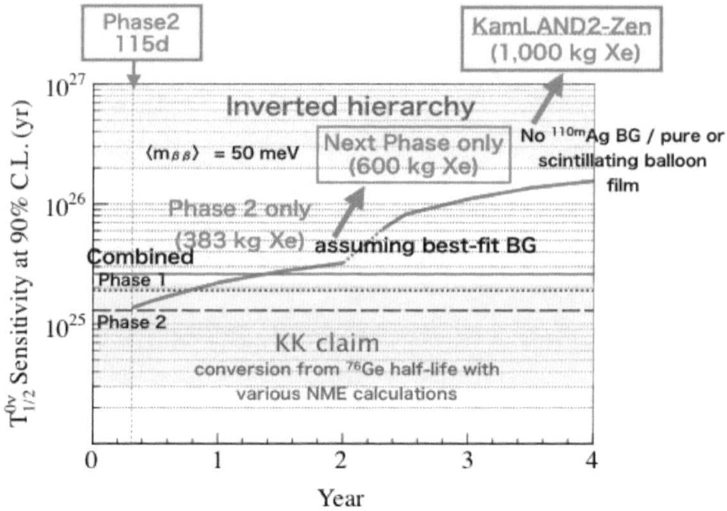

Figure 3: Expected $0\nu\beta\beta$ sensitivities of various phases of KamLAND-Zen.

5 Summary

1. Studies of $0\nu\beta\beta$ can clarify the fundamental problem of the neutrino mass which would open exciting physics.

2. KamLAND-Zen with its unique detector for ^{136}Xe $0\nu\beta\beta$ search conducted Xe-LS purification which reduced the dominant ^{110m}Ag background by a factor 10.

3. Limits at 90%C.L. on the $T^{0\nu}_{1/2}$ obtained from phase-1+phase-2 (115 days) data provides 2.6×10^{25}yr, which corresponds to $\langle m_{\beta\beta}\rangle$ <140-280meV refuting the KK claim for ^{76}Ge.

4. We are now constructing a new larger IB for the next phase with a reduction of contaminants to extend the detector sensitivity into the inverted mass hierarchy region.

5. Various R&D efforts are underway for the future project of KamLAND2-Zen to fully cover the IH region for $\langle m_{\beta\beta}\rangle$ down to $\sim$ 20meV.

Acknowledgments

The KamLAND-Zen experiment is supported by the Grant-in-Aid for Specially Promoted Research under Grant No. 21000001 of the Japanese Ministry of

Education, Culture, Sports, Science and Technology; the World Premier International Research Center Initiative (WPI Initiative), MEXT, Japan; Stiching FOM in the Netherlands; and under the US Department of Energy (DOE) Grant No. DE-AC02-05CH11231, as well as other DOE grants to individual institutions. The Kamioka Mining and Smelting Company has provided service for activities in the mine.

References

[1] J.Schechter and J.W.F.Valle, Phys. Rev. **D 25**, 2951 (1982).
[2] H.V.Klapdor-Kleingrothaus and I.V.Krivosheina, Mod.Phys.Lett.A**21**, 1547(2006).
[3] A.Gando et al., Phys. Rev. **C 85**, 045504 (2012).
[4] A.Gando et al., Phys. Rev. **C 86**, 021601(R) (2012).
[5] A.Gando et al., Phys. Rev. Letters **110**, 062502 (2013).
[6] M.Auger et al., Phys. Rev. Letters **109**, 032505 (2012).

REVIEW OF DOUBLE BETA DECAY EXPERIMENTS

Laurent Simard [a]
LAL, Univ Paris-Sud, CNRS/IN2P3, Université Paris-Saclay, Orsay, France
Institut Universitaire de France, F-75005 Paris, France

Abstract. The search of the neutrinoless double beta decay is currently the only way to test the nature, Dirac or Majorana, of the neutrino. Two different experimental approaches exist : one, where source and detector coïncide, is pure calorimetric. In the other technique, before the measurements of their energies, the two electrons are tracked. The results of the currently data-taking experiments are presented. The expected sensitivities for future experiments are discussed.

1 Majorana neutrinos and double beta decay

As a neutral fermion, it can be identical to its antiparticle (= Majorana particle). The Majorana mass term of neutrino is naturally added in a simple extension of the standard model. It could explain the baryon-antibaryon asymmetry by leptogenesis. The most sensitive process to search for the Majorana nature of the neutrinos is the neutrinoless double beta decay ($0\nu2\beta$), a lepton number violating process ($\Delta L = 2$) : $(A, Z) -> (A, Z+2) + 2e^-$

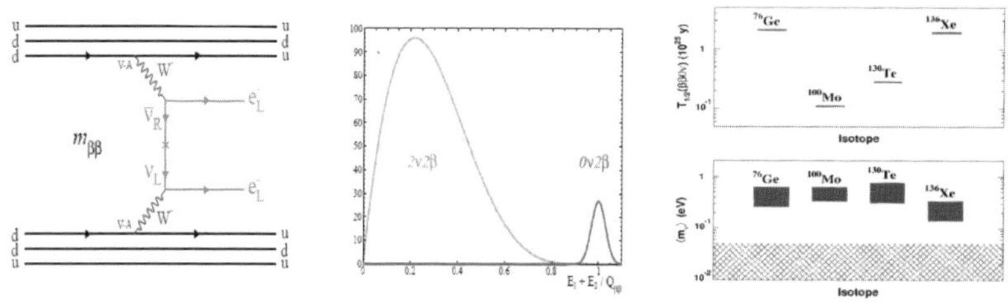

Figure 1: Top : $0\nu2\beta$ mechanism of exchange of a light Majorana neutrino . Middle : 2-electron energy spectra for $\beta\beta$ processes. Right : sensitivities of $\beta\beta$ decay experiments to various isotopes.

For the exchange of a light Majorana neutrino (see left plot of Figure 1) the half-life $T^{0\nu}_{1/2}$ of the process is related to the effective neutrino Majorana mass $m_{\beta\beta}$: $\frac{1}{T^{0\nu}_{1/2}} = G_{0\nu}|M_{0\nu}|^2|m_{\beta\beta}|^2$, where $G_{0\nu}$ is the phase space factor and $M_{0\nu}$ the nuclear matrix element. The process has not been observed so far, and limits on the half-life are tuypically $10^{24} - 10^{25}$ y. Several mechanisms can generate the $0\nu2\beta$ process : light Majorana neutrino exchange $m_{\beta\beta}$, right handed current, Majoron and SUSY. Experimentally, a peak in the 2-electron energy spectrum is expected at $Q_{\beta\beta}$. The current best limits on the $0\nu\beta\beta$ search are given in the right plot of Figure 1.

[a]E-mail: simard@lal.in2p3.fr

2 Recent results or progress on the double beta search

2.1 Germanium detectors

GERDA

The principle is to use bare Ge diodes in liquid argon enriched in ^{76}Ge at 86%. A gradual deployment of the detector strings in the cryostat is performed in Gran Sasso (3800 mwe). The background target is 0.01 count/(keV.kg.y).

The Phase 1 data taking (2011-2013) used 18 kg of ^{76}Ge (8 old semi-coaxials (HdM-IGEX) and 5 new BEGe.) The energy resolution FWHM at $Q_{\beta\beta}$ obtained was [4.2-5.7] keV for coaxials and [2.6-4.0] keV for BEGe. With a 21.6 kg.y exposure, using pulse shape discrimination, the limits sets were : $T^{0\nu}_{1/2} > 2.1\ 10^{25}$y(90%CL) and $|m_{\beta\beta}| < [0.2 - 0.4]$eV [1].

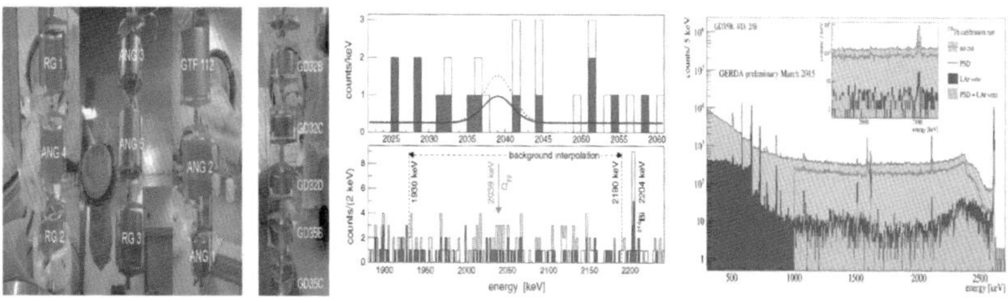

Figure 2: GERDA. Left : Detectors used for phase 1. Middle : Energy spectrum of Phase 1. Right : Energy spectrum with a ^{228}Th source, for Phase2 detectors.

For the phase 2, 50 kg of ^{76}Ge will be used. The expected level of background is 10^{-3}count/(keV.kg.y). With a 100 kg.y exposure the expected sensitivity is $T^{0\nu}_{1/2} > 10^{26}$y for 30 new BEGe diodes due to several improvments : better PSD than for coaxials, use of PSD single site $(0\nu 2\beta)$/multisites (γ-rays) or surface (^{42}K), FWHM $\simeq$ 3 keV at $Q_{\beta\beta}$ and detection of liquid argon scintillation light to veto backgrounds.

A possible extrapolation to higher masses will be estimated. The commissioning has started (spectrum from a ^{228}Th source in Figure 2.1.)

Majorana

The MAJORANA detector consists in enriched HpGe detectors in electroformed copper in Sanford underground lab. The target background level for 1 ton of ^{76}Ge detectors is 1 count/(ROI.kg.yr) for a 4 keV ROI. A 40kg demonstrator is under construction to demonstrate the ability to reach the necessary background level. Early 2015 half of enriched HpGe detectors and some natural detectors (20 kg) are completed. Late 2015 the rest of the HpGe detectors will be implemented.

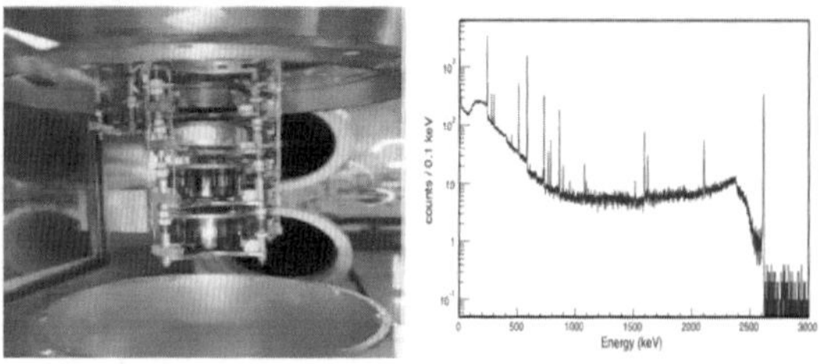

Figure 3: MAJORANA. Middle : One detector. Right : Spectrum obtained with a ^{228}Th source.

2.2 Bolometers

Cuore

The CUORE detector is made of 206 kg of ^{130}Te bolometers. The assembly and cleaning techniques have been improved, as well as the cryostat radiopurity. The target background level is 10^{-2}counts/(keV.kg.y), with a FWHM of 5 keV at $Q_{\beta\beta}$. The expected sensitivity is $T^{0\nu}_{1/2} > 10^{26}$y in 5 years of data taking.

The first tower of CUORE holds 11 kg of ^{130}Te in the same cryostat as CUORICINO, with the same γ background. The improved crystals supports induces a α background reduction. With a 9.8 kg.y exposure the combined results with CUORICINO are : $T^{0\nu}_{1/2} > 4 10^{24}$y(90%CL) and $|m_{\beta\beta}| < [0.27 - 0.76]$eV.

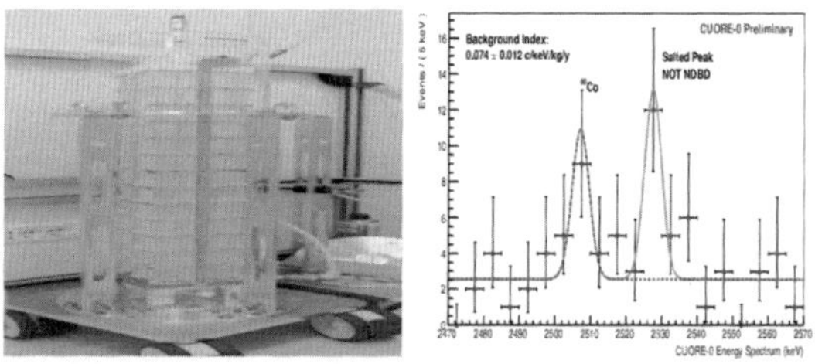

Figure 4: CUORE. Left : One tower. Middle : Spectrum of the CUORE-0 demonstrator.

Lucifer

The Lucifer project consists in 50 ZnSe crystals (18 kg of ^{82}Se). The CUORICINO cryostat will be used after the end of CUORE-0 data taking The expected sensitivity is $T^{0\nu}_{1/2} > 10^{26}$y in 5 years of data taking. The background measurement is ongoing with natural crystals in Gran Sasso. Some difficulties concern the reproducibility during the production of ZnSe crystals.

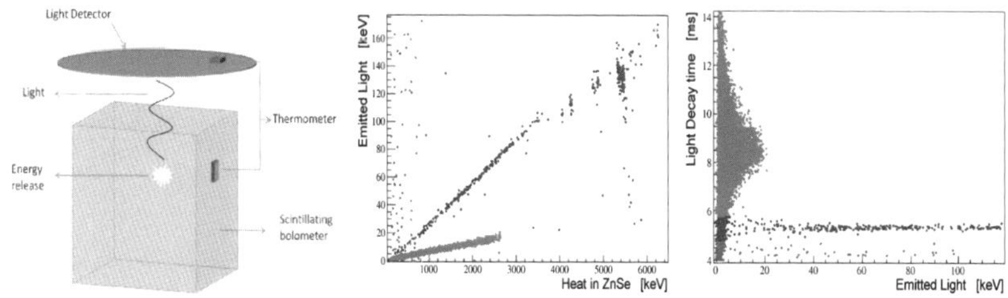

Figure 5: LUCIFER. Left : Principle of scintillating bolometers. Middle : Light versus heat signals. Right : Light decay time vs emitted light.

Lumineu

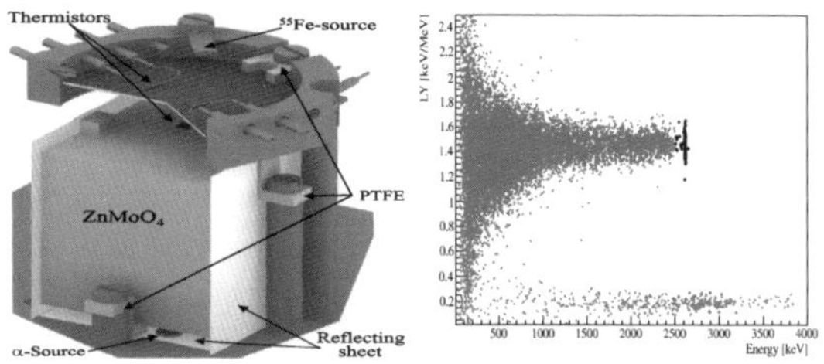

Figure 6: LUMINEU. Left : $Zn^{100}MoO_4$ scintillating bolometer. Right : Light vs heat signals.

The Goal is to build a demonstrator with 40 crystals ($\simeq$ 7 kg of ^{100}Mo). The expected sensitivity is $T^{0\nu}_{1/2} > 1.2\ 10^{25}$y in 2 years. 20 crystals will be placed at Gran Sasso in early 2016 and 30 crystals in LSM in end 2016. A clear α and β/γ discrimination is observed at LSM for the first natural and enriched crystals. The radiopurities of the natural and enriched crystals are very good.

2.3 TPC : EXO-200/nEXO

EXO-200 benefits from the easy and cheap ^{136}Xe enrichment (80%). It hosts 200 kg of liquid xenon in a TPC at WIPP. With a fiducial volume of 79.4 kg of ^{136}Xe, a background level of $(1.7 \pm 0.2)10^{-3}$counts/(keV.kg.y), a FWHM of 89 keV at $Q_{\beta\beta}$, the observed limit is : $T^{0\nu}_{1/2} > 1.1\ 10^{25}$y(90%CL) and $|m_{\beta\beta}| < [0.19 - 0.45]$eV with an exposure of 123.7 kg.y [3].

The nEXO project will be a liquid Xenon TPC in SNOLAB Cryopit. It will host 5 t enriched at 90% in ^{136}Xe. The baseline FWHM 86 keV at $Q_{\beta\beta}$ Scaling EXO-200 background with an improved discrimination, the expected sensitivty is $T^{0\nu}_{1/2}(^{136}Xe) > 4.1\ 10^{27}$y (2.1 10^{28}y with the tagging of the Ba ion) in 10 years of data taking.

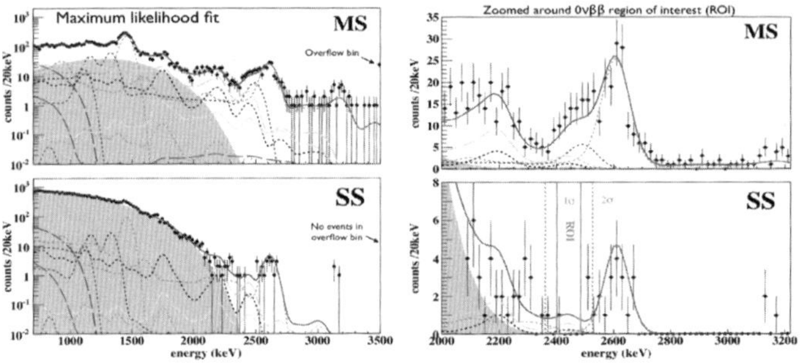

Figure 7: EXO-200. Left : Energy spectrum for single site(top) and multisite (bottom) events. Right : Same plots around $Q_{\beta\beta}$.

2.4 Scintillators :

KamLand-Zen

13 tons of Xe-loaded liquid scintillator are hold in KamLand mini balloon (320 kg of enriched Xe). The energy resolution is 232 keV at $Q_{\beta\beta}$. The first phase (2011-2012) has shown a dominant background from ^{110m}Ag. After purification of the liquid scintillator, the second phase (2013-2014) has demonstrated a more than 10 times reduction of ^{110m}Ag background. The combined limit is : $T^{0\nu}_{1/2} > 2.6\ 10^{25}$y(90%CL) and $|m_{\beta\beta}| < [0.14 - 0.28]$eV with a 198.3 kg.y exposure [4]. In the future, a miniballoon will be rebuilt, the Xe mass increased to 600 kg and the balloon radioactivity reduced.

SNO+

The SNO detector will be filled with 800 kg of ^{130}Te (160 kg within the fiducial volume). The liquid scintillator loaded with Te has a high light yield : FWHM is 265 keV at $Q_{\beta\beta}$. The $T^{2\nu}_{1/2}(^{130}\text{Te})$ is high, inducing a low $2\nu 2\beta$ background. The liquid scintillator must be ultra radiopure in ^{238}U and ^{232}Th. The neutrinos from solar ^{8}B are the ultimate background. The expected sensitivity is $T^{0\nu}_{1/2} > 9.4\ 10^{25}$y(90%CL).

2.5 Tracko-calo : NEMO-3/SuperNEMO

NEMO-3

The NEMO-3 detector was located in the Laboratoire Souterrain de Modane in France under 4800 m.w.e. shielded by 30 cm of borated water and wood, 19 cm of pure iron and radon-free air tent (after 2004). The $0\nu 2\beta$ events are fully reconstructed (individual electron energies, electron angular distribution) and

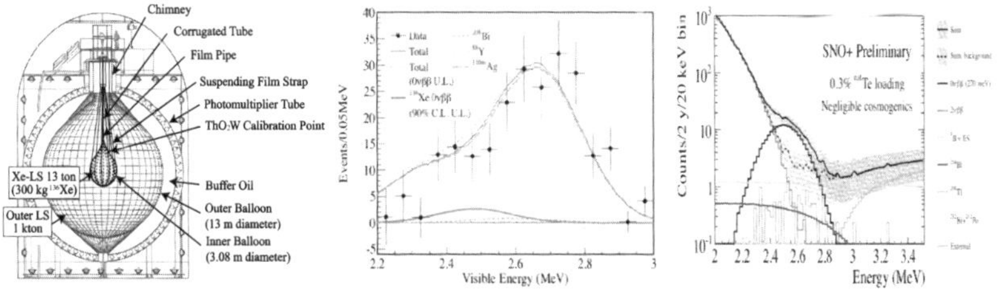

Figure 8: Left : Scheme of KamLand-Zen detector. Middle : KamLand-Zen spectrum around $Q_{\beta\beta}$. Right : Expected SNO+ spectrum around $Q_{\beta\beta}$.

the background powerfully rejected. For ^{100}Mo with an exposure 34.7 kg.y, the limits are [5]: $T^{0\nu}_{1/2} > 1.1\ 10^{24}$y (90%CL) and $|m_{\beta\beta}| < [0.33 - 0.62]$eV

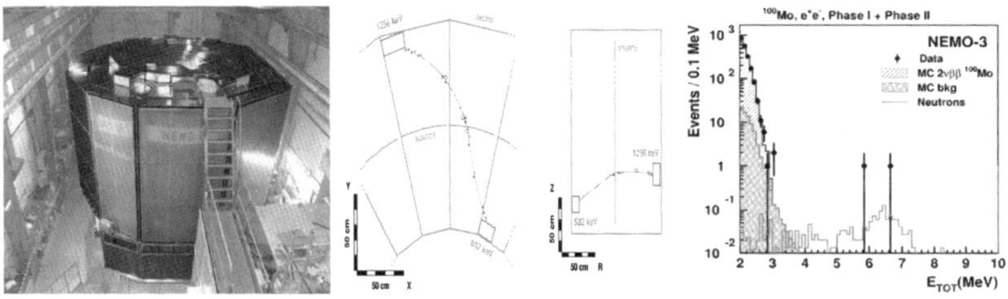

Figure 9: NEMO-3. Left : Photo of the detector at LSM. Middle : Reconstruction of a 2-electron event. Right : 2-electron energy spectrum for ^{100}Mo

No events were observed in the [3.2-10] MeV energy range for a 47kg.y exposure. This is very promising for high-$Q_{\beta\beta}$ isotopes (^{48}Ca, ^{150}Nd, ^{96}Zr). Also limits were sets on right handed currents and supersymmetry.

SuperNEMO

	NEMO-3	SuperNEMO
Isotope	^{100}Mo	^{82}Se (^{150}Nd, ^{48}Ca)
Calorimeter energy resolution FWHM at 1 MeV	[14-17]%	8 %
Source radiopurity : ^{208}Tl (μBq/kg)	$\simeq$ 100	2
Source radiopurity : ^{214}Bi (μBq/kg)	$\simeq$ [60-300]	10
Tracker gas radiopurity : ^{222}Rn (mBq/m^3)	$\simeq$ 5	0.15

The SuperNEMO project should host $\simeq$ 100 kg of $\beta\beta$ isotopes. After a R&D phase the energy resolution was improved by a factor 2 w.r.t. NEMO-3. An automated drift cells production through a wiring robot was developed.A dedicated detector BiPo-3 to measure the radiopurity of 1.4 kg of source at the required sensitivity level was constructed and installed in Canfranc (see figure 2.5). Against Radon background, emanation measurements were done and the gas purification, permeability of materials were studied.

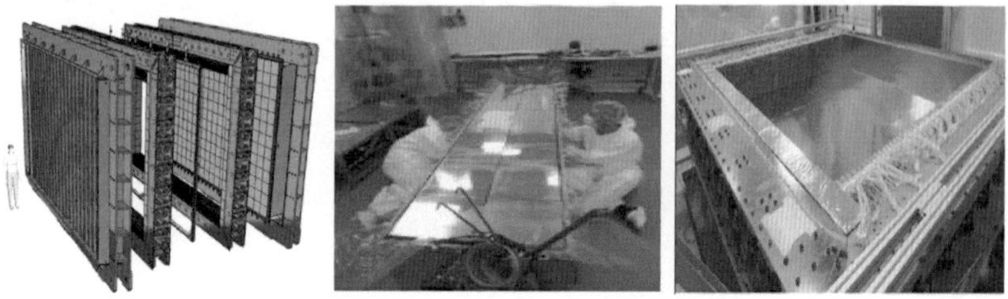

Figure 10: SuperNEMO. Left : Scheme of the tracker, calorimeter of one SuperNEMO module. Middle : BiPo-3 detector at Canfranc. Right : Part of the SuperNEMO demonstrator tracker, during assembly at MSSL.

The first module of SuperNEMO, called SuperNEMO demonstrator, will host 7kg of ^{82}Se in the LSM. It should demonstrate the ability to reach SuperNEMO specifications for background. The expected sensitivity is : $T^{0\nu}_{1/2} > 6.5\ 10^{24}$ y (90%CL), $|m_{\beta\beta}| < [0.2 - 0.4]$ eV with a 17.5 kg.y exposure. The detector is in construction and commissioning expected in 2016.

3 Conclusion

This review is not exhaustive. The recommandation of the NUSAG american committee is to wait for the results of the demonstrators to choose the best technique(s).

References

[1] M. Agostini *et al.*, Phys. Rev. Lett. **111**, 122503 (2013).
[2] K. Alfonso *et al.*, Phys. Rev. Lett. **115**, 102502 (2015),
[3] J. B. Albert *et al.*, Nature **510**, 229 (2014),
[4] A. Gando *et al.*, Phys. Rev. Lett. **110**, 062502 (2013),
[5] R. Arnold *et al,*, Phys. Rev. D **92**, 072011 (2015).

SEARCH FOR $\nu_\mu \to \nu_e$ OSCILLATIONS IN THE OPERA EXPERIMENT

Svetlana Zemskova [a] on behalf of the OPERA collaboration
Dzhelepov Laboratory of Nuclear Problems, Joint Institute for Nuclear Research, 141980 Dubna, Moscow region, Russia

Abstract. The tracking capabilities of the OPERA detector allow to observe $\nu_\mu \to \nu_\tau$ oscillations in appearance mode and to study $\nu_\mu \to \nu_e$ oscillations in the CNGS ν_μ beam. Current results on $\nu_\mu \to \nu_e$ channel in the three-flavour mixing model are presented. The same data allow to constrain the mixing with additional sterile neutrino states. The analysis of the full 2008-2012 OPERA data set is ongoing. Details of the systematic ν_e search will be presented.

1 Introduction

The main goal of the OPERA experiment [1] is a confirmation of $\nu_\mu \to \nu_\tau$ oscillations in appearance mode in the CNGS beam. The detector is located at the Gran Sasso Underground Laboratory (LNGS), 730 km away from the CERN SPS, where the beam is produced. The design of the detector and the tracking capabilities of the OPERA nuclear emulsion allow to study $\nu_\mu \to \nu_e$ oscillations. The energy of the beam ($< E > = 17$ GeV) and the baseline make the OPERA experiment sensitive for the $\Delta m^2 > 0.01\ eV^2$ (LSND [2] and MiniBooNE [3] allowed region) in a search for oscillations involving additional neutrino states.

2 OPERA experiment

The OPERA detector consists of two identical Super Modules (SM), each SM has a target instrumented with tracking detectors (Target Tracker, TT) and a magnetic muon spectrometer. The target contains about 150,000 Emulsion Cloud Chamber modules (ECC or "bricks"). Each ECC weights 8.3 kg and consists of 57 emulsion films interleaved with 56 lead plates. Each brick has a removable doublet of emulsion films (Changeable Sheet or CS) on the downstream face to facilitate the neutrino interaction location.

The TT allow to identify the brick where the interaction occurred. The brick is then extracted from the target and, if the CS analysis confirms the presence of the event in the brick, the 57 emulsion films are developed and sent to the scanning laboratories. A complex analysis at the microscope level is performed [4] and the systematic ν_e search is applied [5]. If the electromagnetic shower is found, the electron track candidate is carefully checked in the first two plates downstream of the vertex. In case of confirmation of an electron

[a]E-mail: `zemskova@numail.jinr.ru`

track attached to the primary vertex, the event is classified as a ν_e interaction. The event energy is reconstructed using the TT data and an energy resolution:

$$\Delta E/E = 0.37 + 0.74/\sqrt{E}, (E \ in \ GeV). \qquad (1)$$

The $\nu_\mu \to \nu_e$ analysis was completed for the 2008-2009 data set, which corresponds to 5.25×10^{19} p.o.t. and to 5255 events registered. In this sample 505 events were not classified as ν_μ CC interactions. Out of them 19 ν_e candidates were found.

3 Oscillation analysis

3.1 Background to $\nu_\mu \to \nu_e$ appearance

Energy cut		20 GeV	30 GeV	No cut
	BG from π^0	0.2	0.2	0.2
BG common to both analysis	BG from $\tau \to e$	0.2	0.3	0.3
	ν_e beam contamination	4.2	7.7	19.4
Total expected BG in the 3-flavour oscillation analysis		4.6	8.2	19.8
BG to non-standard oscillation analysis only	ν_e via 3-flavour oscillation	1.0	1.3	1.4
Total expected BG in the non-standard oscillation analysis		5.6	9.4	21.3
Data		4	6	19

Table 1: Expected and observed number of events for the different energy cuts [5].

The sources of background in the ν_e interaction identification are presented in Table 1. In the analyzed data sample 19.8 ± 2.8 (syst) background ν_e events are expected which is in agreement with the 19 observed ν_e candidates.

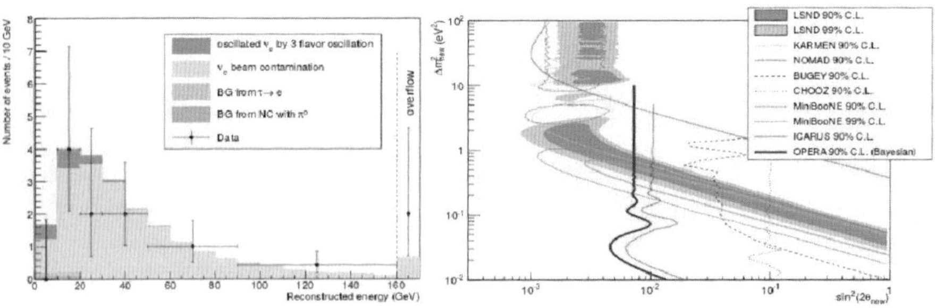

Figure 1: Left plot: Distribution of the reconstructed energy of the ν_e events, and the expected spectrum from the different sources in a stack histogram, normalized to the number of p.o.t. analyzed. Right plot: The exclusion plot in the plane of parameters of the non-standard $\nu_\mu \to \nu_e$ oscillation, in the analysis using the Bayesian method [5].

3.2 Three-flavour mixing scenario

The following parameters are used to compute the expectation in the three-flavour oscillation framework: $\sin^2(2\theta_{13}) = 0.098$, $\sin^2(2\theta_{23}) = 1$, $\Delta m^2{}_{32} =$

$\Delta m^2{}_{31} = 2.32 \times 10^{-3}\ eV^2$, assuming $\delta_{CP} = 0$ and neglecting matter effects. In the whole energy range, 1.4 oscillated ν_e CC events were expected. The simulated and reconstructed energy distributions are presented in Fig. 1.

The number of observed events is compatible with the non-oscillation hypothesis and the upper limit $\sin^2(2\theta_{13}) < 0.44$ is derived at 90% C.L..

3.3 Search for sterile neutrinos

The OPERA data were used in the search for the non-standard oscillations indicated by LSND and MiniBooNE experiments. The conventional approach of expressing the $\nu_\mu \to \nu_e$ oscillation probability in the one mass scale dominance approximation is used:

$$P_{\nu_\mu \to \nu_e} = \sin^2(2\theta_{\text{new}}) \times sin^2(1.27\Delta m^2{}_{\text{new}} L(km)/E(GeV)). \qquad (2)$$

Note however that this approach does not allow a direct comparison between experiments working in different L/E regimes.

The number of observed events and expected backgrounds are represented in Table 1. The curve with the Bayesian upper limit and the results of other experiments, working in different L/E regimes, are shown in Fig. 1.

4 Conclusions and prospectives

The observed number of ν_e interactions in the data sample collected in the 2008-2009 runs (5.25×10^{19} p.o.t.) is compatible with the non-oscillation hypothesis and gives an upper limit in the standard three-flavour framework $\sin^2(2\theta_{13}) < 0.44$ at (90% C.L.). The OPERA results also limit the parameter space available for a non-standard ν_e appearance. It further constrains the still allowed region around $\Delta m^2{}_{new} > 5 \times 10^{-2}\ eV^2$ and for large Δm^2 values the 90% C.L. upper limit on $\sin^2(2\theta_{new})$ reaches 7.2×10^{-3}.

The analysis of the complete 2008-2012 data sample corresponding to 17.97×10^{19} p.o.t. is close to completion. So far 52 ν_e candidates have been found and improvement on the energy reconstruction and background estimation are ongoing.

References

[1] M. Guler et al., the OPERA Collaboration, CERN-SPSC-2000-028, CERN, Geneva, Switzerland (2000).

[2] A. Aguilar-Arevalo et al., the LSND Collaboration, Phys. Rev. **D** 64, 112007 (2001).

[3] A. Aguilar-Arevalo et al., the MiniBooNE Collaboration, arXiv:1207.4809.
[4] N. Agafonova et al., the OPERA Collaboration, New J. Phys. **14**, 033017 (2012).
[5] N. Agafonova et al., the OPERA Collaboration, JHEP **07**, 04 (2013).

SEARCH FOR STERILE NEUTRINO MIXING IN THE $\nu_\mu \to \nu_\tau$ APPEARANCE CHANNEL WITH THE OPERA DETECTOR

Matteo Tenti [a] for the OPERA Collaboration
INFN-CNAF, Viale Berti Pichat 6/2, 40127 Bologna (IT)

Abstract. The OPERA experiment collected data on the CNGS beam from 2008 to 2012. Five oscillated ν_τ CC interaction candidates have been detected so far with an expected background of 0.25 events. The OPERA appearance results are used to derive limits on the mixing parameters of a massive sterile neutrino.

1 Introduction

The neutrino oscillation phenomenon is currently well described by the mixing of three neutrino flavor eigenstates (ν_e, ν_μ and ν_τ) with three mass eigenstates (ν_1, ν_2 and ν_3) [1, 2]. Nevertheless there are several experimental results that cannot be explained within this framework [3]. These anomalies may hint to the existence of sterile neutrino(s) with $\Delta m^2 \sim 1$ eV^2. The OPERA experiment can test the sterile neutrino hypothesis looking for deviations from predictions of the standard three-neutrino model.

The OPERA experiment was designed to perform the direct observation of the appearance of ν_τ in a ν_μ beam identifying, on a event by event basis, the τ lepton produced in the ν_τ Charged Current (CC) interaction. The OPERA detector [4] is composed of two identical supermodules, each consisting of an iron spectrometer downstream of a target section. The target has a mass of about 1.2 kt and a modular structure with approximately 150,000 target units, called bricks. A brick is made of 56 1-mm-thick lead plates, acting as targets, interleaved with 57 nuclear emulsion films, used as micrometric tracking devices: the Emulsion Cloud Chamber (ECC) technique. Bricks are arranged in walls interleaved with planes of scintillator strips forming the Target Tracker (TT). The magnetic spectrometers, consisting of bipolar iron magnets instrumented with Resistive Plate Chambers (RPC) and Precision drift Tubes (PT), are used for the measurement of the muon charge and momentum. For each event, the information provided by the electronic detectors allows to identify the brick containing the neutrino interaction vertex [5]. The full event topology is reconstructed through the measurement of the ECC brick [6]. Finally kinematical and topological cuts are applied to improve the signal-to-background ratio [7].

The OPERA detector was exposed to the ν_μ CERN Neutrinos to Gran Sasso (CNGS) [8] beam from 2008 to 2012. A sample of 19505 neutrino interactions corresponding to 17.97×10^{19} protons on target (pot) have been registered in

[a] E-mail: matteo.tenti@bo.infn.it

the detector fiducial volume. Five ν_τ candidate events were identified, with a number of expected background events equal to 0.25 [7]. The background-only hypothesis is excluded at 5.1σ. The number of expected ν_τ events from the standard three-neutrino model, assuming maximal mixing and $\Delta m^2_{32} = 2.44 \times 10^{-3}$ eV^2, is 2.64, compatible with the number of observed ones.

2 Search for sterile mixing

In this work the addition to the standard 3 neutrino mass eigenstates of a further state with a mass squared difference up to a few eV^2 is considered. The four mass eigenstates are related to the three active neutrinos plus the sterile neutrino by the so-called mixing matrix U. The results of this analysis are given in terms of an effective mixing parameter $\sin 2\theta_{\mu\tau} = 2|U_{\mu 4}||U_{\tau 4}|$, which is the leading mixing term at Short-Base Line (SBL) experiments, allowing a direct comparison with their results.

The number of expected events is evaluated by GLoBES [9,10] as a function of the mixing angles θ_{ij}, the CP-violating phases δ_k and the squared-mass differences Δm^2_{ij}. The parameter Δm^2_{21} is fixed to 7.54×10^{-5} eV^2 and a Gaussian prior, with mean and sigma from global fits [11], is assumed for Δm^2_{31}, i.e. $(2.47 \pm 0.06) \times 10^{-3}$ eV^2 for Normal Hierarchy (N.H.) and $(-2.34 \pm 0.06) \times 10^{-3}$ eV^2 for Inverted Hierarchy (I.H.). The Earth density is assumed to be constant following the PREM model [12,13]. The likelihood $\mathcal{L}$ is defined as:

$$\mathcal{L} = P(n|\mu) \times \pi(\Delta m^2_{31}) \tag{1}$$

where P is a Poissonian depending on the number of expected events μ (which, in turn, depends on the oscillation parameters θ_{ij}, δ_k and Δm^2_{ij}), n the number of observed events and π is the Gaussian prior of the nuisance parameter Δm^2_{31}. The effects of possible systematic errors are negligible because results are dominated by the statistical error. To extract limits on the sterile neutrino mixing parameters, the test statistic $-2\ln\lambda_p$, where λ_p is the profile likelihood ratio, is used.

The test statistic $-2\ln\lambda_p$ is minimized over the nuisance parameters to obtain an exclusion region at 90% Confidence Level (CL) in the $[\Delta m^2_{41}, \sin^2 2\theta_{\mu\tau}]$ parameter plane. The result is shown in Fig. 1. The 90% CL excluded region on Δm^2_{41} extends down to 10^{-2} eV^2 for $\sin^2 2\theta_{\mu\tau} > 0.5$. The analysis is performed assuming $\Delta m^2_{41} > 0$. Since present limits on the sum of neutrino masses from cosmological surveys do not exclude small negative values for Δm^2_{41}, the analysis was repeated following this assumption. The exclusion regions obtained in this way are similar to those of Fig. 1, but with hierarchies exchanged.

At large values of the new squared mass difference ($\Delta m^2_{41} > 1$ eV^2) , the dependence of the probability on Δm^2_{41} vanishes and values of $\sin^2 2\theta_{\mu\tau} > 0.119$ are excluded at 90% CL. At maximal mixing ($\sin^2 2\theta_{\mu\tau} = 1$) the 90% CL upper limit on Δm^2_{41} is 4.7×10^{-3} eV^2 for I.H. while for N.H. values $(0.4 < \Delta m^2_{41} < 2.1) \times 10^{-3}$ eV^2 and $\Delta m^2_{41} > 7.1 \times 10^{-3}$ eV^2 are excluded.

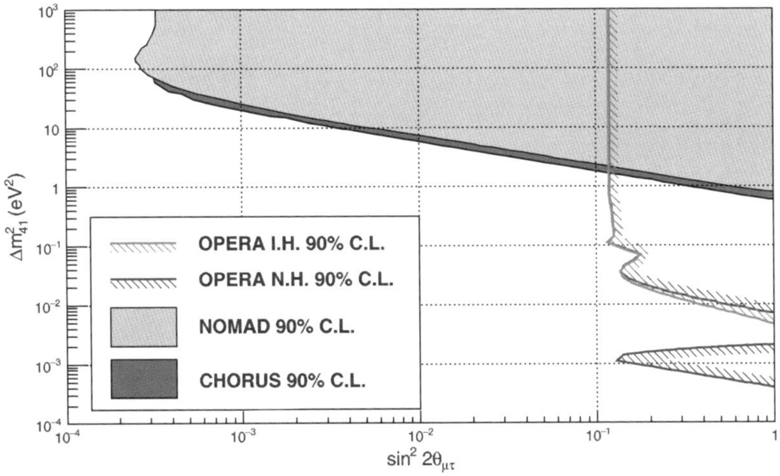

Figure 1: 90% CL excluded region in the [Δm^2_{41}, $\sin^2 2\theta_{\mu\tau}$] plane for normal (blue) and inverted (red) hierarchy. The 90% CL excluded region obtain by CHORUS (dark grey) and NOMAD (grey) are also shown.

3 Conclusions

The OPERA experiment was designed to observe the $\nu_\mu \to \nu_\tau$ oscillation in the so-called "atmospheric sector". Five ν_τ event candidates were currently identified in the OPERA detector exposed to the CNGS beam, with an expected background of 0.25 events, allowing to exclude the background-only hypothesis at 5.1σ. OPERA results have been used to obtain 90% CL exclusion regions for the oscillation parameters of a 3 + 1 sterile neutrino model. The exclusion region in the [Δm^2_{41}, $\sin^2 2\theta_{\mu\tau}$] plane is extended down to 10^{-2} eV^2 for $\sin^2 2\theta_{\mu\tau} > 0.5$ thanks to the peculiar baseline and energy of the OPERA experiment.

References

[1] B. Pontecorvo, Zh. Eksp. Teor. Fiz. **33** (1957) 549; B. Pontecorvo, Zh. Eksp. Teor. Fiz. **34** (1958) 247.

[2] Z. Maki, M. Nakagawa and S. Sakata, Prog. Theor. Phys. **28** (1962) 870.

[3] A. Aguilar-Arevalo et al. [LSND Collaboration], Phys. Rev. D **64** (2001) 112007; A. A. Aguilar-Arevalo et al. [MiniBooNE Collaboration], Phys. Rev. Lett. **110** (2013) 161801; G. Mention et al., Phys. Rev. D **83** (2011) 073006; Th. A. Muller et al., Phys. Rev. C **83** (2011) 054615; C. Giunti and M. Laveder, Phys. Rev. C **83** (2011) 065504.

[4] R. Acquafredda et al. [OPERA Collaboration], JINST **4** (2009) P04018.

[5] N. Agafonova et al. [OPERA Collaboration], New J. Phys. **13** (2011) 053051.

[6] N. Agafonova et al. [OPERA Collaboration], Eur. Phys. J. C **74** (2014) 2986.

[7] N. Agafonova et al. [OPERA Collaboration], Phys. Rev. Lett. **115** (2015) 121802.

[8] R. Baldy et al., CERN-SL-99-034-DI, CERN-SL-99-34-DI, INFN-AE-99-05, INFN-AE-99-5 (1999);
G. Acquistapace et al., CERN-98-02, INFN-AE-98-05, CERN-YELLOW-98-02 (1998).

[9] P. Huber, M. Lindner and W. Winter, Comput. Phys. Commun. **167** (2005) 195.

[10] P. Huber, J. Kopp, M. Lindner, M. Rolinec and W. Winter, Comput. Phys. Commun. **197** (2007) 432-438.

[11] K. A. Olive et al., [Particle Data Group], Chin. Phys. C **38** (2014) 090001.

[12] A. M. Dziewonski and D. L. Anderson, Phys. Earth Planet. Interiors **25** (1981) 297.

[13] F. D. Stacey, Physics of the earth, 2nd ed., Wiley, U.S.A. (1977).

SEARCH FOR STERILE NEUTRINOS IN THE ν_μ DISAPPEARANCE MODE AT FNAL

Matteo Tenti [a] for the NESSiE Collaboration
INFN-CNAF, Viale Berti Pichat 6/2, 40127 Bologna (IT)

Abstract. The NESSiE Collaboration proposed an experiment to definitively clarify the existing tension between the ν_μ disappearance and the ν_e appearance/disappearance at the eV mass scale. That is achieved extending by more than one order of magnitude the range of the investigated mixing angle between standard and sterile neutrinos. To this aim the use of the current FNAL Booster neutrino beam for a Short Baseline experiment was carefully evaluated by considering the use of magnetic spectrometers at two sites, near and far ones. The detector locations were extensively studied, together with the achievable performances of two OPERA like spectrometers.

1 Introduction

The neutrino oscillation phenomenon is currently well described by the mixing of three neutrino flavor eigenstates (ν_e, ν_μ and ν_τ) with three mass eigenstates (ν_1, ν_2 and ν_3) [1,2]. Nevertheless there are several experimental results that cannot be explained within this framework [3]. Notwithstanding the proposal of the existence of sterile neutrinos [4] to explain these anomalies, some tensions [5] on neutrino/antineutrino oscillation results at Short-Baseline (SBL) remain. The main source of tensions corresponds to the absence, mainly at SBL, of any ν_μ disappearance signal in L/E region suggested by the ν_e appearance.

2 The NESSiE Proposal

Since only limited and rather old experimental data [6] are available on searches for ν_μ disappearance at SBL, the NESSiE Collaboration proposed an accurate measurement of ν_μ disappearance using magnetic spectrometers (B = 1.5 T) along the FNAL Booster neutrino beam [7]. The plan is to design, construct and install two spectrometers at two sites, near (at 110 m, on-axis) and far (at 710 m, off-axis, on surface), in line with the FNAL Booster beam. By using very massive detectors (300 tons and 700 tons for the near and far spectrometers, respectively) it would be possible to collect very large event samples in order to improve the disentanglement of systematic effects and to span the oscillation mixing parameter down to till unexplored regions ($\sin^2(2\theta_{new}) \sim 0.01$). By keeping the systematic error at the level of 1-2% for the measurements of the ν_μ interactions, it would be possible to measure:

- ν_μ disappearance in the almost entire available momentum range ($p_\mu \geq$ 500 MeV/c);

[a] E-mail: matteo.tenti@bo.infn.it

- neutrino flux at the Near detector, in the relevant muon momentum range;
- the sign of the muon charge to separate ν_μ from $\bar{\nu}_\mu$ for systematics control.

3 Beam and Site Study

The large uncertainty on the absolute neutrino flux (around 20% for neutrino energies below 1.5 GeV at MiniBooNE site [8]) makes the use of two or more identical detectors at different baselines mandatory when searching for small disappearance phenomena. The ratio of the event rates at the far and near detectors (FNR) as function of neutrino energy is a convenient variable since at first order it benefits from cancellation of systematic uncertainties due to the common effects of proton-target, neutrino cross sections and reconstruction efficiencies. Even in the absence of oscillations, the energy spectra in any two detectors are different leading to a non-flat FNR, especially if the distance of the near detector is comparable with the length of the decay pipe. It is therefore necessary to master the knowledge of the FNR.

All the contributions to the systematic uncertainties on the FNAL Booster neutrino flux were studied in detail by the MiniBooNE collaboration [8]. At first order they factorize out using a double site. However, since their magnitude can limit the FNR accuracy, we focused on the main contribution, which comes from the knowledge of the hadro-production double differential (momentum p, polar angle θ) cross sections in 8 GeV p-Be interactions.

In order to evaluate how the hadro-production uncertainty affects the knowledge of the FNR in our experiment a new beamline simulation was developed. The angular and momentum distribution of pions exiting the Be target were simulated using either FLUKA (2011.2b) or GEANT4 (v4.9.4 p02, QGSP 3.4 physics list). Furthermore the Sanford-Wang empirical formula [9] with parameter (c_i, $i = 1 \ldots 7$) values obtained by a fit of the HARP and E910 data-set in [8], was used:

$$\frac{d^2\sigma}{dpd\Omega} = c_1 p^{c_2}\left(1 - \frac{p}{p_b - 1}\right) e^{-\frac{p^{c_3}}{p_b^{c_4}} - c_5\theta(p - c_6 p_b \cos^{c_7}\theta)} \qquad (1)$$

with p_b being the proton-beam momentum. Six configurations, with different distances (110, 460 and 710 m), either on-axis or off-axis, and different fiducial sizes of the detectors, were studied. The impact of the hadro-production uncertainties on FNR was estimated in two ways:

- given the different hadronic models implemented, using the differences between the FNR spectrum obtained by FLUKA and GEANT4.

- as the relative spread of sampling of the FNR spectrum obtained using the Sanford-Wang parameterization and assuming gaussian distribution for the coefficients c_i, with mean and sigma given by a fit to the HARP-E910 data-set [8]. Parameter correlation was also taken into account.

The configuration, with near detector (1.4×1.4 m^2 transversal section) at 110 m on-axis and far detector (4×4 m^2 transversal section) at 710 off-axis at surface was selected since the systematic uncertainties on FNR remains around 1% in the whole neutrino energy range.

4 Detector Design

Muon magnetic (OPERA like) spectrometers [10], instrumented with RPCs, are used as tracking detectors. To maximize the volume where muons can be identify by range, the two OPERA spectrometers are longitudinally coupled, both at the far and the near sites. The modularity of the OPERA spectrometer allows to cut every single piece at 4/7 of its height, using the bottom part for the far site and the top part for near site. In this way any inaccuracy in geometry and materials will be the same in the two detection sites.

With the proposed setup a very large fraction of muons from CC neutrino interactions is stopped in the spectrometer. For this class of events the momentum is obtained by muon range. For higher energies the muon momentum is determined from the track sagitta measured in the bending plane. The charge of the muon can be determined when its track crosses few RPC planes (≤ 3).

5 Physics Analysis and Results

To determine the sensitivity region that can be explored with an exposure of 6.6×10^{20} p.o.t., corresponding to 3 years of data taking on the FNAL Booster beam, three different analyses exploiting different techniques and approximations are used.

- a Feldman & Cousins technique [11] with ad hoc systematic errors added to the muon momentum distribution;

- a full correlation matrix based on the full Monte Carlo simulation including the reconstruction of the simulated data;

- CLs profile likelihood [7, 12];

For all analyses the two flavor neutrino mixing in the approximation of one mass dominance is considered. The oscillation probability is given by the formula:

$$P = \sin^2(2\theta_{new}) \sin^2(1.27\Delta m^2_{new} L(km)/E(GeV)) \quad (2)$$

where Δm^2_{new} is the mass splitting between a new heavy-neutrino mass-state and the heaviest among the three SM neutrinos, and θ_{new} is the corresponding mixing angle. In the second method, the FNR spectra, in term both of the muon range and the number of crossed planes, are evaluated using smearing matrices obtained by a full Monte Carlo simulation of the detectors.

In this analysis, correlations between the data collected in the Far and Near detectors are taken into account through the covariant matrix of the chosen observable. Since ν_μ disappearance can be observed either by a deficit of events (normalization) or, alternatively, by a distortion of the observable spectrum (shape) both normalization error and shape error (assuming a simple model for the effect on the observable spectrum) are considered. By applying the frequentist method, the χ^2 statistic was used to determine the sensitivity to oscillation parameters. Exploiting the negative-focusing mode of the FNAL Booster beam, a similar study on the search for $\bar{\nu}_\mu$ disappearance, was performed [13]. Unprecedented sensitivities at low mixing angle would be possible by using the charge measurement provided by NESSiE.

As a representative result the sensitivity calculated considering 1% correlated error in the shape is plotted in Fig. 1. The sensitivity computed considering CC and NC events is almost the same as the sensitivity obtained with only CC events and therefore NC background events do not affect the result.

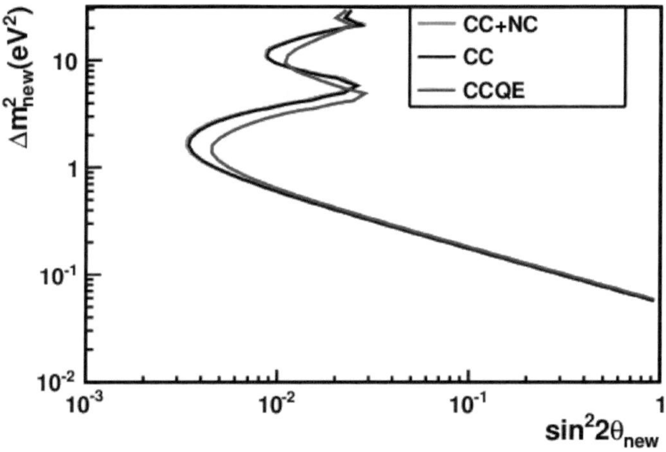

Figure 1: 95% C.L. sensitivity obtained using the muon range for all CC (black) and CC+NC (red) events and for only CCQE events (blue) assuming 3 years of data taking. 1% bin-to-bin correlated error in the shape is considered.

6 Conclusions

Existing anomalies in the experimental neutrino oscillation results may hint to the existence of additional sterile neutrino families. The NESSiE collaboration performed a detailed study of the physics case in order to set a Short-Baseline experiment at the FNAL Booster neutrino beam to exploit the measurement of the charged current events. The best option in terms of physics reach and funding constraints is provided by two spectrometers based on dipoles iron magnets, at the Near and Far sites, located at 110 and 710 m from the FNAL Booster neutrino source, respectively. By the proposed approach systematic errors can be kept at the level of 1-2% for the measurements of the ν_μ and the $\bar{\nu}_\mu$ interactions allowing the exploration of till not explored regions ($\sin^2(2\theta_{new})$ ~ 0.01).

References

[1] B. Pontecorvo, Zh. Eksp. Teor. Fiz. **33** (1957) 549;
B. Pontecorvo, Zh. Eksp. Teor. Fiz. **34** (1958) 247.
Z. Maki, M. Nakagawa and S. Sakata, Prog. Theor. Phys. **28** (1962) 870.

[2] K. A. Olive et al. [Particle Data Group], Chin. Phys. C **38** (2014), 090001. See the review therein: *Neutrino mass, mixing and oscillations* by K. Nakamura and S. T. Petcov.

[3] A. Aguilar-Arevalo et al. [LSND Collaboration], Phys. Rev. D **64** (2001) 112007; A. A. Aguilar-Arevalo et al. [MiniBooNE Collaboration], Phys. Rev. Lett. **110** (2013) 161801; G. Mention et al., Phys. Rev. D **83** (2011) 073006; Th. A. Muller et al., Phys. Rev. C **83** (2011) 054615; C. Giunti and M. Laveder, Phys. Rev. C **83** (2011) 065504.

[4] B. Pontecorvo, Zh. Eksp. Teor. Fiz. **53** (1967) 1717.

[5] J. Kopp, P. A. N. Machado, M. Maltoni, and T. Schwetz, JHEP **05** (2013) 050; T. Schwetz, Nucl. Phys. Proc. Suppl. bf 235-236 (2013) 229-235; C. Giunti et al., Phys. Rev. D **86** (2012) 113014.

[6] F. Dydak et al. [CDHS Collaboration], Phys. Lett. B **134** (1984) 281.

[7] A. Anokhina et al. [NESSiE Collaboration], FNAL-P-1057 (2014), arXiv:1404.2521.

[8] A. A. Aguilar-Arevalo et al. [MiniBooNE Collaboration], Phys. Rev. D **79** (2009) 072002.

[9] J. R. Sanford and C. L. Wang, BNL Internal Report (1967) BNL-11479.

[10] R. Acquafredda et al. [OPERA Collaboration], JINST **4** (2009) P04018.

[11] G. J. Feldman and R. D. Cousins, Phys. Rev. D **57** (1998) 3873.

[12] A. L. Read, J. Phys., G **28** (2002) 2693.

[13] A. Anokhina et al. [NESSiE Collaboration], arXiv:1503.07471v2 (2015)

STATUS OF THE BAIKAL-GVD PROJECT: FIRST CLUSTER DUBNA

A.D. Avrorin[a], A.V. Avrorin[a], V.M. Aynutdinov[a], R. Bannash[g], I.A. Belolaptikov[b], D.Yu. Bogorodsky[c], V.B. Brudanin[b], N.M. Budnev[c], I.A. Danilchenko[a], S.V. Demidov[a], G.V. Domogatsky[a], A.A. Doroshenko[a], A.N. Dyachok[c], Zh.-A.M. Dzhilkibaev[a], S.V. Fialkovsky[e], A.R. Gafarov[c], O.N. Gaponenko[a], K.V. Golubkov[a], T.I. Gress[c], Z. Honz[b], K.G. Kebkal[g], O.G. Kebkal[g], K.V. Konischev[b], A.V. Korobchenko[b], A.P. Koshechkin[a], F.K. Koshel[a], A.V. Kozhin[d], V.F. Kulepov[e], D.A. Kuleshov[a], V.I. Ljashuk[a], M.B. Milenin[e], R.A. Mirgazov[c], E.R. Osipova[d], A.I. Panfilov[a], L.V. Pan'kov[c], E.N. Pliskovsky[b], M.I. Rozanov[f], E.V. Rjabov[c], B.A. Shaybonov[b], A.A. Sheifler[a], M.D. Shelepov[a], A.V. Skurihin[d], A.A. Smagina[b], O.V. Suvorova[a], V.A. Tabolenko[c], B.A. Tarashansky[c], S.A. Yakovlev[g], A.V. Zagorodnikov[c], V.A. Zhukov[a], and V.L. Zurbanov[c]

[a]Institute for Nuclear Research, Moscow, 117312 Russia
[b]Joint Institute for Nuclear Research, Dubna, 141980 Russia
[c]Irkutsk State University, Irkutsk, 664003 Russia
[d]Institute of Nuclear Physics, Moscow State University, Moscow, 119991 Russia
[e]Nizhni Novgorod State Technical University, Nizhni Novgorod, 603950 Russia
[f]St. Petersburg State Marine Technical University, St. Petersburg, 190008 Russia
[g]EvoLogics, Germany
E-mail: [a]

Abstract. In April 2015 the demonstration cluster "Dubna" was deployed and started to take data in Lake Baikal. This array is the first cluster of the cubic kilometer scale Gigaton Volume Detector (Baikal-GVD), which is constructed in Lake Baikal. We review the status and performance of the array.

1 Introduction

The neutrino telescope Baikal Gigaton Volume Detector (GVD) is currently under construction in Lake Baikal [1]. The holy grail of the Baikal-GVD quest for neutrinos of very high energy, more than tens of TeV, is the identification of their astrophysical origin at least with value of the flux measured recently by IceCube [2]. Baikal-GVD is a Cherenkov detector configurated by a three-dimensional lattice of optical modules (OMs) that is photomultiplier tubes housed in transparent pressure spheres, arranged at vertical load-carrying cables to form strings. The telescope has a modular structure and consists of clusters of strings, functionally independent sub-arrays, which are connected to shore by individual electro-optical cables. Each cluster has a central string identical to seven others distant at radius of 60 meters in baseline configuration. The first Baikal-GVD cluster named *Dubna* was deployed and started operation in Lake Baikal in April 2015 according to timeline of the project [1, 3].

[a]E-mail: suvorova@cpc.inr.ac.ru

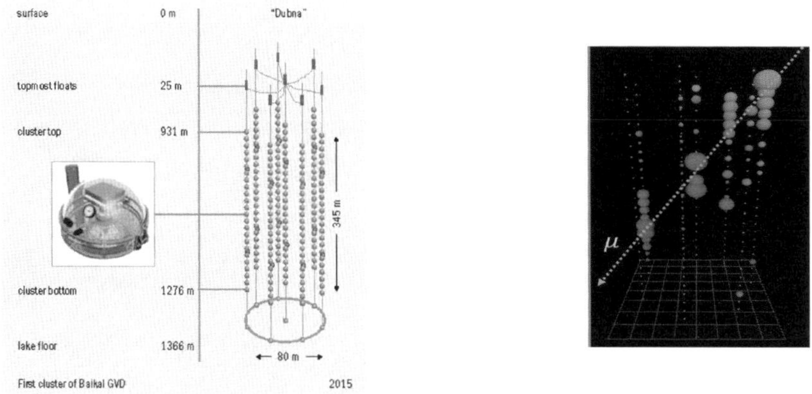

Figure 1: Right: Schematic drawing of the Dubna cluster. Left: Event view of the high-energy down-going muon event observed in June 2015. Each sphere represents an OM. Colors represent the arrival times of the photons where red indicates early and blue late times. The size of the spheres is proportional to the measured number of photoelectrons.

2 Demonstration cluster "Dubna"

The Dubna cluster encloses 1.7 Megatons of fresh Baikal water. The cluster comprises a total of 192 optical modules arranged at eight 345 m long strings, as well as an acoustic positioning system. There is also an instrumentation string with equipment for array calibration and monitoring of environmental parameters. Artistic view of the Dubna cluster with OM view in zoom are shown in Fig. 1 (on left). Each string comprises 24 OMs spaced by 15 m at depths of 900 m to 1250 m below the surface. In 2015 seven side strings have been located at a reduced radius of 40 m around a central one. The reason is to increase the sensitivity to low-energy atmospheric muons and neutrinos which are used for array calibration. In the next year, strings will be moved to the baseline distances.

A section is the basic detection unit of the GVD neutrino telescope. It comprises 12 optical modules and a central module (CeM). Among the basic functions of OMs are detection of the particle radiation; shaping of the output analog pulse for signal transmission to the ADC board; control of the PMT operation modes; calibration and monitoring of the parameters of OM electronic components. The block diagram of a section is published elsewere [3–5]. Each optical module consists of a pressure-resistant glass sphere of 43.2 cm diameter which holds the OM electronics and the PMT which is surrounded by a high permittivity alloy cage for shielding it against the Earth magnetic field. A large photomultiplier tube Hamamatsu R7081-100 with a 10-inch hemispherical photocathode and quantum efficiency up to 35% has been selected as light sensor.

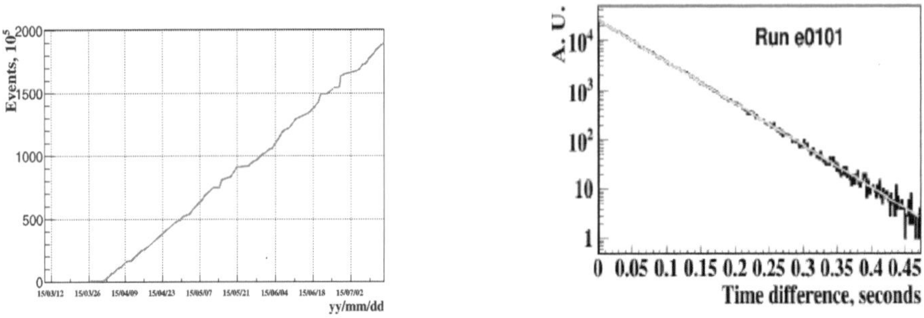

Figure 2: Right:Integrated number of recorded events since April 2015. Left: The time difference between subsequent events is shown for one run.

Besides the PMT, an OM comprises a high voltage power supply unit (HV), a fast two-channel preamplifier, and a controller. For time and amplitude calibration of the measuring channel, two LEDs are installed in the optical module. The OM controller is intended for HV control and monitoring for PMT noise measurements and for time and amplitude calibration. Details of data acquisition, basic controls, methods of calibrations, hard-and soft-ware triggers can be found in [4].

3 Performance and sensitivity

The site chosen for the experiment is in the southern basin of Lake Baikal. Here, the combination of hydrological, hydro-physical, and landscape factors is studied to be optimal for deployment and operation of the neutrino telescope [1]. The water depth is about 1360 m at distances beginning from about of three kilometers from the shore. The water transparency is characterized by an absorption length of about 20–25 m and a scattering length of 30–50 m [1]. The water chemiluminescence is moderate at the detector site.

Since April till June about $1.7{\cdot}10^8$ events have been recorded, while the cluster is operating in several testing and data taking modes. Stability and efficiency of the cluster operation during 2015 are presented in Fig. 2 (on left) by integral rate of general trigger. Quality of data is seen in Fig. 2 (on right) with distribution of time difference between subsequent events for one run. Obtained exponential behavior is consistent with expectation for randomly distributed experimental events. As an illustration, the one event observed in June 2015 is shown in Fig. 1 (on right). The photon distribution of the event is consistent with the pattern of Cherenkov photons from down going high-energy muon.

The first cluster of Baikal-GVD potential to detect astrophysical neutrinos is shown in Fig. 3. We expect about one event per year with $E_{sh} > 100$ TeV

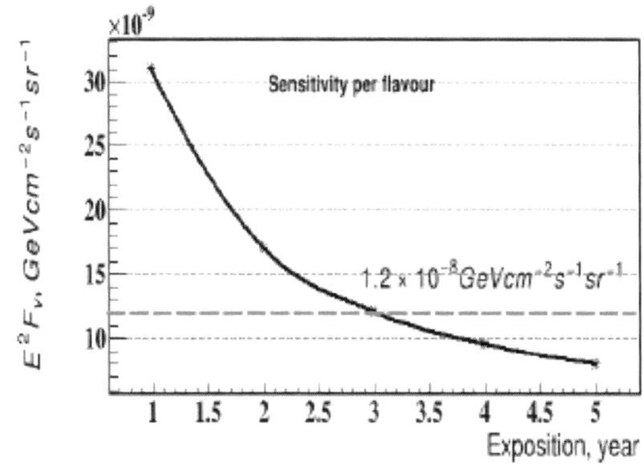

Figure 3: Cluster sensitivity for one flavor neutrino flux with an E^{-2} spectrum as function of the observation years. The long-dashed line indicates the one flavor neutrino flux value obtained by IceCube

from an all-flavor astrophysical flux with the normalization $E^2F=3.6\cdot 10^{-8}$ GeV cm^{-2} s^{-1} sr^{-1} in the cluster, compared to about 10 events in IceCube. Here the search for high-energy neutrinos is based on the selection of cascade events generated by neutrino interactions in the sensitive volume of array. After applying an iterative procedure of vertex reconstruction followed by the rejection of hits contradicting the cascade hypothesis on each iteration stage, events with a final multiplicity of hit OMs $N_{hit} > 20$ are selected as high-energy neutrino events. The accuracy of shower direction reconstruction is about 4 degree (median value), which is substantially better than the 10–15 degree accuracy for IceCube [2]. Three year exposition allows sensitivity at a level of flux value measured by IceCube.

Acknowledgments

This work was supported by the Russian Found for Basic Research (grants 13-02-12221, 14-02-00175, 14-02-00972).

References

[1] V. Aynutdinov et al., NIM **A742**, 82, (2014).
[2] M.G. Aartsen et al.,Science **342**, 1242856, (2013) .
[3] A.D. Avrorin et al., Physics of Particles and Nuclei, **46**, 2, 211 (2015).
[4] A. Avrorin et al., Instr. and Exp. Tech. **57**, 262, (2014).
[5] A.D. Avrorin et al., Proc. of 34th ICRC, Hague, Netherlands, http://pos.sissa.it/archive/conferences/236/1162, (2015).

Physics at Accelerators and Studies in SM and Beyond

SEARCH FOR NEW PHENOMENA AT THE HIGH LUMINOSITY LHC WITH ATLAS

Paul Glaysher [a] (On behalf of the ATLAS Collaboration)
University of Edinburgh

Abstract. ATLAS sensitivity studies into the prospects of measuring Higgs boson properties and performing searches for new phenomena are presented in the context of the High Luminosity Large Hadron Collider. Simulated data provides expected limits on Higgs self-coupling, coupling to dark matter and constraints on discovering further Higgs bosons, with a total integrated luminosity of 3000 fb^{-1} and collisions at $\sqrt{s}$ =14 TeV. Likewise, prospects of supersymmetry and di-jet resonance searches are highlighted.

1 Introduction

The first run of the Large Hadron Collider (LHC) from 2010 - 2012, using proton-proton collisions at $\sqrt{s}$ = 7 and 8 TeV, was highly successful. The highlight was the discovery of a new particle with a mass of 125 GeV compatible with the Standard Model (SM) prediction of the Higgs boson [1,2]. ATLAS [3] has performed initial measurements of the Higgs properties such as its couplings to SM particles [4]. In future LHC runs, with a collision energy of $\sqrt{s}$ = 14 TeV, ATLAS will pursue precision measurements of the Higgs boson properties as well as performing Beyond the Standard Model (BSM) searches. Highlights of the future physics potential of the ATLAS detector at the High Luminosity LHC (HL-LHC) are presented, covering the Higgs sector, supersymmetry and di-jet resonance searches.

The current Run 2 and following Run 3 of the LHC programme is expected to deliver a total integrated luminosity of about 300 fb^{-1} by the year 2022. The peak instantaneous luminosity will be in the range from 2 to 3 $\cdot 10^{34}\text{cm}^2\text{s}^{-1}$. This dataset is foreseen to have an average number of pile-up events per bunch crossing, denoted by PU, of 50 - 60. A planned upgrade of the LHC to the HL-LHC is needed to fully explore rare decay processes, including Higgs boson couplings, and popular models for new phenomena. The HL-LHC will provide 3000 fb^{-1} by 2030, with a peak luminosity of $5{\cdot}10^{34}\text{cm}^2\text{s}^{-1}$ and a value of PU = 140. The increase in PU with instantaneous luminosity is the primary challenge for detector operation at the HL-LHC, degrading the experimental environment, primarily the primary vertex and track reconstruction efficiencies. To mitigate the effect of PU, as well as addressing the issue of radiation damage to the silicon sensors, the ATLAS detector will undergo an extensive upgrade program prior to the HL-LHC running [5,6].

The sensitivity studies presented are performed using particle-level simulated data, to which detector response functions are applied. The detector response

[a]E-mail: `paul.glaysher@cern.ch`

functions are derived from full Geant 4 simulations and are based on an upgraded ATLAS detector, taking into account the impact of future PU scenarios on the detector efficiencies [7, 8].

2 Higgs Physics Goals

2.1 Standard Model Higgs Couplings

The sensitivity to measure the SM Higgs boson couplings were determined for various decay channels, including the yet to be observed $b\bar{b}$ and $\mu\mu$ states. H$\rightarrow b\bar{b}$ suffers from a large background and is expected to be observed in Run 2 [9], while a H $\rightarrow \mu\mu$ observation will require the HL-LHC data, due to the low branching ratio of the process [10]. The expected precision on the couplings is obtained from a fit for each channel using a leading-order tree-level motivated framework [11], under the assumption of a single resonance Higgs boson of mass m_{H} =125 GeV within a narrow width approximation. An overview is given in Ref. [12] for the 300 and 3000 fb^{-1} estimates. The expected precision at which the SM nature of the couplings can be probed with 3000 fb^{-1} is in the 2 - 15 % approximate range depending on the decay channel. BSM searches in the Higgs sector rely on measuring deviations to the SM couplings, their sensitivity is thus determined by the precision of the SM couplings.

2.2 Higgs Self-Coupling

An exciting goal of the HL-LHC is observing di-Higgs boson production, which is sensitive to the Higgs self-coupling. Measuring the self-coupling, λ, will provide the strongest test of assessing the SM nature of the Higgs boson. The expected cross section at NNLO is 40.2 fb for $\sqrt{s}$ = 14 TeV [13]. For this challenging measurement, the most promising signatures come from the final states HH $\rightarrow b\bar{b}\gamma\gamma$, with only 320 expected events for 3000 fb^{-1} but an experimentally clean signature, and HH $\rightarrow b\bar{b}WW$ with 31,000 expected events but subject to large backgrounds. Further, the $b\bar{b}b\bar{b}$ and $b\bar{b}\tau^+\tau^-$ final states are also of interest [14]. Results for HH$\rightarrow b\bar{b}\gamma\gamma$ are shown in Figure 5a. A strong separation of signal and background is achieved through angular and mass cuts. In the case of the HH$\rightarrow b\bar{b}\gamma\gamma$ channel alone, 8.4 signal and approximately 47 background events are selected, assuming a SM coupling λ_{SM}. As shown in Figure 1b, just HH$\rightarrow b\bar{b}\gamma\gamma$ will not be sensitive at the 5 σ discovery level to λ_{SM}, but will be able to rule out large deviations from the SM, namely $\lambda/\lambda_{SM} < -1.3$ and $\lambda/\lambda_{SM} > 8.7$.

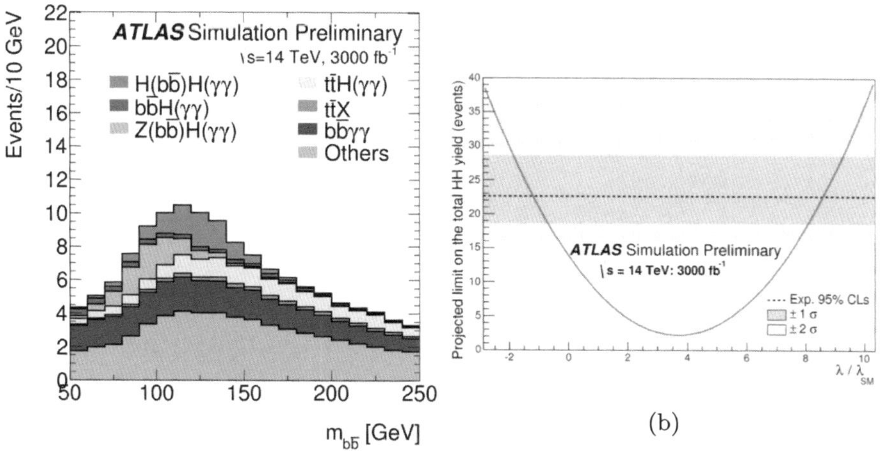

(a) (b)

Figure 1: (a) Expected invariant mass distribution $m_{b\bar{b}}$ in the HH→ $b\bar{b}\gamma\gamma$ channel. Contributions are shown before a mass and angular cut is applied and are normalized to the number of expected events after the full event selection. The background component *Others* is the sum of $c\bar{c}\gamma\gamma$, $b\bar{b}\gamma j$, $b\bar{b}jj$ and $jj\gamma\gamma$, where j signifies a jet. (b) Number of expected HH signal events (red) given as a functions of $\lambda/\lambda_{\rm SM}$. Projected 95% CL limit (black) on excluding HH signal with 1 and 2σ uncertainty bands (green/yellow) for the 3000 fb^{-1} dataset, taken from Ref. [14].

3 Higgs Boson as a Portal to New Phenomena

3.1 Additional Heavy Higgs-like Particles

In several theoretical extension to the SM further Higgs bosons are envisioned. For example, in the Two-Higgs-Doublet Model (2HDM) [15–17], an additional Higgs doublet is considered, resulting in a model with five Higgs bosons: one neutral-CP-odd A, two charged $H^{\pm}$, two neutral CP-even h and H, where h_{125} would be the discovered 125 GeV Higgs boson. Indirect searches for the heavy H through precision measurements of the h_{125} couplings are possible, where the presence of H would be established from deviations of the h_{125} couplings from the SM expectations, in both the couplings to bosons and fermions. The relationship between h_{125} and H is determined by their mixing angle α and tanβ, the ratio of the vacuum expectation values of the two doublets. Likewise, cos$(\beta-\alpha)$ is a measure of the relative coupling strength of H to vector bosons over the SM expectation, such that cos$(\beta-\alpha)$=0 recovers the SM case. For a Minimal Super-symmetric Model (MSSM)-like Type II 2HDM, in which one

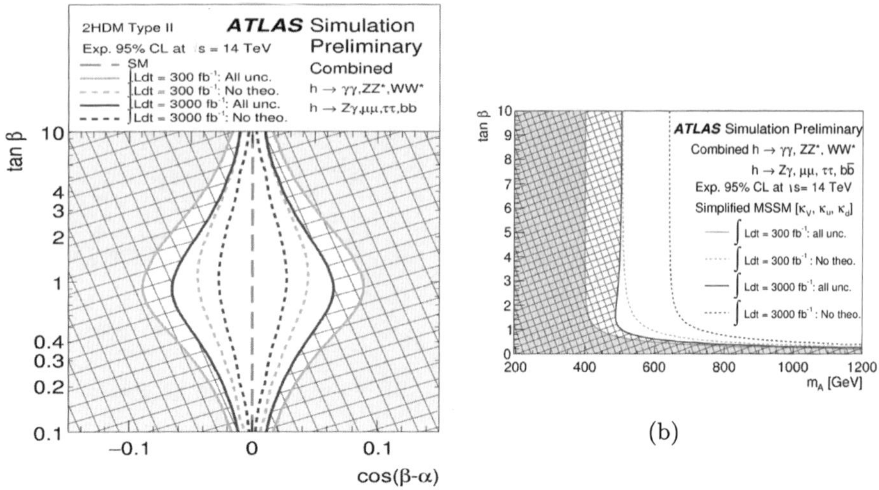

(a) (b)

Figure 2: Excluded regions of (a) the $\cos(\beta-\alpha)$ vs $\tan\beta$ of the Type II 2HDMs and (b) the m_A vs $\tan\beta$ plane for a simplified MSSM scenario. Both are obtained by fits to the measured rates of Higgs boson production and decays, taken from Ref. [18]. The confidence intervals account for a possible relative sign between different couplings in (a). The expected likelihood contours of 95% CL (2σ) are indicated assuming the SM Higgs sector. The light shaded and hashed regions indicate the expected exclusions for 300 and 3000 fb^{-1}, with and without theory uncertainties.

Higgs doublet couples to up-type quarks and the other to down-type quarks and charged leptons, the regions of the $\cos(\beta-\alpha)$ vs $\tan\beta$ plane that are expected to be excluded are shown in Figure 2a. Figure 2b shows a specific example of 2HDM, the MSSM. For a simplified MSSM, where the h_{125} and H mixing is fully determined by m_A and $\tan\beta$, the SM Higgs boson coupling rates κ_V, κ_u and κ_d would be modified by m_A. Although simplified, this model serves as a good benchmark study and is of particular use if no direct SUSY observation is made.

3.2 Higgs Boson to Dark Matter Couplings

One can make the assumption that a WIMP Dark Matter (DM) candidate interacts very weakly with SM particles except for the Higgs boson. The Higgs boson thus decays to DM candidates as well as other particles invisible to the detector, such as neutrinos. By measuring the branching ratio of $H \to$ invisible,

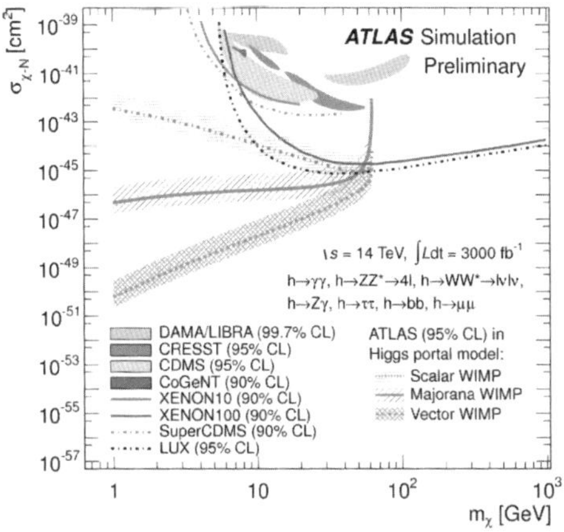

Figure 3: Expected upper limits at 95% CL on the interaction cross section of the WIMP DM candidate as a function of its mass, taken from Ref. [18]. Expected results from ATLAS with 3000 fb^{-1} are shown together with current results from dedicated DM experiments. Scalar, Majorana and Vector WIMP models are considered.

BR_{inv}, limits can be set on the interaction cross section of DM as shown in Figure 3. A 95% CL upper limit of $\text{BR}_{\text{inv}} < 0.13(0.09)$ is expected to be achieved with 3000 fb^{-1}, for a nominal (and improved) systematic uncertainty scenario. This should be compared to the current limit of $\text{BR}_{\text{inv}} < 0.25$ set in Run 1 from direct searches [19]. The prospects for Scalar, Majorana and Vector WIMP models are studied, in all three cases ATLAS at the HL-LHC will provide complementary measurements to those from dedicated DM experiments, in particular for low mass DM candidates.

4 Supersymmetry

Supersymmetry (SUSY) is one of the best motivated and well studied extensions of the SM. The HL-LHC will allow to probe small production cross sections, with expected sensitivity limits on SUSY particle masses in the range of 1-3 TeV. Here one assumes R-parity conserving SUSY and a stable lightest supersymmetric particle. Exclusion and discovery contours are shown for gluino, $\tilde{g}$, and sbottom, $\tilde{b}_1$, searches in Figure 4 as a function of the $\tilde{\chi}_1^0$ mass. Pair-produced $\tilde{g}$ and $\tilde{b}$ are considered which exclusively as $\tilde{g} \to qq\tilde{\chi}_1^0$ or $\tilde{b}_1 \to b\tilde{\chi}_1^0$. Thus, such events are characterised by four or two hadronic jets respectively and a large amount of missing transverse momentum from the undetected $\tilde{\chi}_1^0$.

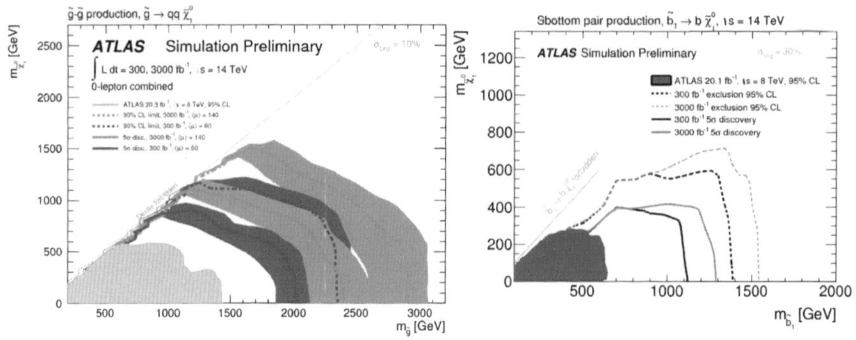

(a) (b)

Figure 4: Exclusion and discovery contours are shown for gluino $\tilde{g}$ (a) and sbottom $\tilde{b_1}$ (b) as a function of the $\tilde{\chi}_1^0$ mass, taken from Ref. [20]. The filled area represents the exclusion limits set in the 8 TeV run, for the $\tilde{g}$ case the blue bands give the 1 σ uncertainty on the production cross section. The grey line is the lower bound of the kinematically forbidden region.

5 Di-Jet Resonance Searches

A number of new phenomena are expected to modify the di-jet mass distribution, creating a 'bump' on the smoothly falling QCD spectrum. For example excited u and d quarks (q*) or quantum black holes (QBH), which decay into two high-momentum hadronic jets. A possible manifestation of quark compositeness, q* are assumed to predominately decay to a qg final state via a gauge boson, with a branching ratio of 0.83. Decays involving W/Z or γ emission are also considered. QBH decays to all allowed two-body final states are simulated by the BlackMax [21] generator, where the QBH production threshold mass is set equal to the Planck scale. Limits set by sensitivity studies on both benchmark models [22] are shown in Figure 5.

6 Summary

With the discovery of the Higgs boson at LHC Run 1 we have entered a new era of particle physics. The unprecedented collision energy of $\sqrt{s}$ =13 - 14 TeV in Run 2 will vastly increase the search-scope for new physics phenomena; but it is the HL-LHC that will offer unique opportunities to fully explore the Higgs sector, and probe for high energy scale New Physics. The prospects of future SM Higgs couplings measurements and the expected sensitivity to BSM physics such as Higgs self-coupling, extra heavy Higgs boson, Higgs coupling to dark matter, supersymmetry and di-jet resonance searches were presented.

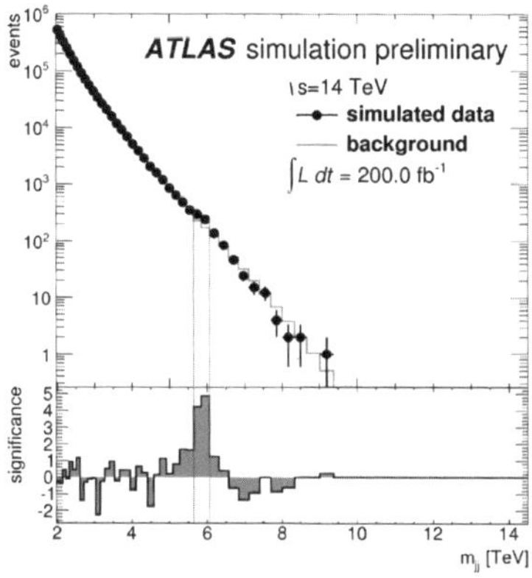

(a)

integrated luminosity [fb^{-1}]	m_{q^*} [TeV]	m_{QBH} [TeV]
0.1	4.0	8.2
1	5.0	8.9
5	5.9	9.2
25	6.6	9.7
300	7.4	10.0
3000	8.0	10.1

(b)

Figure 5: (a) Di-jet spectrum containing a 6 TeV q^* producing a 5 σ excess with 200 fb^{-1}. (b) Expected upper exclusion limits on q^* and QBH masses for a given amount of collected data, taken from Ref. [22].

References

[1] ATLAS Collaboration, Phys.Lett. B716 (2012) 1-29.
[2] CMS Collaboration, Phys. Lett. B 716 (2012) 30.
[3] ATLAS Collaboration, 2008 JINST 3 S08003.
[4] ATLAS Collaboration, Eur.Phys.J. C76 (2016) 6
[5] ATLAS Collaboration, LHCC-I-023, CERN-LHCC-2012-022.
[6] ATLAS Collaboration, LHCC-G-166, CERN-LHCC-2015-020.
[7] ATLAS Collaboration, ATL-PHYS-PUB-2013-004.
[8] ATLAS Collaboration, ATL-PHYS-PUB-2013-009.
[9] ATLAS Collaboration, ATL-PHYS-PUB-2014-011.
[10] ATLAS Collaboration, ATL-PHYS-PUB-2013-014.
[11] LHC Higgs Cross Section Working Group, arXiv:1307.1347 [hep-ph].
[12] ATLAS Collaboration, ATL-PHYS-PUB-2014-016.
[13] de Florian and J. Mazzitelli, Phys. Rev. Lett. 111 (2013) 201801.
[14] ATLAS Collaboration, ATL-PHYS-PUB-2014-019.
[15] G. Branco et al., Phys. Rept. 516 (2012) 1-102.
[16] T. Lee, Phys. Rev. D. 8 (1973) 12261239.
[17] J. F. Gunion and H. E. Haber, Phys. Rev. D. 67 (2003) 075019.
[18] ATLAS Collaboration, ATL-PHYS-PUB-2014-017.
[19] ATLAS Collaboration, JHEP 11 (2015) 206.
[20] ATLAS Collaboration, ATL-PHYS-PUB-2014-010.
[21] D. Dai. et al., Phys. Rev. D 77 (2008) 076007.
[22] ATLAS Collaboration, ATL-PHYS-PUB-2015-004.

CLIC PHYSICS OVERVIEW

Ivanka Božović-Jelisavčić[a] on behalf of the CLICdp collaboration
Vinca Institute of Nuclear Sciences, University of Belgrade 11000 Belgrade, Serbia

Abstract. In this paper, based on the invited talk at the 17th Lomonosov Conference of Elementary Particle Physics, the physics program at the future Compact Linear Collider (CLIC) will be reviewed, with particular emphasis on the Higgs physics studies. It will be demonstrated, on the basis of detailed physics and detector studies carried out at CLIC, that the CLIC is indeed a precision tool for studies both in the Higgs sector and beyond the Standard Model.

1 Introduction

The future Compact Linear Collider (CLIC) offers excellent statistical potential for precision measurements in the QCD-background-free environment of e^+e^- collisions. An energy-staged approach that includes three running center-of-mass energies (380 GeV, 1.4 TeV and 3 TeV) maximizes the potential to explore the Higgs sector in a model-independent way using the favorable properties of Higgsstrahlung - a Higgs production mechanism exploitable at future lepton colliders at center-of-mass energies $\leq$1 TeV. In addition, CLIC offers excellent possibilities to study top physics, including a threshold scan at the lowest center-of-mass energy, as well as to study the Higgs self-coupling and to perform Beyond Standard Model (BSM) searches at the highest energy. Staged implementation along the described lines provides a long-term precision physics program to complement the LHC searches. In Figure 1, cross-sections are given for top-pair production and Higgs production at various center-of-mass energies.

1.1 CLIC accelerator and detector

The CLIC accelerator is based on a novel two-beam scheme, where a high-intensity beam (drive beam) is used to generate RF power for the main beam. Using normal-conducting accelerator structures, the two-beam acceleration provides gradients of 100 MV/m as has been demonstrated at the CTF3 test facility [1]. Small (σ_x=40 nm, σ_y=1 nm) and dense ($\sim 10^9$ particles) bunches lead to strong beamstrahlung effects induced by the electromagnetic fields of the opposite bunches. Consequently, beamstrahlung photons produce hadrons as often as 3.2 interactions per bunch-crossing. With a small bunch separation of 0.5 ns, the above leads to a requirement of 10 ns time stamping in order to cope with the occupancy of the central detectors. At the other hand, Beamstrahlung also deteriorates luminosity spectrum, in particular at high center-of-mass energies, however, it has been shown in [2] that the effect can be controlled at a permille level in the peak region above 80% of the nominal center-of-mass energies. The SiD [3] and ILD [4] detector concepts proposed for the International

[a]E-mail: `ibozovic@vinca.rs`

Linear Collider, have been adapted to the CLIC running conditions. Detector performance is driven by the physics goals requiring: Jet energy resolution better than 3.5% for high-energy jets above 100 GeV, track momentum resolution $\Delta(p_t)/(p_t)^2 \sim 10^{-5}$ GeV^{-1}, impact parameter resolution of the order of a few microns. The above comes respectively from the need to distinguish between Z, W or Higgs bosons, to reconstruct the Higgs boson from the recoil mass of the Z decay products (i.e. muons) and to provide the flavor separation i.e. for the measurement of the Higgs couplings to beauty and charm. Detector hermeticity and the ability to efficiently identify leptons (over 95%) is assumed at all energy stages. In Figure 2 [1], a schematic view of the CLIC detector models is given.

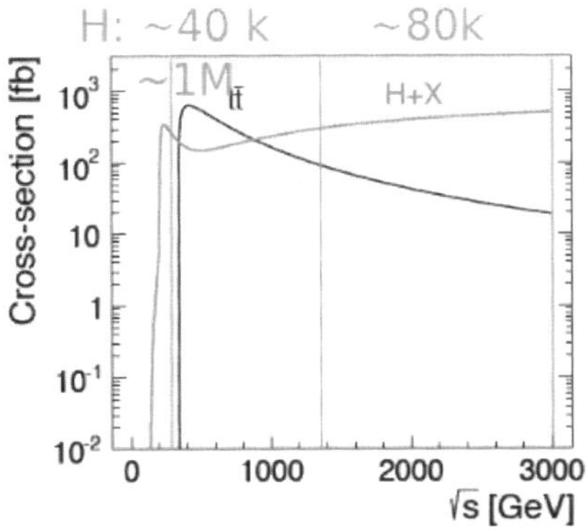

Figure 1: Higgs and top-pair production cross-sections as a function of center-of-mass energy. Approximate statistics of the produced Higgs bosons is also indicated per energy stage. Four years of nominal detector operation are assumed per stage.

2 Higgs studies at CLIC

Higgsstrahlung and WW-fusion are the dominant production mechanisms of the Higgs boson at low and high energies respectively. Combined they lead to $\sim 10^6$ Higgs bosons in four years of nominal detector operation, per stage, assuming 50% data taking efficiency. Depending on the production mechanism, appropriate polarization could eventually double the statistics. Combined study of the two production mechanisms probe the Higgs width and couplings in a model-independent way. This leads to a determination of the Higgs couplings at the level of a percent (except for the rare decays to light particles such as muons or photons), employing all three stages. Assuming that the Higgs total width is constrained by the SM decays, the statistical precision of the Higgs couplings can be improved to a sub-percent level. Details of the combined fit of the Higgs measurements at all energy stages will be discussed in Section 2.2.

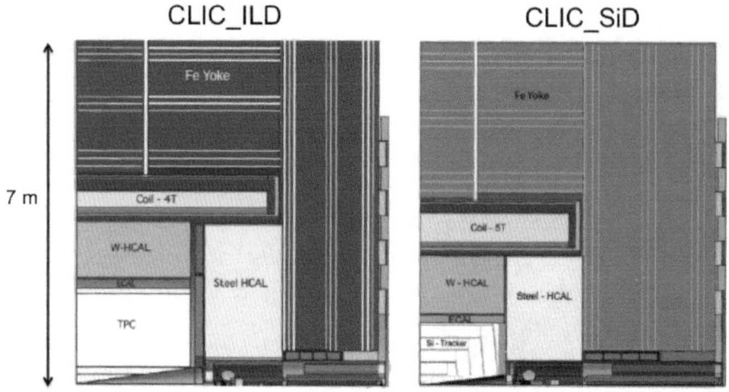

Figure 2: Longitudinal cross section of the top-right quadrant of the CLIC_ILD (left) and CLIC_SiD (right) detector models.

2.1 *Low and high-energy landscape*

The low-energy phase at CLIC, from the perspective of the Higgs studies is primarily motivated by the direct and model-independent measurement of the Higgs coupling to Z that can be obtained from the recoil mass distribution in Higgsstrahlung ($e^+e^- \to ZH, Z \to f\bar{f}, f = e, \mu, q$), with a statistical precision of 0.8%. The g_{HZZ} determination plays a central role in the measurement of the Higgs couplings at future e^+e^- colliders. From the Z decays to leptons, the recoil mass distribution can be used to extract the Higgs mass and the total ZH production cross section with the statistical precision of 110 MeV and 3.8%, respectively. In Figure 3 (left), reconstruction of the Higgs mass from the simulated data is given for HZ, $Z \to \mu^+\mu^-$ decays. In addition, the measurement of the recoil mass distribution from hadronic Z decays provides a direct search for invisible Higgs decays, constraining the branching ratio for $H \to invisible$ decays to 0.57% with 1σ uncertainty.

With the cross section $\sigma_{H\nu\nu}$ scaling with $log(s)$, WW-fusion becomes the dominant Higgs production channel at higher energies. This allows to probe even rare decays (BR$\leq 10^{-3}$) like $H \to \mu^+\mu^-$ (Figure 3 (right)), $H \to Z\gamma$ and $H \to \gamma\gamma$ with a statistical precision of 16%, 42% and 15% respectively. These numbers scale down by a factor ~ 0.7 if an electron polarization of -80% is applied. Excellent performance of tagging algorithms allows heavy flavor separation and the corresponding statistical uncertainty to access $\sigma_{H\nu\nu} \cdot BR(H \to b\bar{b}, c\bar{c}, gg)$ is 0.2%, 2.7% and 1.8% respectively, for the Higgs decays to b, c and gluons, yielding to a statistical precision of 1.5% on the ratio g_{Hcc}/g_{Hbb}. The latter provides a direct test of the Standard Model predictions for the up and down-type quark's couplings to the Higgs boson. Particular relevance of the high energy measurement lies in the ability to access the Higgs self-coupling λ and quartic coupling to W bosons g_{HHWW} with a relative statistical uncertainty of 12% and 3%, respectively. Beam polarization plays important role in this type of measurement

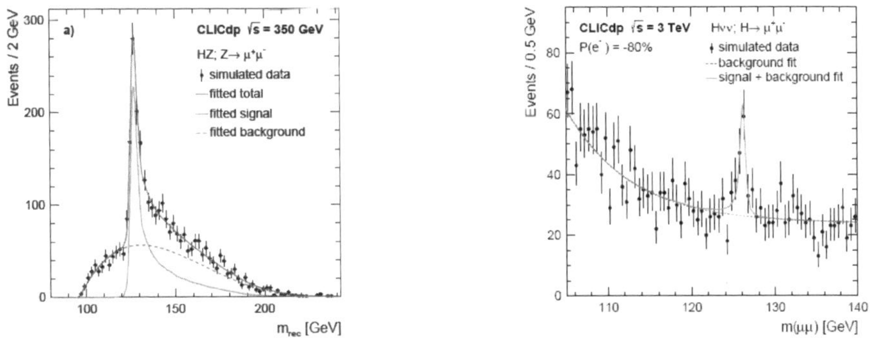

Figure 3: Recoil mass distribution from HZ, Z→ $\mu^+\mu^-$ decays at 350 GeV center-of-mass energy (left); Reconstruction of the di-muon invariant mass in Higgs decays H→ $\mu^+\mu^-$ at 3 TeV center-of-mass-energy (right).

due to the increase of the Higgs production cross-section and suppression of certain types of background.

2.2 *Combined fit of the Higgs measurements*

The full statistics of data (0.5 ab^{-1} at 380 GeV, 1.5 ab^{-1} at 1.4 TeV, 2 ab^{-1} at 3 TeV) can be used in a global fit in order to reach the ultimate precision on the Higgs couplings. Direct access to g_{HZZ}, in a model-independent way, through Higgstrahlung at the lowest center-of-mass energy, allows a fit with minimal theoretical assumptions where the Higgs couplings and the Higgs total width enter as free parameters:

$$\frac{g_{HZZ}{}^2 \cdot g_{Hbb}{}^2}{\Gamma_H} = C_{HZ,H\to b\bar{b}} \tag{1}$$

and, similarly for other processes, fit parameters C_i can be defined where i=1,10. Then, the overall χ^2 can be built:

$$\chi^2 = \sum_{i=1,10} \frac{(C_i/C_i{}^{SM} - 1)^2}{\delta F_i^2} \tag{2}$$

where δF_i stands for the relative statistical uncertainty of the σ·BR measurement for the considered Higgs decay i and σ is the Higgs production cross-section. The above leads to the determination of the Higgs total width with a relative statistical uncertainty of 3.4%, while most of the couplings can be determined at a percent level. On the other hand, assuming that the total (model-dependent) width is determined from the SM branching ratios, a global fit can be performed in a model-dependent way, with the relative partial widths κ_i (i=1,9) as free parameters:

$$\Gamma_H^{md} = \sum_{i=1,9} \kappa_i^2 BR_i^{SM} \qquad\qquad \kappa_i^2 = \Gamma_i/\Gamma_i^{SM} \tag{3}$$

Then, instead of couplings, the corresponding relative partial width is defining C_i in (2), as in example:

$$\frac{\kappa_{HZZ}^2 \cdot \kappa_{Hbb}^2}{\Gamma_H^{md}} = C_{HZ,H\to b\bar{b}} \tag{4}$$

The above, LHC-style approach, leads to a sub-percent precision for most of the couplings. Precision of the Higgs couplings at the three-stage CLIC program is illustrated in Figure 4, for a model-independent and dependent fit.

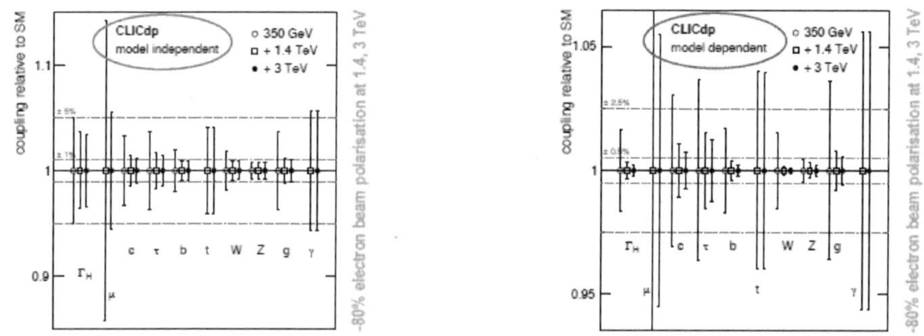

Figure 4: Precision of the Higgs couplings of the three-stage CLIC program for a model-independent (left) and model-dependent fit (right).

3 Top benchmark studies

As an e^+e^- collider, CLIC offers two complementary ways of the top mass measurement. One is by direct reconstruction of the invariant mass of the top decay products and can be performed at arbitrary center-of-mass energy and the other is based on a cross-section scan around the top-pair production threshold. Interpretation of the first method depends on higher order theoretical corrections. The invariant mass approach leads to the statistical uncertainty of 80 MeV, while the threshold scan provides 30 MeV statistical uncertainty, within the total error of 100 MeV dominated by the theoretical uncertainty. Excellent flavor tagging and possibility to separate between jets originating from Z, W and H, enables the measurement of the top Yukawa coupling at 1.4 TeV centre-of-mass energy. With electron polarization of -80%, it is possible to achieve a relative statistical uncertainty of the top Yukawa coupling below 4% [5]. In Figure 5 (left) [5], the top threshold scan is illustrated for 174 GeV top mass.

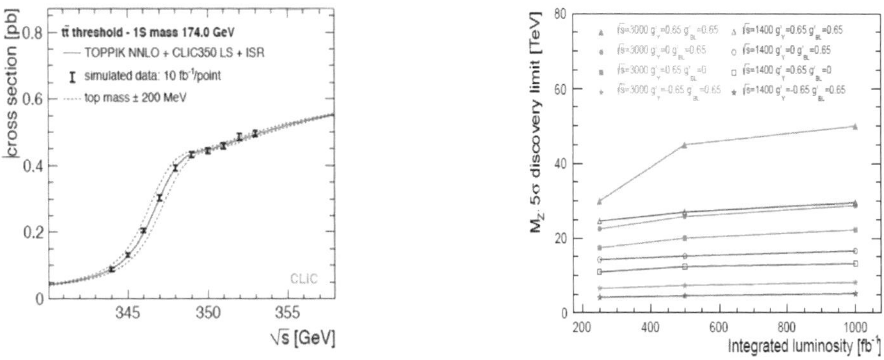

Figure 5: Illustration of the top-pair production cross-section sensitivity to the top mass performed around the production threshold (left); 5σ discovery limit on Z$'$ mass as a function of integrated luminosity (right).

4 BSM searches

At CLIC, potential signatures of BSM physics can be searched for either through direct reconstruction of new particles, with an approximate mass reach of $\sqrt{s}/2$, and indirectly, in a model-dependent way, using evidence of deviations in precision observables (like cross-sections, FB and LR asymmetries, etc.). Indirect searches extend the sensitivity for BSM physics beyond the direct kinematic reach of the machine. As showed in [1], mass measurements at a percent level or better are possible for most particles in the accessible SUSY scenarios. In indirect searches, fermion-pair production can be used to test extended gauge theories (i.e. with Z$'$) where sensitive observables receive higher order corrections due to the appearance of new gauge bosons. Figure 5 (right) [5] illustrates 5 σ discovery limits on Z$'$ mass, as a function of integrated luminosity. It is clear that the sensitivity is extended far beyond the operational center-of-mass energy. The same is true for probing the fundamental scale in models with extra dimensions and for many other indirect searches.

5 Complementarity with the LHC program

The capacity for precision studies of either SM processes, or particles discovered at LHC, together with a stand-alone discovery reach, makes CLIC complementary to the LHC physics program. The ability to measure the Higgs couplings to a percent level or better is, in principal, more precise than what can be done by LHC or its high luminosity upgrade HL-LHC. This is of particular importance for measurement of the Higgs self-coupling λ, defining the shape of the Higgs potential. The same is true for probing the Higgs compositeness - very similar in phenomenology to the SM Higgs. In the later case, the only difference comes from the relative correction proportional to ξ where $\xi = (v/f)^2$ and v is the vacuum expectation value of the SM Higgs and f

is the scale of compositeness. Figure 6 [5] illustrates current constraints and the reach on ξ at LHC and CLIC. With the double-Higgs production included, the estimated reach at CLIC is around ξ=0.002 corresponding to the scale of compositeness of 70 TeV. For comparison, the estimates for HL-LHC reach on the scale of compositeness f, with an integrated luminosity of 3000 fb^{-1}, are $f \leq$1.2 TeV at 95% CL [6].

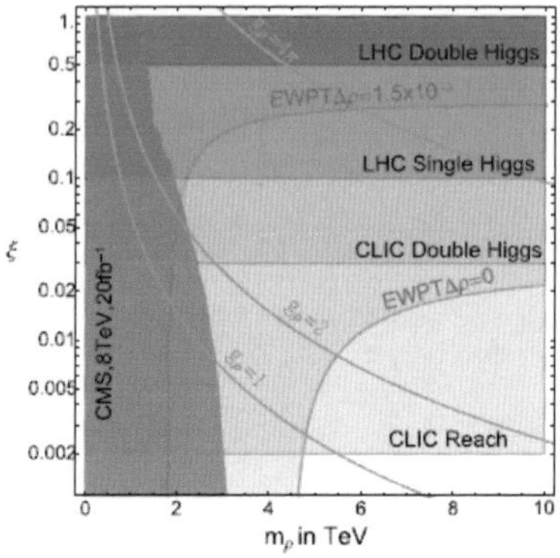

Figure 6: Summary plot on the current constraints and prospects on Higgs compositeness probe. Correction factor to the strength of the Higgs interactions ξ is given as function of the mass of the vector resonance of the composite theory.

6 Summary

The CLIC accelerator is an attractive option for a future e^+e^- collider, with the demonstrated feasibility through the extensive simulation and prototyping, comprehending physics and accelerator and detector R&D [1]. Staged implementation results in a broad physics program, from precision studies of Higgs physics to BSM probes. The most of Higgs couplings can be probed at a percent or sub-percent level. Rare Higgs decays to the lightest particles such as muons and photons can be accessed with a statistical precision of several percent, while the Higgs self-coupling can be determined with 12% uncertainty at the highest energy. Composite Higgs can be indirectly probed up to the scale of 70 TeV. A scan of the top mass at the production threshold allows the top mass to be determinated with a statistical uncertainty smaller than the theoretical error. Employment of beam polarisation enables determination of the top Yukawa coupling below 4%. BSM physics can be accessed through direct and indirect searches. The first ones give access to SUSY masses at a percent level, in most of the available scenarios, while the indirect searches extend the energy scale far beyond the kinematic reach of the machine (as in the case of extended gauge theories, theories with extra-dimensions, etc.) The foreseen physics program at CLIC extends and complements the physics studies planned for LHC and HL-LHC.

Acknowledgments

Vinca Institute activity at CLIC/CLICdp has been supported by the Ministry of Education, Science and Technological Development of the Republic of Serbia, through the project OI171012.

References

[1] L. Linssen et al. (editors), Physics and Detectors at CLIC: CLIC Conceptual Design Report, ANL-HEP-TR-12-01, CERN-2012-003, DESY 12-008, KEK Report 2011-7, CERN, 2012.
[2] S. Lukic, I. Bozovic Jelisavcic, M. Pandurovic, I. Smiljanic, JINST **8** P05008, (2013).
[3] H. Aihara et al. (editors), SiD Letter of Intent, SLAC-R-989, SLAC, 2009.
[4] T. Abe et al., The International Large Detector: Letter of Intent, DESY 2009-87, DESY, 2010.
[5] H. Abramowicz et al., Physics at the CLIC e+e- Linear Collider Input to the Snowmass Process 2013, 2013, arXiv:1307.5288.
[6] P. Glaysher, Proceedings of the 17th Lomonosov Conference on Elementary Particle Physics, August 20–26, 2015, Moscow, Russia, Ed. by A. Studenikin, World Scientific, Singapore.

PHYSICS AT THE FCC-ee

David d'Enterria[a]
CERN, EP Department, 1211 Geneva, Switzerland

Abstract. The physics program accessible in e^+e^- collisions at the Future Circular Collider (FCC-ee) is summarized. The FCC-ee aims at collecting multi-ab^{-1} integrated luminosities in e^+e^- at $\sqrt{s}$ = 90, 160, 240, and 350 GeV, yielding 10^{12} Z bosons, 10^8 W^+W^- pairs, 10^6 Higgs bosons and $4 \cdot 10^5$ top-quark pairs per year. Such huge data samples combined with a $\mathcal{O}$(100 keV) c.m. energy uncertainty will allow for Standard Model measurements with unparalleled precision and searches for new physics in regions not probed so far. The FCC-ee will be able to (i) indirectly discover new particles coupling to the Higgs and electroweak bosons up to scales $\Lambda \approx 7$ and 100 TeV; (ii) perform competitive SUSY tests at the loop level in regions beyond the LHC reach; and (iii) achieve the best potential in direct collider searches for dark matter and sterile neutrinos with masses below 60 GeV.

1 Introduction

Today, the understanding of particle physics is theoretically encoded in the Standard Model (SM), a renormalizable quantum field theory –unifying quantum mechanics and special relativity– that describes the fundamental interactions (except gravity) via a local SU(3)× SU(2)×U(1) gauge-symmetry group. The three gauge-symmetry terms give rise to the strong, weak and electromagnetic forces, while the particles fall into different representations of these symmetry groups. The SM Lagrangian (without neutrino masses) contains 19 free parameters to be determined experimentally: 3 gauge couplings, 9 Higgs–fermion Yukawa couplings, 3 mixing-angles, 2 Charge-Parity (CP) phases, and 2 Higgs boson couplings. Despite its tremendous success to accurately and consistently describe all phenomena observed at particle accelerators so far –including the recent experimental confirmation of the existence of its last missing piece, the Higgs boson– the theory is not complete and has several outstanding open questions to solve:

1. Dark matter (DM): The SM describes only ~4% of the universe energy budget, the rest being in unknown DM (and dark energy) forms, pointing to the existence of new massive particles (such as e.g. SUSY partners, axions, heavy ν's...).

2. Flavour problem: The huge dominance of matter over antimatter in the universe cannot be explained by the known SM sources of CP violation. More generally, the SM fails to explain the rationale behind the observed pattern of fermion families masses and flavour mixings.

[a]E-mail: dde@cern.ch

3. Neutrino masses: The generation of non-zero neutrino masses, called for by the observation of their flavour oscillations, is beyond the SM (BSM) and requires new particles such as right-handed ν's.

4. Hierarchy (or "fine tuning", "naturalness") problem: Quadratically - divergent virtual SM corrections affect the running of the Higgs boson mass between the widely separated electroweak and Planck scales, calling for new (e.g. SUSY) particles to stabilize such "untamed" quantum corrections.

5. Strong CP problem: The absence of CP-violation in QCD is naturally explained by postulating a (Peccei-Quinn) symmetry that gives rise to a new particle (axion).

6. Other fundamental SM issues include the current inability to explain dark energy, the cosmological constant, the origin of inflation, or gauge-gravity unification.

Many (or all) such fundamental questions may likely *not* be fully answered through the study of p-p collisions at the LHC. Despite their lower center-of-mass energies, e^+e^- colliders feature several advantages compared to their hadronic counterparts in terms of new physics studies: (i) direct model - independent searches for new particles coupling to Z^*/γ^* with masses up to $m \approx \sqrt{s}/2$; (ii) signals and backgrounds with very precisely known theoretical QCD and electroweak corrections with uncertainties (well) below 1% in many cases; and (iii) very clean experimental environment with final states free of "holes" or "blind spots" typical of p-p searches involving new particles with difficult decay modes (e.g. in certain SUSY scenarios with soft final-states and/or without missing transverse energy). Thus, combined with high-luminosities, an e^+e^- collider can provide access to studies with permil-level precision δX, allowing one to place indirect constraints on BSM physics – appearing as virtual corrections to well-controlled processes– up to very-high energies $\Lambda \propto (1\text{ TeV})/\sqrt{\delta X}$. Plans exist to build future circular (FCC-ee [1], CEPC [2]) and/or linear (ILC [3], CLIC [4]) e^+e^- colliders (Fig. 1).

The advantages of circular over linear e^+e^- machines are (i) much larger luminosity below $\sqrt{s} \approx 400$ GeV (thanks to much higher collision rates, adding continuous top-up injection to compensate for luminosity burnoff); (ii) the possibility to have several interaction points (IPs); and (iii) a very precise measurement of the beam energy $E_{\rm beam}$ through resonant transverse depolarization [5]. Linear e^+e^- colliders, on the other hand, feature (i) much larger $\sqrt{s}$ reach (circular colliders are not competitive above $\sqrt{s} \approx 400$ GeV due to synchroton radiation scaling as $E^4_{\rm beam}/R$), and (ii) easier longitudinal beam polarization. At the FCC (with a radius R = 80–100 km), e^+e^- collisions

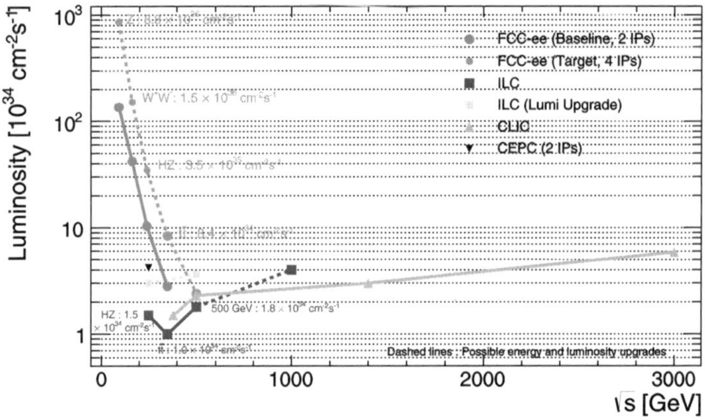

Figure 1: Target luminosities as a function of center-of-mass energy, in the range $\sqrt{s} \approx 90$–3000 GeV, for future circular (FCC-ee, CEPC) and linear (ILC, CLIC) e^+e^- colliders.

present clear advantages with respect to the older LEP ($\times 10^4$ more bunches, and $\delta E_{\rm beam} = \pm 0.1$ MeV compared to ± 2 MeV) and the ILC (crab-waist optics scheme, up to 4 IPs) yielding luminosities $\times(10^4$–$10)$ larger than both machines in the $\sqrt{s} = 90$–350 GeV range [6]. Table 1 lists the target FCC-ee luminosities and associated total number of events at each $\sqrt{s}$. They have been obtained for the relevant cross sections including initial state radiation and smearings due to beam-energy spreads: $\sigma_{\rm e^+e^-\to Z} = 43$ nb, $\sigma_{\rm e^+e^-\to H} = 0.29$ fb, $\sigma_{\rm e^+e^-\to W^+W^-} = 4$ pb, $\sigma_{\rm e^+e^-\to HZ} = 200$ fb, $\sigma_{\rm e^+e^-\to t\bar{t}} = 0.5$ pb and $\sigma_{\rm VV\to H} = 30$ fb. The completion of the FCC-ee core physics program (described in the next sections) requires $\sim$10 years of running.

$\sqrt{s}$ (GeV):	90 (Z)	125 (eeH)	160 (WW)	240 (HZ)	350 ($t\bar{t}$)	350 (WW→H)
$\mathscr{L}$/IP (cm^{-2} s^{-1})	$2.2{\cdot}10^{36}$	$1.1{\cdot}10^{36}$	$3.8{\cdot}10^{35}$	$8.7{\cdot}10^{34}$	$2.1{\cdot}10^{34}$	$2.1{\cdot}10^{34}$
$\mathscr{L}_{\rm int}$ (ab^{-1}/yr/IP)	22	11	3.8	0.87	0.21	0.21
Events/year (4 IPs)	$3.7{\cdot}10^{12}$	$1.2{\cdot}10^{4}$	$6.1{\cdot}10^{7}$	$7.0{\cdot}10^{5}$	$4.2{\cdot}10^{5}$	$2.5{\cdot}10^{4}$
Years needed (4 IPs)	2.5	1.5	1	3	0.5	3

Table 1: Target luminosities, events/yr, and years needed to complete the W, Z, H and top programs at FCC-ee. $\mathscr{L} = 10^{35}$ cm^{-2} s^{-1} corresponds to $\mathscr{L}_{\rm int} = 1$ ab^{-1}/yr for 1 yr $= 10^7$ s.

2 Indirect constraints on BSM via high-precision W, Z, top physics

Among the main goals of the FCC-ee is to collect multi-ab^{-1} at $\sqrt{s} \approx 91$ GeV (Z pole), 160 GeV (WW threshold), and 350 GeV ($t\bar{t}$ threshold) in order to measure key properties of the W and Z bosons and the top-quark, as well as other fundamental SM parameters, with unprecedented precision. The combination of huge data samples available at each $\sqrt{s}$ and the exquisite control of the c.m. energy (at the ±100 keV level) leading to very accurate energy threshold scans, allows the experimental precision of many SM parameters to be improved by a factor better than 25 with respect to the current state of the art (Table 2) [7]. Some FCC-ee experimental precision targets are ±100 keV for m_Z, ±500 keV for m_W, ±10 MeV for m_t, a relative statistical uncertainty of the order of $3 \cdot 10^{-5}$ for the QED α coupling (through $e^+e^- \to \mu^+\mu^-$ forward-backward asymmetries above and below the Z peak) [8], one-permil for the QCD coupling α_s (through hadronic Z and W decays) [9], and one-permil on the electroweak top couplings $F^{\gamma t, Z t}_{1V,2V,1A}$ (through angular distributions in $e^+e^- \to t\bar{t} \to \ell\nu q\bar{q}b\bar{b}$) [10]. In many cases, the dominant uncertainty will be of theoretical origin, and developments in the calculations are needed in order to match the expected experimental uncertainty.

Observable	Measurement	Current precision	FCC-ee stat.	Possible syst.	Challenge
m_Z (MeV)	Z lineshape	91187.5 ± 2.1	0.005	< 0.1	QED corr.
Γ_Z (MeV)	Z lineshape	2495.2 ± 2.3	0.008	< 0.1	QED corr.
R_ℓ	Z peak	20.767 ± 0.025	0.0001	< 0.001	QED corr.
R_b	Z peak	0.21629 ± 0.00066	0.000003	< 0.00006	$g \to b\bar{b}$
N_ν	Z peak	2.984 ± 0.008	0.00004	0.004	Lumi meas.
N_ν	$e^+e^- \to \gamma Z(\text{inv.})$	2.92 ± 0.05	0.0008	< 0.001	–
$A^{\mu\mu}_{FB}$	Z peak	0.0171 ± 0.0010	0.000004	< 0.00001	E_{beam} meas.
$\alpha_s(m_Z)$	$R_\ell, \sigma_{had}, \Gamma_Z$	0.1190 ± 0.0025	0.000001	0.00015	New physics
$1/\alpha_{QED}(m_Z)$	$A^{\mu\mu}_{FB}$ around Z peak	128.952 ± 0.014	0.004	0.002	EW corr.
m_W (MeV)	WW threshold scan	80385 ± 15	0.3	< 1	QED corr.
$\alpha_s(m_W)$	Γ_W, B^W_{had}	$B^W_{had} = 67.41 \pm 0.27$	0.00018	0.00015	CKM matrix
m_t (MeV)	$t\bar{t}$ threshold scan	173200 ± 900	10	10	QCD
Γ_t (MeV)	$t\bar{t}$ threshold scan	1410^{+290}_{-150}	12	?	$\alpha_s(m_Z)$
y_t	$t\bar{t}$ threshold scan	$\mu = 2.5 \pm 1.05$	13%	?	$\alpha_s(m_Z)$
$F^{\gamma t, Z t}_{1V,2V,1A}$	$d\sigma^{t\bar{t}}/dx\, d\cos(\theta)$	4%–20% (LHC-14 TeV)	(0.1–2.2)%	(0.01–100)%	–

Table 2: Examples of achievable precisions in representative Z, W and top measurements.

None of these measurements can be carried out at the LHC (or other e^+e^- machines) with such a level of precision. Physics beyond the SM can thereby be indirectly probed through loop corrections induced by possible new heavy particles [11]. Figure 2 shows limits on the W-mass vs. top-mass plane (left), and on the subset of leptonic dimension-6 operators of a model-independent Effective Field Theory of the SM parametrizing possible new physics (right) [12]. Such measurements impose unrivaled constraints on new weakly-coupled physics. Whereas LEP electroweak precision tests (EWPT) bound any BSM physics at scales $\Lambda \gtrsim 7$ TeV, FCC-ee would reach up to $\Lambda \approx 100$ TeV for some operators.

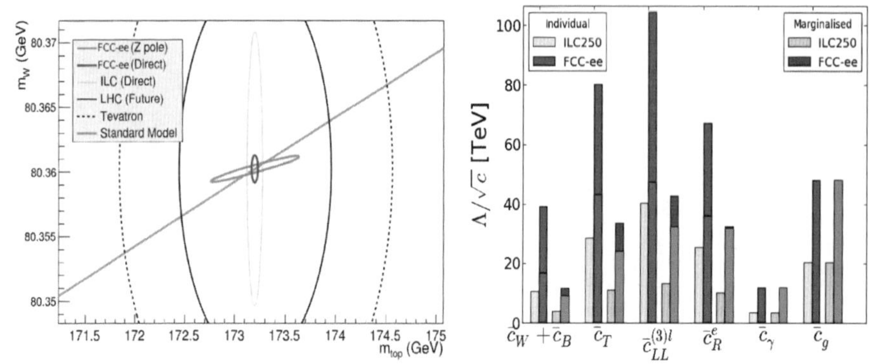

Figure 2: Left: Comparisons of 68% C.L. limits in the m_t–m_W plane at various colliders [1]. Right: 95% C.L. limits at FCC-ee(250 GeV) for leptonic operators sensitive to EWPT [12].

3 Indirect constraints on BSM via high-precision Higgs physics

The Higgs sector of the SM can be probed with a unique precision with a high-luminosity lepton collider. In the range of FCC-ee energies, Higgs production peaks at $\sqrt{s} \approx 240$ GeV dominated by Higgsstrahlung ($e^+e^- \to$ HZ), with some sensitivity at $\sqrt{s} = 350$ GeV also to vector-boson-fusion ($VV \to$ H $e^+e^-, \nu\nu$) and the top Yukawa coupling y_t (via $e^+e^- \to t\bar{t}$ with a virtual Higgs exchanged among the top quarks). The target total number of Higgs produced at the FCC-ee (4 IPs combined, all years) amounts to 2.1 million at 240 GeV, 75 000 in $VV \to$ H at 350 GeV, and 19 000 in s-channel $e^+e^- \to$ H at $\sqrt{s} = 125$ GeV (Table 1). Unique Higgs physics topics are open to study with such large data samples:

- High-precision model-independent determination of the Higgs couplings, total width, and exotic and invisible decays (Fig. 3 right) [1].
- Higgs self-coupling through loop corrections in HZ production [13].
- First-generation Yukawa couplings: (u,d,s) via exclusive decays H $\to$ Vγ (V $= \rho, \omega, \phi$) [14], and electron via resonant $e^+e^- \to$ H at $\sqrt{s} = m_H$ [15].

The recoil mass method in $e^+e^- \to$ HZ is unique to lepton colliders, and allows for an accurate tagging of Higgs events (Fig. 3, left) irrespective of their decay mode. It provides, in particular, a high-precision ($\pm 0.4\%$) measurement of $\sigma_{e^+e^- \to HZ}$ and, therefore, of g^2_{HZ}. From the measured value of $\sigma_{e^+e^- \to H(XX)Z} \propto \Gamma_{H \to XX}$ and the different known decays fractions $\Gamma_{H \to XX}$, one can then obtain the total Higgs boson width with $\mathcal{O}(1\%)$ uncertainty combining different final states. The H Z($\ell^+\ell^-$) final state can be used to directly measure the invisible decay width of the Higgs boson in events where its decay products escape undetected, by analyzing the distribution of the mass recoiling against the lepton pair. The Higgs boson invisible branching fraction can be measured with an absolute precision of 0.2%. If not observed, a 0.5% upper limit (95% C.L.) can be set on this branching ratio [1].

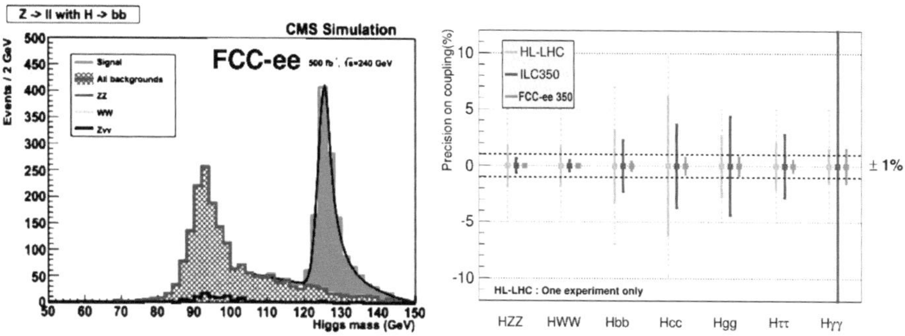

Figure 3: Left: Distribution of recoil mass against $Z \to \ell\ell$ in the $e^+e^- \to$ HZ process with H$\to b\bar{b}$. Right: Comparison of expected relative uncertainties for the Higgs boson couplings at future facilities [1].

In addition, loop corrections to the Higgsstrahlung cross sections at different center-of-mass energies are sensitive to the Higgs self-coupling. The effect is tiny but visible at FCC-ee thanks to the extreme precision achievable on the g_{HZ} coupling. Indirect (model-dependent) limits on the trilinear g_{HHH} can be set with a $\mathcal{O}(70\%)$ uncertainty, comparable to that expected at HL-LHC [13]. The large Higgs data samples available also open up the study of exotic (e.g. flavour-violating Higgs) and very rare SM decays. The Higgs couplings to first- and second-generation fermions, which may reveal new dynamics on the flavour structure of the SM [16], can be accessed via exclusive decays $\mathrm{H} \to \mathrm{V}\gamma$, for $V = \rho, \omega, \phi$, with sensitivity to the u,d,s quark Yukawas [14]. The $\mathrm{H} \to \rho\gamma$ channel appears the most promising with $\mathcal{O}(50)$ events expected. The low mass of the electron translates into a tiny $\mathrm{H} \to \mathrm{e}^+\mathrm{e}^-$ branching ratio $\mathrm{BR}_{\mathrm{e}^+\mathrm{e}^-} = 5 \cdot 10^{-9}$ which precludes any experimental observation of this decay mode and, thereby a determination of the electron Yukawa coupling. The resonant s-channel production, despite its small cross section [17], is not completely hopeless and preliminary studies indicate that could be observed at the 3σ-level with 90 ab^{-1} at $\sqrt{\mathrm{s}} = 125$ GeV with a c.m. energy spread commensurate with the Higgs boson width itself ($\approx$4 MeV, requiring beam monochromatization) [15].

In summary, the FCC-ee provides the smallest uncertainties for the measurements of Higgs boson couplings to gauge bosons and fermions (Fig. 3, right). Since any deviation δg_{HXX}, relative to the SM value $g^{\mathrm{SM}}_{\mathrm{HXX}}$, can be approximately translated into BSM scale limits through the expression: $\Lambda \gtrsim (1\,\text{TeV})/\sqrt{(\delta g_{\text{HXX}}/g^{\text{SM}}_{\text{HXX}})/5\%}$. The expected 0.15% precision for the most precise coupling, g_{HZZ}, would thus set competitive bounds, $\Lambda \gtrsim 7$ TeV, on any new physics coupled to the scalar sector of the SM.

4 Indirect constraints on supersymmetry

Supersymmetry has many appealing features as a candidate framework for BSM physics, solving the naturalness problem (through light stop contributions to

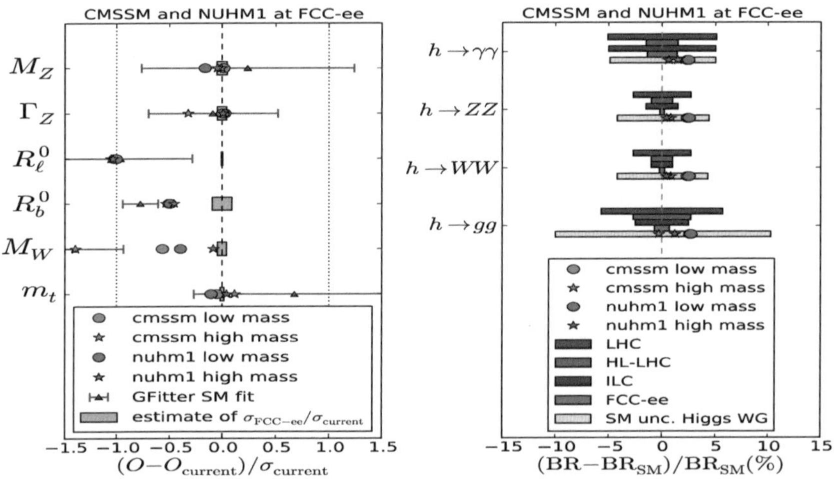

Figure 4: Comparison of estimated precision in EWPT (left) and Higgs branching ratios (right) measurements: Current results (error bars, left; green boxes, right), low- and high-mass best-fit CMSSM and NUHM1 points (circles and stars) and future prospects (bars) [18].

the running of the Higgs boson mass), stabilizing the electroweak vacuum, providing dark matter candidate(s) (the stable lightest SUSY particle, if R-parity is conserved), predicting grand unification of gauge interactions, and connecting to string theory where SUSY is preserved. Since stops and higgsinos carry electroweak quantum numbers, e^+e^- colliders have better sensitivity to those than to coloured sparticles (squarks, gluinos). The precision electroweak and top (Section 2) and Higgs (Section 3) studies at the FCC-ee not only impose generic constraints on new physics scales at multi-TeV energies, but can also be interpreted in terms of sensitivity to broad classes of SUSY models (such as the Constrained MSSM, Non-Universal Higgs Masses, or natural SUSY) [18, 19]. Figure 4 compares the precision on EWPT (left) and Higgs (right) observables at the LHC, ILC and FCC-ee with the deviations from their SM values expected for the low- and high-mass CMSSM and NUHM1 best-fit points [1, 18]. It is clear that the FCC-ee has the best ability to distinguish these models from the Standard Model. In particular, the high-mass CMSSM points in Fig. 4 provide an example of SUSY model which likely lies beyond the LHC reach, featuring narrow strips where stop-neutralino coannihilation is important, or focus-point strips at higher values of the ratio $m_0/m_{1/2}$. The EWPT and Higgs measurements at the FCC-ee would be able to probe both types of narrow parameter-space strips that extend to large sparticle masses, and indirectly determine CMSSM parameters also in such a pessimistic scenario [18].

5 Direct BSM physics constraints: dark matter & right-handed ν's

The impact of the FCC-ee goes beyond indirect BSM studies and has also a strong discovery potential in direct searches of other key BSM extensions such as dark matter (DM) [20] and right-handed neutrinos [21], by exploiting the possibility to measure very rare decays of the Z and H bosons into such

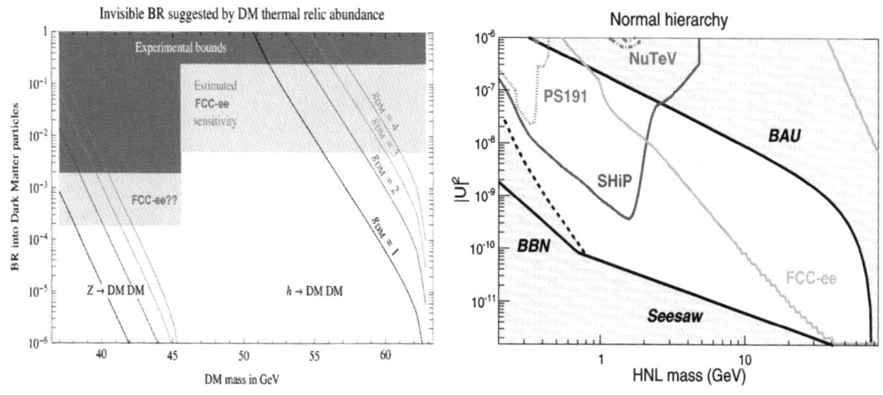

Figure 5: Regions of FCC-ee sensitivity for: (i) Rare Z and H decays into DM pairs in the $BR_{Z,H\to DM\,DM}$ vs. m_{DM} plane (left) [20], and (ii) sterile neutrinos as a function of their mass and mixing to light neutrinos (normal hierarchy) for 10^{13} Z decays (right) [21].

new particles. As a matter of fact, measurements of the invisible Z and H widths at the FCC-ee provide the best collider options to test DM lighter than $m_{Z,H}/2$ that couples via SM mediators. Figure 5 (left) shows the limits in the plane (branching ratio, DM mass) for the decays $Z, H \to DM\,DM$. Similarly, Fig. 5 (right) shows the unrivaled sensitivity of searches for high-mass sterile neutrinos via decays $Z \to N\nu_i$ (with $N \to W^*\ell, Z^*\nu_j$) as a function of their mass and mixing to light neutrinos (normal hierarchy) [21].

6 Summary

The FCC-ee has a unique program of searches for new physics via high-precision studies of the W, Z, and Higgs bosons, and the top quark with uncertainties at the permil level or below, thanks to the huge luminosities $\mathcal{O}(1-100)$ ab^{-1} (for 4 interaction points) and the exquisite control of the beam energy in the range $\sqrt{s} \approx 90$–350 GeV. By searching for tiny deviations with respect to the SM predictions in a rich set of measurements, BSM physics sensitivities reach scales as large as $\Lambda \approx 7, 100$ TeV for new particles coupling to the scalar and electroweak SM sectors, respectively. When interpreted in terms of broader classes of SUSY models, such precision electroweak and Higgs observables provide indirect tests of supersymmetry in regions which are beyond the LHC reach. Last but not least, the FCC-ee covers also direct BSM searches, through very rare and invisible Higgs and Z bosons decays which provide the best collider options to test dark matter and sterile neutrinos with masses up to $m_{\rm DM,HNL} \approx 60$ GeV.

References

[1] M. Bicer *et al.* [TLEP Design Study Working Group], JHEP **1401**, 164 (2014); [arXiv:1308.6176 [hep-ex]].

[2] M. Ahmad *et al.* [CEPC-SPPC Study Group], IHEP-CEPC-DR-2015-01, IHEP-TH-2015-01, HEP-EP-2015-01 (2015).

[3] T. Behnke *et al.* [ILC Collaboration], arXiv:1306.6327 [physics.acc-ph].

[4] L. Linssen, A. Miyamoto, M. Stanitzki and H. Weerts, doi:10.5170/CERN-2012-003; [arXiv:1202.5940 [physics.ins-det]].

[5] M. Koratzinos, A. Blondel, E. Gianfelice-Wendt and F. Zimmermann, arXiv:1506.00933 [physics.acc-ph].

[6] F. Zimmermann, 1st FCC annual workshop, Washington, March 2015; `https://indico.cern.ch/event/340703/session/109/contribution/190/`

[7] R. Tenchini, arXiv:1412.2928 [hep-ex].

[8] P. Janot, JHEP **1602**, 053 (2015); arXiv:1512.05544 [hep-ph].

[9] D. d'Enterria *et al.*, arXiv:1512.05194 [hep-ph].

[10] P. Janot, JHEP **1504**, 182 (2015); [arXiv:1503.01325 [hep-ph]].

[11] J. Fan, M. Reece and L. T. Wang, JHEP **1509**, 196 (2015); [arXiv:1411.1054 [hep-ph]].

[12] J. Ellis and T. You, arXiv:1510.04561 [hep-ph].

[13] M. McCullough, Phys. Rev. D **90**, 015001 (2014) arXiv:1312.3322.

[14] A. L. Kagan, G. Perez, F. Petriello, Y. Soreq, S. Stoynev and J. Zupan, Phys. Rev. Lett. **114**, 101802 (2015); [arXiv:1406.1722 [hep-ph]].

[15] D. d'Enterria, G. Wojcik, R. Aleksan, 8th FCC-ee Physics Workshop, LPNHE-Paris, Oct. 2014, `https://indico.cern.ch/event/337673/session/6/contribution/20`

[16] D. Ghosh, R. S. Gupta and G. Perez, arXiv:1508.01501 [hep-ph].

[17] S. Jadach and R. A. Kycia, Phys. Lett. B **755** 58 (2016); arXiv:1509.02406 [hep-ph].

[18] O. Buchmueller, M. Citron, J. Ellis, *et al.*, Eur. Phys. J. C **75**, 469 (2015); [arXiv:1505.04702 [hep-ph]].

[19] J. Fan, M. Reece and L. T. Wang, JHEP **1508**, 152 (2015); [arXiv:1412.3107 [hep-ph]].

[20] A. De Simone, G. F. Giudice and A. Strumia, JHEP **1406**, 081 (2014) [arXiv:1402.6287 [hep-ph]]; A. Strumia, 9th FCC-ee Physics Workshop, SNS-Pisa, Feb. 2015,
`http://www.sns.it/eventi/9th-fcc-ee-physics-workshop`

[21] A. Blondel *et al.* [FCC-ee study Team Collaboration], arXiv:1411.5230.

FREE ELECTRON LASERS: PROGRAMS AND CHALLENGES

Giuseppe Dattoli [a], Federico Nguyen [b], Ivan P. Spassovsky [c]
ENEA-Nuclear Fusion Department, Via E. Fermi 45 – 00044 Frascati (Roma), Italy

Abstract. Free Electron Laser devices have provided a breakthrough within the framework of the sources of coherent radiation. They are the result of multiple technological achievements, including the development of accelerators able to provide high brightness electron beam. We describe the main features of these revolutionary devices and address some potential novel applications.

1 Introduction

The Free Electron Laser (FEL) can be considered a laser, even though the underlying emission process does not occur in an atomic or a molecular system, with population inversion, but it is provided by a relativistic electron beam, passing through the magnetic field of an undulator [1–5]. Relativistic particles moving in curved path lose energy emitting synchrotron radiation. An intrinsic limiting factor to accelerate light particles at high energy with circular machines, synchrotron radiation turned to have an increasingly important role in many areas of basic and applied sciences. It provides, indeed, high intensity flux of photons on a broad band of frequencies, which cannot be covered with conventional sources.

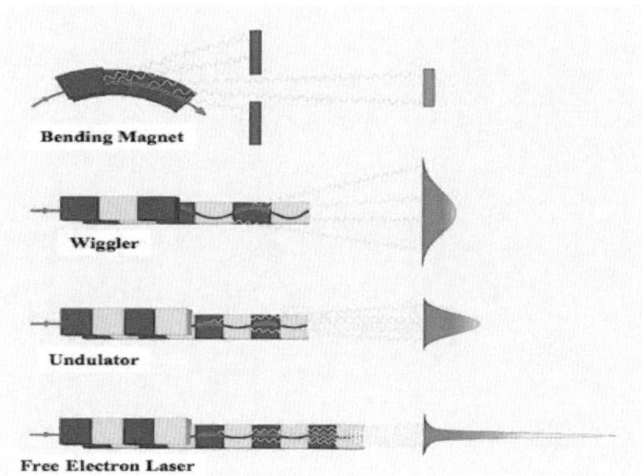

Figure 1: Different schemes used to produce synchrotron radiation.

Figure 1 summarizes different schemes through which synchrotron radiation is produced. In bending magnets, electrons move along a circular path and

[a]E-mail: giuseppe.dattoli@enea.it
[b]E-mail: federico.nguyen@enea.it
[c]E-mail: ivan.spassovsky@enea.it

produce a smooth spectrum. The wiggler is a sequence of $2N$ bending magnets with alternate polarities. The outgoing relevant radiation is not different from that of a bending magnet, aside for the N-fold enhancement of the intensity. The undulator is an N-period structure with $2N$ alternating poles, where the electrons perform small transverse periodic oscillations. Radiation emitted in any point of the undulator interferes constructively and the result is that it is squeezed into a discrete spectrum.

In a FEL, the kinetic energy of a relativistic electron beam is transformed into an intense beam of electromagnetic radiation by wiggling the electrons transversely in a long undulator. FELs are getting the main source of tunable, intense and coherent photons of either ultra-short time resolution or ultra-fine energy resolution, from the far infrared (IR) to the hard X-ray regime. Conventional laser light sources are indeed limited to spectroscopic techniques in the infrared, visible and near ultraviolet (UV) range excluding all the measurements in need for photons of energy larger than a few eV.

2 FEL Basic Elements

A charged particle beam is associated to a power equal to the product of its energy with its current P(MW) = I(A)E(MeV). A 20 MeV beam with a current of 2 A has a power of 40 MW. Such power is delivered to the beam by the accelerator. From now on, we will refer to linear accelerators where the acceleration is performed in radiofrequency cavities. The electron beam power corresponds in the case of free electron lasers to the pump power in the case of conventional lasers. Free electrons passing through magnetic structures like in Fig. 1 produce flashes of radiation. Synchrotron light sources rely on this process. We consider the case of a magnetic field associated to an undulator:

$$\vec{B} = \left(0, B_0 \sin\frac{2\pi z}{\lambda_u}, 0\right) \tag{1}$$

such field is oscillating in the vertical direction with a peak value B_0 and periodicity λ_u. It can be obtained with two series of magnets with alternate N-S orientation. The Lorentz force due to the undulator field introduces a transverse component in the electron motion, initially only longitudinal. After one undulator period, the average transverse and longitudinal velocities are:

$$\beta_T \simeq \frac{K}{\sqrt{2}\,\gamma}, \quad \beta_L \simeq 1 - \frac{1}{2\gamma^{*2}} \tag{2}$$

where the undulator parameter K is introduced to take into account the effect of the transverse motion on the longitudinal velocity and γ^* is an average Lorentz factor over the period λ_u:

$$\gamma^* \simeq \frac{\gamma}{\sqrt{1+\frac{K^2}{2}}}, \quad K = \frac{eB_0\lambda_u}{2\pi m_e c^2}$$

The electrons moving in the undulator emit radiation and this process can be viewed as a sort of spontaneous emission. After one undulator period, the emitted radiation has slept ahead of the electron beam by a quantity

$$\delta \simeq (1-\beta_L)\lambda_u = \frac{\lambda_u}{2\gamma^2}\left(1+\frac{K^2}{2}\right) \tag{3}$$

Since δ represents a phase advance of the electromagnetic wave, constructive interference with the radiation emitted at the next undulator period, requires the condition $\delta = n\lambda$ n being an integer number. From these considerations, the definition of the undulator radiation wavelength goes as:

$$\lambda_n = \frac{\lambda_u}{2n\gamma^2}\left(1+\frac{K^2}{2}\right) \tag{4}$$

where $n = 1$ represents the fundamental wavelength, while $n > 1$ is related to the higher harmonics. The radiation pulse after N undulator periods has a structure of a step function with a length $N\delta$, corresponding to a pulse duration $\Delta\tau \sim N\delta/c$. The spectral distribution is given by the Fourier transform of this pulse and the spectral width $\Delta\omega$ results defined from the Parseval relation $\Delta\tau\Delta\omega \simeq \pi$, and we obtain the relative bandwidth

$$\frac{\Delta\omega}{\omega} = \frac{1}{2N} \tag{5}$$

The spectral intensity profile of radiation emitted by a single undulator is just:

$$f(\nu) \propto \left[\frac{\sin\frac{\nu}{2}}{\frac{\nu}{2}}\right]^2, \quad \nu = 2\pi N\frac{\omega-\omega_0}{\omega_0} \tag{6}$$

where ν is the dimensionless detuning parameter, used to define the spectral profile with respect to the central reference wavelength. The FEL interaction between electrons and radiation is a combination of two competing effects: electron energy loss with amplification of the photon energy, and absorption of stimulating photons with subsequent energy transfer to the electrons. If these effects of stimulated emission and absorption are independent, the gain $G(\nu) \propto f(\nu_+) - f(\nu_-)$ describes the stimulating field net energy increase.

FELs can be arranged [6] in three different modes, shown in Fig. 2: the self-amplified spontaneous emission (SASE), the seeded high-gain amplifier, and the oscillator. In the first two modes, the FELs operate in the high gain regime while the oscillators operate in the low gain regime.

3 High Gain Free Electron Lasers

In the SASE mode of operation, the electron beam amplifies the spontaneous undulator radiation emitted in the early portion of the undulator. While in

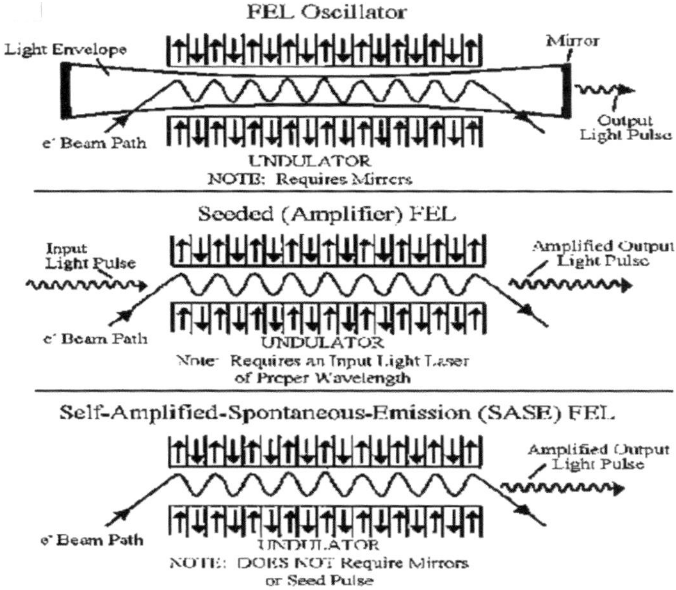

Figure 2: Different FEL modes of operation (courtesy of Z. Huang).

the seeded high gain amplifier mode, the electron beam amplifies the externally injected seed radiation.

As the electrons and the radiation co-propagate through the undulator, the radiation slips through the electron beam in accordance with the resonance condition, so that the radiation intensity grows exponentially along the undulator. The scaling behaviour of a high gain FEL amplifier is characterized by the Pierce parameter ρ:

$$\rho \propto \sqrt[3]{\frac{I_{peak}\lambda_u^2 K^2}{I_A\, \varepsilon\, \beta}} \tag{7}$$

where I_{peak} is the peak current, $I_A \simeq 17$ kA is the Alfvén current, and the electron beam envelope is characterized by the emittance ε and the betatron function β. The radiation intensity grows as e^{z/L_G} and the gain length L_G is:

$$L_G = \frac{\lambda_u}{4\pi\sqrt{3}\,\rho} \tag{8}$$

Figure 3 shows the main steps in high gain operating FELs. Similar to the eigenmodes of the free-space propagation of a radiation field, there is an infinite number of FEL eigenmodes. Each couples independently to the electron beam and has different growth rates or gain lengths. The mode with the largest

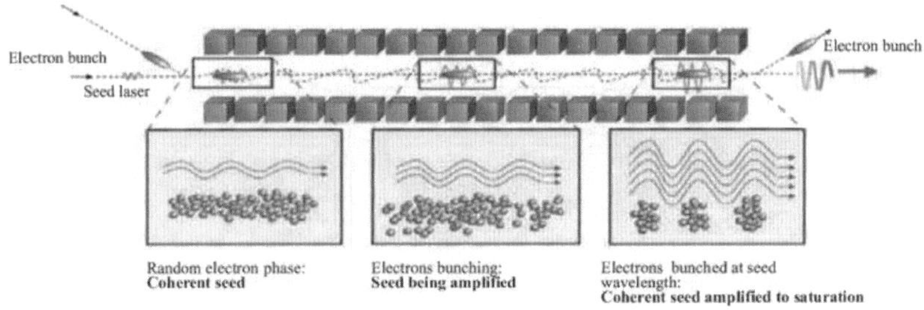

Figure 3: High gain seeded mechanism of FEL operation.

growth rate dominates after several gain lengths and the radiation field becomes transversely coherent. At saturation, gain guiding vanishes and the electron beam radiates into multiple modes. Typically, the fundamental FEL eigenmode is similar to that of free-space propagation and the resulting reduction in transverse coherence at saturation is negligible. At saturation, the bandwidth of the FEL radiation almost coincides with the energy spread of the electron beam:

$$\frac{\delta\omega}{\omega} \approx \frac{\sigma_E}{E} \approx \rho \tag{9}$$

SASE has been important in the development of hard X-ray FELs because amplifiers and oscillators are difficult in this spectral region due to the absence of appropriate seed radiation for the former and of suitable high-reflectivity mirrors for the latter. With the advances in FEL and laser harmonic generation techniques, the soft X-ray region can be reached with the help of a seeded harmonic generation amplifier scheme.

4 Low Gain FELs: Oscillators

In an oscillator, one of several optical pulses are trapped within an optical cavity, and the gain occurs each time the trapped optical pulse meets and travels together with an electron bunch through an undulator located within the optical cavity. FEL oscillators are feasible from the IR to hard X-ray regions by employing suitable reflectors for the cavity, which could be multilayer dielectrics, metals at grazing incidence, or Bragg crystals as appropriate for the wavelength region in which the FEL operates. Figure 4 shows the operation scheme of an oscillator FEL. The main features are similar to those of a conventional oscillator Laser: the spontaneous emission process (from the electron beam in an optical cavity) and the amplification of this radiation throughout a stimulated emission mechanism. During the interaction between electrons and radiation, the electron beam gets modulated in energy which results in space

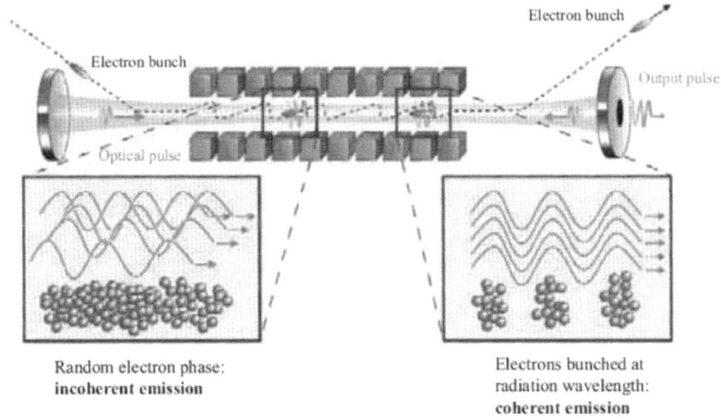

Figure 4: Oscillator mode of FEL operation.

modulation – bunching – at the radiation wavelength. This bunching allows the production of coherent radiation.

In this case, the intracavity radiation pulse increases exponentially with a concomitant reduction in gain. The system will reach saturation when $[1 + G(I)]\ R \approx 1$, where $G(I)$ is the intensity-dependent gain and R is the total round-trip reflectivity.

Thanks to high reflectivity mirrors and to the coherent high harmonics generation, oscillator FEL operation gets possible in the vacuum ultraviolet region, namely at $\mathcal{O}(100\text{ nm})$ order of magnitude.

5 Prospects and Challenges

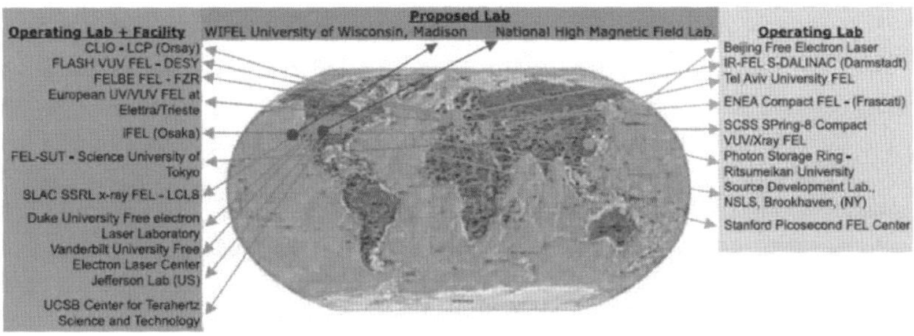

Figure 5: FEL sources worldwide scenario.

Figure 5 shows the worldwide scenario of FEL sources. The engineering issues in the design of FEL devices covers a mosaic of different technologies such as magnetic materials for undulators and high quality electron beams.

According to the specific goals of the FEL users, different aspects are studied with more accuracy. FEL devices operating in the oscillator configuration demand for a breakthrough in the technology of mirror manufacturing, able to confine high intensity radiation to a broad wavelength spectrum. At the same time, high performance accelerators are needed to provide beams with low angular dispersion, very good energy resolution and brightness as well as further progress is needed on undulators with high on-axis field and small period.

Furthermore, interest is rapidly growing towards the application of FEL beams in many areas of biomedical sciences. The challenge is to probe living matter with finer length, time and energy resolution, where biological systems can be observed at their femto-second time, nanometer distance and milli-eV energy scale. In view of this, the next generation of light sources should have high transverse and time coherence, pulse duration down to femtoseconds, high brightness and peak power, wide range and easy tuning of photon energy and polarization. FELs are unique in order to achieve these goals.

Acknowledgments

The Authors express their sincere appreciation to Prof. A. Studenikin for his invitation at the Lomonosov Conference. G. D. acknowledges the collaboration with Dr. K. Zhukovsky who provided new and interesting avenues in the analysis of the emission process in exotic undulators.

References

[1] C.A. Brau, *"Free-Electron Lasers"* (Academic Press, Oxford, 1990).
[2] G. Dattoli, A. Renieri, and A. Torre, *"Lectures on Free Electron Laser Theory and Related Topics"* (World Scientific, Singapore, 1993).
[3] E.L. Saldin, E.A. Schneidmiller, and M.V. Yurkov, *"The Physics of Free Electron Lasers"* (Springer-Verlag, Berlin, 1999).
[4] F. Ciocci, G. Dattoli, A. Torre, and A. Renieri, *"Insertion Devices for Synchrotron Radiation and Free Electron Laser"* (World Scientific, Singapore, 2000).
[5] G. Dattoli, M. Del Franco, M. Labat, P.L. Ottaviani, and S. Pagnutti, in *"Free Electron Lasers"*, Ed. by S. Varro (ISBN: 978-953-51-0279-3, InTech, 2012).
[6] W. A. Barletta *et al.*, Nucl. Instrum. Meth. A **618**, 69 (2010).

MEASUREMENTS OF THE PROPERTIES OF THE HIGGS BOSON USING THE ATLAS DETECTOR

Marco Filipuzzi [a]
DESY, Hamburg, Germany
On behalf of the ATLAS Collaboration

Abstract. A review of the Higgs coupling measurements and spin/CP tests with up to 25 fb^{-1} of $\sqrt{s}$ = 7 and 8 TeV proton-proton collision data collected by the ATLAS detector at the LHC in 2011 and 2012 is presented. No significant deviation from Standard Model expectations is observed.

1 Introduction

In July 2012, the ATLAS [1] and CMS Collaborations announced the discovery of a new particle [2, 3] at the Large Hadron Collider (LHC) at CERN in the context of the search for the Higgs boson of the Standard Model of Particle Physics (SM). Since then, with an increased dataset, improved detector understanding and refined analysis strategies, a wide range of properties of this new particle has been measured and tested for possible deviations from SM predictions. In these proceedings, tests of the spin and parity of the new particle performed by the ATLAS experiment will be presented in the first section, while measurements of the coupling strengths between the new particle and the gauge bosons and fermions will be presented in the second part.

2 Spin and parity of the Higgs boson in diboson decays

The study of the spin and parity properties presented in these proceeding is based on the $H \to \gamma\gamma$, $H \to ZZ^* \to 4l$ and $H \to WW^* \to e\nu\mu\nu$ decay channels and their combination [4].

In the SM, the Higgs boson is a CP-even scalar particle, $J^P = 0^+$. Theories of physics beyond the SM (BSM) often require an extended Higgs sector featuring several neutral Higgs bosons. Such cases may include CP-mixing in the Higgs boson interactions, which could result in observable differences in the kinematics of the final-state particles produced in their decays. Therefore the SM hypothesis $J^P = 0^+$ is compared to alternative spin-0 models: a pseudoscalar boson $J^P = 0^-$ and a BSM scalar boson $J^P = 0_h^+$. Models with $J^P = 2^+$ with universal and non-universal couplings [5,6] are also considered. In these tests of fixed spin and parity hypotheses it is assumed that the resonance decay involves only one CP eigenstate.

An effective field theory (EFT) approach is adopted to describe the interaction between the new particle and the SM vector bosons. The possible BSM scenarios considered for the spin and parity of the boson are: spin-2, pure BSM

[a]E-mail: marco.filipuzzi@cern.ch

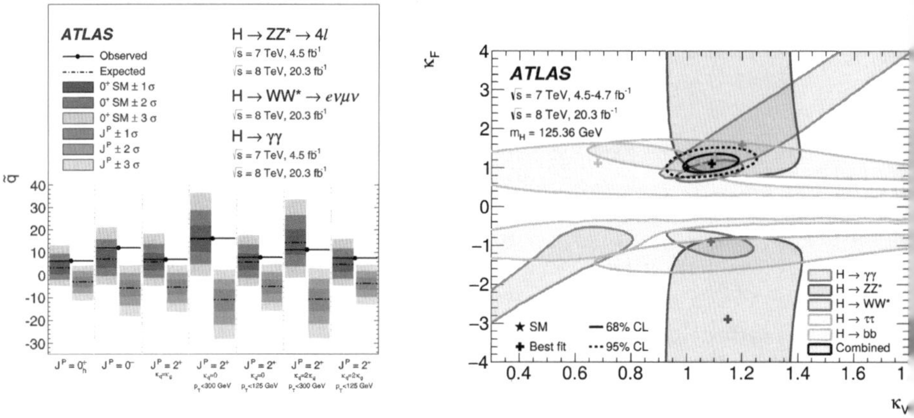

Figure 1: (left) Distributions of the test statistics $\tilde{q}$ for the SM Higgs boson and for the J^P alternative hypotheses [4]. (right) Results of fits for the two-parameter benchmark model that probes different coupling-strength scale factors for fermions and vector bosons, assuming only SM contributions to the total width: individual channels and their combination including 68% and 95% CL contours [8].

spin-0 CP-even or CP-odd Higgs boson and a mixture of the SM spin-0 state and a BSM spin-0 CP-even or CP-odd state.

The analyses rely on discriminant observables chosen to be sensitive to the spin and parity of the Higgs boson. In the $H \to \gamma\gamma$ channel the kinematic variables used in this analysis are the production angle of the two photons, measured in the Collins-Soper frame [7] and the diphoton transverse momentum $p_T^{\gamma\gamma}$. For the $H \to WW^*$ case a Boosted Decision Tree (BDT) algorithm based on m_{ll}, p_T^{ll}, $\Delta\phi_{ll}$ and m_T (see [4]), is used and trained for each specific hypothesis tested. Finally, for the $H \to ZZ^* \to 4l$ analysis the full kinematic information is available from the measured decay products.

The test statistic $\tilde{q}$ used to distinguish between two spin-parity hypotheses is based on the ratio between the profiled likelihoods evaluated under the SM and the alternative spin-parity hypothesis. All the hypotheses tested are summarized in Figure 1 (left). Expected ± 1, ± 2 and ± 3 σ regions of the test statistic are shown for the SM Higgs boson (blue) and for the alternative J^P hypotheses (red) together with the observed $\tilde{q}$ value (black points). All tested alternative hypotheses are rejected at more than 99% confidence level (CL) in favour of the SM hypothesis.

Limits are set on the admixture of a BSM spin-0 CP-even or a spin-0 CP-odd component to a SM-like spin-0 state as described in [4], and are compatible with the SM predictions.

3 Higgs boson production/decay rates and coupling strengths

This section presents the combined results from the analyses of the $H \to \gamma\gamma$, ZZ^*, WW^*, $\tau\tau$, $b\bar{b}$, $\mu\mu$ and $Z\gamma$ decay modes and of the constraints on $t\bar{t}H$ and off-shell Higgs boson production [8]. The general approach of the analysis is to divide events in distinct categories to maximize the sensitivity to different production modes and to perform a global fit to account for cross-contaminations between categories and uncertainty correlations between different channels.

For a measurement of Higgs boson coupling strengths, production and decay modes cannot be treated independently, as most observed processes involve at least two Higgs boson coupling strengths. Following the leading-order framework recommended in Ref. [5], the Higgs boson coupling-strength scale factors κ_j are measured for selected benchmark models. All scale factors are defined such that a value of $\kappa_j = 1$ corresponds to the best available SM prediction. Only modifications of the absolute values of the coupling strengths are considered, while the tensor structure of the couplings is assumed to be the same as in the SM, meaning that the observed state is assumed to be a CP-even scalar.

Under the assumptions that only SM particles contribute to the $gg \to H$, $H \to \gamma\gamma$, $H \to Z\gamma$ and $gg \to ZH$ loops and that no undetected or invisible Higgs boson decays exist, a benchmark coupling model that allows for different Higgs boson coupling strengths to fermions and bosons is considered. The coupling-strength scale factors are defined in this model as κ_F for all fermions and κ_V for all vector bosons. The results for this benchmark model are shown in Figure 1 (right). The measured values of κ_F and κ_V are compatible with their SM expectation and the two-dimensional compatibility of the SM hypothesis with the best-fit point is 41%. The best-fit values and uncertainties are:

$$\kappa_F = 1.11 \pm 0.16 \left[^{+0.12}_{-0.11}(\text{stat})^{+0.10}_{-0.09}(\text{syst})^{+0.06}_{-0.05}(\text{theo})\right]$$

$$\kappa_V = 1.09 \pm 0.07 \left[^{+0.05}_{-0.05}(\text{stat})^{+0.03}_{-0.03}(\text{syst})^{+0.04}_{-0.03}(\text{theo})\right].$$

Additional benchmark models testing fermion couplings, BSM contributions and a generic model are considered in Ref. [8]. Results for these measurements, performed for different assumptions, are compatible with the SM predictions.

References

[1] ATLAS Collaboration, JINST. **3**, S08003 (2008).
[2] ATLAS Collaboration, Phys. Lett. **B716**, 1-29 (2012).
[3] CMS Collaboration, Phys. Lett. **B716**, 30-61 (2012).
[4] ATLAS Collaboration, Eur. Phys. J. **C75**, (2015) 476.
[5] LHC Higgs Cross Section Working Group, CERN-2013-004 (2013).

[6] P. Artoisenet et al., JHEP **1311** (2013) 043.
[7] J. C. Collins and D. E. Soper, Phys. Rev. **D 16** (1977) 2219.
[8] ATLAS Collaboration, arXiv:1507.04548 [hep-ex] submitted to Eur. Phys. J (2015) .

HIGGS TO TAUTAU CHANNELS WITH CMS

Konstantin Androsov[a] on behalf of the CMS collaboration
University of Siena and INFN Pisa, Italy

Abstract. A brief overview of the searches for a neutral Higgs boson using $\tau\tau$ decay channels with events recorded by the CMS experiment in 2011 and 2012 is given. It includes three searches, using similar analysis techniques: the searches for the standard model Higgs boson and the MSSM Higgs boson decaying to a pair of τ, and the search for a neutral scalar heavy Higgs H decaying to a pair of 125 GeV Higgs bosons in the final state $bb\tau\tau$.

1 Introduction

After the discovery at the Large Hadron Collider (LHC) of a Higgs boson with a mass around 125 GeV [1, 2], direct measurements of the Higgs couplings to fermions should provide further verification of the standard model (SM) consistency. On the other hand, the discovered Higgs boson might belong to a Higgs sector predicted by models beyond the SM (BSM) and direct searches for an additional scalar boson can test BSM predictions.

The CMS detector [3] has good τ reconstruction and identification performance [4], which is exploited in analyses where a Higgs boson is decaying to a pair of τ. This report summarizes three analyses performed by CMS using p-p collisions data collected in 2011 and 2012 with a center of mass energy, respectively, of 7 and 8 TeV, corresponding to 4.9 and 19.7 fb^{-1}. These analyses are:

- Search for the SM H decaying into $\tau\tau$ [5] (SM $H \to \tau\tau$).
- Search for an additional neutral Higgs boson predicted by the Minimal Supersymmetric extension of the SM (MSSM) [6, 7] (MSSM $H \to \tau\tau$).
- Search for a heavy scalar boson H decaying to a pair of 125 GeV Higgs bosons hh, in the final states $hh \to bb\tau\tau$, predicted by the MSSM and the two-Higgs-doublet models (2HDM) [8] ($H \to hh \to bb\tau\tau$)[b].

These analyses exploit very similar techniques to reconstruct a Higgs decaying in $\tau\tau$ channels.

2 Event selection

A common baseline selection, depending on the final state, is used in all the three analyses. In the SM and MSSM $H \to \tau\tau$ searches, all possible di-τ final states are analyzed, but $\tau_e\tau_e$ final state[c] for the MSSM analysis. In the

[a]E-mail: `konstantin.androsov@cern.ch`

[b]In the $H \to hh \to bb\tau\tau$ analysis only data collected in 2012 is used.

[c]By τ_e and τ_μ the leptonic τ decays: $\tau^- \to l^-\bar{\nu}_l\nu_\tau$, where $l = e, \mu$ is an electron or a muon, or the charge conjugate state are indicated.

$H \to hh \to bb\tau\tau$ search, the three most sensitive final states are analyzed[d]: $bb\tau_e\tau_h$, $bb\tau_\mu\tau_h$ and $bb\tau_h\tau_h$.

The main backgrounds for these analyses are electroweak processes $Z \to ll$ ($l = e, \mu, \tau$) + jets, W + jets, di-boson and $t\bar{t}$, and the QCD multi-jets processes. Combinations of data-driven and MC background techniques are used to model shape and normalization of these backgrounds. In each analysis, to improve the signal sensitivity, selected events were split into orthogonal categories, where additional criteria are used, depending on category, to reduce the contributions from the main backgrounds.

The most discriminant criteria are [5, 7, 8]: a requirement on the transverse mass, $m_T \equiv \sqrt{2\left(|\boldsymbol{p}_T^l||\boldsymbol{E}_T^{miss}| - \boldsymbol{p}_T^l\boldsymbol{E}_T^{miss}\right)}$, of a lepton candidate is used to reduce W+jets background; a veto on the presence of the third lepton is used to reduce $Z \to ll$ background; a requirement on the $L_T^{LL'} \equiv |\boldsymbol{p}_T^L| + |\boldsymbol{p}|_T^{L'}$ variable allows to reduce Z+jets background and a requirement on the $P_\zeta \equiv \left(\boldsymbol{p}_T^e + \boldsymbol{p}_T^\mu + \boldsymbol{E}_T^{miss}\right)\boldsymbol{\zeta}/|\boldsymbol{\zeta}|$ variable reduces $t\bar{t}$ background. In addition, multi-variate (MVA) discriminators, based on kinematic variables, are used to reduce dominant backgrounds.

3 Signal extraction and results

In the SM and MSSM $H \to \tau\tau$ analyses to distinguish the Higgs boson signal from the background, the τ lepton pair invariant mass, $m_{\tau\tau}$, is used to extract the signal. The $m_{\tau\tau}$ resolution is improved by using a dynamic likelihood-based algorithm, SVfit [9], which combines the kinematic information of the visible products of the τ candidates decays with the missing transverse energy.

To extract the signal in the $H \to hh \to bb\tau\tau$ analysis, the mass of the heavy Higgs boson candidate, m_H, is used. To improve the sensitivity, events where $m_{\tau\tau}$ or the b jet candidate pair invariant mass, m_{bb}, are significantly incompatibles with the 125 GeV hypothesis, are excluded by applying a mass window selection. Then, the m_H resolution is improved by applying a kinematic fit, where lighter Higgs boson candidates masses are constrained to be 125 GeV.

The results of the MSSM $H \to \tau\tau$ and $H \to hh \to bb\tau\tau$ searches are shown in Figure 1. No evidence of a signal was found. For both analyses, the 95% CL upper limit on cross section times branching fraction for the production of a neutral scalar boson was set. The model-dependent corresponding exclusion limits were set for different MSSM benchmark scenarios in the MSSM $H \to \tau\tau$ analysis and for the MSSM low $\tan\beta$ scenario and 2HDM type-II model parameter space in the $H \to hh \to bb\tau\tau$ analysis.

Figure 2 shows the the SM $H \to \tau\tau$ search results. The measured significance of this process is 3.2σ and the expected significance is 3.7σ.

---

[d]By τ_h all possible hadronic decays of τ^- or their charge conjugates are indicated.

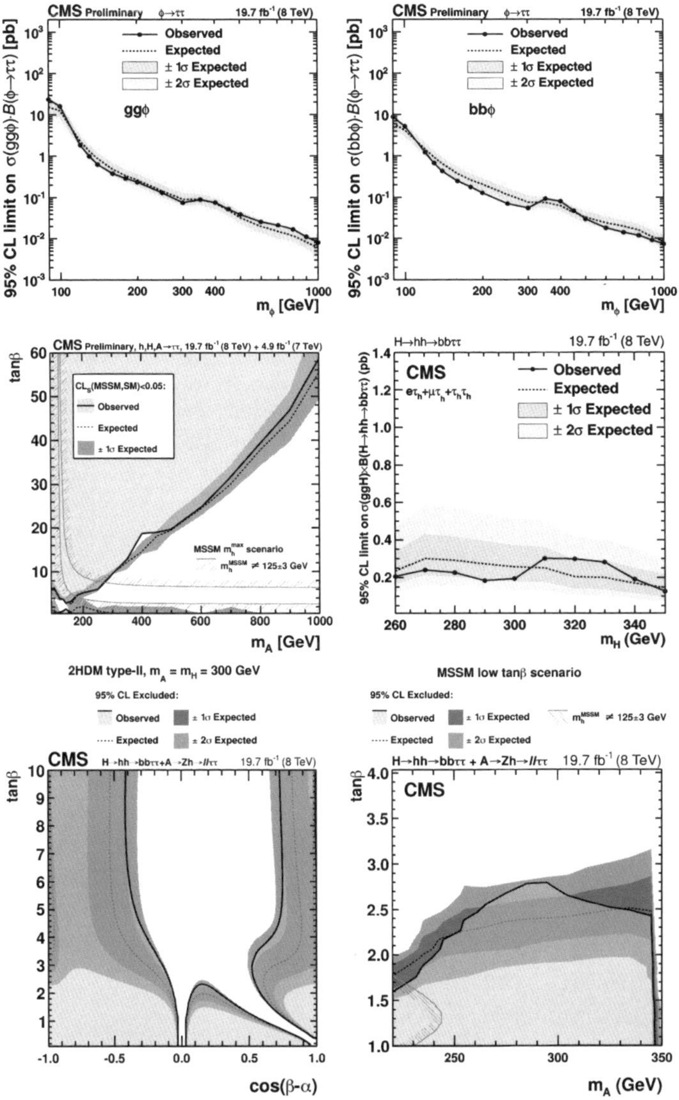

Figure 1: 95% CL upper limits on $\sigma(gg\phi) \cdot \mathcal{B}(\phi \to \tau\tau)$ *(top left)* and $\sigma(bb\phi) \cdot \mathcal{B}(\phi \to \tau\tau)$ *(top right)* for the production of an additional neutral scalar ϕ that decays into τ pair; region in $m_A - \tan\beta$ parameter space excluded at 95% CL in the m_h^{max} MSSM benchmark scenario *(middle left)* [7]. 95% CL upper limits on $\sigma(ggH) \cdot \mathcal{B}(H \to hh \to bb\tau\tau)$ *(middle right)* for the production of an additional neutral scalar H that decays into pair of 125 GeV Higgs bosons with the final state $bb\tau\tau$; the 95% CL exclusion region in the $m_A - \tan\beta$ parameter space for the low-$\tan\beta$ MSSM scenario *(bottom left)* and the 95% CL exclusion regions in the $\cos(\beta - \alpha)$ vs. $\tan\beta$ plane of 2HDM Type II model for $m_A = m_H = 300$ GeV *(bottom right)*, combining the results of the $H \to hh \to bb\tau\tau$ and $A \to Zh \to ll\tau\tau$ analysis [8].

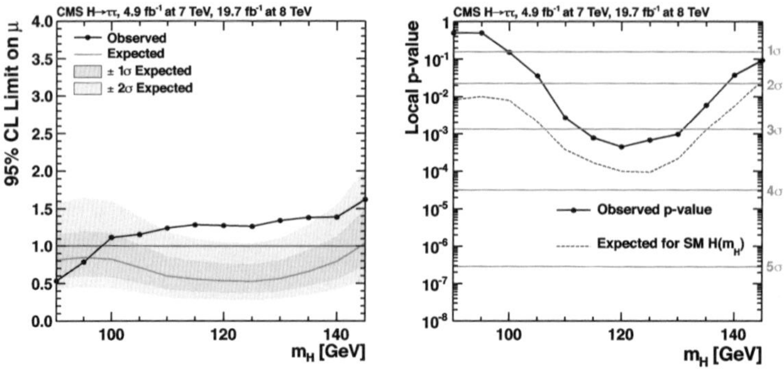

Figure 2: Observed 95% CL upper limit on the signal strength parameter μ, together with the expected limit obtained in the background-only hypothesis *(left)*, and local p-value and significance as a function of the SM Higgs boson mass hypothesis *(right)* obtained by the SM $H \to \tau\tau$ search [5].

4 Conclusions

A brief overview of three Higgs boson searches with $\tau\tau$ channels, using the data collected by CMS in 2011-2012, was given. The MSSM $H \to \tau\tau$ and the $H \to hh \to bb\tau\tau$ searches found no evidence of a signal, while the SM $H \to \tau\tau$ search shows 3.2σ (3.7σ) measured (expected) signal significance.

References

[1] S. Chatrchyan *et al.* [CMS Collaboration], Phys. Lett. B **716** (2012) 30 doi:10.1016/j.physletb.2012.08.021 [arXiv:1207.7235 [hep-ex]].

[2] G. Aad *et al.* [ATLAS Collaboration], Phys. Lett. B **716** (2012) 1 doi:10.1016/j.physletb.2012.08.020 [arXiv:1207.7214 [hep-ex]].

[3] S. Chatrchyan *et al.* [CMS Collaboration], JINST **3** (2008) S08004. doi:10.1088/1748-0221/3/08/S08004

[4] V. Khachatryan *et al.* [CMS Collaboration], arXiv:1510.07488 [physics.ins-det].

[5] S. Chatrchyan *et al.* [CMS Collaboration], JHEP **05** (2014) 104 doi:10.1007/JHEP05(2014)104 [arXiv:1401.5041 [hep-ex]].

[6] V. Khachatryan *et al.* [CMS Collaboration], JHEP **10** (2014) 160 doi:10.1007/JHEP10(2014)160 [arXiv:1408.3316 [hep-ex]].

[7] CMS Collaboration [CMS Collaboration], CMS-PAS-HIG-14-029.

[8] V. Khachatryan *et al.* [CMS Collaboration], arXiv:1510.01181 [hep-ex].

[9] L. Bianchini, J. Conway, E. K. Friis and C. Veelken, J. Phys. Conf. Ser. **513** (2014) 022035. doi:10.1088/1742-6596/513/2/022035

Recent Results from the Fermilab Tevatron Experiments

Phillip Gutierrez [a] for the D0 and CDF Collaborations
University of Oklahoma, Norman, Oklahoma, USA

Abstract. We present a review of recent results from the Fermilab Tevatron CDF and D0 experiments covering a wide range of topics including combined results.

1 Introduction

The Fermilab $p\bar{p}$ Tevatron collider commenced operation in 1985 with the first collisions observed by the CDF experiment in 1986. The official start of run 1 at $\sqrt{s} = 1.8$ TeV, which included both the CDF and D0 experiments, occurred in 1992 and proceeded through 1996 accumulating ≈ 120 pb^{-1} of integrated luminosity. This was sufficient to lead to the discovery of the top quark in 1995 by CDF and D0 [1, 2]. Following a period of four years in which the two detectors underwent upgrades, run 2 officially started in 2001 at $\sqrt{s} = 1.96$ TeV. A large number of results have been produced from the ≈ 10 fb^{-1} of integrated luminosity collected, with the highlights being the discovery of $B_s^0\bar{B}_s^0$ oscillations and the electroweak production of the top quark, and observing evidence for the Higgs boson in the $b\bar{b}$ final state. Run 2 concluded in 2011, and since then both collaborations continue to analysis their data producing new results. In this presentation, we discuss a selection of recent results produced independently and jointly by the two collaborations during the year 2015 using their full datasets.

2 Higgs Boson *Spin-Parity*

In 2013, the Tevatron collaborations published their combined Higgs searches in the mass range 90 to 200 GeV [3]. The searches were performed assuming standard model (SM) production mechanisms and decay modes. The combined analysis yielded a three standard deviation excess in the mass range 120 to 125 GeV where the SM signal is dominated by associated vector boson production with the vector boson (V) decaying to leptons and the Higgs boson to $b\bar{b}$. The mass from this search is consistent with the LHC discovery [4, 5], and the Tevatron evidence in the $b\bar{b}$ final state [6].

The authors of Ref. [7] suggested the use of Tevatron data to test models of Higgs bosons having exotic spin-parity (J^P) in the $p\bar{p} \to VH$ channel with $H \to b\bar{b}$ and leptonic final states of the V. CDF and D0 examined the possibility of $J^P = 0^-$, 2^+ and the SM prediction of 0^+. It is important to note that the $J^P = 0^-$ and 2^+ states have a 50% greater acceptance than the SM decay in

[a] E-mail: pgutierrez@ou.edu

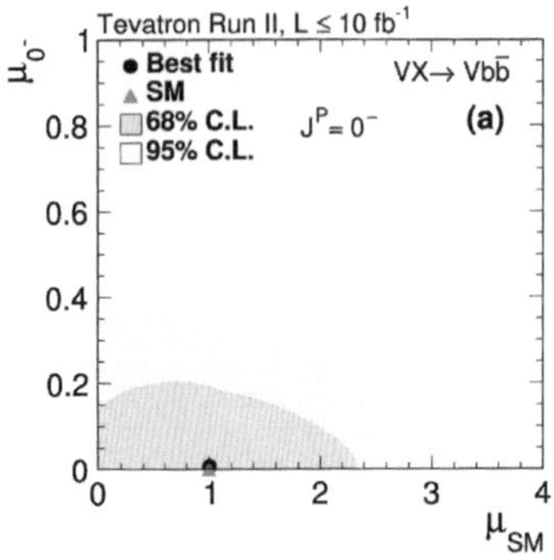

Figure 1: The signal strength μ of the SM prediction $J^P = 0^+$ versus $J^P = 0^-$.

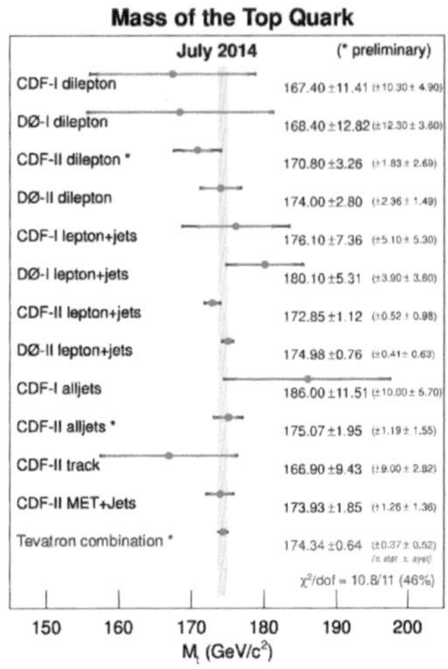

Figure 2: Recent Tevatron top quark mass measurements.

the $Z \to \nu\bar{\nu}$ final state. For production near threshold, the $Vb\bar{b}$ invariant mass is sensitive to the spin-parity.

Combining results from earlier D0 and CDF measurements assuming the SM Higgs production rate and decay to $b\bar{b}$, the exotic $J^P = 0^-$ and 2^+ are excluded at 5 and 4.9 standard deviations, respectively (see Fig. 1). If an admixture of SM 0^+ and either 0^- or 2^+ J^P states is assumed, the fraction of the exotic component at 95% confidence level was found to be < 0.36 for both cases [8].

3 Top Quark

3.1 Mass

The top quark mass is an important parameter in the SM. The masses of the top quark, W boson, and Higgs boson are interrelated through loop corrections and determine the stability of the electroweak vacuum [9]. The most recent Tevatron mass combination, which includes the then most accurate single measurement, the D0 lepton plus jets result $m_t = 174.98 \pm 0.76$ GeV [10], yields a mass of $m_t = 174.34 \pm 0.64$ GeV, corresponding to a relative precision of 0.37% [11] (see Fig. 2).

Both CDF and D0 have produced new mass measurements in the di-lepton final state. The di-lepton final state has very small backgrounds, but since the W from each top quark decay produces a neutrino, the jet energy scale (JES) can not be calibrated directly on the data used in the mass measurement. This leads to relatively large systematic uncertainties. To circumvent this problem, both collaborations perform two independent measurements that are then combined. Both collaborations use a variant of the neutrino weighting method, which is sensitive to the mass. CDF then performs a second measurement using the scalar product of the leptons and the two leading jets, which is insensitive to the JES. The two measurements are then combined as follows: $m_t^{\rm hyb} = w m_t^{\rm reco} + (1-w) m_t^{\rm alt}$, where the value of w was optimized on Monte Carlo. Templates were then used to fit to the $m_t^{\rm hyb}$ distribution to extract the mass $m_t = 171.5 \pm 1.9({\rm stat}) \pm 2.5({\rm syst})$ GeV [12]

D0 improves the precision of the neutrino weighting method by correcting the JES using the $t\bar{t} \to \ell$+jets sample, where the W mass can be constrained. Using this technique, the measured mass is $m_t = 173.3 \pm 1.4({\rm stat}) \pm 0.8({\rm syst})$ GeV [13].

3.2 Forward-Backward Asymmetry

At leading order, the production rapidity distributions of t and $\bar{t}$ should be symmetric about $y = 0$. At second order, there are Feynman diagrams that interfere to produce an asymmetry. Since these diagrams contain loops, it is possible that the asymmetry can be enhanced due to non-SM effects. Some early measurements, in fact, gave an approximate 3 standard deviation asymmetry difference relative to the SM prediction at large $t\bar{t}$ invariant mass, which led to much speculation. Subsequent results in the $\ell +$ jets channel [14, 15] were closer to SM predictions. Two new measurements in the di-lepton final state [16, 17] that complement the $\ell +$ jets measurements are presented here.

The measurement of the forward-backward asymmetry $A_{\rm FB}^{t\bar{t}} \propto N(\Delta y > 0) - N(\Delta y < 0)$ is performed in the $t\bar{t}$ rest frame rapidity given by the variable $\Delta y = y_t - y_{\bar{t}}$. In order to extract the parton-level forward-backward asymmetry ($A_{\rm FB}^{t\bar{t}}$) from the observed kinematics, both experiments reconstruct $A_{\rm FB}^{t\bar{t}}$ from the characteristics of the observed $t\bar{t}$ event. D0 applies the same matrix element method used to extract m_t [10], while CDF uses a likelihood-algorithm in which they scan over the allowed kinematic range deriving a probability density function of the kinematics. The results of these measurements along with those of the earlier lepton plus jets measurements are presented in Fig. 3.

In addition, the D0 analysis is performed to simultaneously extract $A_{\rm FB}^{t\bar{t}}$ and the analyzing power times polarization ($\kappa^{\pm} P_n^{\pm}$), where $P_n^{\pm}$ the polarization of the t ($\bar{t}$) relative to the quantization axis n defined as the direction of the proton beam, and $\kappa^{\pm}$ is the analyzing power of the lepton $\ell^{\pm}$, which is predicated to be ≈ 1 to 0.1%. This result is presented in Fig. 4.

3.3 Single Top Quark Production

With the discovery of s-channel electroweak production of the top quark in 2014, the Tevatron collaborations sought to improve the precision of the t- and $s + t$-channel measurements by combining their results. Both collaborations

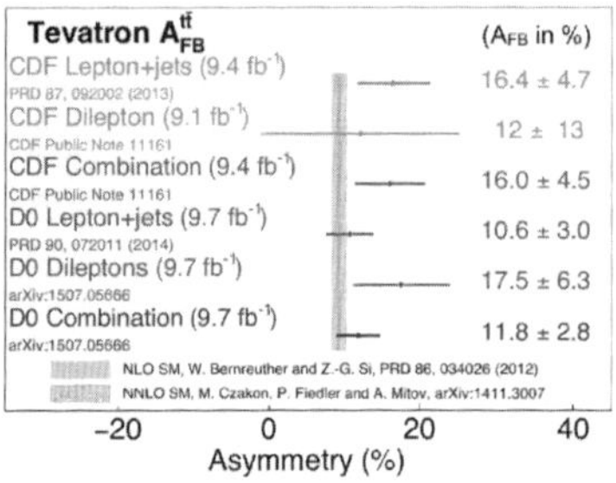

Figure 3: The latest Tevatron $A_{\mathrm{FB}}^{t\bar{t}}$ measurements.

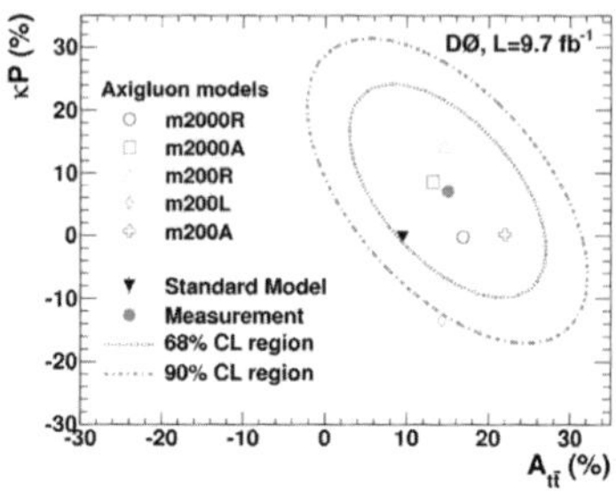

Figure 4: The D0 polarization measurement.

split their analyses into two measurements: one optimized on s-channel production and the second on the t-channel. To extract the signal, CDF applies a neural network discriminant. D0 uses a combination of boosted decision trees, matrix element method, and Bayesian neural networks, and then combines these discriminants into a single Bayesian neural network discriminant. The s-channel analyses follow similar methods, but are specifically optimized for the s-channel topology [18, 19].

To derive the final combined cross section measurements, likelihood fits to the various discriminants are performed and are then combined through a Bayesian analyses taking a uniform prior probability density over positive values of the signal cross sections. The resulting 2-dimensional posterior probability density is a function of σ_s and σ_t, from which the cross sections, $\sigma_t = 2.25^{+0.29}_{-0.31}$ pb and $\sigma_s = 1.29^{+0.26}_{-0.24}$ pb, are extracted [20]

Furthermore, in the SM the cross section for electroweak production of top quarks is directly proportional to $|V_{tb}|^2$. This allows for a direct measurement of $|V_{tb}|^2$ without assuming unitarity or number of fermion families. Following the procedure outlined above, the posterior probability density is obtained assuming a uniform prior probability in the range $[0, \infty]$, which results in an expected value of $|V_{tb}|^2 = 1.04^{+0.12}_{-0.10}$. If one restricts $|V_{tb}| < 1$, then a 95% lower limit of $|V_{tb}|^2 > 0.84$ is achieved.

4 *b* Quark

4.1 Forward-Backward Asymmetry

Following up on the $t\bar{t}$ asymmetry measurement, both D0 and CDF perform a similar measurement on $b\bar{b}$. D0, which has the ability to independently reverse the solenoid field of the inner tracker and the toroid of the muon detector, can remove detector asymmetries to first order. Using a sample of $B^{\pm} \rightarrow$

$J/\psi(\mu^+\mu^-)K^\pm$, D0 derives the asymmetry in $B^\pm$ production from the $J/\psi K^\pm$ invariant mass distribution through a log likelihood fit in bins of p_T and η. The result is given as a function of both p_T and η with the result independent of both and consistent with zero. The asymmetry calculated over all p_T is $A_{\mathrm{FB}} = -0.24 \pm 0.41(\mathrm{stat}) \pm 0.19(\mathrm{syst})$ for $\langle p_T \rangle = 12.9$ GeV [21]. Even though the data are inconsistent with the predication of MC@NLO, the author of Ref. [22] finds good theoretical agreement with the D0 result.

CDF has performed the forward-backward asymmetry measurement using b tagged jets containing a soft muon, which provides the flavor tag. The measurement is performed in the invariant mass of the $b\bar{b}$ jets in the range 150 GeV to 350 GeV [23]. Even though this measurement suffers from dilution due to mixing and cascade decays (for example $b \to c$, which inverts the lepton sign), it gives good agreement with the theory prediction of [22].

4.2 Violation CPT & Lorentz Invariance

D0 has observed a 3.6 standard deviation discrepancy in CP violation relative to the SM prediction in the same sign di-muon forward-backward asymmetry [24]. A possible interpretation of this result could be due to CPT violation such that the oscillation probability of the B_s^0 and $\bar{B}_s^0$ back to their original states are different, $P(B_s^0 \to B_s^0) \neq P(\bar{B}_s^0 \to \bar{B}_s^0)$, due to a flavor dependent vector field Δa_μ that selects a direction in space-time [25, 26]. For CPT violation to produce the observed di-muon asymmetry, the quantity $\Delta a_T - 0.396\Delta a_z$ must be of order 10^{-12} GeV, where Δa_T and Δa_Z are the transverse and longitudinal components of Δa_μ, respectively.

The D0 collaboration has searched for such an effect by measuring the time dependent asymmetry $B_s^0 \to \mu^\pm D_s^\mp$ relative to the sidereal time. As the Fermilab Tevatron changes its direction relative to this preferred direction, the forward-backward muon asymmetry, varies over a sidereal day. The observed 95% confidence interval is $(-0.8 < \Delta a_T - 0.396\Delta a_z < 3.9) \times 10^{-13}$ GeV [27].

5 Summary

In addition to the results presented, new results in QCD and measurements of the electroweak parameters were also published in 2015. An estimated 40 additional analysis are in the pipeline including both Tevatron combinations and results from the individual experiments on a variety of topics including the W boson mass measurement.

Acknowledgments

The author thanks the College of Arts and Sciences and the Homer L. Dodge Department of Physics and Astronomy at the University of Oklahoma, and the

United States Department of Energy for providing travel funds to present at this conference. The author also thanks the organizers for their hospitality.

References

[1] F. Abe et al., (CDF Collaboration), "Observation of Top Quark Production in $\overline{p}p$ Collisions with the Collider Detector at Fermilab", Phys. Rev. Lett. **74**, 2626 (1995).

[2] S. Abachi et al., (D0 Collaboration), "Observation of the Top Quark", Phys. Rev. Lett. **74**, 2632 (1995).

[3] T. Aaltonen et al., (CDF and D0 Collaborations), "Higgs boson studies at the Tevatron", Phys. Rev. D **88**, 052014 (2013).

[4] G. Aad et al., (ATLAS Collaboration), "Observation of a new particle in the search for the Standard Model Higgs boson with the ATLAS detector at the LHC", Physics Letters B **716**, 1 (2012).

[5] S. Chatrchyan et al., (CMS Collaboration), "Observation of a new boson at a mass of 125 GeV with the CMS experiment at the LHC", Physics Letters B **716**, 30 (2012).

[6] T. Aaltonen et al., (CDF and D0 Collaborations), "Evidence for a Particle Produced in Association with Weak Bosons and Decaying to a Bottom-Antibottom Quark Pair in Higgs Boson Searches at the Tevatron", Phys. Rev. Lett. **109**, 071804 (2012).

[7] J. Ellis et al., "A fast track towards the Higgs spin and parity", Journal of High Energy Physics **2012**, 134 (2012).

[8] T. Aaltonen et al., (CDF and D0 Collaborations), "Tevatron Constraints on Models of the Higgs Boson with Exotic Spin and Parity Using Decays to Bottom-Antibottom Quark Pairs", Phys. Rev. Lett. **114**, 151802 (2015).

[9] S. Alekhin, A. Djouadi, and S. Moch, "The top quark and Higgs boson masses and the stability of the electroweak vacuum", Physics Letters B **716**, 214 (2012).

[10] V. M. Abazov et al., (D0 Collaboration), "Precision measurement of the top-quark mass in lepton+jets final states", Phys. Rev. D **91**, 112003 (2015).

[11] T. Aaltonen et al., (CDF and D0 Collaborations), "Combination of CDF and D0 results on the mass of the top quark using up to 9.7 fb^{-1} at the Tevatron", (2014), arXiv:`1407.2682 [hep-ex]`.

[12] T. Aaltonen et al., "Measurement of the top-quark mass in the $t\bar{t}$ dilepton channel using the full CDF Run II data set", Phys. Rev. D **92**, 032003 (2015).

[13] V. Abazov et al., (D0 Collaboration), "Precise measurement of the top quark mass in dilepton decays using optimized neutrino weighting", Physics Letters B **752**, 18 (2016).

[14] T. Aaltonen et al., (CDF Collaboration), "Measurement of the top quark forward-backward production asymmetry and its dependence on event kinematic properties", Phys. Rev. D **87**, 092002 (2013).

[15] V. M. Abazov et al., (D0 Collaboration), "Measurement of the forward-backward asymmetry in the distribution of leptons in $t\bar{t}$ events in the lepton + jets channel", Phys. Rev. D **90**, 072001 (2014).

[16] T. Aaltonen et al., (CDF Collaboration), "Measurement of the Forward-Backward Asymmetry of Top-Quark Pairs in the Dilepton Final State and Combination at CDF", CDF/ANAL/TOP/PUB/11161 (2015).

[17] V. M. Abazov et al., (D0 Collaboration), "Simultaneous measurement of forward-backward asymmetry and top polarization in dilepton final states from $t\bar{t}$ production at the Tevatron", Phys. Rev. D **92**, 052007 (2015).

[18] V. Abazov et al., (D0 Collaboration), "Evidence for s-channel single top quark production in $p\bar{p}$ collisions at $\sqrt{s} = 1.96$ TeV", Physics Letters B **726**, 656 (2013).

[19] T. Aaltonen et al., (CDF Collaboration), "Observation of s-Channel Production of Single Top Quarks at the Tevatron", Phys. Rev. Lett. **112**, 231803 (2014).

[20] T. Aaltonen et al., (CDF and D0 Collaborations), "Tevatron Combination of Single-Top-Quark Cross Sections and Determination of the Magnitude of the Cabibbo-Kobayashi-Maskawa Matrix Element V_{tb}", Phys. Rev. Lett. **115**, 152003 (2015).

[21] V. M. Abazov et al., (D0 Collaboration), "Measurement of the Forward-Backward Asymmetry in the Production of $B^{\pm}$ Mesons in $p\bar{p}$ Collisions at $\sqrt{s} = 1.96$ TeV", Phys. Rev. Lett. **114**, 051803 (2015).

[22] C. W. Murphy, "Bottom-quark forward-backward and charge asymmetries at hadron colliders", Phys. Rev. D **92**, 054003 (2015).

[23] T. Aaltonen et al., (CDF Collaboration), "First measurement of the forward-backward asymmetry in bottom-quark pair production at high mass", Phys. Rev. D **92**, 032006 (2015).

[24] V. M. Abazov et al., (D0 Collaboration), "Study of CP-violating charge asymmetries of single muons and like-sign dimuons in $p\bar{p}$ collisions", Phys. Rev. D **89**, 012002 (2014).

[25] V. A. Kostelecký and S. Samuel, "Spontaneous breaking of Lorentz symmetry in string theory", Phys. Rev. D **39**, 683 (1989).

[26] V. A. Kostelecký and R. Potting, "CPT and strings", Nuclear Physics B **359**, 545 (1991).
[27] V. M. Abazov et al., "Search for Violation of CPT and Lorentz Invariance in B_s^0 Meson Oscillations", Phys. Rev. Lett. **115**, 161601 (2015).

MEASUREMENT OF THE TWO-PHOTON EXCHANGE IN THE ELASTIC $e^{\pm}p$ SCATTERING AT THE OLYMPUS EXPERIMENT AT DESY

Dmitry Khaneft [a] for the OLYMPUS Collaboration
Johannes Gutenberg-Universitä Mainz, Mainz, Germany

Abstract. The OLYMPUS experiment aims at measuring the two-photon exchange amplitude in the elastic scattering of electrons off protons. This amplitude comprises a possible explanation for the significant, experimental discrepancy in the determination of the ratio of the proton electric to magnetic form factor, R = G_E/G_M, extracted using Rosenbluth separation method and polarization transfer technique. A measurement of the elastic scattering cross section ratio $\sigma_{e^+p}/\sigma_{e^-p}$ will render a direct measurement of the two-photon exchange amplitude. The OLYMPUS experiment was carried out at DESY Hamburg, Germany using 2.01 GeV electron and positron beams of the DORIS storage ring incident on an internal hydrogen gas target. Multiple independent luminosity monitors were operated in parallel to the main spectrometer during data taking to allow for a precise relative luminosity measurement. Approximately 4.45 $\mathrm{fb}^{-}1$ of data were collected. The current status of the OLYMPUS experiment will be discussed.

1 Introduction

A significant discrepancy have been observed by two types of experiments that have measured the ratio of proton form factors, G_E/G_M [1]. The results obtained by the Rosenbluth separation method and the polarization transfer technique are comparable in the lower Q^2 range, while the difference becomes more pronounced for measurements at higher energies. Fig. 1 shows the ratio of proton form factors, as a function of Q^2, obtained from Rosenbluth separation method in unpolarized experiments and data obtained using polarization techniques. Also shown are the ratios from fits [2] of the form factors to the world data set. Two-photon exchange is believed to be the reason of the discrepancy. The ratio of electron-proton to positron-proton elastic cross sections, if affected by the two-photon exchange, will deviate from unity. Previous measurements shown the deviation of the ratio from unity but the large uncertainties do not allow precise determination of the two-photon exchange contribution.

2 The OLYMPUS Experiment

The OLYMPUS experiment [15] seeks to measure two-photon exchange contribution over a range of four-momentum transfer of 0.4 $\mathrm{GeV}^2/c^2 < Q^2 <$ 2.2 GeV^2/c^2 through a measurement of the ratio of electron-proton to positron-proton elastic scattering cross sections with a total uncertainty of less than 1%. In the OLYMPUS kinematic region the cross section ratio is expected to deviate

[a]E-mail: `khaneftd@kph.uni-mainz.de`

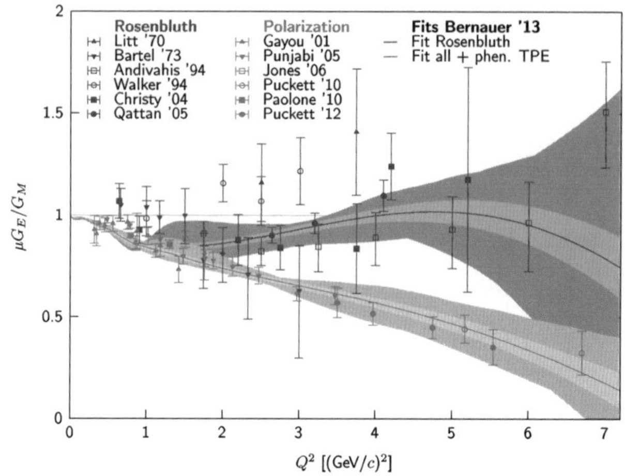

Figure 1: The ratio of proton form factors, $\mu_p G_E^p/G_M^p$, as a function of Q^2 from (unpolarized) Rosenbluth measurements [3–8] is inconsistent with recent data collected using polarization techniques [9–14]. Also shown are the ratios from fits [2] of the form factors to the world data set. The light shaded bands show statistical uncertainty and the dark shaded bands show model uncertainty added linearly.

from unity by 5-10%, if the form factor discrepancy between Rosenbluth and recoil polarization measurements is caused by two-photon exchange. Experimental data as well as some theoretical predictions on the ratio of positron to electron elastic scattering cross sections at a beam energy of 2 GeV are shown in Fig. 2a.

The OLYMPUS experiment was equipped with an eight-coil toroidal magnet and detectors located in two horizontal sectors on both sides of the beam line (see Fig. 2b). Each side consisted of drift chambers for particle tracking and a set of time-of-flight scintillator bars for triggering and measurements of energy deposition, particle position, and timing. Two luminosity monitors were used: symmetric Møller/Bhabha (SYMB) detector and gas electron multiplier (GEM) detectors interleaved with multi-wire proportional chambers (MWPCs). The SYMB consisted of two Cherenkov calorimeters placed at the symmetric Møller/Bhabha scattering angle (1.29° at the OLYMPUS beam energy). Three pairs of GEM detectors and MWPCs were placed symmetrically in each sector at 12°. They measured elastically scattered leptons in coincidence with recoil protons detected in the opposite sector drift chamber.

The measurements have been performed at the DORIS storage ring at DESY which provided both electron and positron beams at 2.01 GeV, alternating between beam species on a daily basis. The toroidal magnet current was set to 5000 A to allow particle charge and momentum determination from the cur-

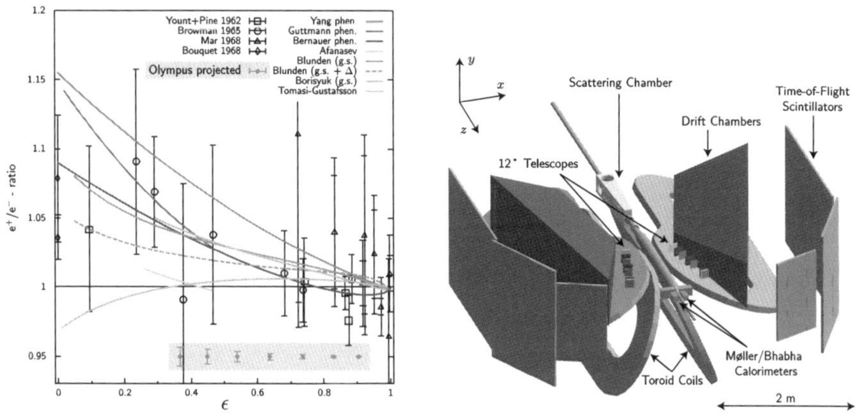

(a) (b)

Figure 2: (a) The ratio of positron to electron elastic scattering cross-sections at a beam energy of 2 GeV as a function of ϵ showing phenomenological predictions [2, 16, 17], a number of theoretical predictions on hard two-photon exchange [18–24], and existing ratio data [25–28]. In addition the projected OLYMPUS uncertainties are shown. (b) A model of the OLYMPUS experiment without the top four magnet coils to show the detectors.

vature of their tracks and to remove low-energy, charged background particles from the tracking detectors. Around 10% of the data was taken with the negative magnet polarity which will provide a check on systematic uncertainties. Leptons were scattered off the internal hydrogen gas target [29]. Scattered leptons and recoil protons were detected in coincidence over a wide range of scattering angles ($25° \leq \theta \leq 75°$, $-15° \leq \phi \leq 15°$).

To summarize, the OLYMPUS experiment collected approximately 4.45 fb^{-1} of data for electron and positron elastic scattering from hydrogen. It will provide a precise determination of the two-photon exchange contribution by measuring the ratio of electron-proton to positron-proton elastic cross sections.

References

[1] Pierre A.M. Guichon and M Vanderhaeghen. *Physical Review Letters*, 91:142303, 2003.

[2] J. C. Bernauer et al. *Phys. Rev.*, C90(1):015206, 2014.

[3] J. Litt, G. Buschhorn, D.H. Coward, H. Destaebler, L.W. Mo, R.E. Taylor, B.C. Barish, S.C. Loken, J. Pine, J.I. Friedman, G.C. Hartmann, and H.W. Kendall. *Physics Letters B*, 31(1):40 – 44, 1970.

[4] W. Bartel, F.-W. Bsser, W.-R. Dix, R. Felst, D. Harms, H. Krehbiel, P.E. Kuhlmann, J. McElroy, J. Meyer, and G. Weber. *Nuclear Physics B*, 58(2):429 – 475, 1973.
[5] L. Andivahis et al. *Phys. Rev. D*, 50:5491–5517, 1994.
[6] R. C. Walker et al. *Phys. Rev. D*, 49:5671–5689, 1994.
[7] M. E. Christy et al. *Phys. Rev. C*, 70:015206, 2004.
[8] I. A. Qattan et al. *Phys. Rev. Lett.*, 94:142301, 2005.
[9] O. Gayou et al. *Phys. Rev. C*, 64:038202, 2001.
[10] V. Punjabi et al. *Phys. Rev. C*, 71:069902, 2005.
[11] M. K. Jones et al. *Phys. Rev. C*, 74:035201, 2006.
[12] A. J. R. Puckett et al. *Phys. Rev. Lett.*, 104:242301, 2010.
[13] M. Paolone et al. *Phys. Rev. Lett.*, 105:072001, 2010.
[14] A. J. R. Puckett et al. *Phys. Rev. C*, 85:045203, 2012.
[15] R Milner, D Hasell, M Kohl, U Schneekloth, N Akopov, et al. *Nuclear Instruments and Methods in Physics Research Section A*, 741:1–17, 2014.
[16] Yu Chun Chen, Chung Wen Kao, and Shin Nan Yang. *Physics Letters B*, 652(56):269 – 274, 2007.
[17] J. Guttmann, N. Kivel, M. Meziane, and M. Vanderhaeghen. *The European Physical Journal A - Hadrons and Nuclei*, 47:1–5, 2011. 10.1140/epja/i2011-11077-4.
[18] P. G. Blunden, W. Melnitchouk, and J. A. Tjon. *Phys. Rev. Lett.*, 91:142304, 2003.
[19] Y.-C. Chen, A. Afanasev, S. J. Brodsky, C. E. Carlson, and M. Vanderhaeghen. *Phys. Rev. Lett.*, 93:122301, 2004.
[20] Andrei V. Afanasev, Stanley J. Brodsky, Carl E. Carlson, Yu-Chun Chen, and Marc Vanderhaeghen. *Phys. Rev. D*, 72:013008, 2005.
[21] P. G. Blunden, W. Melnitchouk, and J. A. Tjon. *Phys. Rev. C*, 72:034612, 2005.
[22] S. Kondratyuk, P. G. Blunden, W. Melnitchouk, and J. A. Tjon. *Phys. Rev. Lett.*, 95:172503, 2005.
[23] Dmitry Borisyuk and Alexander Kobushkin. *Phys. Rev. C*, 74:065203, 2006.
[24] E. Tomasi-Gustafsson, M. Osipenko, E. A. Kuraev, and Yu. M. Bystritskiy. *Physics of Atomic Nuclei*, 76(8):937–946, 2013.
[25] D. Yount and J. Pine. *Phys. Rev.*, 128:1842–1849, 1962.
[26] A. Browman, F. Liu, and C. Schaerf. *Phys. Rev.*, 139:B1079–B1085, 1965.
[27] Jerry Mar, Barry C. Barish, Jerome Pine, David H. Coward, H. DeStaebler, J. Litt, A. Minten, Richard E. Taylor, and M. Breidenbach. *Phys. Rev. Lett.*, 21:482–484, 1968.

[28] B. Bouquet, D. Benaksas, B. Grossette, B. Jean-Marie, G. Parrour, J.P. Poux, and R. Tchapoutian. *Physics Letters B*, 26(3):178 – 180, 1968.

[29] J.C. Bernauer, V. Carassiti, G. Ciullo, B.S. Henderson, E. Ihloff, J. Kelsey, P. Lenisa, R. Milner, A. Schmidt, and M. Statera. *Nuclear Instruments and Methods in Physics Research Section A*, 755:20 – 27, 2014.

ELECTROWEAK MEASUREMENTS WITH THE ATLAS DETECTOR

Krasnopevtsev D. [a]
on behalf of the ATLAS Collaboration
National Research Nuclear University MEPhI,115409, Russia, Moscow

Abstract. ATLAS measurements of multi-boson production processes involving combinations of W, Z and isolated photons at 8 TeV are summarized. Standard Model cross section are measured with high precision by ATLAS and are compared to world averages. Production processes sensitive to vector-boson fusion and vector-boson scattering are also presented and used for the triple and quartic gauge boson couplings limits setting.

1 Introduction

Precision electroweak (EWK) measurements at the Large Hadron Collider (LHC) aim to validate Standard Model (SM) in new energy scale, to improve accuracy of known SM parameters and to estimate new physical contributions. The triple (TGC) and quartic (QGC) gauge boson couplings can be tested at the LHC through studies of multiboson production or through studies of vector boson fusion (VBF) and vector boson scattering (VBS) processes. This proceeding reports on recent ATLAS [1] studies using 20.3 fb^{-1} of proton-proton collisions at a center of mass energy of 8 TeV. In the following, the most up-to-date measurements of 4 lepton and $W\gamma\gamma$ cross sections will be reviewed, along with that of vector boson fusion with Z and vector boson scattering with WW. Studies of these processes are motivated by both theoretical and experimental considerations, as will be illustrated in the following sections.

2 Electroweak measurements

2.1 Inclusive 4 lepton production

The main goal of this study is to test validity of the SM through the interplay of QCD and EW effects for different 4 lepton (4l) productions [2]. The following productions could be observed at the LHC: qq originated 4l production, 4l from higgs decays, non-resonant ZZ to 4 leptons production. Leptonic channel is very clean. Background contribution is less than 10% and the impact of background uncertainty to the cross section is less than 1%. Combined cross section results are in agreement with SM predictions within statistical and systematic uncertainties Fig. 1(a). Discrepancies between signal strength and current LO prediction are interpreted as missing higher order correction to $gg \to 4l$. This correction is extracted from data in $M_{4l} > 180$ GeV region: $\mu_{gg} = 2.4 \pm 1.0(stat.) \pm 0.5(syst.) \pm 0.8(theory)$.

[a]E-mail: Dimitriy.Krasnopevtsev@cern.ch

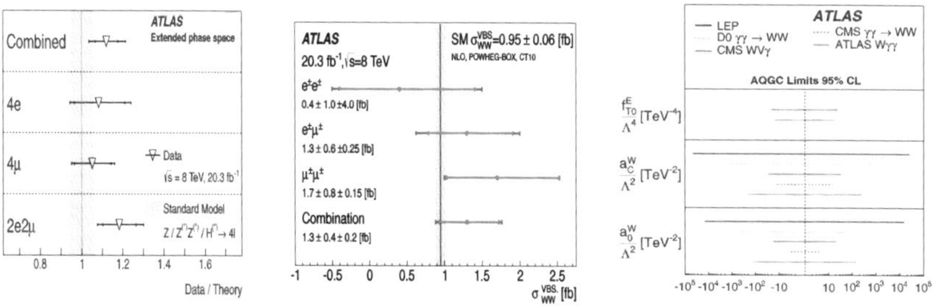

(a) 4 lepton (b) WW VBS (c) W$\gamma\gamma$

Figure 1: (a) The measured 4 lepton cross sections in extended phase space divided by the expected inclusive cross sections. The results per channel and for the three-channel combinations in the extended phase space [3]; (b) The measured cross sections for WW VBS region compared to the predicted standard model cross section [6]; (c) Comparison of the 95% CL limits on f^E_{T0}/Λ^4, a^W_C/Λ^2, and a^W_0/Λ^2 obtained by W$\gamma\gamma$ analysis in ATLAS and other different experiments [8].

2.2 Vector boson fusion Z

Vector boson fusion Z signal events also contain leptons as final state but they are accompanied by at least two high p_T jets [4]. Strong production is dominant for this study and VBF process is a component of EWK Zjj production. EWK components could be extracted using jets invariant mass distribution M_{jj} and better p_T balance of the Zjj system. EWK results are in a good agreement with Powheg predictions and show that background-only hypothesis rejected with significance above 5 σ:
$\sigma_{exp} = 54.7 \pm 4.6(stat)^{+9.8}_{-10.4}(sys) \pm 1.5(lumi)\ fb$,
$\sigma_{theory} = 46.1 \pm 0.2(stat)^{+0.3}_{-0.2}(scale) \pm 0.8(PDF) \pm 0.5(model)\ fb$.
Electroweak Zjj production cross section results are used to test of anomalous triple gauge coupling (aTGC) limits on λ_Z and $g_{1,Z}$ parameters.

2.3 Vector boson scattering WW

VBS WW production is observed for the first time with a significance of 3.6 standard deviations and is sensitive to the mechanism of electroweak symmetry breaking as well as beyond the standard model physics [5]. Events with two reconstructed same-charge leptons and two or more jets are analyzed. Measured cross-sections are in agreement with theory prediction Fig. 1(b). Anomalous quartic gauge coupling (aQGC) limits on $\alpha 4$ and $\alpha 5$ parameters are set at 95% confidence level.

2.4 $W\gamma\gamma$ production

$W\gamma\gamma$ production [7] is the first measurements of triboson process in ATLAS. Signal events contain two isolated photons and W decayed into leptons. The largest backgrounds are jets faking photons and leptons and are estimated using data-driven techniques. Combined results show a statistical significance above three standard deviations for SM $W\gamma\gamma$ process. Cross section is higher by 1.9 σ than SM NLO prediction in inclusive case. Better agreement is observed for the exclusive case. ATLAS anomalous quartic couplings limits for f^E_{T0}, a^W_C, and a^W_0 are set for exclusive case and are better or similar to LEP and D0 collaborations Fig. 1(c).

3 Future electroweak measurements with ATLAS

A new run of the LHC has started recently during summer 2015 at an increased center of mass energy of 13 TeV. Recent EWK studies include the first $W^{\pm}$ and Z production cross sections measured by ATLAS at the highest centre-of-mass energy ever available from a collider and show agreement with theoretical calculations based on NNLO QCD [9]. A variety of multi-boson channels studies have been started and more EWK measurement with ATLAS detector are expected to come soon.

Acknowledgments

The reported study was supported by RFBR, research project No. 14-02-31074.

References

[1] ATLAS Collaboration, JINST **3**, S08003 (2008)
[2] ATLAS collaboration, submitted to Physics Letters B, http://arxiv.org/abs/1509.07844.
[3] https://atlas.web.cern.ch/Atlas/GROUPS/PHYSICS/PAPERS/STDM-2014-15/
[4] ATLAS collaboration, JHEP **04**, 031 (2014).
[5] ATLAS collaboration, Phys. Rev. Lett. **113**, 14 (2014).
[6] https://atlas.web.cern.ch/Atlas/GROUPS/PHYSICS/PAPERS/STDM-2013-06/
[7] ATLAS collaboration, Phys. Rev. Lett. **115**, 3 (2015).
[8] https://atlas.web.cern.ch/Atlas/GROUPS/PHYSICS/PAPERS/STDM-2013-05/
[9] ATLAS collaboration, ATLAS-CONF-2015-039, https://cdsweb.cern.ch/record/2045487

TOP PHYSICS IN ATLAS

Roger Naranjo [a]
on behalf of the ATLAS collaboration
DESY, Hamburg, Germany.

Abstract. These proceedings summarize the latest measurements on top production, top properties and searches using the ATLAS detector at the LHC. The measurements are performed on pp collision data with a center of mass energy $\sqrt{s} = 7, 8$ and 13 TeV.

1 Introduction

The top quark is the heaviest known elementary particle of the Standard Model (SM). It has a life-time shorter than the hadronization time and offers a window to search for new physics. With large numbers of top events produced at LHC in Run-1, it is possible to study its properties in detail. These proceedings present measurements of top production, top properties and searches at the center of mass energy of 7, 8 and 13 TeV by the ATLAS experiment [1].

2 Top production measurements

2.1 Top pair production

The ATLAS experiment has performed measurements of the inclusive top-antitop production cross section of $t\bar{t}$ production at center of mass energy of 7, 8 and 13 TeV in the dilepton channel [2,4]. The measurement at 8 TeV represents the most precise measurement to date, even more precise than next-to-next-to-leading order (NNLO) predictions [3]. At 13 TeV, the cross section is found to be $\sigma_{t\bar{t}} = 825 \pm 49$ (stat) $\pm$ 60(syst) $\pm$ 83(lumi) pb. In the three analyses, a simultaneous fit of the cross section, b-jet reconstruction and tagging efficiency is performed in the $e\mu$ channel. All the measurements are in agreement with SM predictions, as shown in Figure 1a.

New physics could affect the shape of the differential cross section as a function of top kinematics. In this conference, measurements performed in two different topologies were presented: the resolved and the boosted topologies, which are optimized for tops with transverse momentum $p_{\mathrm{T}} < 300$ and and $p_{\mathrm{T}} > 300$ GeV respectively. In the resolved topology, the measurement of the top-antitop differential cross section is performed as a function of the mass, p_{T} and rapidity of the top pair system at 7 TeV [5]. These measurements are performed in a fiducial region closely matching the detector acceptance and using tops observables based on stable particles in the $e\mu$ channel.

[a]E-mail: `roger.naranjo@desy.de`

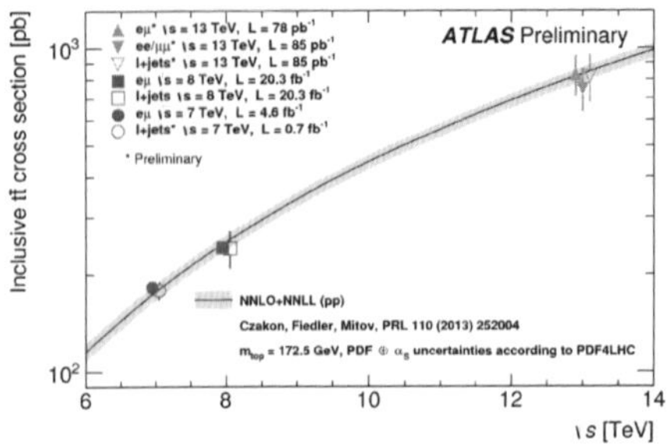

(a)

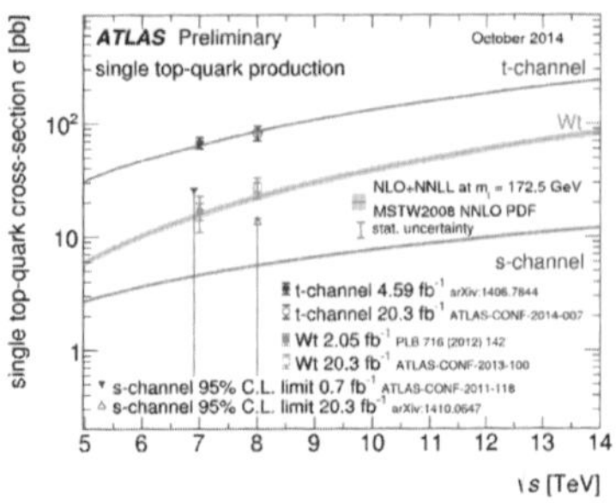

(b)

Figure 1: (a) Cross section for $t\bar{t}$ pair production in pp collisions as a function of centre-of-mass energy. The results at $\sqrt{s}$ = 13, 8 and 7 TeV are compared to the theory predictions [3]. (b) Summary of ATLAS measurements of the single top production cross sections in various channels as a function of the center of mass energy compared to theoretical calculations [8].

The observed spectra are softer than the Monte Carlo (MC) predictions. In the boosted topology, the differential cross section as a function of the p_{T} of the $t\bar{t}$ system is measured in the semileptonic decay channel at 8 TeV [6]. The measurements are performed in both a fiducial and total phase space. The measurements are also found to be softer than the MC predictions.

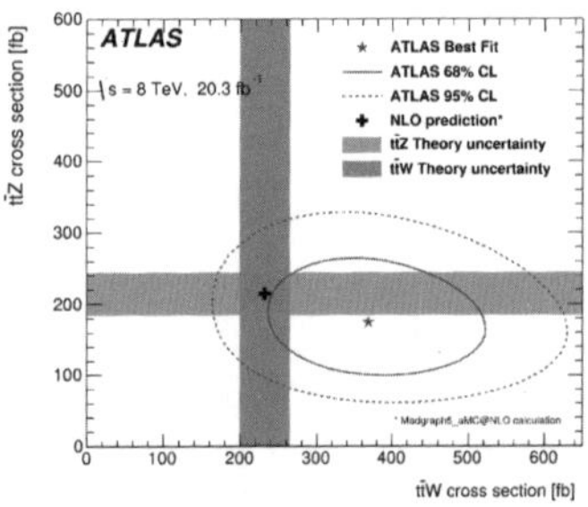

Figure 2: The result of the simultaneous fit to the $t\bar{t}W$ and $t\bar{t}Z$ cross sections along with the 68% and 95% CL uncertainty contours [7].

2.2 Single top production

Several measurements of the inclusive single top cross section in the s, t and Wt channel at 7 and 8 TeV have been performed by ATLAS. The latest 8 TeV results were presented in this conference. Figure 1b shows a summary of these measurements [8]. The inclusive cross sections for t-channel is found to be $\sigma_t = 82.6 \pm 1.2\,(\text{stat.}) \pm 11.4\,(\text{syst.}) \pm 3.1\,(\text{PDF}) \pm 2.3\,(\text{lumi.})$ pb, showing a good agreement with the NLO+NNLL[b] predictions [9], as well as the inclusive measurement in the Wt channel where the measured value is $\sigma_t = 23.0 \pm 1.3\,(\text{stat.}) \pm^{+3.2}_{3.5}\,(\text{syst.}) \pm 1.1\,(\text{lumi.})$ pb [10]. An upper limit for the s-channel cross section was established [11].

[b]Next-to-leading order plus next-to-next-to-leading-logarithm

2.3 Top pair production in association with vector bosons

The $t\bar{t}W$ and $t\bar{t}Z$ cross sections were measured in final states with two, three or four leptons [7]. In this analysis, a simultaneous fit is performed in 20 signal and control regions. The control regions are defined by varying the number jets and lepton charge in the event. The results agree with the SM prediction as shown in Figure 2.

3 Top quark properties

3.1 Mass

Measurements of the top mass, performed in $t\bar{t}$ events in the semileptonic and the dileptonic channels, were presented [12]. The analysis in the semileptonic channel uses a 3-dimensional template fit which determines simultaneously the top quark mass, the jet and b-jet energy scale factor. This measurement is the most precise mass measurement performed by ATLAS to date. The dilepton channel uses a one dimensional template fit and the combined result is found to be $m_{top} = 172.99 \pm 0.48(\text{stat.}) \pm 0.78(\text{syst.})$ GeV.

In single top decays, the measurement is performed through a one dimensional template fit using the invariant mass of the lepton and b-jet, in order to reduce the modelling uncertainties [13]. Neural networks are used to optimize the purity of the selection to 50%. The measured value is found to be $m_{\text{top}} = 172.2 \pm 0.7(\text{stat.}) \pm 2.0(\text{syst.})\,\text{GeV}$.

Moreover, a measurement of the pole mass using the normalized differential cross section for $t\bar{t}$ + 1-jet as a function of the inverse of the invariant mass of system was performed [14]. The selected events are identified using the lepton+jets top-quark pair decay channel, where lepton refers to either an electron or a muon. The measured value is found to be $m_{\text{top}} = 173.7 \pm 1.5(\text{stat.}) \pm 1.4(\text{syst.})^{+1.0}_{-0.5}(\text{theory})\,\text{GeV}$.

3.2 Spin correlation

Top pairs produced via the strong interaction are produced unpolarized at leading order. Their spins are correlated and information transferred to decay products. A measurement of the $t\bar{t}$ spin correlation at 8 TeV was performed [15]. In this analysis, the $t\bar{t}$ spin correlation is extracted from dilepton events by using the difference in the azimuthal angle of the two selected leptons. The measured value is consistent with the SM predictions. Moreover, the measurement of the angular distribution of the decay products is sensitive to super symmetric top squark (stop) pair production. The top squarks with masses between the top quark mass 172.5 GeV and 191 GeV are excluded at the 95% confidence level.

4 Searches

In this conference, limits in searches for flavor changing neutral currents (FCNC) were presented [16]. FCNC are predicted by the SM but are largely suppressed by the GIM mechanism. In Beyond the Standard Model (BSM) models the FCNC are enhanced, giving a window to search for new physics. One search is performed in $t\bar{t}$ events with one top quark decaying to a Z-boson and a light quark and the other to a W-boson and a b-quark. Both bosons are required to decay leptonically. No evidence of FCNC is found and an observed upper limit is established. Another search for FCNC is performed in single top production via $gu/c \to t$ with the top decaying to a Z-boson and a light quark [16]. Only the leptonic channel is considered and upper limits are established on the cross section times branching ratio and on the coupling constants of the processes.

5 Conclusions

ATLAS has performed multiple measurements in the top physics domain. The measurements of cross sections and properties are in agreement with the SM predictions. Some top quark properties were used to exclude BSM models and limits were established for FCNC searches.

References

[1] ATLAS Collaboration, 2008 JINST 3 S08003.
[2] ATLAS collaboration, Eur. Phys. J. C 74 (2014), no. 10, 3109 .
[3] M. Czakon, P. Fiedler and A. Mitov, Phys. Rev. Lett. 110 (2013), 252004.
[4] ATLAS collaboration, ATLAS-CONF-2015-033.
[5] ATLAS collaboration, JHEP 1506 (2015), 100.
[6] ATLAS collaboration, arXiv:1510.03818 [hep-ex].
[7] ATLAS collaboration, JHEP 1511 (2015), 172.
[8] ATLAS collaboration. Top public results. https://twiki.cern.ch/twiki/bin/view/AtlasPublic/TopPublicResults
[9] ATLAS collaboration, ATLAS-CONF-2014-007.
[10] ATLAS collaboration, ATLAS-CONF-2014-100.
[11] ATLAS collaboration, Phys. Lett. B 740 (2015), 118.
[12] ATLAS collaboration, Eur. Phys. J. C 75 (2015), no. 7, 330.
[13] ATLAS collaboration, ATLAS-CONF-2014-055.
[14] ATLAS collaboration, JHEP 1510 (2015), 121.
[15] ATLAS collaboration, Phys. Rev. Lett. 114 (2015), no. 14, 142001.
[16] ATLAS collaboration, arXiv:1508.05796 [hep-ex].
[17] ATLAS collaboration, arXiv:1509.00294 [hep-ex].

SUPERSYMMETRY SEARCHES IN ATLAS

Elena Romero Adam [a] on behalf of the ATLAS Collaboration
Instituto de Física Corpuscular (IFIC-CSIC), Calle Catedratico Jose Beltran, 2, E-46980 Paterna (Valencia), Spain

Abstract. Weak scale supersymmetry remains one of the best motivated and studied Standard Model extensions. This contribution summarises recent ATLAS results for searches for supersymmetric particles with the LHC Run 1 data at $\sqrt{s} =$ 8 TeV. A sensitivity study for the $\sqrt{s} = 13$ TeV data is also briefly presented.

1 Introduction

Supersymmetry (SUSY) [1] is a generalisation of space-time symmetries that predicts new bosonic partners for the fermions and new fermionic partners for the bosons of the Standard Model (SM). The physical superpartners of the SM particles would be: scalar partners of quarks and leptons (squarks ($\tilde{q}$) and sleptons ($\tilde{\ell}$)), fermionic partners of gauge and Higgs bosons (gluinos ($\tilde{g}$), charginos and neutralinos). The charginos ($\tilde{\chi}^{\pm}$) and neutralinos ($\tilde{\chi}^{0}$) are mixtures of the bino, winos and higgsinos, collectively referred to as the electroweakinos, that are superpartners of the U(1), SU(2) gauge bosons and the Higgs bosons, respectively. Their mass eigenstates are referred to as $\tilde{\chi}_i^{\pm}$ ($i = 1, 2$) and $\tilde{\chi}_j^0$ ($j = 1, 2, 3, 4$) in order of increasing mass. Since no superpartners of any of the SM particles have been observed, SUSY, if realised in nature, must be a broken symmetry, therefore SUSY particles (called sparticles) must be heavier than their SM partners and could be produced in the Large Hadron Collider (LHC) [2].

Several classes of phenomenological and simplified models covering different combinations of physics objects in the final state have been studied in ATLAS [3] with the LHC Run 1 data. Here just a few examples of the rich variety of results at $\sqrt{s}$ =8 TeV, using a total integrated luminosity of about 20 fb^{-1}, will be shown giving emphasis to the most recent results.

This document is organised as follows: Section 2 will present strong production searches, Section 3 will be dedicated to 3rd generation squarks production and Section 4 to electroweak production. In those sections only RPC searches will be considered, leaving RPV searches for Section 5. Long-lived particles searches will not be considered in this contribution but can be found in Refs. [4, 5].

2 Strong Production Searches

This Section presents several results from inclusive searches for gluinos and first- and second-generation squarks.

[a]E-mail: `Elena.Romero.Adam@cern.ch`

First, a simplified model with one-step decays is considered. In this case the pair-produced squarks decay via the $\tilde{\chi}_1^{\pm}$ into a W boson and the $\tilde{\chi}_1^0$. The free parameters are $m_{\tilde{q}}$ and $m_{\tilde{\chi}_1^0}$ with $m_{\tilde{\chi}_1^{\pm}} = (m_{\tilde{q}} + m_{\tilde{\chi}_1^0})/2$. This model was probed with two independent analyses: the 0-lepton (which requires the absence of isolated electrons or muons, at least two to at least six jets and significant $E_{\mathrm{T}}^{\mathrm{miss}}$) and the 1-lepton (requiring at least one isolated electron or muon, jets and high $E_{\mathrm{T}}^{\mathrm{miss}}$). The 95% CL exclusion limits obtained with the results of the two analyses separately as well as the statistical combination of both can be seen in Fig. 1. Details about the analyses and studies of other models can be seen in Ref. [6].

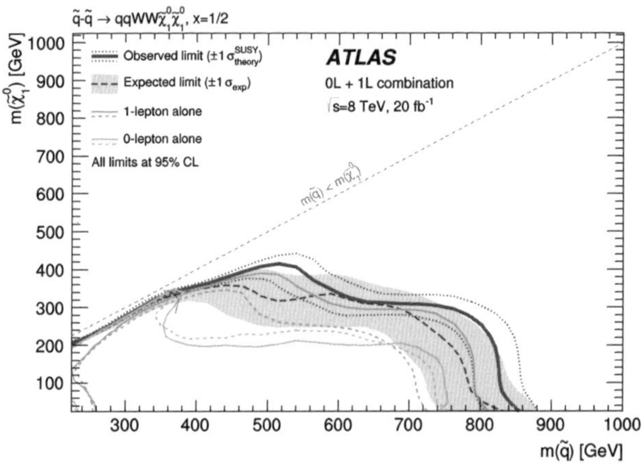

Figure 1: Observed and expected exclusion limits for simplified models of squark-pair production with one-step decays via the $\tilde{\chi}_1^{\pm}$ into a W boson and the $\tilde{\chi}_1^0$. The area below the observed limit is excluded at 95% CL. The yellow band includes all experimental uncertainties; the red dotted lines indicate the theory uncertainty on the cross-section. Taken from Ref. [6]

Another of the many interesting searches performed in the frame of the strong production searches is the Z+MET analysis [7], where events are required to contain at least two same-flavoured leptons (electrons or muons), with opposite charge, coming from the decay of a Z boson, at least two jets and high $E_{\mathrm{T}}^{\mathrm{miss}}$. After using data-driven methods to estimate all major backgrounds, which are thoroughly cross-checked with other methods (including data-driven, semi data-driven and MC), the total background estimation is 10.6 ± 3.2 events, whereas the observed number of events is 29. This corresponds to a significance of 3.0 σ. Figure 2 shows the number of observed events contrasted with the estimated background for all the regions considered in the analysis. This plot is especially interesting because it shows that the excess is not an effect of the incorrect estimation of any backgrounds. As it can be seen, in all the regions but the Signal Region (SR) the observed and estimated number of events are in very good agreement. The question if this excess is a hint of new physics or

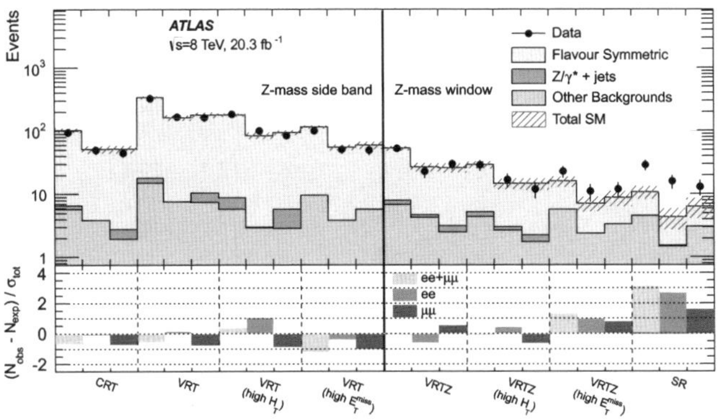

Figure 2: Summary of the agreement between data and prediction in the regions included in the Z+MET analysis. The detailed definition of all the regions can be found in Ref. [7]

a statistical fluctuation remains to be answered by the LHC Run 2 data. These results have been interpreted in a General Gauge Mediated (GGM) model and exclusion limits have been set [7].

3 Searches for Direct Pair Production of 3$^{\rm rd}$ Generation Squarks

Several searches targeting specific stop and sbottom final-state topologies can be found in Ref. [8]. Here an example will be shown where a series of simplified models is considered. It includes direct stop pair production as the only SUSY production process, and assumes that no supersymmetric particle other than the $\tilde{t}_1$ itself and the LSP, taken to be the lightest neutralino $\tilde{\chi}_1^0$, is involved in the decay. Under this assumption, there is little model dependence left in the stop phenomenology. The stop decay modes are defined mainly by the mass separation $\Delta m(\tilde{t}_1, \tilde{\chi}_1^0)$ between the stop and the neutralino. Figure 3 shows the 95% CL exclusion limits obtained in the $(m_{\tilde{t}_1}, m_{\tilde{\chi}_1^0})$ plane by the relevant analyses listed in Ref. [8].

4 Electroweak Production Searches

The ATLAS results for electroweakino searches at 8 TeV in the framework of simplified models are summarised in Fig. 4 in the (m($\tilde{\chi}_1^{\pm}$,$\tilde{\chi}_2^0$), m($\tilde{\chi}_1^0$)) plane. Each of the $\tilde{\chi}_1^{\pm}/\tilde{\chi}_2^0/\tilde{\chi}_3^0$ decays considered in the plot is assumed to have 100% branching fraction, and the production cross-section is for pure wino $\tilde{\chi}_1^+\tilde{\chi}_1^-$ and $\tilde{\chi}_1^{\pm}\tilde{\chi}_2^0$, and pure higgsino $\tilde{\chi}_2^0\tilde{\chi}_3^0$. The limits for $\tilde{\chi}_1^+\tilde{\chi}_1^-$ and $\tilde{\chi}_1^{\pm}\tilde{\chi}_2^0$ production with SM boson mediated decays are summarised in Fig. 4 (left). The limits for $\tilde{\chi}_1^+\tilde{\chi}_1^-$, $\tilde{\chi}_1^{\pm}\tilde{\chi}_2^0$ and $\tilde{\chi}_2^0\tilde{\chi}_3^0$ production with $\tilde{\ell}$-mediated decays are summarised in Fig. 4 (right). More details about the analyses that derived to these limits can be found in Ref. [9].

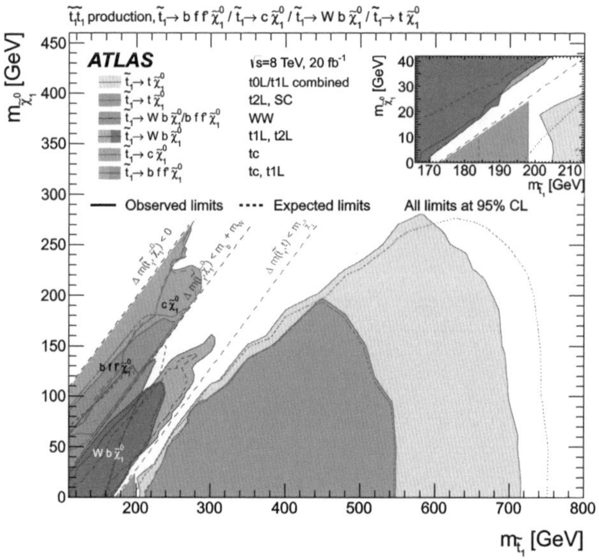

Figure 3: Summary of searches for direct stop pair production in models where no supersymmetric particle other than the $\tilde{t}_1$ and the $\tilde{\chi}_1^0$ is involved in the $\tilde{t}_1$ decay. The 95% CL exclusion limits are shown in the $(m_{\tilde{t}_1}, m_{\tilde{\chi}_1^0})$ mass plane. The dashed and solid lines show the expected and observed limits respectively, including all uncertainties except the theoretical signal cross-section uncertainty (PDF and scale). Four decay modes are considered separately with a branching ratio of 100%. Taken from Ref. [8]

5 R-parity-violating Searches

Introducing RPV [10] into supersymmetric models can significantly weaken mass and cross-section limits from collider experiments and also provide a rich phenomenology. Most relevant is the fact that the lightest supersymmetric particle (LSP) is unstable and decays to SM particles rather than escaping unseen as predicted by models that conserve R-parity. This contribution shows the phenomenological minimal supersymmetric Standard Model (pMSSM) with R-parity violation through bilinear terms (bRPV).

The 95% CL lower limits obtained on the μ (the higgsino mass parameter) and $m_{\tilde{q}_{L,L}}$ (the mass parameter for left-handed top and bottom squarks) parameters in natural SUSY with bRPV [10] are shown in Fig. 5.

6 Run1 SUSY Searches in Summary

This section summarises the combined sensitivity and constraints from 22 separate ATLAS analyses of the Run 1 LHC dataset. The interpretation of those

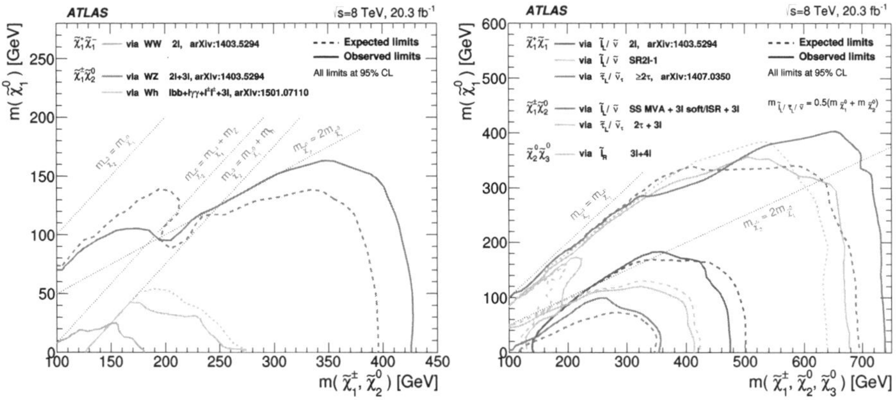

Figure 4: The 95% CL exclusion limits on $\tilde{\chi}_1^+\tilde{\chi}_2^-$, $\tilde{\chi}_1^\pm\tilde{\chi}_2^0$ and $\tilde{\chi}_2^0\tilde{\chi}_3^0$ production with (left) SM-boson-mediated decays and (right) $\tilde{\ell}$-mediated decays as a function of the $\tilde{\chi}_1^\pm$, $\tilde{\chi}_2^0$ and $\tilde{\chi}_1^0$ masses. The production cross-section is for pure wino $\tilde{\chi}_1^+\tilde{\chi}_2^-$ and $\tilde{\chi}_1^\pm\tilde{\chi}_2^0$, and pure higgsino $\tilde{\chi}_2^0\tilde{\chi}_3^0$. Taken from Ref. [9]

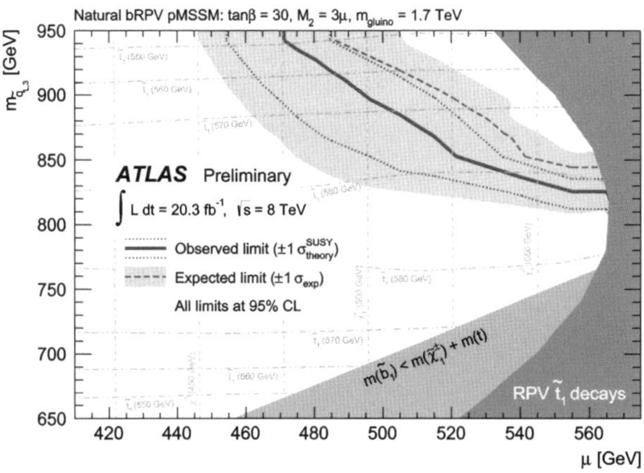

Figure 5: Observed (solid) and expected (dashed) 95% CL exclusion limit contours for the natural bRPV model. The region below and to the left of the solid line is excluded. The yellow band shows the effect of varying the experimental systematic uncertainties by $\pm 1\sigma$, while the dotted lines show the effect of the theoretical cross-section uncertainty. The other shaded areas at high μ are not explored in this analysis, see Ref. [10] for details.

results is done here within the wider framework of the pMSSM, where the over a hundred parameters of the MSSM are reduced to 19 applying a series of assumptions motivated by either experimental constraints or general features of possible SUSY breaking mechanisms. In this section the model is assumed to conserve R-parity and the LSP is required to be the lightest neutralino. A total of 310,327 model points are selected, each of which satisfies constraints from previous collider searches, precision measurements, cold dark matter energy density measurements and direct dark matter searches. More details about this study can be found in Ref. [11].

The impact of the ATLAS Run 1 searches on this model space is presented in Fig. 6, showing their overall effect in constraining such supersymmetric models. The plot shows the fraction of model points excluded for each sparticle as a function of the sparticle mass.

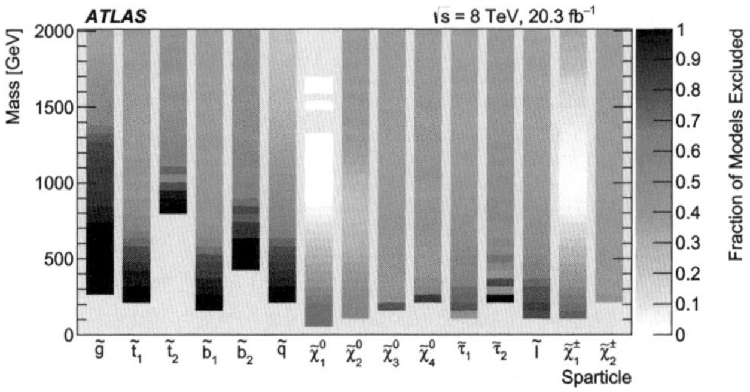

Figure 6: Each vertical bar is a 1D projection of the fraction of model points excluded, with colour coding representing the fraction of model points excluded for each sparticle. Taken from Ref. [11].

7 What's next?

This section describes the study of the expected sensitivity to pair-produced gluinos in final states with exactly one isolated lepton (an electron or a muon) and large missing transverse momentum. The sensitivity measure is the signal significance calculated based on the discovery p_0 values from the background-only hypothesis test for a counting experiment. The uncertainty on the SM background expectation is taken to be at the same level as the one observed in Run 1 (25%). Various assumptions on the integrated luminosity are evaluated. More details on this study and several others can be found in Ref. [12].

Figure 7 shows the best p_0-value obtained in the optimisation for the simplified model with gluino pair production and decay via a $\tilde{\chi}_1^{\pm}$ and two quarks to a $\tilde{\chi}_1^0$ and a W boson [12].

The LHC Run 2 has already started and several performance checks in preparation for SUSY searches have been made with an integrated luminosity about 78 pb^{-1} and $\sqrt{s} = 13$ TeV. These studies show good agreement between data and MC simulations.

8 Conclusion

Many SUSY searches have been performed, with Run 1 data, targeting different production processes and covering a wide range of final states. No significant

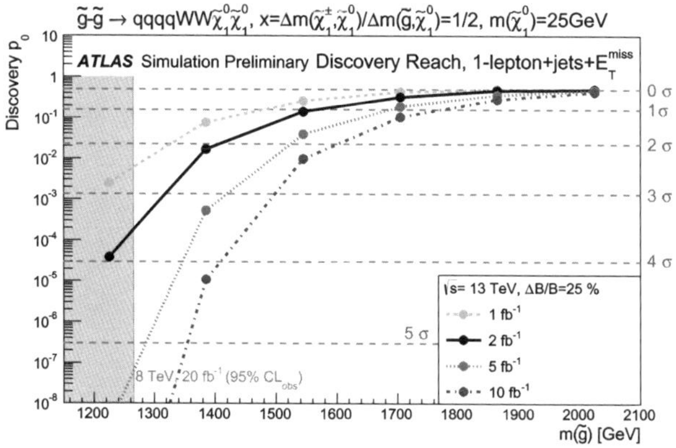

Figure 7: The best p_0-value obtained in the optimisation is shown for the simplified model with gluino pair production and decay via a chargino and two quarks to a neutralino and a W boson. Different integrated luminosities were assumed in the optimisation: 1, 2, 5 or 10 fb^{-1}. The plot assumes a total uncertainty of 25% on the background yield. Taken from Ref. [12]

excesses have been observed and 95% CL exclusion limits have been set for a large variety of models. However, an interesting 3 σ excess has been observed in the Z+MET [7] analysis, which will be further investigated with Run 2 data.

The LHC Run 2, which has already started, presents great potential for the discovery of new Physics due to its higher center-of-mass energy and luminosity.

Acknowledgments

The author acknowledges the support from the ATLAS project FPA2012-39055-C02-01 and the Centro de Excelencia Severo Ochoa SEV-2014-0398.

References

[1] S.P. Martin, arXiv:hep-ph/9709356 (1997).
[2] L. Evans and P. Bryant (editors) 2008 JINST 3 S08001.
[3] ATLAS Collaboration, JINST 3 (2008) S08003
[4] ATLAS Collaboration, Eur. Phys. J. C 75 (2015) 407 [arXiv:1506.05332].
[5] ATLAS Collaboration, JHEP 1501 (2015) 068 [arXiv:1411.6795].
[6] ATLAS Collaboration, JHEP 1510 (2015) 054 [arXiv:1507.05525].
[7] ATLAS Collaboration, Eur. Phys. J. C 75 (2015) 318 [arXiv:1503.03290].
[8] ATLAS Collaboration, Eur. Phys. J. C 75 (2015) 510 [arXiv:1506.08616].
[9] ATLAS Collaboration, arXiv:1509.07152 [hep-ex].

[10] ATLAS Collaboration, ATLAS-CONF-2015-018, http://cds.cern.ch/record/2017303.
[11] ATLAS Collaboration, JHEP 1510 (2015) 134 [arXiv:1508.06608].
[12] ATLAS Collaboration, ATL-PHYS-PUB-2015-005, http://cds.cern.ch/record/2002608.

SEARCH FOR EXOTIC PHYSICS BEYOND THE STANDARD MODEL WITH THE ATLAS DETECTOR

Yury Smirnov [a] on behalf of the ATLAS Collaboration
National Research Nuclear University "MEPhI" (Moscow Engineering Physics Institute), 115409 Moscow, Russia

Abstract. A summary is given of non-SUSY searches for New Physics with the ATLAS detector at the LHC. Shown results use a data sample collected with a center-of-mass energy of $\sqrt{s} = 8$ TeV and an integrated luminosity of around 20 fb^{-1} in proton-proton collisions. Four recent searches using leptons, photons, missing transverse momentum, and jets are presented. No significant deviations from Standard Model expectations are observed, hence new limits on a wide set of predictions for several Standard Model extensions are set.

1 Introduction

The Standard Model (SM), the most successful theory of particle physics, has its own obvious limits: it does not address phenomena such as gravity and hierarchy problem, dark matter, non-zero neutrino mass and some others. Plenty of SM extensions were proposed to answer these questions. This paper covers a variety of relevant searches for exotic phenomena performed with the ATLAS detector [1] at the CERN Large Hadron Collider.

2 Heavy Majorana neutrinos search

A unique feature of neutrinos (with respect to other SM fermions) is that neutrinos could be their own antiparticles, the so-called Majorana fermions. If this is so indeed, then the unusually low mass scale of the light neutrinos could be generated via a seesaw mechanism [2], implying the existence of a yet unobserved heavy Majorana neutrino. The search [3] was conducted for events with exactly two leptons with the same sign of an electric charge and at least two high-p_{T} jets (see Fig. 1). No significant excess of events was observed above the SM background. Limits were set on the cross-section times branching ratio, the mixing between heavy neutrinos and SM ones as a function of the heavy neutrino mass and, finally, on heavy Majorana neutrino masses.

3 High-mass diphoton resonances search

The diphoton channel is important for searches for physics beyond the SM since new high-mass states decaying to two photons are predicted in numerous SM extensions. One popular example of theories predicting a new high-mass diphoton resonance is the Randall-Sundrum (RS) model [4]. This model attempts to

[a] E-mail: `Yury.Smirnov@cern.ch`

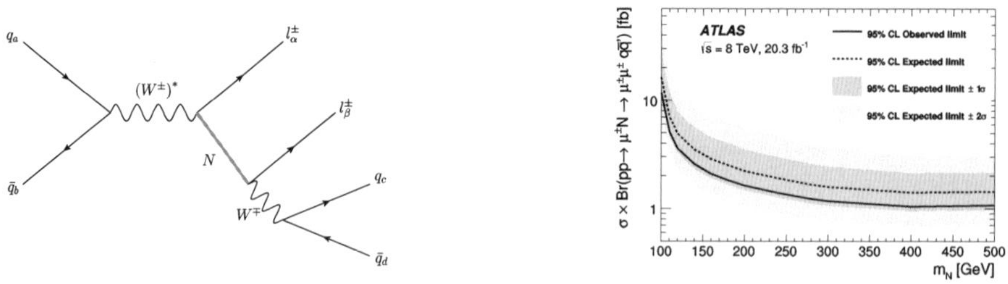

Figure 1: (left) The tree-level diagram for the production of a heavy Majorana neutrino (denoted by N); (right) Observed and expected 95% confidence level (CL) upper limits on the cross-section times branching ratio for the production of heavy Majorana neutrinos as a function of the heavy neutrino mass in the $\mu^{\pm}\mu^{\pm}$ channel. Taken from Ref. [3].

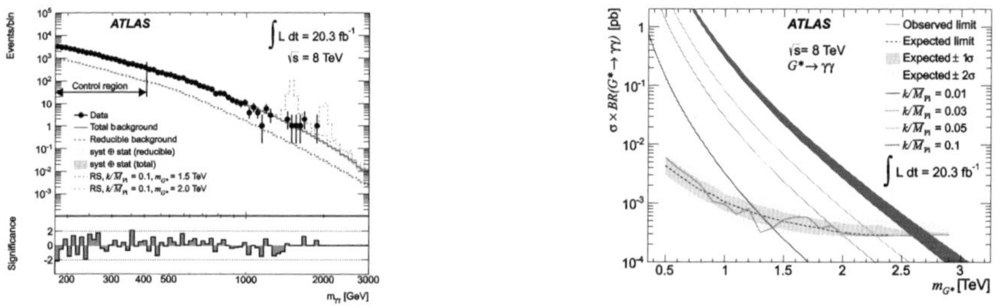

Figure 2: (left) Observed invariant mass distribution of the selected diphoton events, SM background prediction and two examples of signal predictions; (right) Expected and observed upper limits on $\sigma \times BR(G^* \to \gamma\gamma)$ at 95% CL as a function of the assumed value of the graviton mass. Also, the curves show the RS model predictions as a function of m_{G^*}. Taken from Ref. [5].

solve the so-called hierarchy problem of the SM (the fact that the electroweak scale is 16 orders of magnitude smaller than the Planck scale) by postulating an existence of Kaluza-Klein tower of graviton excitations G^* decaying to two photons. In this search, events with two isolated high-E_{T} photons were sought for [5], with dominant background coming from SM processes with prompt $\gamma\gamma$ production (see Fig. 2). No significant excess over the expected background was observed, and the results were interpreted in terms of upper limits on the production cross-section times branching ratio as a function of the resonance mass.

4 Multi-charged particles search

Several theories beyond the SM predict particles possessing an electric charge above the elementary charge e: the almost-commutative model [6], the walking technicolor model [7] and the left-right symmetric model [8]. Any observation of particles predicted by the first two models could have implications for

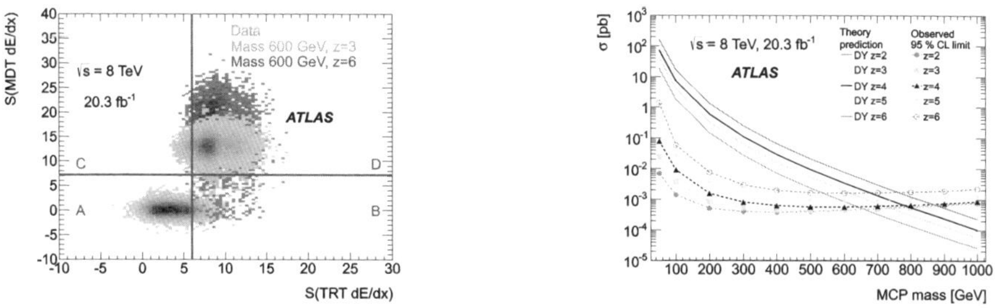

Figure 3: (left) Distributions of ionization estimators in two ATLAS subsystems for data and two simulated signal samples with multi-charged particles; (right) Observed 95% CL cross-section upper limits and theoretical cross-sections as functions of the sought particles mass. Taken from Ref. [9].

the formation of composite dark matter, while particles from the third model could shed some light on new symmetries beyond SM. This analysis sought for particles traversing the entire ATLAS detector, leaving an abnormally large ionization signal along their tracks [9]. Again, the observed results were consistent with the expected background (see Fig. 3), and they were interpreted as the limits on production cross-section and mass of the multi-charged particles.

5 Search for long-lived neutral particles decaying into lepton jets

Several possible extensions of the SM predict the existence of a hidden sector, which is weakly coupled to the visible one [10]. Some unstable hidden states may be produced at colliders and decay back to SM particles with sizeable branching fractions. If the lightest unstable hidden states have masses in the MeV to GeV range, they would decay mainly to leptons and possibly light mesons. This search studied a case where the two sectors couple via a vector portal, in which a light hidden photon (dark photon, $\gamma_{\rm d}$) mixes kinetically with the SM photon [11]. If the dark photon is the lightest state in the hidden sector, it decays back to SM particles. Due to their small mass, these particles are typically produced with a large boost and, due to their weak interactions, can have non-negligible lifetime. As a result, collimated jet-like structures containing pairs of electrons and/or muons and/or charged pions ("lepton jets", LJs) may have been produced far from the primary interaction vertex of the event (see Fig. 4).

No excess with respect to the background expectation was observed, and the upper limits were set on the $\sigma \times BR$ for non-SM Higgs boson decays to LJs, $H \to 2\gamma_{\rm d} + X$ and $H \to 4\gamma_{\rm d} + X$, as a function of the long-lived particle mean lifetime. Also, the results were interpreted as exclusion contours in the kinetic mixing parameter ϵ vs. $\gamma_{\rm d}$ mass plane and significantly improved the constraints from other experiments (see Fig. 5).

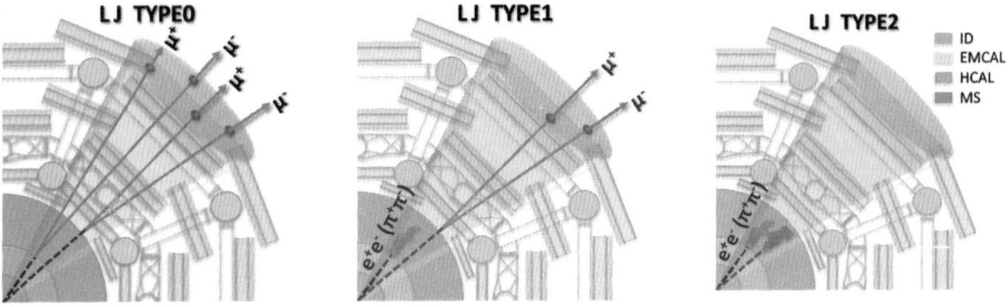

Figure 4: Schematic picture of the LJ classification according to the $\gamma_{\rm d}$ decay final states: left TYPE0 LJ (muons only), center TYPE1 LJ (muons and jets), right TYPE2 LJ (jets only). Taken from Ref. [11].

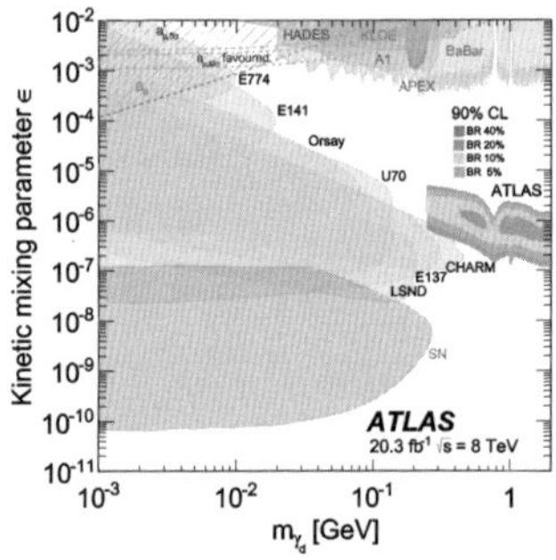

Figure 5: Parameter space exclusion plot for dark photons as a function of the $\gamma_{\rm d}$ mass and of the kinetic mixing parameter ϵ. Shown are existing 90% CL exclusion regions from several previous searches, as well as the exclusion limits from the present search. Taken from Ref. [11].

References

[1] ATLAS Collaboration, JINST **3** S08003 (2008).
[2] R. N. Mohapatra and G. Senjanović, Phys. Rev. **D 23**, 165 (1981).
[3] ATLAS Collaboration, JHEP07 (2015) 162.
[4] L. Randall and R. Sundrum, Phys. Rev. Lett. **83**, 3370 (1999).
[5] ATLAS Collaboration, Phys. Rev. **D 92**, no. 3, 032004 (2015).
[6] C. A. Stephan, J. Phys. A **39**, 9657 (2006).
[7] F. Sannino and K. Tuomine, Phys. Rev. **D 71**, 051901 (2005).
[8] R. N. Mohapatra and J. C. Pati, Phys. Rev. **D 11**, 566 (1975).
[9] ATLAS Collaboration, Eur. Phys. J. C **75**, no. 8, 362 (2015).
[10] M. J. Strassler and K. M. Zurek, Phys. Lett. B **651**, 374 (2007).
[11] ATLAS Collaboration, JHEP11 (2014) 088.

TOTEM PHYSICS SEARCHES AND FORWARD PHYSICS AT THE LHC

Christophe Royon [a]
on behalf of the TOTEM Collaboration
University of Kansas, Lawrence, USA
Institute of Physics, Academy of Science, Prague, Czech Republic
Nuclear Physics Institute (PAN), Cracow, Poland

Abstract. We will describe some of the future measurements to be performed by the TOTEM collaboration on soft and hard diffraction in order to understand better the structure of the Pomeron. We will also describe the prospects concerning the search for quartic $\gamma\gamma\gamma\gamma$ anomalous couplings.

1 The TOTEM detectors and their acceptance

The TOTEM collaboration installed some roman pots located at about 200-220 meters from the CMS interaction point on both sides of CMS in order to measure intact protons after collisions [1]. In addition, there were some roman pot detectors located at 147 m but they were moved to 220 m during the 2015 shutdown. The TOTEM collaboration also installed the $T1$ and $T2$ forward station in the very forward region of CMS that are sensitive to the $3 < |\eta| < 6.5$ region in pseudo-rapidity. These detectors can be used as veto for total cross section and soft diffraction measurements. The TOTEM pots alone can be used to measure soft diffraction as well as total, elastic and inelastic cross sections.

It is possible to collect combined data between CMS and TOTEM [1]. The CMS-TOTEM collaboration aims at combining the data using the vertical roman pots of TOTEM and the CMS detector. The acceptance in (ξ, t) for high $\beta^* \sim 90$m is given in Fig. 1, left (t is the transfered energy squared at the proton vertex and ξ is the proton momentum fraction carried by the Pomeron). We see that we have a good acceptance in a large domain in ξ and thus in diffractive mass. The luminosity accumulated in special high-β^* is rather low and the high-β^* measurements are restricted to processes with a high enough cross sections. The CMS-TOTEM-Precision Proton Spectrometer (CT-PPS) aims at taking data using the horizontal pots of TOTEM and the CMS detector at high luminosity for the standard low-β^* lattice parameters of the LHC. The acceptance is shown in Fig. 1, right. We see that we have a good acceptance for rather high values of ξ up to 0.15, so at high diffractive masses. For these runs, the luminosity is quite high and we are sensitive to processes with low cross section. The situation is thus complementary to the high-β^* running scenario.

[a]E-mail: `christophe.royon@cern.ch`

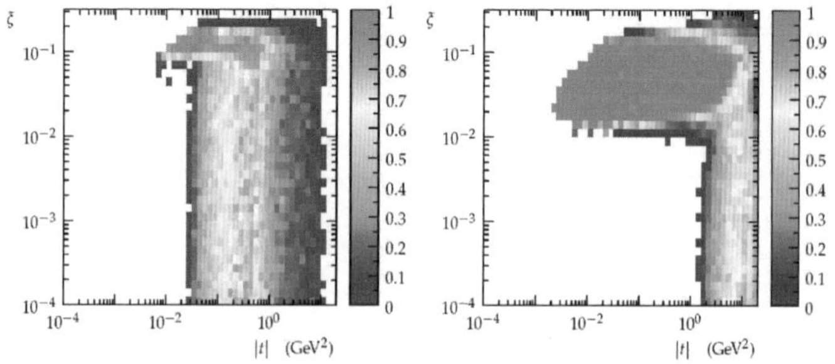

Figure 1: Acceptance in (t,ξ) of the roman pots detectors for high-β^* (left) and low-β^* (right).

2 Measurements at low and medium luminosities

2.1 TOTEM measurements on soft diffraction and total cross section

Many data have already been accumulated at 8 TeV with a high value of β^* of 1 km. This optics allowed TOTEM to reach very low values of $|t|$ down to 6. 10^{-4} GeV^2 since the detectors could approach the beam at 3σ. This low value of $|t|$ allows accessing the elastic scattering in the Coulomb-Nuclear region and a measurement of ρ is thus possible [2]. Much luminosity with billions of events have also been accumulated recently at 13 TeV and new measurements are expected soon. Special high β^* =2.5 km optics is also expected to be achieved in 2016 or 2017, allowing TOTEM to reach even lower values of t. Using the same soft diffrsction data, it is is also possible to look for resonances in soft diffraction and for glueballs as we will see when we discuss exclusive diffraction.

2.2 Mesaurement at medium luminosities: Hard diffraction

In this section, we will discuss briefly how we can assess the gluon and quark contents in the Pomeron at the LHC, and also the search for BFKL resummation effects.

Inclusive diffraction measurements at the LHC We discuss potential measurements at the LHC that can constrain the Pomeron structure. The Pomeron structure in terms of quarks and gluons has been derived from QCD fits at HERA and it is possible to probe this structure and the QCD evolution at the LHC in a completely new kinematical domain. All the following studies have been performed using the Forward Physics Monte Carlo (FPMC), a generator that has been designed to study forward physics, especially at the LHC [3]. It aims to provide a variety of diffractive processes in one common framework, *i.e.* single diffraction, double pomeron exchange, central exclusive production and two-photon exchange.

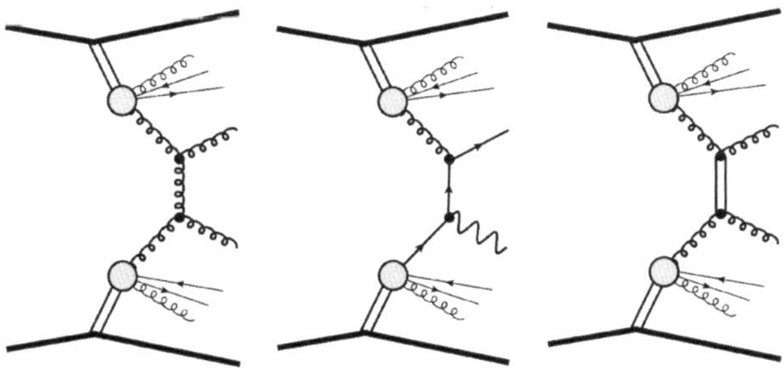

Figure 2: Inclusive diffractive diagrams. From left to right: jet production in inclusive double pomeron exchange, γ+jet production in DPE, jet gap jet events

Dijet production in double Pomeron exchanges processes and sensitivity to the gluon density in the pomeron

One can first probe if the Pomeron is universal between *ep* and *pp* colliders, or in other words, if we are sensitive to the same object at HERA and the LHC using as an example dijet production in single diffractive and double pomeron exchange at the LHC. It is possible to assess the gluon and quark densities using the dijet and $\gamma + jet$ productions. The different diagrams of the processes that can be studied at the LHC are shown in Fig. 2, namely double pomeron exchange (DPE) production of dijets (left), of γ+jet (middle), sensitive respectively to the gluon and quark contents of the Pomeron, and the jet gap jet events (right).

The dijet production in DPE events at the LHC is sensitive to the gluon density in the Pomeron. In order to quantify how well we are sensitive to the Pomeron structure in terms of gluon density at the LHC, we display in Fig. 3, the dijet mass fraction, the ratio of the dijet mass to the total diffractive mass [4,5]. The central black line displays the cross section value for the gluon density in the Pomeron measured at HERA including an additional survival probability of 0.03. The yellow band shows the effect of the 20% uncertainty on the gluon density taking into account the normalisation uncertainties. The dashed curves display how the dijet cross section at the LHC is sensitive to the gluon density distribution especially at high β. For this sake, we multiply the gluon density in the Pomeron from HERA by $(1-\beta)^{\nu}$ where ν varies between -1 and 1. When ν is equal to -1 (resp. 1), the gluon density is enhanced (resp, decreased) at high β. We note that the curves corresponding to the different values of ν are much more separated at high values of the dijet mass fraction, meaning that this observable is indeed sensitive to the gluon density at high β.

Sensitivity to the Pomeron structure in quarks using $\gamma+$jet events and W asymmetry

The QCD diffractive fits performed at HERA assumed that $u = d = s =$

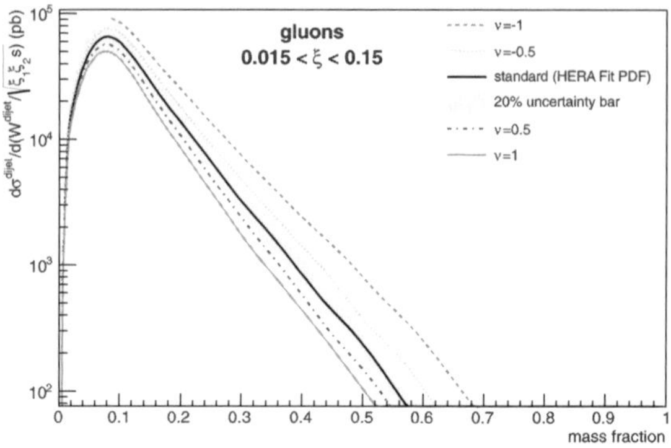

Figure 3: DPE di-jet mass fraction distribution. The different curves correspond to different modifications of the Pomeron gluon density extracted from HERA data (see text).

$\bar{u} = \bar{d} = \bar{s}$, since data were not sensitive to the difference between the different quark component in the Pomeron. On the contrary, measuring the γ+jet to the dijet cross section ratios as a function of the diffractive mass M allows to distinguish between different assumptions on the quark content of the Pomeron. For instance, varying d/u between 0.25 and 4 leads to a variation of the cross section ratio by a factor 2.5.

In addition, it is possible to use the W asymmetry in single diffractive W production [6]. Typically, the muon asymmetry is directly sensitive to the quark content of the pomeron and varies by a factor 6 at low ξ between the assumptions $u/d = 2$ or $u/d = 1/2$ for the quark content in the pomeron.

Jet gap jet production in double Pomeron exchanges processes

The observation of gap between jets in double Pomeron exchange events [8,9] is ideal to probe the BFKL [7] evolution equation. We predict that the ratio of jet-gap-jet with respect with "standard" dijet production in DPE events will be of the order of 20%, which would represented a good test of the BFKL evolution equation.

2.3 Exclusive processes in diffraction

The advantage of the exclusive diffractive and photon exchange processes is that all particles can be measured in the final state. Both protons can be measured in CMS-TOTEM/CT-PPS and the produced particles (jets, vector mesons, Z boson....) in CMS, and there is no energy losses such as in the pomeron remnants. It is thus possible to reconstruct the properties of the object produced exclusively (via photon and gluon exchanges) from the tagged proton since the system is completely constrained. It is worth mentioning that it is also possible to constrain the background by requesting the matching between the information of the two protons and the produced object, and thus, central exclusive production is a potential channel for beyond standard modelphysics at high masses.

For instance, the CMS-TOTEM experiment performed extensive studies of possible measurements of exclusive states at high β^*. It is worth mentioning that the search for glueball states and the probe of the low x gluon density down to $x \sim 10^{-4}$ will be possible. It is possible to discover or exclude glueballs at low masses since the 1-10 GeV region in masses can be probed diffractively ($\xi \sim 10^{-4} - 10^{-3}$), ensuring pure gluonic exchanges. It is then possible to check the $f_0(1500)$ or $f_0(1710)$ glueball candidates (lattice calculations predict a 0++ glueball at 1.7 GeV with a ~100 MeV uncertainty, favoring the $f_0(1710)$ candidate). The simulation of a possible signal of $f_0(1710) \to \rho^0\rho^0$ and of the non resonant $\rho^0\rho^0$ background including the CMS tracker performance leading to a 20-30 MeV resolution was performed and a luminosity of ~ 0.06 pb^{-1} is needed for a 7 σ signal. About 0.6 pb^{-1} is needed for decay characterisation In addition, it is possible to perform a spin analysis of $f_0(1710) \to \rho^0\rho^0 \to 4\pi$ to determine if $J = 0$ or 2: as an example, a measurement of the polar angle of the $\pi^+\pi^-$ pair for the ρ candidate needs about 5 pb^{-1}.

In addition, using the same data set, the measurement of the cross section times branching ratio for the three $\chi_{C,0,1,2}$ states will be performed allowing a comparison with the results of the LHCb measurement.

3 Search for quartic $\gamma\gamma\gamma\gamma$ anomalous coupling events in diffraction

In this section, we will describe briefly the search for $\gamma\gamma\gamma\gamma$ quartic anomalous couplings using proton tagging at the LHC that regained interest after the excess observed in diphoton production by the CMS and ATLAS experiments [10, 11] (the search for $WW\gamma\gamma$ and $ZZ\gamma\gamma$ couplings is given in [12]).

3.1 Theoretical motivations

Four-photon (4γ) interactions through diphoton production via photon fusion with intact outgoing protons are considered. The pure photon dimension-eight operators read

$$\mathcal{L}_{4\gamma} = \zeta_1^\gamma F_{\mu\nu}F^{\mu\nu}F_{\rho\sigma}F^{\rho\sigma} + \zeta_2^\gamma F_{\mu\nu}F^{\nu\rho}F_{\rho\lambda}F^{\lambda\mu} \tag{1}$$

and they can induce the $\gamma\gamma\gamma\gamma$ process, highly suppressed in the SM [13]. We discuss here possible new physics contributions to $\zeta_{1,2}^\gamma$ that can be probed and discovered at the LHC using the forward proton detectors.

Loops of heavy charged particles contribute to the 4γ couplings [13] as $\zeta_i^\gamma = \alpha_{\rm em}^2 Q^4 m^{-4} N c_{i,s}$, where $c_{1,s}$ is related to the spin of the heavy particle of mass

Cut / Process	Signal (full)	Signal with/without f.f. (EFT)	Excl.	DPE	DY dijet+ pile up	$\gamma\gamma$+ pile up
$[0.015 < \xi_{1,2} < 0.15$, $p_{T1,(2)} > 200$ (100) GeV]	65.	18. (187.)	0.13	0.2	1.6	2968
$m_{\gamma\gamma} > 600$ GeV	64.	17. (186.)	0.10	0	0.2	1023
$[p_{T2}/p_{T1} > 0.95$, $\|\Delta\phi\| > \pi - 0.01]$	64.	17. (186.)	0.10	0	0	80.2
$\sqrt{\xi_1\xi_2 s} = m_{\gamma\gamma} \pm 3\%$	61.	12. (175.)	0.09	0	0	2.8
$\|y_{\gamma\gamma} - y_{pp}\| < 0.03$	60.	16. (169.)	0.09	0	0	0

Table 1: Number of signal for $Q_{\text{eff}} = 4$, $m = 340$ GeV and background events after various selections for an integrated luminosity of 300 fb^{-1} and $\mu = 50$ at $\sqrt{s} = 14$ TeV. Values obtained using the corresponding EFT couplings with and without form factors are also displayed. Excl. stands for exclusive backgrounds and DPE for double pomeron exchange backgrounds.

m running in the loop and Q its electric charge. These couplings scale as $\sim Q^4$ and are enhanced in presence of particles with large charges. For a 500 GeV vector (fermion) resonance with $Q = 3$ (4), large couplings ζ_i^γ of the order of $10^{-13} - 10^{-14}$ Gev^{-4} can be reached.

Beyond perturbative contributions to ζ_i^γ from charged particles, non-renormalizable interactions of neutral particles are also present in common extensions of the SM. Such theories can contain scalar, pseudo-scalar and spin-2 resonances that couple to the photon and generate the 4γ couplings by tree-level exchange as $\zeta_i^\gamma = (f_s\, m)^{-2}\, d_{i,s}$, where $d_{1,s}$ is related to the spin of the particle.

3.2 Experimental sensitivity to quartic four photon couplings

The $\gamma\gamma\gamma\gamma$ process can be probed via the detection of two intact protons in the forward proton detectors and two energetic photons in the corresponding electromagnetic calorimeters.

The anomalous $\gamma\gamma\gamma\gamma$ process has been implemented in the Forward Physics Monte Carlo (FPMC) generator [3]. The FPMC generator was also used to simulate the background processes giving rise to two intact protons accompanied by two photons, electrons or jets that can mimic the photon signal. Those include exclusive SM production of $\gamma\gamma\gamma\gamma$ via lepton and quark boxes and $\gamma\gamma \to e^+e^-$. The central exclusive production of $\gamma\gamma$ via two-gluon exchange, was simulated using ExHuME. This series of backgrounds is called "Exclusive" in Table 1. FPMC was also used to produce $\gamma\gamma$, Higgs to $\gamma\gamma$ and dijet productions via double pomeron exchange (called DPE background in Table 1 and Fig. 4). Such backgrounds tend to be softer than the signal and can

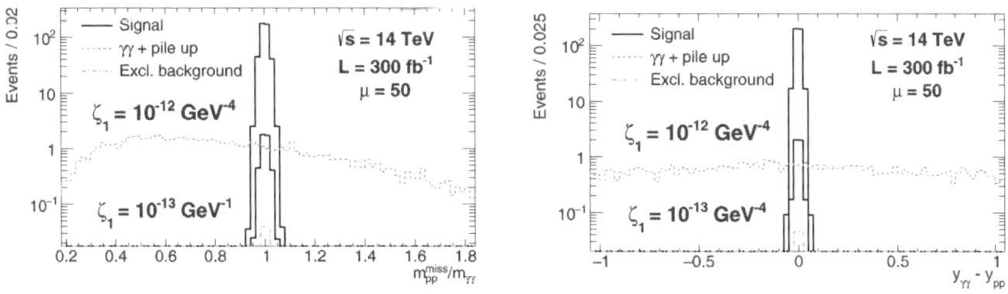

Figure 4: Diphoton to missing proton mass ratio (left) and rapidity difference (right) distributions for signal and backgrounds.

be suppressed with requirements on the transverse momenta of the photons and the diphoton invariant mass. In addition, the final-state photons of the signal are typically back-to-back and have about the same transverse momenta. Requiring a large azimuthal angle $|\Delta\phi| > \pi - 0.01$ between the two photons and a ratio $p_{T,2}/p_{T,1} > 0.95$ greatly reduces the contribution of non-exclusive processes.

Additional background processes include the quark and gluon-initiated production of two photons, two jets and Drell-Yan processes leading to two electrons. The two intact protons arise from pile-up interactions (these backgrounds are called $\gamma\gamma$ + pile-up and DY, dijet + pile-up in Table 1). The pile-up background is further suppressed by requiring the proton missing invariant mass to match the diphoton invariant mass within the expected resolution and the diphoton system rapidity and the rapidity of the two protons to be similar.

The number of expected signal and background events passing respective selections is shown in Table 1 for an integrated luminosity of 300 fb^{-1} for a center-of-mass energy of 14 TeV. Exploiting the full event kinematics with the forward proton detectors allows to completely suppress the background with a signal selection efficiency after the acceptance cuts exceeding 70%. Tagging the protons is absolutely needed to suppress the $\gamma\gamma$ + pile-up events. Further background reduction is even possible by requiring the photons and the protons to originate from the same vertex that provides an additional rejection factor of 40 for 50 pile-up interactions, showing the large margin on the background suppression. A similar study at a higher pile-up of 200 was performed and led to a very small background. The sensitivity extends up to $7 \cdot 10^{-15}$ GeV^{-4} allowing us to probe further the models of new physics described above.

We also performed a full amplitude calculation in Ref. [13] that avoids the dependence on the choice of form factors needed in order to avoid quadratic divergences of scattering amplitudes. Sensitivities were found to be similar leading to possible discoveries of vector or fermions at high masses and high effective charges.

Recently, the interest of diphoton production in exclusive events regained a considerable interest due to the possible excess seen by the ATLAS and CMS collaborations in diphoton cross sections at about 750 GeV [11], By tagging the protons in the final state, it will be possible to select photon-induced processes only without any significant background [10].

In this short report, we presented present and potential measurements to be performed by the TOTEM and CMS collaborations related to a better understanding of the pomeron structure in terms of QCD and to exploratory physics where unprecedented sensitivities can be reached on quartic anomalous $\gamma\gamma\gamma\gamma$ couplings.

References

[1] TOTEM Coll., CERN-LHCC-2014- 020; CMS and TOTEM Coll., CERN-LHCC-2014-021.

[2] TOTEM Coll., Europhysics Lett. **101** (2013) 21004; Europhysics Lett. **101** (2013) 21003; Europhysics Lett. **101** (2013) 21002; Europhysics Lett. **96** (2011) 21002; Europhysics Lett. **95** (2011) 41001; Nucl. Phys. B **899** (2015) 527; Phys. Rev. Lett. **111**, 012001 (2013).

[3] M. Boonekamp et al., e-Print: arXiv:1102.2531;

[4] C, Marquet, C. Royon, M. Saimpert, D. Werder, Phys.Rev. D **88** (2013) 7, 074029.

[5] O. Kepka, C. Royon, Phys.Rev. D**76** (2007) 034012; CDF Coll., Phys. Rev. D **77**, (2008) 052004.

[6] A. Chuinard, C. Royon, Rafal Staszewski, preprint arXiv:1510.04218.

[7] V. S. Fadin, E. A. Kuraev, L. N. Lipatov, Phys. Lett. B**60** (1975) 50; I. I. Balitsky, L. N. Lipatov, Sov.J.Nucl.Phys. **28** (1978) 822;

[8] C. Marquet, C. Royon, M. Trzebinski, R. Zlebcik, Phys.Rev. D**87** (2013) 3, 034010; O. Kepka, C. Marquet, C. Royon, Phys. Rev. D**83** (2011) 034036.

[9] C. Marquet, C. Royon, Phys. Rev. D**79** (2009) 034028; O. Kepka, C. Royon, C. Marquet, R. Peschanski, Eur. Phys .J. C**55** (2008) 259-272; Phys. Lett. B**655** (2007) 236-240; H. Navelet, R. Peschanski, C. Royon, S. Wallon, Phys. Lett. B**385** (1996) 357; H. Navelet, R. Peschanski, C. Royon, Phys. Lett. B366 (1996) 329.

[10] S. Fichet, G. von Gersdorff, C. Royon, preprint : arXiv:1512.05751; S. Fichet, G. von Gersdorff, C. Royon, preprint : arXiv:1601.01712.

[11] CMS Coll., CMS-PAS-EXO-15-004; ATLAS Coll., ATLAS-CONF-2015-081.

[12] E. Chapon, O. Kepka, C. Royon, Phys. Rev. **D81** (2010) 074003; O. Kepka and C. Royon, Phys. Rev. D **78** (2008) 073005.

[13] S. Fichet, G. von Gersdorff, B. Lenzi, C. Royon, M. Saimpert, JHEP 1502 (2015) 165; S. Fichet, G. von Gersdorff, O. Kepka, B. Lenzi, C. Royon, M. Saimpert, Phys.Rev. D**89** (2014) 114004.

LHC OPTICS AND ELASTIC SCATTERING MEASURED BY THE TOTEM EXPERIMENT

F. Nemes, for the TOTEM Collaboration[1,2 a]

[1] *CERN, Geneva, Switzerland*
[2] *Wigner RCP, H-1121 Budapest XII, Konkoly-Thege 29-33, Hungary*

Abstract. The TOTEM experiment at the LHC has measured proton-proton elastic scattering in dedicated runs at $\sqrt{s} = 7$ and 8 TeV centre-of-mass LHC energies. The proton-proton total cross-section $\sigma_{\rm tot}$ has been derived for both energies using a luminosity independent method. TOTEM has excluded a purely exponential differential cross-section for elastic proton-proton scattering with significance greater than 7σ in the $|t|$ range from 0.027 to 0.2 GeV^2 at $\sqrt{s} = 8$ TeV.

1 Introduction

The TOTEM (TOTal cross section, Elastic scattering and diffraction dissociation Measurement at the LHC) experiment has been designed to measure the total proton-proton (pp) cross-section, elastic scattering and diffractive processes at the LHC [1], see Fig. 1.

The experimental apparatus of TOTEM is composed of three subdetectors: the Roman Pots (RP), the T1 and T2 inelastic forward telescopes. The detectors are placed symmetrically on both sides of the Interaction Point 5 (IP5), which is shared with the CMS experiment.

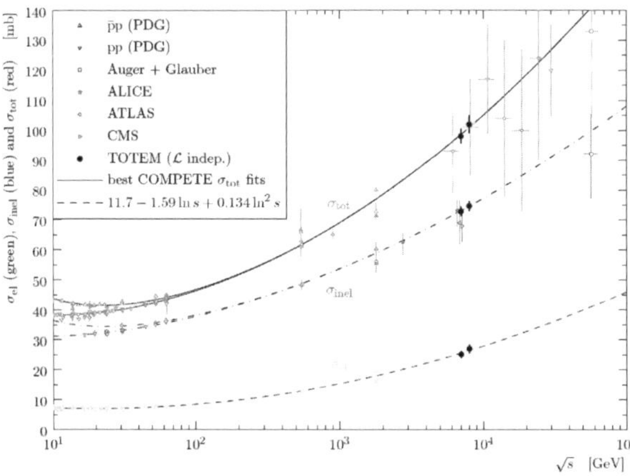

Figure 1: A compilation of total, inelastic and elastic pp cross-section measurements. The black points indicate the TOTEM measurements at $\sqrt{s} = 7$ and 8 TeV using the luminosity independent method [2–7].

[a]E-mail: `fnemes@cern.ch`

The RPs are moveable beam-pipe insertions, hosting edgeless silicon detectors to detect leading protons scattered at very small angles. In order to maximize the acceptance of the experiment for elastically scattered protons, the RPs are able to approach the beam center to a transverse distance as small as 1 mm. The alignment of the RPs is optimized by reconstructing common tracks going through the overlap between the vertical and horizontal RPs [1,8].

Before the LHC long shutdown one (LS1) the RPs, used for measurements, were located at distances of 215–220 m from IP5 [1]. The actual layout, i.e., after the LHC LS1, is different in RP location and quantity. The RP stations previously installed at ±147 m, from IP5, have been relocated to ±210 m. Moreover, two newly designed horizontal RPs have been installed between the two units of the ±220 m station [9, 10].

2 Elastic scattering and total cross-section σ_{tot} measurements

For each tagged elastic event the four-momentum transfer squared t is reconstructed using the LHC optical functions, characterized with the so-called betatron amplitude at IP5 β^* [1]. The TOTEM experiment developed a novel experimental method to estimate the optical functions at the RP locations, using the measured elastically scattered protons [11–14], see Fig. 2.

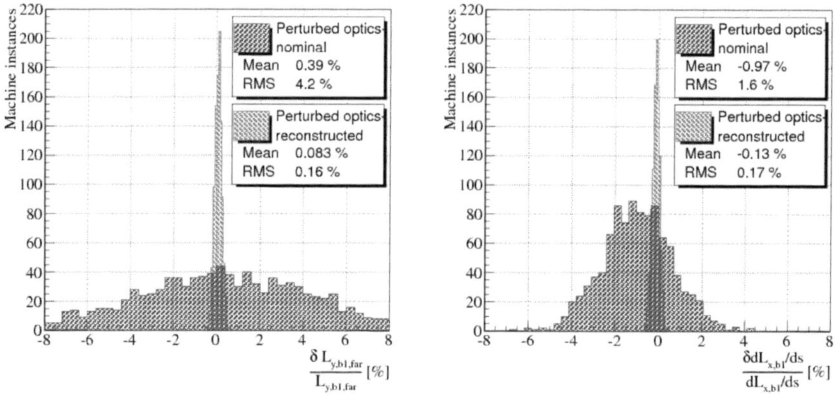

Figure 2: The MC error distribution of $\beta^* = 3.5$ m optical functions L_y and dL_x/ds for Beam 1 at $\sqrt{s} = 7$ TeV, before and after optics estimation.

The total inelastic rate N_{inel}, measured by the T1 and T2 telescopes, and the total nuclear elastic rate N_{el} with its extrapolation to zero four-momentum transfer squared $t = 0$ are combined with the optical theorem to obtain the total cross-section in a luminosity, $\mathcal{L}$, independent way

$$\sigma_{\text{tot}} = \frac{16\pi}{1+\rho^2} \cdot \left.\frac{\mathrm{d}N_{\text{el}}}{\mathrm{d}t}\right|_{t=0} \cdot (N_{\text{el}} + N_{\text{inel}})^{-1}. \tag{1}$$

The measured ratio of the elastic and inelastic rates $N_{\text{el}}/N_{\text{inel}}$ allows for the determination of the elastic and inelastic cross-sections as well [1, 14].

The TOTEM experiment determined the total pp cross-section at $\sqrt{s}$ = 7 TeV using the luminosity independent method [2], which was shown to be consistent with the total cross-sections measured in independent ways, see Table 1. The elastic and inelastic cross-sections were found to be $\sigma_{\text{el}} = 25.1 \pm 1.1$ mb and $\sigma_{\text{inel}} = 72.9 \pm 1.5$ mb.

The measurement was repeated at $\sqrt{s} = 8$ TeV, yielding to $\sigma_{\text{tot}} = 101.7 \pm 2.9$ mb, $\sigma_{\text{el}} = 27.1 \pm 1.4$ mb and $\sigma_{\text{inel}} = 74.7 \pm 1.7$ mb [3]. A compilation of the results is shown in Fig. 1, which also demonstrates that the observed cross-sections are in agreement with low-energy data and cosmic ray results as well [4–7].

Method	$\mathcal{L}$ independent [2]	[12]	[13]	[13]
σ_{tot} [mb]	98.0 ± 2.5	98.3 ± 2.8	98.6 ± 2.2	99.1 ± 4.3

Table 1: The total cross-section σ_{tot} results measured by the TOTEM experiment at $\sqrt{s}$ = 7 TeV with four different methods.

Thanks to a very-high statistics $\beta^* = 90$ m data set at $\sqrt{s} = 8$ TeV energy, the TOTEM experiment has excluded a purely exponential elastic pp differential cross-section [15]. The exclusion's significance is greater than 7σ in the $|t|$ range from 0.027 to 0.2 GeV^2, see Fig. 3. Using refined parametrizations for the extrapolation to the optical point, $t = 0$, yields total cross-section values $\sigma_{\text{tot}} = 101.5 \pm 2.1$ mb and $\sigma_{\text{tot}} = 101.9 \pm 2.1$ mb, compatible with the previous TOTEM measurement. The TOTEM experiment performed its first measurement of elastic scattering in the Coulomb-nuclear interference (CNI) region [16]. The data have been collected at $\sqrt{s} = 8$ TeV with a special beam optics of $\beta^* = 1000$ m. The ρ parameter was for the first time at LHC extracted via the Coulomb-nuclear interference, and was found to be $\rho = 0.12 \pm 0.03$.

3 Conclusions

The TOTEM experiment has measured elastic pp scattering at $\sqrt{s}$ =7 and 8 TeV. The total, elastic and inelastic cross-sections have been derived for both energies using a luminosity independent method. TOTEM has excluded a purely exponential differential cross-section for elastic pp scattering with significance greater than 7σ in the $|t|$ range from 0.027 to 0.2 GeV^2 at $\sqrt{s}$ =

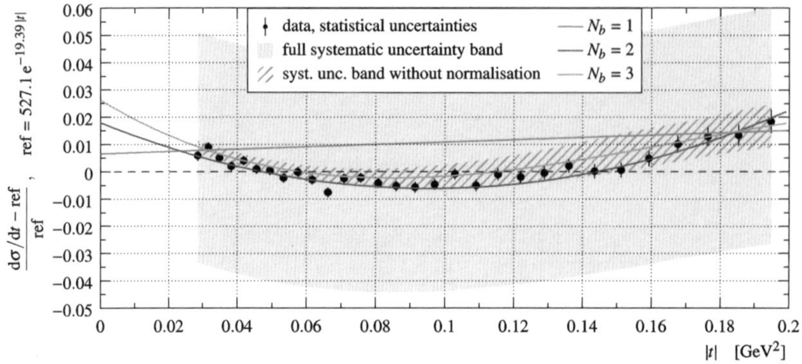

Figure 3: Differential cross-section measured at $\sqrt{s} = 8$ TeV LHC energy plotted as relative difference from a reference exponential. The black dots represent data points with statistical uncertainty bars [15].

8 TeV. Using $\beta^* = 1000$ m optics at $\sqrt{s} = 8$ TeV energy the ρ parameter was for the first time at LHC extracted via the Coulomb-nuclear interference.

4 Acknowledgement

This work was supported by the institutions involved in the TOTEM collaboration and partially also by NSF (US), the Magnus Ehrnrooth Foundation (Finland), the Waldemar von Frenckell Foundation (Finland), the Academy of Finland, the Finnish Academy of Science and Letters (The Vilho, Yrjo and Kalle Vaisa la Fund) as well as by the Hungarian OTKA grant NK 101438.

References

[1] G. Anelli *et al.* [TOTEM Coll.], JINST **3** (2008) S08007.
[2] G. Antchev *et al.* [TOTEM Coll.], Europhys. Lett. **101**, 21004 (2013).
[3] G. Antchev *et al.* [TOTEM Coll.], Phys. Rev. Lett. **111** (2013) 1, 012001.
[4] B. Abelev *et al.* [ALICE Coll.], Eur. Phys. J. C **73** (2013) 6, 2456
[5] S. Chatrchyan *et al.* [CMS Coll.], Phys. Lett. B **722** (2013) 5
[6] ATLAS Collaboration, Nature Commun. **2** (2011) 463.
[7] K. A. Olive *et al.* [PDG Coll.], Chin. Phys. C **38** (2014) 090001.
[8] G. Antchev *et al.* [TOTEM Coll.], Europhys. Lett. **95** (2011) 41001
[9] TOTEM Coll. [TOTEM Coll.], CERN-LHCC-2013-009, LHCC-P-007.
[10] M. G. Albrow [CMS and TOTEM Coll.], PoS DIS **2015** (2015) 064.
[11] G. Antchev *et al.* [TOTEM Coll.], New J. Phys. **16** (2014) 103041
[12] G. Antchev *et al.* [TOTEM Coll.], Europhys. Lett. **96** (2011) 21002

[13] G. Antchev *et al.* [TOTEM Coll.], Europhys. Lett. **101** (2013) 21002.
[14] G. Antchev *et al.* [TOTEM Coll.], Europhys. Lett. **101** (2013) 21003.
[15] G. Antchev *et al.* [TOTEM Coll.], Nucl. Phys. B **899** (2015) 527
[16] G. Antchev *et al.* [TOTEM Coll.], CERN-PH-EP-2015-325

DIFFRACTIVE STUDIES AT THE LHC. A CONCEPTUAL SURVEY [a]

Vladimir Petrov [b]
Institute for High Energy Physics,
NRC "Kurchatov Institute", 142281 Protvino, Russia

Abstract. This is a short survey of the gross features of the hadron diffraction processes observed at the LHC energies with brief theoretical comments to them.

Diffractive studies at the LHC have quickly grown to quite a vast and a very special and diversified domain of high-energy research. In these studies, while keeping a tight relation with the study of the structure of hadrons, we use the high collision energy not to penetrate to the smaller distances, but on the contrary, to the region of greater and greater spatial and temporal scales. I limit myself mostly to elastic proton-proton scattering and touch the status of inelastic diffraction (which, together with elastic scattering, form a quasi self-enclosed "gentleman's set")only in a very cursory way.

1 Experiment

Let me start with experimental findings in the studies of diffraction processes at the LHC and our evaluation of their significance.

1.1 ***Total, elastic and inelastic cross-sections continue to grow at the LHC energies*** *[1].*

Is this conclusion trivial or little informative? Some authors, before the LHC experiments, expressed (e.g. on the basis of cosmic ray data) a firm belief that cross-sections would grow at the LHC as well. One can easily inspect that CR data without the LHC points fully admit the onset of the constancy of the total *pp* cross-sections. Moreover, even the rise of the total *pp* cross-sections still fairly admits a decrease or constancy of the elastic or inelastic *pp* cross-sections (not of both, sure). So the LHC results are in no way trivial or little important. With a nice precision they show that *pp* cross-sections continue to remain in the growth regime which began at the ISR at the energies three hundred times lower than at the LHC. (Earlier the growth was observed for $\sigma_{tot}^{K^+p}$ at the cms energy 6 GeV at the Serpukhov proton synchrocyclotron.)

---

[a]The talk at the 17th Lomonosov Conference on Elementary Particle Physics, Moscow State University, Moscow, 20 26 August, 2015.

[b]E-mail: `vladimir.petrov@ihep.ru`

1.2 ***The transverse interaction region size(embodied in the forward slope B(s)) continues to grow and achieves*** $\approx 1.3fm$ ***at 7 TeV.*** *[2]*

However, the most of this size is still due to the fixed proper sizes of the protons ($\sqrt{\langle b^2\rangle|_{nucleon}} \approx 0.66fm.$) Actually, it is a bit less than two radii of colliding protons. This can be interpreted as an evidence that at the LHC region of energies the colliding protons barely stop to overlap leaving more space for the Pomeron exchange "in free space" [3] , [4]. The last circumstance can incite a more rapid growth [5] of the forward slope (which seems to follow a Regge-like linear log growth up top the Tevatron) but this effect isnt still quite clearly seen [4].

1.3 ***The t-dependence at*** $0.02GeV^2 \leq -t \leq 0.20GeV^2$ ***and cms energy 8 TeV deviates from a simple exponential of the type*** $exp(B(s)t)$. *[6]*

Such a deviation has been repeatedly observed and discussed at lower energies. Theoretical explanation is not unique: from the formal influence of the two-pion threshold in the Pomeron trajectory to possible oscillations of unclear origin. No satisfactory interpretation of this observaion is suggested by now.

1.4 ***Diffractive pattern exhibited by the differental cross-section shows that the dip value of*** $-t$ ***goes on closer to zero and the very dip gets shallower.*** *[2]*

Such an evolution of the dip (related to the size of the scatterer/absorber as borrowed from optics) is being routinely interpreted as the growing interaction radius. However we still lack the theoretically feasible relationship of this phenomenon with the evolution of the forward slope or with any other measure of the size of "effective scatterer". Early attempts (motivated by the hypothesis of " geometrical scaling") to universally parametrize the dip position as $|t_{DIP}| \sim 1/\sigma_{tot}$ unfortunately failed.

The very diffractive pattern still demonstrates only one clear diffractive minimum and one secondary maximum. This is in a stark contrast with optical diffraction off an opaque disc. On one hand it could mean that analogous disc is not just similar to a simple opaque disc with the sharp edge. On the other hand some models show that diffraction minima and maxima will become more and more visible and closer to a typical optical picture. This is to happen, however, at unreasonably high energies. This feature (inherent to Regge-eikonal models)seems to contradict to some hints that the depth of the dip diminishes with the energy growth and may fairly become a shoulder. From this perspective the observed features of the diffractive pattern could be interpreted as one more indication that even at the LHC we are still very far from the "Asymptopia" (a hypothetical energy region where the evolution of cross-sections can be considered as "truly asymptotic").

1.5 Some preliminary data are available on the measurement of the ratio $\rho = ReT(s,0)/ImT(s,0) = ctgArgT(s,0)$. *[7]*

It seems that experimentalists (and even some theorists) begin to be aware of the inevitable model dependence of the parameter ρ. Physical interpretation of one or another value and behaviour of this quantity is not known. At the moment we cannot say what is a fundamental physical difference between the values of ρ, say, 0.12 or 0.14 . It would be much more informative to know the phase of the scattering amplitude at non-zero t [8]. We could, in this case, to estimate, e.g. the effective (both transverse and longitudinal)extent of the interaction region . As far as I can judge, the experimental prospects for that are close to zero. In general, the Coulomb-nuclear interference is but an example of the holography of hadronic beams. I use the term "holography" in its normal meaning, not in the sense of crazy "almighty AdS/CFT correspondence". Would we be able to arrange the interference of two monochromatic proton beams as it is made in optical holography we would have a 3D image of the interaction region. I read something about magnetic mirrors for particle beams which deviate them without loss of energy but it is too far from my knowledge and expertise. So, things go as they go.

1.6 Inelastic diffraction [9]

In compare with elastic scattering the data on inelastic diffraction are relatively poor though recent measurements by TOTEM of SD and DD cross-section well supplement the available material.There is one problem I would like to focus: the definition of what we can consider not just a process with a rapidity gap bracketed between multi-particle states but something which could be qualified as "diffraction dissociation" of one or both incoming hadrons. It stems from the classical consideration by Good and Walker who formulated conditions for observation not just a generic multi-particle state but a state which is a materialization of virtual quantum fluctuations of an incoming hadron. Such a condition refers to a very high momentum of the incoming particle and, thus, is not, generally, Lorentz invariant. E.g. if the incoming hadron is in its rest frame it's quite problematic to unravel its "proper" fluctuations from the vacuum ones. In contrast, in the frame where the incoming hadron moves fast vacuum fluctuations live much shorter time than the proper ones [12]. But which momentum we should call "fast enough" to be sure that we deal exactly with diffractive dissociation? The problem of definition of diffractive dissociation was addressed e.g. in Refs. [10], [11] ,though with no irrevocable conclusions. My trial suggestion for the SD case could be the demand that in the cms frame the supposed products of the incoming hadron dissociation shouldn't occupy more that half of the available rapidity space. This imposes quite severe limits on the missing mass

$$M_X^2 \leq m\sqrt{s}$$

where $m = O(1GeV)$. This gives for the LHC at 13 TeV $M_X^2 \leq 120GeV^2$ or $M_X \leq 11GeV$. Higher missing mass include not only products of the incoming hadron dissociation but also the products of generic Pomeron-proton interaction. Such a criterion does not look justified for the double diffractive dissociation because it admits a zero rapidity gap. One thing seems necessary: the products of supposed dissociation should carry away a finite fraction of the parent hadron momentum. This argument actually goes back to celebrated papers on limiting fragmentation [13]. From the conceptual viewpoint the studies of the inelastic diffraction should be a source of information on the properties of Reggeons and their interactions [14]. Unfortunately, due to the lack of time I can't say anything about extremely interesting subject of the central diffraction which allows to study Reggeon interactions and interplay between hard and soft subprocesses [15]. Ongoing measurements of the CMS-TOTEM collaboration in this field seem very promising.

2 Theory

2.1 QCD

If to start *ab ovo* , i.e. from the official theory of strong interaction, QCD, we have to state that practioners of this theory still are not able to produce experimentally falsifiable (in the sense of the distinguished Sir Karl Popper) predictions for any of measurable quantities in the field of hadron diffraction: total or differential cross-sections or else. We know that thorough and labour-consuming investigations of Regge trajectories in QCD were being conducted during almost 40 years. Unfortunately, not only the Regge trajectories were not obtained but even a single real number, such as the Pomeron intercept, is still not calculated in a satisfactory way. The only more or less feasible result is that the Pomeron intercept does not depend on the QCD gauge coupling [16] and, probably, we have to deal with infinitely many Pomerons [17].This property is inherent to asymptotically free field theories. One interesting finding which corroborates with the mentioned independence of Regge intercepts on the QCD coupling was made in [18]: the Odderon intercept is exactly 1. It follows immediately that the Odderon decouples from nucleons at $t = 0$. To some extent this can explain elusiveness of the Odderon in low - t measurements. We have to avow, however, that these conclusions are important and interesting but too poor to be subjects of direct experimental check.

The general trend in the development of theory is often related with so-called "Reggeon field theory", an effective framework with Reggeons as basic field-theoretic entities in a *sui generis* 2+1 space-"time" where the role of time is played by rapidity while the space is the impact parameter space. One can develop in this peculiar field theory (with some anti-Hermitean vertices) a renormalization group technique. On this way, one can hope that (provided that some important physical parameters would be given by the microscopic theory (QCD)) it would be possible to get more or less concrete physical predictions for observable quantities. Indeed, some interesting results were obtained in 1970's [19]. Recently, after a long period of oblivion, a new interest to this method is awakened [20]. Unfortunately, there is quite a significant arbitrariness in this approach which still limits results mostly by the formalism developments. On the other hand, we do not see anything that can preclude a direct derivation of predictions for the hadron diffraction from QCD, though it seems clear that being tightly related with the large and growing characteristic distances this derivation cannot escape the confinement problem.Although, it is quite probable that for practical calculations at these distances an effective field theory with colourless basic elements ("QCD strings" etc)will be more relevant than dealing directly with confined quarks and gluons.

At present we can only hope for the future (desirably more rapid) progress in this area which would reduce the gap between theorists and experimentalists, who during the mentioned 40 years have gone far ahead both in techniques and results.

2.2 ***Models***

What remains? The models. They differ in the amount of the QCD terminology (basically of mimicking character) and in the degree of clarity of formulation of their fundamental premises. The latter is very important to be able to understand the meaning of both success or failure of one or another model. Though, if a model would demonstrate an undoubted success then it could be the reason to find physical interpretation of its ideological requisite with possible far reaching consequences. Alas, the very first TOTEM results on the differential cross-sections at 7 TeV ruled out all the models on the market (see e.g. an excellent review of models up to 2009 in [21]). Sure, this sorrow circumstance incited the authors of the failed models to undertake, later or sooner, due modifications and improvements and to get ready for the next comparison. This occurs not for the first and, most likely, not for the last time. What disappoints is that seemingly nobody of the authors tried to draw physically meaningful conclusions from their failure. Description prevails over explanation. It would be very good if somebody could break this boring scenario.

3 Conclusions

To summarize all above said I would like to stress again that in the field of diffractive studies there is a significant lag of theory from experiment. Recent findings of the LHC collaborations (mostly TOTEM) give a new impetus for improvement of the model description and making the latter more substantiated by the fundamental QCD premises. Unfortunately the accumulated data on diffractive processes is insufficient for resolving of one important and long debated hypothesis about universality of strong interactions at high energies. It seems that it is true for the proton - proton and proton - anti-proton case but clearly from the point of view of pure QCD these particles are too close. That is why I believe many support the appeal to our experimental colleagues to reflect on possibilities to obtain secondary beams and to provide us with the data on $\pi p, Kp$ interactions in the TeV region of energies. We wait with impatience the data from the 13 TeV run as well.

Acknowledgments

I am indebted to my colleagues A. A. Godizov, V. V. Ezhela, A. V. Kisselev, R. A. Ryutin and A. P. Samokhin for fruitful and very helpful discussions on the issues raised in this talk. I am thankful to the organizers of the 17th Lomonosov Conference for invitation to give this talk.

References

[1] G. Antchev et al. (TOTEM Collaboration), EPL **96** (2011)21002 .
[2] G. Antchev et al. (TOTEM Collaboration), EPL **101** (2013)21002 .
[3] J. Pumplin. Phys.Lett.**B289**(1992)449;
[4] V. A. Petrov in *"Particle and Astroparticle Physics, Gravitation and Cosmology: Predictions, Observations and New Projects"* (Proceedings of the 30th International Workshop on High Energy Physics(IHEP 2014), June 23-27, 2014, Protvino, Russia.World Scientific, Singapore. P.110-120.
[5] V.A. Schegelsky and M.G. Ryskin, Phys. Rev. **D85**,094024 (2012); V.A. Okorokov. Adv.High Energy Phys. **2015** (2015) 914170.
[6] G. Antchev et al. (TOTEM Collaboration), Nucl. Phys. **B 899** (2015) 527-546
[7] G. Antchev et al. (TOTEM Collaboration), CERN-PH-EP-2015-325.
[8] J. Prochazka,M.V. Lokajicek and V. Kundrat. arXiv:1509.05343
[9] G. Antchev et al. (TOTEM Collaboration), Phys. Rev. Lett. **111** (2013) 262001
[10] V. Khoze et al. Eur.Phys.J. **C69** (2010) 85-93.
[11] M. Kuusela et al. Int.J.Mod.Phys. **A25** (2010) 1615-1647
[12] V. N. Gribov. hep-ph/0006158.
[13] J. Benecke et al. Phys.Rev. **188** (1969) 2159-2169; R. P. Feynman. Phys.Rev.Lett. **23** (1969) 1415-1417.
[14] V. A. Petrov and R. A. Ryutin. arXiv:1507.07869.
[15] R. A. Ryutin. Phys.J. **C74** (2014) 11, 3162.

[16] V. A. Petrov. hep-ph/0603103.
[17] C. Lovelace. Phys.Lett. **B55** (1975) 187; D. Heckathorn. Phys.Rev. **D18** (1978) 1286; L. N. Lipatov. Sov.Phys.JETP **63** (1986) 904.
[18] J. Bartels, L. N. Lipatov and G. P. Vacca, Phys. Lett. **B477**(2000)178.
[19] M. Moshe. Phys.Rep.**37C**(1978)255.
[20] J. Bartels et al.AIP Conf.Proc. **1654** (2015) 070003.
[21] R. Fiore, L. Jenkovszky et al.Int.J.Mod.Phys. **A24** (2009) 2551.

KAON FEMTOSCOPY MEASUREMENTS IN THE SELEX EXPERIMENT

Grigory Nigmatkulov [a] (on behalf of the SELEX Collaboration)
Department of Experimental Nuclear Physics and Cosmophysics, National Research Nuclear University MEPhI (Moscow Engineering Physics Institute), 115409, Moscow, Russia

Abstract. We present the measurement of the two-kaon correlation functions using the 600 GeV/c Σ^- beam from the SELEX (E781) experiment at the Fermilab Tevatron. The femtoscopic analysis was performed using the three-dimensional Bertsch-Pratt decomposition. The dependence of the correlation functions on the average transverse pair momentum was investigated.

1 Introduction

In order to understand properties of the bulk matter production it is important to know the dynamics of the particle collision. In particular, spatio-temporal aspects of the reaction need to be experimentally addressed. The two-particle correlations at small relative momentum (correlation femtoscopy) carry the information about the size and lifetime of the subatomic reactions [1]. During the last three decades correlation femtoscopy was intensively used as a tool for the precise measurement of the spatio-temporal properties of the particle emitting source [2]. The femtoscopic measurements usually performed for pions. The measurements of charged kaons may provide addition information about the size, orientation and dynamical timescales of the emission region.

2 Two-particle correlation function

Due to the quantum statistical correlations, the spatio-temporal characteristics of the particle emitting source are encoded in the two-particle relative momentum distributions. In order to extract this information Kopylov and Podgoretsky [3] proposed to use two-particle correlation function:

$$C(\mathbf{q}) = \frac{A(\mathbf{q})}{B(\mathbf{q})}, \tag{1}$$

where $C(\mathbf{q})$ is the correlation function and $A(\mathbf{q})$ is the measured two-particle distribution of the pair momentum difference $\mathbf{q} = \mathbf{p_1} - \mathbf{p_2}$. The identical particles that are produced in the same event are influenced by the quantum statistical correlations. The $B(\mathbf{q})$ term is the reference relative momentum distribution, where the quantum statistical correlations are absent. In this analysis the event mixing technique, when particle pairs are constructed by using particles from different events, was used to construct the uncorrelated

[a]E-mail: `ganigmatkulov@mephi.ru; nigmatkulov@gmail.com`

reference sample. The femtoscopic radii are directly proportional to the width of the enhancement of two-meson correlation function at low relative momentum.

The dynamical properties of the emitting source can be studied via the measurement of the average transverse pair momentum dependence, $k_T = |\mathbf{p_{T,1}} + \mathbf{p_{T,2}}|/2$, of the correlation function [4]. In case of the limited statistics, the correlation function is constructed in the one-dimensional quantity:

$$Q_{inv} = \sqrt{(\mathbf{p_1} - \mathbf{p_2})^2 - (E_1 - E_2)^2}, \tag{2}$$

where E_1 and E_2 are the energies of the first and the second particle from the pair, respectively. The one-dimensional femtoscopic analysis allows to extract only the spacial size of the emitting source. In order to extract the space and time properties of the particle emitting region, the three-dimensional decomposition of the relative momentum $\mathbf{q}$ can be performed according to the Bertsch-Pratt "out-side-long" parametrization [5, 6]. In this parametrization, the "long" direction is parallel to the beam, the "out" direction is that of the pair transverse momentum, and the "side" direction is perpendicular to these two. We will use the q_{out}, q_{side} and q_{long} quantities to indicate the projections of the relative pair momentum in these directions.

In the current analysis, the Bertsch-Pratt decomposition of the two-particle relative momentum is performed in the longitudinally co-moving system (LCMS), in which the longitudinal momentum of the pair, $p_{l,1} + p_{l,2}$, vanishes [7].

3 Femtoscopic correlations of charged kaons

The charged kaon correlation functions are measured in 600 GeV/c $\Sigma^- C(Cu)$ interactions recorded by the SELEX (E781) detector [8]. We use the same event, single- and pair-particle selection criteria that were used in Ref. [9]. The two-kaon relative momentum, $\mathbf{q}$, is decomposed according to the Bertsch-Pratt parametrization in the LCMS. Figure 1 shows the correlation function projections in "out", "side" and "long" directions (from left to right). To project onto one $\mathbf{q}$ component, the others are integrated over the range $0 - 0.3$ GeV/c.

The circles, squares and triangles represent the correlation functions measured for (0.00–0.30), (0.30–0.55) and (0.55–1.00) GeV/c average transverse pair momentum ranges. The peaks at low $\mathbf{q}$ correspond to the enhancement due to the quantum statistical correlations. The width of the this enhancement describe the source radius for the given decomposition direction. The increase of the correlation function width with increasing k_T corresponds to the decrease of the source radii for the given direction. The deviation of the correlation functions from unity at high relative pair momenta is coming from the non-femtoscopic correlations. This non-femtoscopic correlations may be

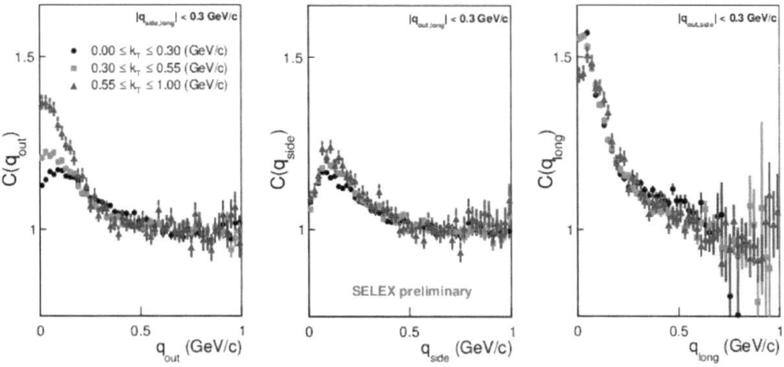

Figure 1: The charged kaon correlation functions measured in $\Sigma^- C(Cu)$ interactions. The circles, squares and triangles represent correlation functions measured for (0.00–0.30), (0.30–0.55) and (0.55–1.00) GeV/c k_T ranges. For the projection on one of the $\mathbf{q}$ directions, integration in other two directions was done over the range (0.00-0.30) GeV/c.

caused, for instance, by the energy and momentum conservation induced correlations [10]. It is observed that for each k_T interval the correlation functions in the longitudinal direction have a smaller width compare to the sideward and outward directions that corresponds to the bigger source radii in the direction along the beam axis.

4 Conclusions

The three-dimensional charged kaon correlation functions are measured using 600 GeV/c Σ^- beam in the SELEX experiment. The average transverse pair momentum dependence of the kaon correlation functions is investigated. The correlation function width in the longitudinal direction is smaller compare to the widths measured in outward and sideward directions for the (0.00–0.30), (0.30–0.55) and (0.55–1.00) GeV/c k_T intervals. This reflects that the emitting source size along the beam direction is bigger than the source sizes in the transverse to the beam directions.

Acknowledgments

The authors thank ITEP and National Research Nuclear University MEPhI (Moscow Engineering Physics Institute) for providing computing powers and support for data analysis and simulations.

References

[1] W. Bauer, C.K. Gelbke, S. Pratt, Annu. Rev. Nucl. Part. Sci. **42**, 77 (1992).

[2] M.A. Lisa, S. Pratt, R. Soltz, U. Wiedemann, Ann. Rev. Nucl. Sci. **55**, 357 (2005).

[3] G.I. Kopylov, M.I. Podgoretsky, Sov. J. Nucl. Phys. **15**, 219 (1972).

[4] S.V. Akkelin, Y.M. Sinyukov, Phys. Lett. **B 356**, 525 (1995).

[5] S. Pratt, Phys. Rev. Lett. **53**, 1219 (1983).

[6] G. Bertsch, M. Gong, M. Tohyama, Phys. Rev. **C 37**, 1896 (1988).

[7] S. Pratt, T. Csorgo, J. Zimanyi, Phys. Rev. **C 42**, 2646 (1990).

[8] J.S. Russ *et al.*, in "High-energy physics" (Proceedings of the 29th International Conference, ICHEP'98, July 23-29, 1998, Vancouver, Canada), Ed. by A. Astbury, D. Axen, J. Robinson, World Scientific, Singapore, V. II, 1259, 1998.

[9] G.A. Nigmatkulov *et al.*, Phys. Lett. **B 753**, 458 (2016).

[10] Z. Chajecki, M.A. Lisa, Phys. Rev. **C 78**, 064903 (2008).

Astroparticle Physics and Cosmology

THE TRANSIENT HIGH ENERGY SKY

Neil Gehrels [a]
Astroparticle Physics Division, NASA/Goddard Space Flight Center, Greenbelt, MD 20771, USA
John K. Cannizzo [b]
CRESST/Joint Center for Astrophysics, UMBC, Baltimore, MD 21250, USA

Abstract. High energy transients make up a diverse and exotic class of objects, from terrestrial lightning to γ-ray bursts at cosmological distances. In this review we provide a detailed look at look at some of the more exciting transients observed over the last few years by *Swift* and other high energy missions.

1 A Potpourri of High Energy Transients

Table 1 indicates the wide range of classes of high energy transients, along with their associated energies. In this section we preview the topic by examining two completely different classes of high energy transients, one terrestrial (γ-ray flashes) and one galactic (soft gamma repeaters).

Terrestrial γ-ray flashes (TGFs) were first seen by the Burst and Transient Source Experiment (BATSE) on the *Compton Gamma-Ray Observatory.* They are thought to be due to decaying electric fields above thunderclouds after a lightning discharge. Relativistic electrons interact with nuclei of atoms in the atmosphere and produce γ-rays via bremsstrahlung. A process known as runaway electron avalanche is thought to be relevant, but the details are uncertain (Gurevich et al. 1992; Dwyer 2003). Subsequent observations by the *Reuven Ramaty High Energy Solar Spectroscopic Imager* (*RHESSI*) have revealed TGFs with much higher energies, and show that ~500 TGFs occur per day, which is a small fraction of the total number of daily lightning strikes on Earth ($\sim 3 - 4 \times 10^6$). This estimate excludes effects relating to beaming and atmospheric obscuration of low altitude TGFs. The *Fermi*/GBM is currently detecting TGFs at a rate of ~ 100 yr^{-1}; some are even detected by *Fermi*/LAT during special Earth-pointed observations. This high rate has held since 2009 when the BGOs were included in the triggering algorithm. Perhaps the most spectacular discovery has been that of 511 keV emission from e^+/e^- annihilation in TGFs (Briggs et al. 2011). It is unsettling to consider the fact that terrestrial processes are capable of producing antimatter in significant quantities.

Soft gamma repeaters (SGRs) emit bursts of γ-rays and X-rays which are thought to be due to the rearrangement of powerful magnetic fields in magnetars – pulsars with magnetic fields of $\sim 10^{15}$ G. The first one seen was the "March 5th Event" from 1979 which was observed by two Soviet interplanetary

[a] E-mail: neil.gehrels@nasa.gov
[b] E-mail: john.k.cannizzo@nasa.gov

Table 1: X-ray/γ-ray Transients

Source	Energy Source	E(γ-ray)
lightning	E field	10^{10} erg
γ-ray burst	gravity	10^{52} erg
magnetar	B field	10^{44} erg
tidal disruption	gravity	10^{52} erg
stellar flare	B field	10^{32} erg
supernova/nova	nuclear	10^{49} erg
accreting BH/NS	gravity	10^{36} erg s^{-1}
AGN	gravity	10^{43} erg s^{-1}

spacecraft *Venera 11* and *Venera 12* (Mazets et al. 1979). This event, SGR 0526-66, was localized to the SN remnant N49 in the LMC. Given a distance of ~50 kpc, the isotropic equivalent energy emitted was $\sim 5 \times 10^{44}$ erg, compared to $\sim 10^{41}$ erg for a typical SGR burst. A ~8 s periodicity thought to be the NS spin period is plainly evident in the data. Thompson & Duncan (1995) present an extensive model of the March 5th event and other SGRs as magnetars, i.e., $B \simeq 10^{15}$ G. They argue that the March 5 burst was due to a large-scale readjustment of the stellar magnetic field, while the more standard SGR bursts are caused by the release of magnetic stresses within a more localized patch of the crust.

On Aug 27, 1998 a second giant flare from an SGR was seen. SGR 1900+14 became the brightest extra-solar system γ-ray source ever. The 5.16 s spin period of the pulsar could be easily seen directly in the light curve (Kouveliotou et al. 1999). The flare was so bright that it produced 5.16 s ionization modulations in the upper atmosphere of the Earth. Thompson & Duncan (2001) argue that the extremely high luminosity $L \gtrsim 10^6 L_{\rm Edd}$ during the initial ~0.5 s spike in SGR 1900+14 demands $B \gtrsim 10^{15}$ G. Shortly after the launch of *Swift*, on 27 Dec 2004, a giant γ-ray flare was seen from SGR 1806-20 with a peak flux of ~5 erg cm^{-2} s^{-1}. SGR bursts are much brighter than ordinary X-ray bursts, which are due to thermonuclear flashes of accumulated hydrogen on the surface of a NS ($L \sim 10^3 - 10^4 L_{\rm Edd}$ vs. $L \sim L_{\rm Edd}$), and they also have harder spectra than ordinary X-ray bursts.

2 V404 Cygni - Galactic Black Hole Transient

The microquasar V404 Cyg (also designated GS 2023+338) is a low-mass X-ray binary (LMXB) with a 6.5 d orbital period and a mass function $f = 6.3 M_\odot$ (Casares, Charles, & Naylor 1992). Combining the mass function with spectroscopic data yields a central black hole mass of $\sim 12 M_\odot$. Radio parallax yields a distance 2.39 ± 0.14 kpc (Miller-Jones et al. 2009). Its unusually hard and bright quiescent spectrum may indicate the presence of an advection dominated

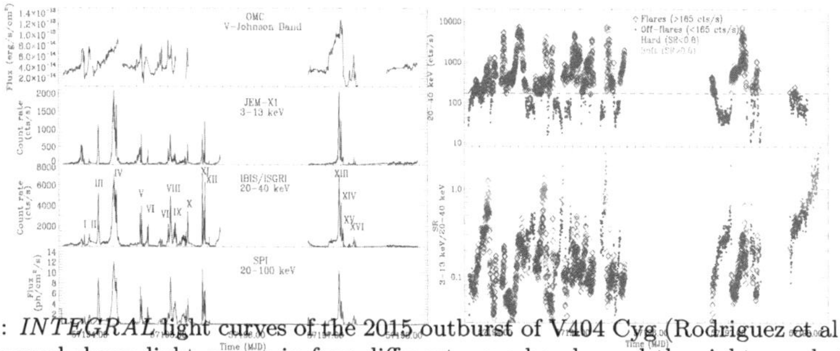

Figure 1: *INTEGRAL* light curves of the 2015 outburst of V404 Cyg (Rodriguez et al. 2015). The left panel shows light curves in four different wave bands, and the right panel gives the 20–40 keV count rate (*top*) and f(3–13 keV)/f(20–40 keV) softness ratio (*bottom*). The dashed horizontal line indicates 1 Crab (20–40 keV). MJD 57193 corresponds to 2015 June 20.

accretion flow, or ADAF (Narayan, Barret, & McClintock 1997). V404 Cyg displays infrequent outbursting behavior, thought to be due to the accretion disk limit cycle mechanism, wherein long periods of inactivity correspond to times of mass accretion within the accretion disk, followed by brief periods of activity during which some fraction of the stored mass is accreted onto the central BH (Cannizzo, Chen, & Livio 1995; King & Ritter 1998; Dubus, Hameury, Lasota 2001). Due to a combination of larger dimensions (i.e., wider binaries) plus lower secondary mass transfer rates, the quiescent intervals for LMXBs can span decades, in contrast to weeks or months for cataclysmic variables, which undergo the same type of storage limit cycle accretion disk instability (Warner 1987). Also, a few LMXBs, such as A0620-00, exhibit smooth FRED-type outbursts (fast rise-exponential decay), whereas others, such as V404 Cyg, show chaotic outbursting behavior, perhaps indicating jetted emission. During the course of its outburst cycle, an LMXB traverses a familiar "turtle diagram" in hardness ratio vs. flux, cycling through various combinations of high-low and hard-soft states (Esin, McClintock, & Narayan 1997; Körding et al. 2008).

After 25 years of quiescence, V404 Cyg became active in 2015 June (Figure 1, from Rodriguez et al. 2015). Historically it was known to exhibit extremely bright and variable outbursts in all wavelengths. It is a prime candidate for exploring the connection between accretion and ejection, which require simultaneous observations at several wavelengths. Rodriguez et al. (2015) present the results of long, nearly uninterrupted *INTEGRAL* observations and obtain light curves from the optical through soft γ-rays. V404 Cyg was extremely variable in all wavebands, exhibiting 18 flares exceeding 6 Crab (20–40 keV) in $<$3 d. Some of the optical flares occurred in conjunction with the X-ray flares, while others were delayed $\gtrsim$10 min. Rodriguez et al. (2015) suggest that the first kind were associated with X-ray reprocessing by an accretion disk or the companion star, while the latter were due to plasma ejections, also seen in the radio. King et al. (2015) analyze two *Chandra* exposures of V404 Cyg. They

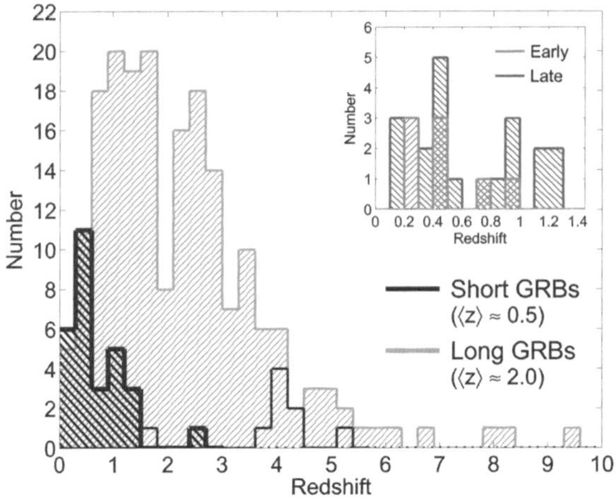

Figure 2: The redshift distribution of sGRBs (black) and lGRBs (gray) (Berger 2014). Open histogram indicates z upper limits based on the lack of a Lyman-α break in afterglow and/or host galaxy optical detections. Inset shows z distribution of sGRBs by host galaxy type.

propose that strong P-Cygni profiles seen in high flux states are due to winds associated with the disruption of the outer accretion disk by radiation pressure as the luminosity approaches Eddington.

3 Short vs. Long GRBs

Gamma-ray bursts (GRBs) are intense flashes of radiation produced at cosmological distances $z \simeq 2$. GRBs come in two primary flavors, long and short, with the dividing point being roughly 2 s (Kouveliotou et al. 1993). A further division can be made spectrally according to their hardness ratio (i.e., ratio of high to low energies). The redshift range is from about 0.2 to 2 for *Swift* short GRBs (sGRBs), with a mean of about 0.4. For *Swift* long GRBs (lGRBs) the range is between about 0.009 and 8.2, with a mean of about 2.3 (Figure 2, from Berger 2014). For sGRBs near enough to be studied in detail, no accompanying supernova is seen. By contrast, nearby lGRBs do have associated supernovae. SGRBs are found in both star-forming and non-star-forming galaxies; lGRBs are found in star-forming galaxies – typically dwarf irregulars. To date *Swift* has found $\gtrsim$90 sGRBs and $\gtrsim$900 lGRBs.

The typical energy release is $\sim 10^{49} - 10^{50}$ erg for sGRBs and $\sim 10^{50} - 10^{51}$ erg for lGRBs. These ranges are based on observed isotropic-equivalent energies of $\sim 10^{50}$ erg for sGRBs and $\sim 10^{53}$ erg for lGRBs, and estimates for jet beaming for each class, $\theta_{\rm j} \sim 5°$ for lGRBs and $\theta_{\rm j} \sim 5 - 10°$ for sGRBs (Burrows et al.

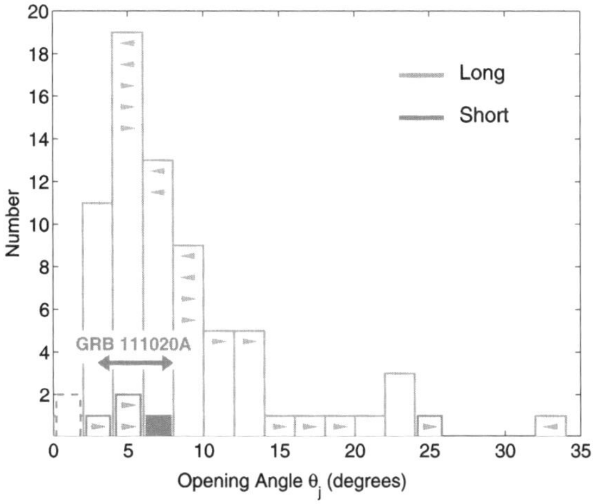

Figure 3: Distribution of opening angles for lGRBs (orange) and sGRBs (blue) (Fong et al. 2012). Arrows represent upper and lower limits.

2006; Grupe et al. 2006; Fong et al. 2012). Beaming angles for sGRBs are still highly uncertain (Figure 3, from Fong et al. 2012). The $L_X/E_{\gamma-\mathrm{iso}}$ values at 11 hr post-GRB are similar between lGRBs and sGRBs (Gehrels et al. 2008; Nysewander et al. 2009). The sGRBs have weaker X-ray afterglows, a mean value of $\sim 7\times10^{-10}$ erg cm^{-2} s^{-1} versus $\sim 3\times10^{-9}$ erg cm^{-2} s^{-1} for lGRBs. Although many of the details are uncertain, the two mechanisms are thought to be collapsars for lGRBs and merging neutron stars for sGRBs (Gehrels, Ramirez-Ruiz, & Fox 2009). LGRBs are intrinsically very bright, the brightest explosions in the Universe. The highest redshift lGRBs were far above detector threshold. SGRBs, by contrast, constitute a flux-limited sample, and are seen primarily nearby, $z \lesssim 1$.

4 Short GRBs: The Future

What does the future hold for sGRBs? In the next few years as we enter into the detection era for LIGO we will witness opportunities for a LIGO–*Swift* synergism brought about by the detection of NS-BH and NS-NS mergers. By the end of this decade we should have at hand a bonanza of data from GW (gravity wave)/EM (electromagnetic) counterpart detections. It will be rare to catch a jetted afterglow in concert with a GW signal, therefore the great hope is for a "kilonova" associated with the supernova-like transient powered by this radioactive decay of material ejected from the NS during a NS-BH or NS-NS merger. Kilonovae are expected to exhibit a lingering red afterglow (Kasen et al. 2013; Barnes & Kasen 2013; Grossman et al. 2014).

Observational confirmation of such an event would be important given that this mechanism may be the predominant source of stable $r-$process elements in the universe (Freiburghaus, Rosswog, & Thielemann 1999; Goriely, Bauswein, & Janka 2011). The biggest problem is that, since the kilonova emission is weak, it could usually be masked by the normal afterglow. GRB 130603B might be the first detected kilonova (Tanvir et al. 2013; Berger, Fong, & Chornock 2013). It was a short GRB at $z = 0.356$ with a duration ~0.2 s in the BAT. Tanvir et al. (2013) present optical and near-infrared observations that provide evidence for an accompanying kilonova.

5 Conclusion

The sky is rich in transients of many types. *Swift* is exploring the transient sky with unprecedented sensitivity and coverage. Every year brings new discoveries in time domain science. Explosion mechanisms range from gravitational collapse to nuclear burning to $B-$field reconnection. The future is bright, with new time domain capability across the electromagnetic spectrum, plus wide-field GW, neutrino, and cosmic ray detectors coming on board.

References

[1] Barnes J., Kasen D., 2013. ApJ, 775, 18.
[2] Berger E., 2014. ARA&A, 52, 43.
[3] Berger E., Fong W., Chornock R., 2013. ApJ, 774, L23.
[4] Briggs M.S., Connaughton V., Wilson-Hodge C., et al., 2011. Geophys. Res. Lett., 38, L02808.
[5] Cannizzo J.K., Chen W., Livio M., 1995. ApJ, 454, 880.
[6] Casares J., Charles P.A., Naylor T., 1992. Nature, 355, 614.
[7] Dubus G., Hameury J.-M., Lasota J.-P., 2001. A&A, 373, 251.
[8] Dwyer J.R., 2003. Geophys. Res. Lett, 30, 2055.
[9] Esin A.A., McClintock J.E., Narayan R., 1997. ApJ, 489, 865.
[10] Fong W., Berger E., Margutti R., et al., 2012. ApJ, 756, 189.
[11] Freiburghaus C., Rosswog S., Thielemann F.-K., 1999. ApJL, 525, L121.
[12] Gehrels N., Barthelmy S.D., Burrows D.N., et al., 2008. ApJ, 689, 1161.
[13] Gehrels N., Ramirez-Ruiz E., Fox D.B., 2009. ARA&A, 47, 567.
[14] Goriely S., Bauswein A., Janka H.-T., 2011. ApJL, 738, L32.
[15] Grossman D., Korobkin O., Rosswog S., Piran T., 2014. MNRAS, 439, 757.
[16] Grupe D., Burrows D.N., Patel S.K., Kouveliotou C., Zhang B., Mészáros P., Wijers R.A.M., Gehrels N., 2006. ApJ, 653, 462.

[17] Gurevich A.V., Milikh G.M., Roussel-Dupré R., 1992. Phys. Lett. A, 165, 463.
[18] King A.L., Miller J.M., Raymond J., Reynolds M.T., Morningstar W., 2015. ApJL, 813, L37.
[19] King A.R., Ritter H., 1998. MNRAS, 293, L42.
[20] Körding E., Rupen M., Knigge C., Fender R., Dhawan V., Templeton M., Muxlow T., 2008. Science, 320, 1318.
[21] Kouveliotou C., Meegan C.A., Fishman G.J., Bhat N.P., Briggs M.S., Koshut T.M., Paciesas W.S., Pendleton G.N., 1993. ApJL, 413, L101.
[22] Kouveliotou C., Strohmayer T., Hurley K., et al., 1999. ApJL, 510, L115.
[23] Mazets E.P., Golentskii S.V., Ilinskii V.N., Aptekar R.L., Guryan Iu.A., 1979. Nature, 282, 587.
[24] Miller-Jones J.C.A., Jonker P.G., Dhawan V., Brisken W., Rupen M.P., Nelemans G., Gallo E., 2009. ApJL, 706, L230.
[25] Narayan R., Barret D., McClintock J.E., 1997. ApJ, 482, 448.
[26] Nysewander M., Fruchter A.S., Pe'er A., 2009. ApJ, 701, 824.
[27] Rodriguez J., Cadolle Bel M., Alfonso-Garzón J., et al., 2015. A&A, 581, L9
[28] Rothschild R.E., Kulkarni S.R., Lingenfelter R.E., 1994. Nature, 368, 432.
[29] Rothschild R.E., Lingenfelter R.E., Seward F.D., Vancura O., 1993. *The Compton Gamma Ray Observatory; American Institute of Physics* ed. M. Friedlander, N. Gehrels, R. J. Macomb, 280, 808.
[30] Soderberg A.M., Chandra P., Frail D., 2008. GCN, 7506, 1.
[31] Soderberg A.M., Kulkarni S.R., Nakar E., et al., 2006. Nature, 442, 1014.
[32] Tanvir N.R., Fox D.B., Levan A.J., et al., 2009. Nature, 461, 1254.
[33] Tanvir N.R., Levan A.J., Fruchter A.S., Hjorth J., Hounsell R.A., Wiersema K., Tunnicliffe R.L., 2013. Nature, 500, 547.
[34] Thompson C., Duncan R.C., 1995. MNRAS, 275, 255.
[35] Thompson C., Duncan R.C., 2001. ApJ, 561, 908.
[36] Warner B., 1987. MNRAS, 227, 23.

THE PLANCK LEGACY: REINFORCING THE CASE FOR A STANDARD MODEL OF COSMOLOGY (ΛCDM)

Nazzareno Mandolesi[1,5,2 a], Diego Molinari[2,1 b], Alessandro Gruppuso [1,3 c], Carlo Burigana[1,2,3 d] and Paolo Natoli[2,4,1 e] (on behalf of *Planck* Collaboration)

[1] *INAF–IASF [f] Bologna, Via Piero Gobetti 101, I-40129 Bologna, Italy*

[2] *Dipartimento di Fisica e Scienze della Terra, Università degli Studi di Ferrara, Via Giuseppe Saragat 1, I-44122 Ferrara, Italy*

[3] *INFN [g], Sezione di Bologna, Via Irnerio 46, I-40126, Bologna, Italy*

[4] *Agenzia Spaziale Italiana Science Data Center, Via del Politecnico snc, 00133, Roma, Italy*

[5] *Agenzia Spaziale Italiana, Viale Liegi 26, Roma, Italy*

Abstract. We present a brief review of the main results of the *Planck* 2015 release describing the new calibration of the data, the maps delivered in temperature and, for the first time, in polarization, the cosmological parameters and the lensing potential. In addition, we present estimates of the Galactic foregrounds in polarization: future satellite experiments will have the challenge to remove foregrounds with great accuracy to be able to measure tensor-to-scalar ratios less than ~ 0.01.

1 Introduction

Cosmic Microwave Background (CMB) radiation, whose first prediction can be attributed to R Alpher, R. Herman and G. Gamov in 1948 [1], is a blackbody radiation at T=2.72548 ± 0.00057 K [2] extremely uniform across the sky. It is the relic radiation emerged at the time the nuclei and electrons recombined to form neutral hydrogen, when the Universe was $\simeq 380,000$ years old [18] and the mean free path of the photons became larger than the Universe itself, corresponding to the so-called Last Scattering Surface (LSS). CMB tiny temperature and polarization anisotropies encode a wealth of cosmological information.

The *Planck* mission, launched and operated by ESA, represent a third-generation satellite devoted to the measurement of the all-sky temperature and polarization anisotropies of the CMB with unprecedented precision, to test the standard cosmological model and investigate the early universe physics. *Planck* observed the microwave sky continuously from the 12th of August 2009 to the 23rd of October 2013 in 9 bands between 30 GHz and 857 THz with angular resolutions between 30 and 5 arcmin, respectively, and a sensitivity of $\Delta T/T_{CMB} \sim 2 \times 10^{-6}$ [3]. The satellite is equipped with two instruments: the Low Frequency Instrument (LFI), with pseudo-correlation temperature and

[a] E-mail: mandolesi@iasfbo.inaf.it

[b] E-mail: molinari@iasfbo.inaf.it

[c] E-mail: gruppuso@iasfbo.inaf.it

[d] E-mail: burigana@iasfbo.inaf.it

[e] E-mail: paolo.natoli@gmail.com

[f] Istituto Nazionale di Astrofisica – Istituto di Astrofisica Spaziale e Fisica Cosmica

[g] Istituto Nazionale di Fisica Nucleare

Experiment	Amplitude μK_{CMB}	Galactic coordinates l [deg]	b [deg]
LFI	3365.5 ± 3.0	264.01 ± 0.05	48.26 ± 0.02
HFI	3364.5 ± 1.0	263.94 ± 0.02	48.21 ± 0.008
Planck 2015 nominal	3364.5 ± 2.0	264.00 ± 0.03	48.24 ± 0.02
WMAP	3355 ± 8	263.99 ± 0.14	48.26 ± 0.03

Table 1: LFI, HFI, and WMAP measurements of the Solar dipole [5].

polarization sensitive radiometers at 30, 44, 70 GHz, operated for about 48 months; the High Frequency Instrument (HFI), with bolometers at 100, 143, 217, 353 GHz in temperature and polarization and at 545 and 857 GHz in temperature only, operated for about 30 months. In the first cosmological release, delivered in 2013 [4], the first 15 months of temperature data have been analysed. A second cosmological release in 2015 [5] involved the full mission data in temperature and, for the first time, including the polarization data.

We present here a brief review of the main results of the *Planck* 2015 release.

2 Calibration

In the *Planck* 2013 release, the photometric calibration of the data has been performed using the Solar dipole, that is the dipole induced in the observed CMB pattern by the motion of the Solar System barycentre with respect to the LSS. In the *Planck* 2015 release, instead, the photometric calibration has been performed using the orbital dipole. This is the modulation of the Solar dipole induced by the orbital motion of the satellite around the Solar System barycentre. Using this method, an independent measurement of the Solar dipole for each detector was extracted and used in the *Planck* calibration pipeline. Since the orbital motion of the satellite is very well known, the measurement of the orbital dipole is one of the most accurate calibration methods.

The amplitude of the orbital dipole is one order of magnitude smaller than the Solar dipole and therefore it is important to have a low noise and a good control of the systematic effects. A measurement of the Solar dipole gives an a posteriori check on calibration within the two *Planck* instruments. This is fundamental for a joint analysis of the data, and for the comparison with other CMB experiments. In Table 1 we report the amplitude and direction of the Solar dipole for the LFI [6] and HFI [7] in comparison with WMAP [8]. The *Planck* nominal dipole is the result of the combination of the LFI and HFI measurements and it is used to subtract the dipole from the frequency maps. The measurements of the Solar dipole independently performed by *Planck* and WMAP agree within 0.28% in amplitude and to better than 2 arcmin in direction, proving the good consistency and calibrations of these experiments.

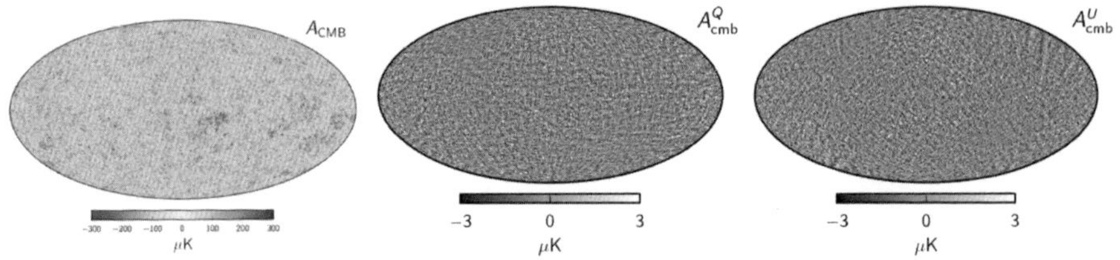

Figure 1: (*left*) CMB total intensity component separated map from a joint analysis of the *Planck*, WMAP and 408 MHz observations and Stokes Q (*center*) and U (*right*) amplitude maps results of the component separation of the only *Planck* data [9, 10].

3 Maps and power spectrum

Four different component separation methods have been considered in the *Planck* collaboration to separate the contribution of the CMB signal from all the astrophysical contaminants emitted by our Galaxy or of extragalactic origin [9]. The use of multiple methods is needed to have a robustness check of the products of component separation. The methods are: SMICA [11] (independent component analysis of power spectra), NILC [12] (needlet-based internal linear combination), Sevem [13] (template fitting) and Commander [14] (pixel-based parameter and template fitting with Gibbs sampling). All the methods have been applied to the *Planck* data producing four sets of cleaned CMB maps in temperature and polarization. An example of the results is shown in Fig. 1. SMICA and Commander methods produce also maps of the foreground components, useful for astrophysical studies.

The temperature maps delivered by the four methods are very consistent with each other and are expected to give equivalently robust results when used for cosmological analyses, e.g. tests of isotropy, Gaussianity and estimation of the lensing effect. The *Planck* 2015 release is characterised by the first release of polarization data and maps. The cleaned polarization CMB maps represent an important improvement in terms of coverage, angular resolution, and sensitivity with respect to previous results. However, the maps still show the presence of important anomalous features at large angular scales, due to systematic effects in the input frequency maps between 100 and 217 GHz. Since their analysis is still on-going, the first multipoles ($\ell < 30$) have been filtered out from the polarization maps delivered in 2015.

In the assumption that the fluctuations of the anisotropy pattern are purely Gaussian, all the statistical information present in the maps can be encoded in the two point correlation function or in its harmonic transform, the Angular Power Spectrum (APS). *Planck* analyses [16, 17] did not find any significant deviation from Gaussianity supporting this assumption. In Fig. 2 we show the APS in temperature, polarization and cross-correlation extracted by the 2015 *Planck* likelihood tool [15] consisting in an hybrid approach, based on a direct calculation of the large angular scale ($\ell < 30$) likelihood, and pseudo-spectral

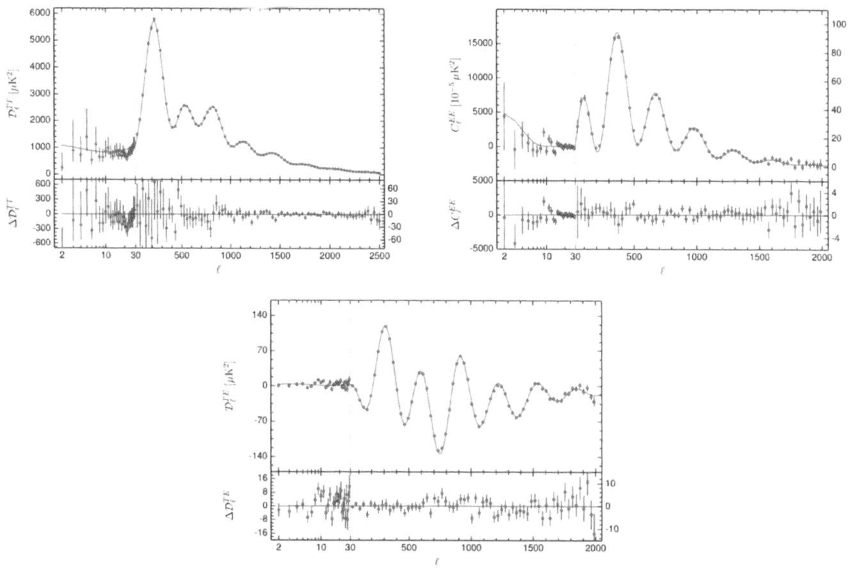

Figure 2: The *Planck* 2015 angular power spectra in temperature (*top left*) and polarization, EE (*top right*) and TE (*bottom*) and their residuals with respect to the theoretical model shown in red obtained from only temperature and large scales ($\ell < 30$) polarization data [15].

estimates at small angular scales. In Fig. 2 the red curve is the theoretical ΛCDM model based on the *Planck* 2015 best fit of only temperature data and low multipole polarization data: the level of agreement with the EE and TE spectra measured at intermediate and small angular scales is impressive.

4 Cosmological parameters

The *Planck* 2015 likelihood, based on the full mission data, is used to explore the cosmological parameter space with a Markov Chain Monte Carlo (MCMC) method in order to extract the values and uncertainties of the 6 parameters of the ΛCDM model [18]. The results, considering different sets of data, are shown in Table 2 including also derived parameters. They represent a great confirmation of the standard cosmological model. They are in excellent agreement with those of the previous release and, with the addition of external datasets such as measurements of the Baryon Acoustic Oscillations (BAO), type Ia supernovae observations collected in the Joint Light-curve Analysis (JLA) and Hubble constant (H0) measurements, these results represent, to date, the best estimates of all the parameters of the ΛCDM model.

The same approach can be applied to models that are extensions of the standard cosmological model obtaining e.g. tight constraints on the curvature of the

Parameter	TT+lowP	TT+lowP+lensing	TT,TE,EE+lowP	TT+lowP+lensing+ext
$\Omega_b h^2$	0.02222 ± 0.00023	0.02226 ± 0.00023	0.02225 ± 0.00016	0.02227 ± 0.00020
$\Omega_c h^2$	0.1197 ± 0.0022	0.1186 ± 0.0020	0.1198 ± 0.0015	0.01184 ± 0.0012
$100\theta_{MC}$	1.04085 ± 0.00047	1.04103 ± 0.00046	1.04077 ± 0.00032	1.04106 ± 0.00041
τ	0.078 ± 0.019	0.066 ± 0.016	0.079 ± 0.017	0.067 ± 0.013
$ln(10^{10} A_s)$	3.089 ± 0.036	3.062 ± 0.029	3.094 ± 0.034	3.064 ± 0.024
n_s	0.9655 ± 0.0062	0.9677 ± 0.0060	0.9645 ± 0.0049	0.9681 ± 0.0044
H_0	67.31 ± 0.96	67.81 ± 0.92	67.27 ± 0.66	67.90 ± 0.55
Ω_m	0.315 ± 0.013	0.308 ± 0.012	0.3156 ± 0.0091	0.3065 ± 0.0072
σ_8	0.829 ± 0.014	0.8149 ± 0.0093	0.831 ± 0.013	0.8154 ± 0.0090
$10^9 A_s e^{-2\tau}$	1.880 ± 0.014	1.874 ± 0.013	1.882 ± 0.0012	1.873 ± 0.011

Table 2: Parameter 68% confidence limits for the base ΛCDM model from *Planck* CMB power spectra, in combination with lensing reconstruction ("lensing") and external data ("ext",i.e. BAO+JLA+H0) [18].

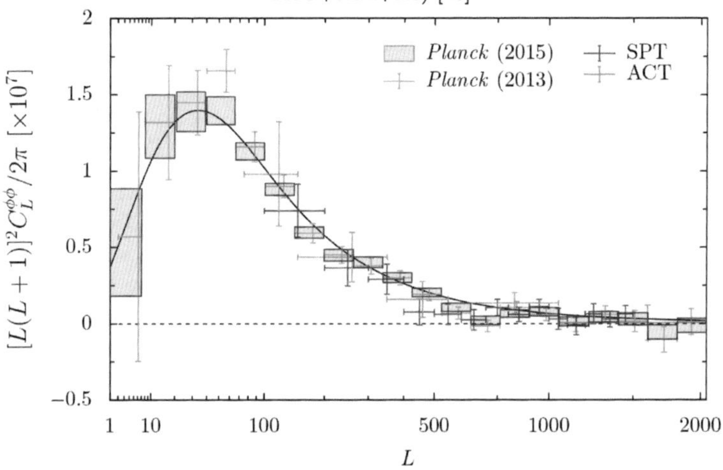

Figure 3: *Planck* 2015 full-mission lensing potential power spectrum measurement compared with the previous release (*Planck* 2013) and with the results of the South Pole telescope (SPT) and the Atacama Cosmology Telescope (ACT) [19].

Universe ($\Omega_K = 0.0008^{+0.0040}_{-0.0039}$), confirming excellent consistency with flatness, the total sum of the neutrino masses ($\sum m_\nu[ev] < 0.194$), the total number of neutrino species ($N_{eff} = 3.04 \pm 0.33$), the ratio between the amplitudes of tensor and scalar perturbations ($r_{0.002} < 0.113$) and the w parameter of Dark Energy equation of state ($w = -1.019^{+0.075}_{-0.080}$).

5 Lensing

During their journey from the last scattering surface to us, the CMB photons experienced small deviations in their path due to the gravitational effect of the matter structures present in the Universe. These deviations cause small but distinctive statistical signatures onto the observed CMB fluctuations. Their

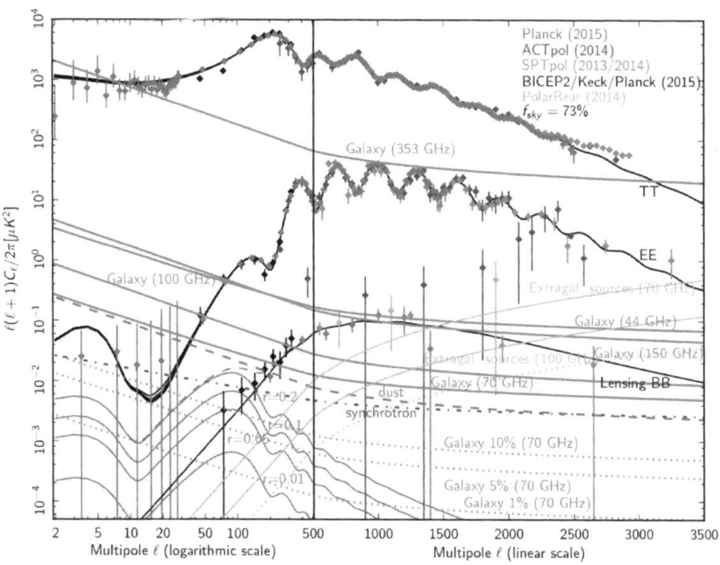

Figure 4: To date combined view of the CMB APS measurements in temperature and polarization of the most recent CMB experiments. They are compared with the total foreground emission (synchrotron plus dust) in polarization at different frequencies. The dots red lines, representing various levels of the Galactic foreground residuals at 70 GHz after an hypothetical component separation, are compared to the primordial B modes for various tensor-to-scalar ratios (green lines).

analysis proves the lensing potential, $C_\ell^{\phi\phi}$. Using the SMICA CMB cleaned map and applying five different quadratic estimators based on the correlations of the CMB temperature and polarization, *Planck* measured the CMB lensing potential with unprecedented precision [19]. The broadband amplitude of $C_\ell^{\phi\phi}$, shown in Fig. 3, is now known to better than 2.5% accuracy, with a detection of the lensing effect at more than 40σ. These results allowed to form a full-sky reconstruction of the projected mass distribution. Moreover, lensing B-modes are detected at 10σ, with both a correlation analysis with the Cosmic Infrared Background and the TTEB 4-point correlation function.

6 Conclusions and future

The *Planck* mission provide extremely precise data for both cosmology, allowing the best confirmation of the standard cosmological model, and astrophysics, with its big reservoir of data crucial for scientific analyses of next years. However, there are still some opened questions. One of the most important is the measurement of the primordial B modes that would be a key confirmation of the inflationary scenario. Remarkably, B mode amplitude proves the energy scale of inflation. However, a measurement of the B modes demands not only big technological efforts, because of the need of e.g. thousands of more stable and sensitive detectors, but it is also an astrophysical and data analysis challenge, because of the presence of foreground emissions that must be removed with great accuracy. In Fig. 4 we show the measurements of the CMB APS in temperature and polarization made by the most recent CMB experiments. In addition, we show the total Galactic foreground contamination in polarization at different frequencies as the sum of synchrotron and dust emissions, when we apply a suitable mask [10] excluding the 27% of the sky. The results show that the frequencies around 70 GHz are the cleanest. For future experiments targeted to detect and characterize the primordial B modes for small tensor-to-scalar ratios, $r \lesssim 0.01$, also foregrounds have to be understood and removed to a very precise level. Depending on the value of r, the APS amplitude of foreground residuals has to be less than 1% with respect to the original foreground signals, requiring a good characterization of also subdominant astrophysical emission components and features through a larger set of observational frequencies.

Acknowledgments

We acknowledge the use of the NASA Legacy Archive for Microwave Background Data Analysis and of the ESA *Planck* Legacy Archive. The ASI/INAF Agreement 2014-024-R.0 for the *Planck* LFI Activity of Phase E2 is acknowledged. Some results of this paper have been derived using HEALPix [20] package.

References

[1] R. Alpher, H. Bethe, and G. Gamow, Phys.Rev., **73**, 803-804 (1948).
[2] D. J. Fixsen, The Astrophys. J., **707**, 916 (2009).
[3] J.A. Tauber, N. Mandolesi, J.-L. Puget, et al. A&A, **520**, A1 (2010).
[4] Planck Collaboration, A&A, **571**, A1 (2014).
[5] Planck Collaboration, arXiv:1502.01582 (2015).
[6] Planck Collaboration, arXiv:1505.08022 (2015).
[7] Planck Collaboration, arXiv:1502.01587 (2015).
[8] G. Hinshaw, J. L. Weiland, R. S. Hill et al., The Astrophys. J. Supplement, **180**, 225 (2009).
[9] Planck Collaboration, arXiv:1502.05956 (2015).
[10] Planck Collaboration, arXiv:1502.01588 (2015).
[11] J.-F. Cardoso, M. Le Jeune, J. Delabrouille, et al. IEEE Journal of Selected Topics in Signal Processing, **2**, 735 (2008).
[12] J. Delabrouille, J.-F. Cardoso, M. Le Jeune et al. A&A, **493**, 835 (2009).
[13] R. Fernández-Cobos, P. Vielva, R. B. Barreiro and E. Martínez-González MNRAS, **420**, 2162 (2012).
[14] H. K. Eriksen, J. B. Jewell, C. Dickinson, et al., The Astrophys. J., **676**, 10 (2008).

[15] Planck Collaboration, arXiv:1507.02704 (2015).
[16] Planck Collaboration, arXiv:1502.01592 (2015).
[17] Planck Collaboration, arXiv:1506.07135 (2015).
[18] Planck Collaboration, arXiv:1502.01589 (2015).
[19] Planck Collaboration, arXiv:1502.01591 (2015).
[20] K.M. Gorski, E. Hivon, A.J. Banday, et al., The Astrophys. J., **622**, 759 (2005).

RECENT RESULTS FROM DAMA/LIBRA AND PERSPECTIVES

R. Bernabei, P. Belli [a], S. d'Angelo [b], A. Di Marco, F. Montecchia [c]
Dip. di Fisica, Univ. "Tor Vergata" and
INFN sez. Roma "Tor Vergata", I-00133 Rome, Italy
A. d'Angelo, A. Incicchitti, D. Prosperi [b]
Dip. di Fisica, Univ. "La Sapienza" and
INFN sez. Roma, I-00185 Rome, Italy
F. Cappella, V. Caracciolo, R. Cerulli
INFN, Laboratori Nazionali del Gran Sasso, I-67100 Assergi (AQ), Italy
C.J. Dai, H.L. He, H.H. Kuang, X.H. Ma,X .D. Sheng, R.G. Wang, Z.P. Ye [d]
IHEP, Key Laboratory of Particle Astrophysics, 100049 Beijing, China

Abstract. The DAMA/LIBRA experiment ($\sim$ 250 kg sensitive mass composed by highly radio-pure NaI(Tl)) is in data taking in the underground Laboratory of Gran Sasso (LNGS). The so called DAMA/LIBRA–phase1 and the former DAMA/NaI experiment ($\sim$ 100 kg of highly radio-pure NaI(Tl)) collected data for 14 independent annual cycles, exploiting the model-independent Dark Matter (DM) annual modulation signature (total exposure 1.33 ton $\times$ yr). A DM annual modulation effect has been observed at 9.3 σ C.L., supporting the presence of DM particles in the galactic halo. No systematic or side reaction able to mimic the observed DM annual modulation has been found or suggested by anyone. Recent analyses on possible diurnal effects, on the Earth shadowing effect and on possible interpretation in terms of Asymmetric Mirror DM will be mentioned. At present DAMA/LIBRA is running in its phase2 with increased sensitivity after an upgrade of the experiment.

1 Introduction

The DAMA project is dedicated to the development and use of low background scintillators for underground physics. The main experiment is DAMA/LIBRA [1–15] that, after the pioneering DAMA/NaI [8, 16, 17], is further investigating the presence of DM particles in the galactic halo by exploiting the model independent DM annual modulation signature [18].

Because of the Earth's revolution around the Sun, which is moving in the Galaxy, the flux of DM particles impinging a terrestrial detector is expected to be maximum around $\simeq$ June 2nd when the projection of the Earth orbital velocity on the Sun velocity with respect to the Galactic frame is maximum, and minimum around $\simeq$ December 2nd when the two velocities are opposite. This effect, known as DM annual modulation signature, is very effective because the signal induced by DM particles must simultaneously satisfy many requirements:

[a] E-mail: pierluigi.belli@roma2.infn.it
[b] deceased
[c] also Dip. di Ingegneria Civile e Ingegneria Informatica, Università di Roma "Tor Vergata", I-00133 Rome, Italy
[d] also University of Jing Gangshan, Jiangxi, China

the rate must contain a component modulated according to a cosine function (1) with one year period (2) and a phase peaked roughly at $\simeq$ June 2nd (3); the modulation must only be present in a well-defined low energy range (4); it must apply only to those events in which just one detector of many actually "fires" (*single-hit* events), since the DM particle multi-interaction probability is negligible (5); the modulation amplitude in the region of maximal sensitivity must be $\simeq$ 7% for usually adopted halo distributions (6), but it can be larger (even up to $\simeq$ 30%) in case of some possible scenarios. This signature is model independent and might be mimicked only by systematic effects or side reactions able to simultaneously satisfy all the requirements given above; no one is available [1–4, 7, 8, 12, 13, 16, 17, 19].

The full description of the DAMA/LIBRA set-up and performances during the phase1 and phase2 (presently running) and other related arguments have been discussed in details in Refs. [1–4, 6–8, 13] and references therein.

2 DM annual modulation results

The data of DAMA/LIBRA–phase1 correspond to 1.04 ton $\times$ yr collected in 7 annual cycles; when including also the data of the DAMA/NaI experiment the total exposure is 1.33 ton $\times$ yr collected in 14 annual cycles. In order to investigate the presence of an annual modulation with proper features in the data, many analyses have been carried out. All these analyses point out the presence of an annual modulation satisfying all the requirements of the signature [2–4,8]. In Fig. 1, as example, it is plotted the time behaviour of the experimental residual rate of the *single-hit* scintillation events for DAMA/LIBRA–phase1 in the (2–6) keV energy interval. When fitting the *single-hit* residual rate of DAMA/LIBRA–phase1 together with the DAMA/NaI ones, with the function: $A\cos\omega(t-t_0)$, considering a period $T=\frac{2\pi}{\omega}=1$ yr and a phase $t_0=152.5$ day (June 2nd) as expected by the DM annual modulation signature, the following modulation amplitude is obtained: $A=(0.0110\pm0.0012)$ cpd/kg/keV, corresponding to 9.2 σ C.L.. When the period, and the phase are kept free in the fitting procedure, the modulation amplitude is (0.0112 ± 0.0012) cpd/kg/keV (9.3 σ C.L.), the period $T=(0.998\pm0.002)$ year and the phase $t_0=(144\pm7)$ day, values well in agreement with expectations for a DM annual modulation signal. In particular, the phase is consistent with about June 2nd and is fully consistent with the value independently determined by Maximum Likelihood analysis [4]. The run test and the χ^2 test on the data have shown that the modulation amplitudes singularly calculated for each annual cycle of DAMA/NaI and DAMA/LIBRA–phase1 are normally fluctuating around their best fit values [2–4].

No modulation has been found in any possible source of systematics or side reactions; thus, cautious upper limits on possible contributions to the

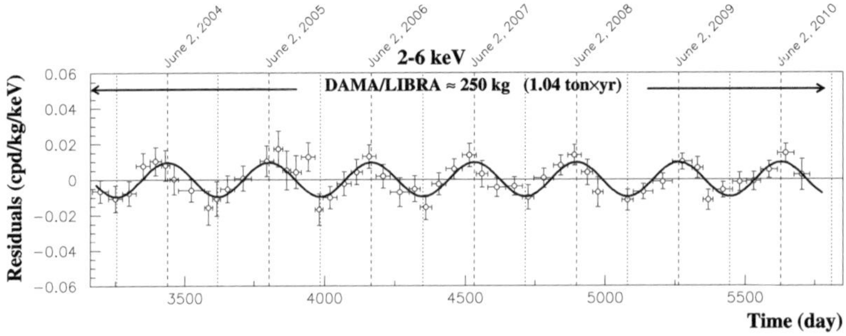

Figure 1: Experimental residual rate of the *single-hit* scintillation events measured by DAMA/LIBRA–phase1 in the (2–6) keV energy interval as a function of the time. The superimposed curve is the cosinusoidal function behaviour $A\cos\omega(t - t_0)$ with a period $T = \frac{2\pi}{\omega} = 1$ yr, a phase $t_0 = 152.5$ day (June 2nd) and modulation amplitude, A, equal to the central values obtained by best fit on the data points of the entire DAMA/LIBRA–phase1. The dashed vertical lines correspond to the maximum expected for the DM signal (June 2nd), while the dotted vertical lines correspond to the minimum.

DAMA/LIBRA–phase1 measured modulation amplitude have been obtained (see Refs. [2–4]). It is worth noting that they do not quantitatively account for the measured modulation amplitudes, and also are not able to simultaneously satisfy all the many requirements of the signature. Similar analyses have also been performed for the DAMA/NaI data [16,17].

In particular, in Ref. [7,13] a simple and intuitive way why the neutrons, the muons and the solar neutrinos cannot give any significant contribution to the DAMA annual modulation results is outlined. Other arguments can be found in Refs. [1–4,7,8,12,13,16,17,19]. In conclusion, DAMA give model-independent evidence (at 9.3σ C.L. over 14 independent annual cycles) for the presence of DM particles in the galactic halo.

As regards comparisons, we recall that no direct model independent comparison is possible in the field when different target materials and/or approaches are used; the same is for the strongly model dependent indirect searches. In particular, the DAMA model independent evidence is compatible with a wide set of scenarios regarding the nature of the DM candidate and related astrophysical, nuclear and particle Physics; for examples some given scenarios and parameters are discussed e.g. in Refs. [2,8,16] and references therein. Further large literature is available on the topics. In conclusion, both negative results and possible positive hints reported in literature can be compatible with the DAMA model-independent DM annual modulation results in various scenarios considering also the existing experimental and theoretical uncertainties. Moreover, scenarios also exist for which the DAMA approach is favoured.

Three recent analyses on the DAMA/LIBRA–phase1 data are summarized in the following. In the first one the presence of possible diurnal effects has been investigated [12], while in the second one the so called "Earth Shadow Effect" has been considered [14]. Finally, the annual modulation result has been interpreted in terms of Asymmetric Mirror DM [15]. These items were extensively studied in the quoted references, here the results are shortly addressed.

3 Diurnal modulation

The results obtained by investigating the presence of possible diurnal variation in the low-energy *single-hit* scintillation events collected by DAMA/LIBRA–phase1 have been analysed in terms of a DM second order model-independent effect due to the Earth diurnal rotation around its axis [12].

This daily modulation of the rate on the sidereal time – expected when taking into account the contribution of the Earth rotation velocity – also presents some specific peculiarities. In particular the interest in this signature is also that the ratio R_{dy} of this DM diurnal modulation amplitude over the DM annual modulation amplitude is a model independent constant at given latitude; considering the LNGS latitude, it is $R_{dy} = \frac{S_d}{S_m} \simeq 0.016$. Thus, taking into account the DM annual modulation effect pointed out by DAMA/LIBRA–phase1 for *single-hit* events in the low energy region, the expected value of the diurnal modulation amplitude for the (2–6) keV energy interval is $\simeq 1.5\times 10^{-4}$ cpd/kg/keV. Fig. 2 shows the time behaviour of the experimental residual rates of *single-hit* events both as a function of solar (*left*) and of sidereal (*right*) time, in the (2–6) keV interval [12].

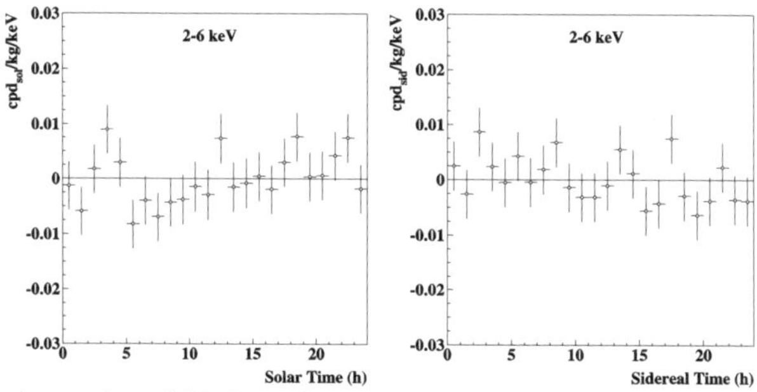

Figure 2: Experimental model-independent diurnal residual rate of the *single-hit* scintillation events, measured by DAMA/LIBRA–phase1 in the (2–6) keV energy intervals as a function of the hour of the solar (*left*) and sidereal (*right*) day. See Ref. [12] for details.

No diurnal variation with a significance of 95% C.L. is found at the reached level of sensitivity [12]. In conclusion the presence of any significant diurnal variation and of time structures can be excluded at the reached level of sensitivity. In addition, considering the (2–6) keV energy interval the obtained upper limit on the DM diurnal modulation amplitude is 1.2×10^{-3} cpd/kg/keV (90% C.L.) [12]; thus, the effect of DM diurnal modulation, expected because of the Earth diurnal motion on the basis of the DAMA DM annual modulation results, is out the present sensitivity [12]. DAMA/LIBRA–phase2, presently running, with a lower software energy threshold [6] can also offer the possibility to increase sensitivity to such an effect.

4 Daily effect on the sidereal time due to the shadow of the Earth

The results obtained in the investigation of possible diurnal effects for low-energy *single-hit* scintillation events of DAMA/LIBRA–phase1 have been analysed in terms of Earth Shadow Effect, a model-dependent effect that is expected for DM candidates inducing nuclear recoils and having high cross-section (σ_n) with ordinary matter [14].

In fact a diurnal variation of the low energy rate could be expected for these specific candidates, because during the sidereal day the Earth shields a given detector with a variable thickness, eclipsing the wind of DM particles. The induced effect should be a daily variation of their velocity distribution, and therefore of the signal rate measured deep underground. However, this effect is very small and would be appreciable only in case of high cross-section spin independent coupled candidates. Such candidates must constitute a little fraction (ξ) of the Galactic dark halo in order to fulfil the positive DAMA result on annual modulation. By the fact, this analysis decouples ξ from σ_n. Considering the measured DM annual modulation effect and the absence – at the present level of sensitivity – of diurnal effects, the analysis selects allowed regions in the three-dimensional space: ξ, σ_n and DM particle mass in some model scenarios; for details see Ref. [14].

5 Asymmetric mirror DM

The model independent annual modulation effect observed by the DAMA experiments has also been investigated in terms of a mirror-type dark matter candidates in some scenarios [15].

In the framework of asymmetric mirror matter, the DM originates from hidden (or shadow) gauge sectors which have particles and interaction content similar to that of ordinary particles. It is assumed that the mirror parity is spontaneously broken and the electroweak symmetry breaking scale v' in the mirror sector is much larger than that in the Standard Model, $v = 174$ GeV. In this case, the mirror world becomes a heavier and deformed copy of our world, with mirror particle masses scaled in different ways with respect to the masses of the ordinary particles. Then, in this scenario dark matter would exist in the form of mirror hydrogen composed of mirror proton and electron, with mass of about 5 GeV which is a rather interesting mass range for dark matter particles.

The data analysis in the Mirror DM model framework allows the determination of the $\sqrt{f}\epsilon$ parameter (where f is the fraction of DM in the Galaxy in form of mirror atoms and ϵ is the coupling constant). In the analysis several uncertainties on the astrophysical, particle physics and nuclear physics models have been taken into account in the calculation. The obtained values of the $\sqrt{f}\epsilon$ parameter in the case of mirror hydrogen atom ranges between 7.7×10^{-10} to 1.1×10^{-7}; they are well compatible with cosmological bounds [15].

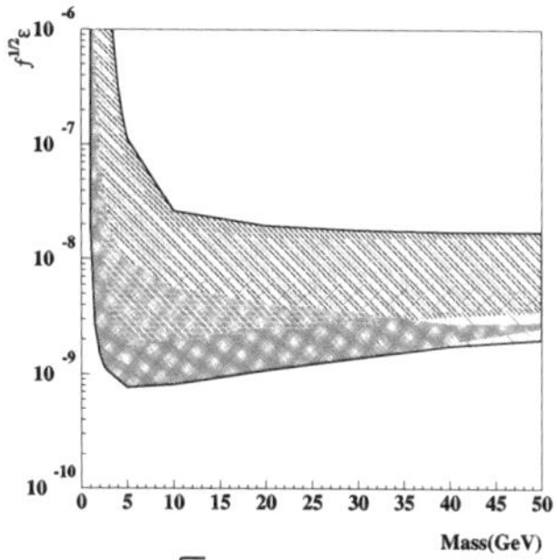

Figure 3: Allowed regions for the $\sqrt{f\epsilon}$ parameter as function of mirror hydrogen mass, obtained by marginalizing all the models for each considered scenario. The allowed intervals identify the $\sqrt{f\epsilon}$ values corresponding to C.L. larger than 5σ from the *null hypothesis*, that is $\sqrt{f\epsilon} = 0$. The allowed regions corresponding to five different scenarios are depicted in different hatching; the black line is the overall boundary; for details see Ref. [15].

In addition, releasing the assumption $M_{A'} \simeq 5m_p$, the allowed regions for the $\sqrt{f}\epsilon$ parameter as function of $M_{A'}$, mirror hydrogen mass, obtained by marginalizing all the models for each considered scenario, are shown in Fig. 3. Five scenarios are reported with different hatching of the allowed regions.

6 Conclusions and perspectives

The cumulative exposure with ultra low background NaI(Tl) target by the former DAMA/NaI and DAMA/LIBRA–phase1 is 1.33 ton $\times$ yr (orders of magnitude larger than those available in the field) giving a model-independent positive evidence at 9.3 σ C.L. for the presence of DM candidates in the galactic halo with full sensitivity to many kinds of astrophysical, nuclear and particle Physics scenarios. Other rare processes have also been searched for by DAMA/LIBRA-phase1; see for details Refs. [9–11].

At the end of 2010 an important upgrade has been performed when all the PMTs have been replaced with new ones having higher Quantum Efficiency (QE). In this new configuration a software energy threshold below 2 keV have been reached [6]. DAMA/LIBRA is thus in its phase2 and after optimization periods it is continuously running with higher sensitivity.

DAMA/LIBRA–phase2 is continuously running in order: (1) to increase the experimental sensitivity thanks to the lower software energy threshold of the experiment; (2) to improve the corollary investigation on the nature of the DM particle and related astrophysical, nuclear and particle physics arguments; (3)

to investigate other signal features; (4) to investigate rare processes other than DM with high sensitivity.

Future improvements to increase the sensitivity of the set-up can be considered by using high QE and ultra-low background PMTs directly coupled to the NaI(Tl) crystals. In this way a further large improvement in the light collection and a further lowering of the software energy threshold could be obtained.

Finally, for completeness, we also mention that low background $ZnWO_4$ crystal scintillators have recently been proposed within the DAMA collaboration for the study of the directionality of DM candidates inducing nuclear recoils. The features and performances of such anisotropic scintillators are very promising [20] and are under exploration.

Acknowledgments

We would like to thank the INR-Kiev group, led by F.A. Danevich, for their contribution to the development of the low background $ZnWO_4$ crystal scintillators for DM directionality and A. Addazi and Z. Berezhiani for their contribution to the paper on Asymmetric Mirror DM.

References

[1] R. Bernabei et al., Nucl. Instr. and Meth. **A 592**, 297 (2008).
[2] R. Bernabei et al., Eur. Phys. J. **C 56**, 333 (2008).
[3] R. Bernabei et al., Eur. Phys. J. **C 67**, 39 (2010).
[4] R. Bernabei et al., Eur. Phys. J. **C 73**, 2648 (2013).
[5] P. Belli et al., Phys. Rev. **D 84**, 055014 (2011) .
[6] R. Bernabei et al., J. of Instr. **7**, P03009 (2012).
[7] R. Bernabei et al., Eur. Phys. J. **C 72**, 2064 (2012).
[8] R. Bernabei et al., Int. J. of Mod. Phys. **A 28**, 1330022 (2013).
[9] R. Bernabei et al., Eur. Phys. J. **C 62**, 327 (2009).
[10] R. Bernabei et al., Eur. Phys. J. **C 72**, 1920 (2012).
[11] R. Bernabei et al., Eur. Phys. J. **A 49**, 64 (2013).
[12] R. Bernabei et al., Eur. Phys. J. **C 74**, 2827 (2014).
[13] R. Bernabei et al., Eur. Phys. J. **C 74**, 3196 (2014).
[14] R. Bernabei et al., Eur. Phys. J. **C 75**, 239 (2015).
[15] A. Addazi et al., Eur. Phys. J. **C 75**, 400 (2015).
[16] R. Bernabei el al., La Rivista del Nuovo Cimento **26 n.1**, 1-73 (2003).
[17] R. Bernabei et al., Int. J. Mod. Phys. **D 13**, 2127 (2004).
[18] K.A. Drukier et al., Phys. Rev. **D 33**, 3495 (1986); K. Freese et al., Phys. Rev. **D 37**, 3388 (1988).
[19] R. Bernabei et al., Eur. Phys. J. **C 18**, 283 (2000).
[20] F. Cappella et al., Eur. Phys. J. **C 73**, 2276 (2013).

ASTROPHYSICS IN THE MEV RANGE

Aldo Morselli [a]
INFN Roma Tor Vergata

Abstract. High-energy phenomena in the cosmos, and in particular processes leading to the emission of gamma- rays in the energy range 10 MeV - 100 GeV, play a very special role in the understanding of our Universe. This energy range is indeed associated with non-thermal phenomena and challenging particle acceleration processes. The Universe can be thought as a context where fundamental physics, relativistic processes, strong gravity regimes, and plasma instabilities can be explored in a way that is not possible to reproduce in our laboratories. High-energy astrophysics and atmospheric plasma physics are indeed not esoteric subjects, but are strongly linked with our daily life. Understanding cosmic high-energy processes has a large impact on our theories and laboratories applications. The technology involved in detecting gamma-rays is challenging and drives our ability to develop improved instruments for a large variety of applications.
The energy range between 1 and 100 MeV is an experimentally very difficult range and remained uncovered since the time of COMPTEL. New instruments can address all astrophysics issues left open by the current generation of instruments. In particular, the breakthrough angular resolution in the energy range 10 MeV - 1 GeV is crucial to resolve patchy and complex features of diffuse sources in the Galaxy and in the Galactic Centre as well as increasing the point source sensitivity. This instrument addresses scientific topics of great interest to the community, with particular emphasis on multifrequency correlation studies involving radio, optical, IR, X-ray, soft gamma-ray and TeV emission. The possibility to study not only the pair production regime but also the Compton regime with this kind of detectors is currently under investigation and it is another possible very interesting breaktrought.
One of the most important goal is the detection of gamma rays and cosmic rays from the annihilation or decay of dark matter particles. This is a promising method for identifying dark matter, understanding its intrinsic properties, and mapping its distribution in the universe. Based on N-body simulations the largest γ-ray signal from DM annihilation is expected from the centre of the Galaxy. In the same region a large γ-ray background is produced by bright discrete sources and the cosmic-rays interacting with the interstellar gas and the photons fields but the DM-induced gamma-ray emission is expected to be so large there that the search is still worthwhile. A good angular resolution below 100 MeV is crucial and I will review the effort in this direction.

Astrophysical searches for dark matter (DM) are a fundamental part of the experimental efforts to explore the dark sector. The strategy is to search for DM annihilation products in preferred regions of the sky, i.e., those with the highest expected DM concentrations and still close enough to yield high DM-induced fluxes at the Earth. For that reason, the Galactic Center (GC), nearby dwarf spheroidal galaxy (dSphs) satellites of the Milky Way, as well as local galaxy clusters are thought to be among the most promising objects for DM searches. In particular, dSphs represent very attractive targets because they are highly DM-dominated systems and are expected to be free from any other

[a]E-mail: aldo.morselli@roma2.infn.it

astrophysical gamma-ray emitters that might contaminate any potential DM signal. Although the expected signal cannot be as large as that from the GC, dSphs may produce a larger signal-to-noise (S/N) ratio. This fact allows us to place very competitive upper limits on the gamma-ray signal from DM annihilation [1–3], using data collected by the Large Area Telescope (LAT) onboard the Fermi gamma-ray observatory [4]. These are often referred to as the most stringent limits on DM annihilation cross-section obtained so far.

Despite these interesting limits derived from dSphs, the GC is still expected to be the brightest source of DM annihilations in the gamma-ray sky by several orders of magnitude. Although several astrophysical processes at work in the crowded GC region make it extremely difficult to disentangle the DM signal from conventional emissions, the DM-induced gamma-ray emission is expected to be so large there that the search is still worthwhile. Furthermore, the DM density in the GC may be larger than what is typically obtained in N-body cosmological simulations. Ordinary matter (baryons) dominates the central region of our Galaxy [5]. Thus, baryons may significantly affect the DM distribution. As baryons collapse and move to the center they increase the gravitational potential, which in turn forces the DM to contract and increase its density. This is a known and qualitatively well understood physical process [6,8,9]. It is also observed in many cosmological simulations that include hydrodynamics and star formation [10–15]. If this is the only effect of baryons, then the expected annihilation signal will substantially increase [5, 7].

A preliminary analysis of *Fermi* LAT observations of the GC region was presented in [20], [21], [22] and then analized in [23].

These results produced a lot of activity outside the Fermi collaboration with claims of evidence for dark matter in the Galactic Center (i.e. [24] [25] and references therein).

This possibility was already considered in the analisys of the EGRET galactic center excess [26] but there are other possible explanations, e.g., past activity of the Galactic Center [27], [28] or a population of millisecond pulsars around the Galactic Center [29].

So, how to discriminate between different hypotesis ?

Are we seeing dark matter with the Fermi-lat in a region around the Milky Way center? Maybe yes, but we can't be sure as far as we don't understand the background at the level needed for disentangle a DM-induced γ-ray flux in this interesting region. It would be really important to have a new experiment with better angular resolution at energies below 100 MeV that can point to the Galactic Center for a relevant fraction of time.

In [30] for example we analyze in detail the constraints that can be obtained for generic DM candidates from *Fermi*-LAT inner Galaxy gamma-ray measurements assuming some specific (and well motivated) DM distributions. The approach is conservative, requiring simply that the expected DM signal does not exceed the gamma-ray emission observed by the *Fermi*-LAT in an optimized region around the GC. The region is chosen in such a way that the S/N ratio is maximized. This kind of analysis, without modeling of the astrophysical background, was also carried out by the *Fermi*-LAT collaboration to constrain DM models from Galactic halo observations [31].

There we also illustrate the case $\langle \sigma v \rangle = 3 \times 10^{-26}$ cm^3 s^{-1}, which corresponds to the value of the annihilation cross-section associated to the correct thermal relic abundance for a Weakly Interacting Massive Particle (WIMP) whose annihilation is dominated by the s-wave (velocity-independent) contribution and thus, $\Omega_{DM}\ h^2 \approx 3 \times 10^{-27}$ cm^3 s^{-1} $\langle \sigma v \rangle^{-1} \approx 0.1$ [33]. For comparison, the constraints are given considering only the contribution from prompt gamma rays and the total contribution from prompt plus ICS gamma rays.

First, it is worth noting that if the DM density follows an Einasto, NFW or Burkert profile, the upper limits on the annihilation cross section are above the value of the thermal cross-section for any annihilation channel. Nevertheless, the situation is drastically different when we consider the DM compression due to baryonic infall in the inner region of the Galaxy. As pointed out in Ref. [5], the effect of the baryonic adiabatic compression might be crucial for indirect DM searches, as it increases by several orders of magnitude the gamma-ray flux from DM annihilation in the inner regions, and therefore the DM detectability. Indeed, by adopting the NFW$_c$ profile and for a $b\bar{b}$, $\tau^+\tau^-$ and W^+W^- channel, the thermal annihilation cross-section is already reached for a DM mass of 680, 530 and 490 GeV, respectively. For the $\mu^+\mu^-$ channel the effect of the prompt gamma rays is less important since generally fewer photons are produced in the FSR compared to the hadronic decays of the other channels. (For the W^+W^- which is open when $m_{DM} \gtrsim 90$ GeV, the $W^\pm$ decays produce a large number of photons, especially at high energy). Notice that the lower bound associated with prompt gamma rays for $\mu^+\mu^-$ is 100 GeV compared to about 500–700 GeV in the other channels. Thus the ICS is important in this case, also due to the relatively harder $e^\pm$ spectrum [32]. We can see that for $B_0 = 1$ μG the lower bound on the DM mass turns out to be 358 GeV and for $B_0 = 10$ μG the bound is 157 GeV, using the MIN diffusion model. For MED and MAX diffusion models the values turn out to be 404, 171 GeV and 439, 179 GeV, respectively. As discussed in [30], when the magnetic field is stronger the energy of the injected $e^\pm$ is more efficiently liberated in the form of microwaves, resulting in a softer gamma-ray spectrum, and producing therefore lower constraints . Therefore, we have shown that in those cases in which the ICS component is dominant (for heavy WIMP masses in general), the variation of the magnetic field can significantly alter the expected gamma-ray fluxes from the inner regions of the Galaxy.

Although the above results can be interpreted in general as implying that vanilla WIMP models and contracted DM profiles are incompatible with the

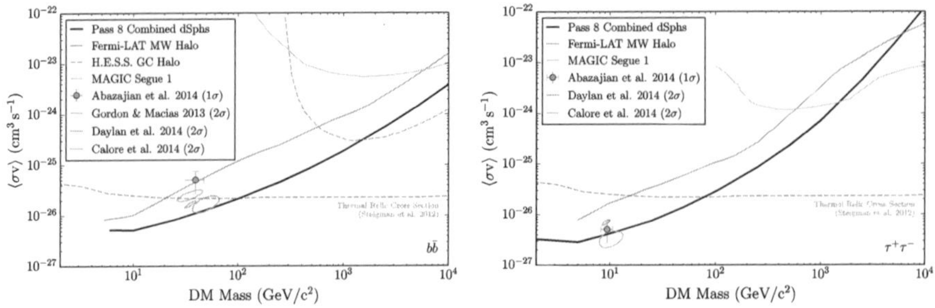

Figure 1: Comparison of constraints on the DM annihilation cross section for the $\bar{b}b(left)$ and $\bar{\tau}\tau$ (right) channels [35] with previously published constraints from LAT analysis of the Milky Way halo (3σ limit) [31], 112 hours of observations of the Galactic Center with H.E.S.S. [16], and 157.9 hours of observations of Segue 1 with MAGIC [17]. Closed contours and the marker with error bars show the best-fit cross section and mass from several interpretations of the Galactic center excess [24].

Fermi data, one should keep in mind that if one works in the framework of a specific particle physics model this conclusion might in principle be avoided in some regions of the parameter space. For example, the final state can be a combination of the annihilation channels presented here, as in supersymmetry where the lightest neutralino annihilation modes are 70% $\bar{b}b$ – 30% $\tau^+\tau^-$ for a Bino DM, and 100% W^+W^- for a Wino DM (or for a Higgs-portal model). More importantly, the value of the annihilation cross section in the Galactic halo might be smaller than 3×10^{-26} cm^3 s^{-1} for a DM candidate that is thermally produced. For example, in the early Universe coannihilation channels can also contribute to $\langle \sigma v \rangle$. Also, DM particles whose annihilation in the early Universe is dominated by p-wave (velocity-dependent) contributions would have a smaller value of $\langle \sigma v \rangle$ in the Galactic halo, where the DM velocity is much smaller than at the time of freeze-out, and can therefore escape the constraints presented here. These two effects can in fact occur in some regions of the parameter space of well motivated models for particle DM, such as the neutralino. In this sense, the results derived above for pure annihilation channels can be interpreted as limiting cases that give an idea of what can happen in realistic scenarios.

A new version of the event-level reconstruction and analysis framework (called Pass 8) is now implemented from the Fermi LAT collaboration. With this new analysis software we can increase the efficiency of the instrument at high energy and have a data set based on independent event analysis thus gaining a better control of the systematic effects [34].

A new analysis of the dSphs with the use of Pass 8 begin to constrain some of the preferred parameter space for a DM interpretation of a gamma-ray excess in the Galactic center region.

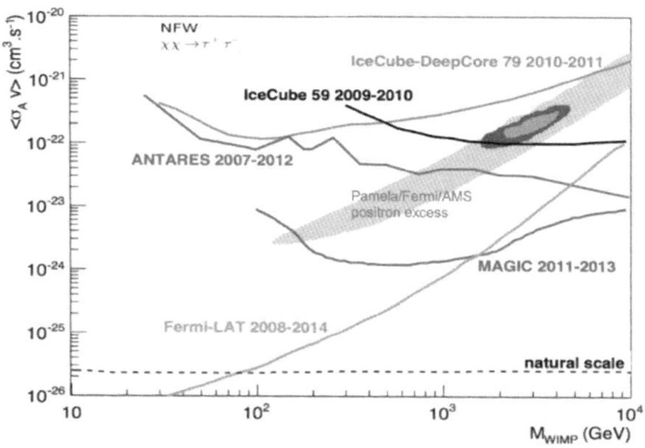

Figure 2: Comparison of constraints on the DM annihilation cross section for the $\overline{\tau}\tau$ channel [35] with Antares [18], IceCube-DeepCore [19] and MAGIC [17]

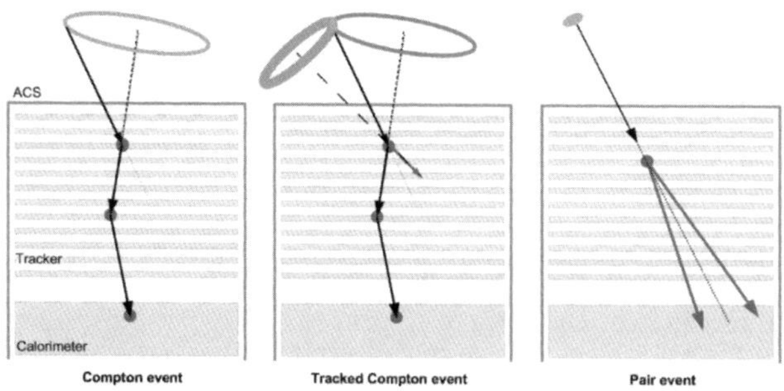

Figure 3: Representative event topologies for Compton events without (left) and with electron tracking (center) and for a pair event (right panel) inside ASTROGAM .

As shown in 1, for interpretations assuming a $\bar{b}b$ final state, the best-fit models lie in a region of parameter space slightly above the 95% CL upper limit from this analysis, with an annihilation cross section in the range of (1-3)$\times 10^{-26}$ cm^3 s^{-1} and m_{DM} between 25 and 50 GeV. However, uncertainties in the structure of the Galactic DM distribution can significantly enlarge the best-fit regions of $\langle\sigma v\rangle$ channel, and m$_{DM}$. Figure 2 shows a comparison of constraints on the DM annihilation cross section for the $\overline{\tau}\tau$ channel [35] with Antares [18], IceCube-DeepCore [19] and MAGIC [17]. One can see that the Fermi limits are the best limits below 2 TeV.

At lower energies a new instrument like Gamma-Light [36] , ASTROGAM [37] or Pangu [38] can really improve these results.

The ASTROGAM instrument is based on double-sided Silicon detectors coupled to front-end-electronics capable of acquiring analog information on energy deposition in the range 20-1000 keV with high efficiency and high signal-to-noise. Both Compton events induced by photons in the range 0.3-30 MeV and pair production events in the 30 MeV-30 GeV range can be detected by the ASTROGAM Tracker equipped with a Calorimeter and an Anticoincidence system. Figure 3 shows representative topologies for Compton and pair events. For Compton events, point interactions of the gamma ray in tracker and calorimeter produce spatially resolved energy deposits, which have to be reconstructed in sequence using the redundant kinematic information from multiple interactions. Once the sequence is established, two sets of information are used for imaging: the total energy and the energy deposit in the first interaction measure the first Compton scatter angle. The combination with the direction of the scattered photon from the vertices of the first and second interactions generates a ring on the sky containing the source direction. Multiple photons from the same source enable a full deconvolution of the image, using probabilistic techniques. For energetic Compton scatters (above 1 MeV), measurement of the track of the scattered electron becomes possible, resulting in a reduction of the event ring to an arc, hence further improving event reconstruction. Compton scattering depends on polarization of the incoming photon, hence careful statistical analysis of the photons for a strong (e.g., transient) source yields a measurement of the degree of polarization of its high-energy emission. Pair events produce two main tracks from the electron and positron at small opening angle. Tracking of the initial opening angle and the plane spanned by electron and positron enables direct back-projection of the source. Multiple scattering in the tracker material (or any intervening passive materials) leads to broadening of the tracks and limits the angular resolution at low energies. The nuclear recoil taking up an unmeasured momentum results in a small uncertainty, usually negligible compared to instrumental effects. The energy of the gamma ray is measured using the calorimeter. Polarization information in the pair domain is given by the azimuthal orientation of the electron-positron plane.

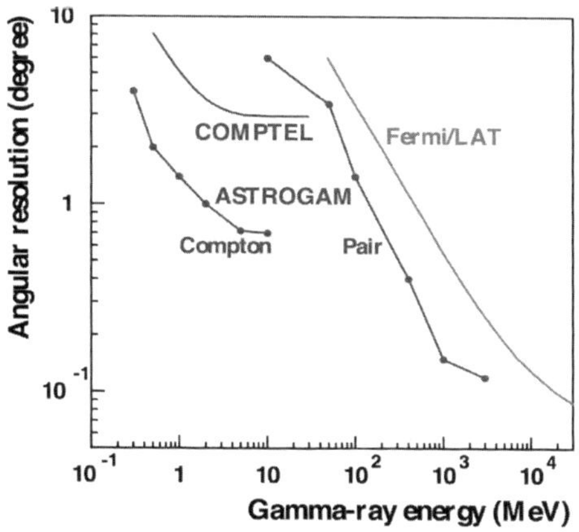

Figure 4: Point Spread Function (PSF, 68% containment radius) of the ASTROGAM gamma-ray detector. For comparison, we show the Fermi-LAT Pass7 PSF and the COMPTEL instrument. In the Compton domain, the performance of ASTROGAM and COMPTEL is the FWHM of the angular resolution measure (ARM) .

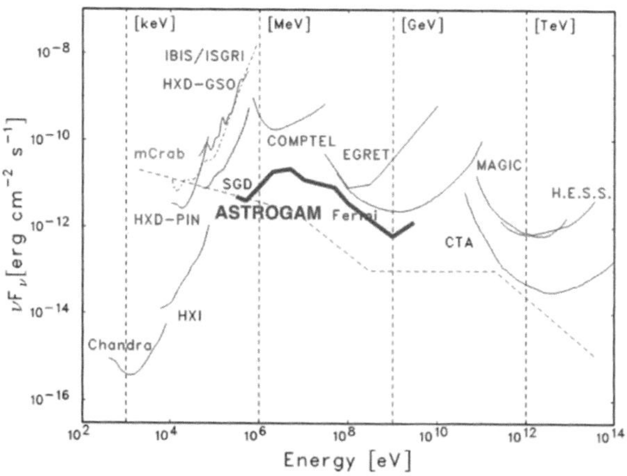

Figure 5: Point source continuum sensitivity of different X and γ-ray instruments compared with ASTROGAM.

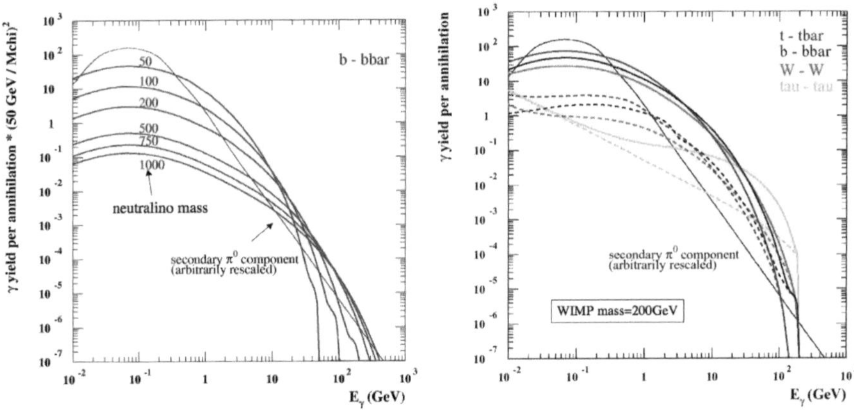

Figure 6: Differential energy spectra per annihilation for a few sample annihilation channels and a fixed WIMP mass (200 GeV) (left) and differential γ-ray energy spectra per annihilation for a fixed annihilation channel ($\bar{b}b(left)$) and for a few sample values of WIMP masses [26]. For comparison we also show the emissivity, with an arbitrarily rescaled normalization, from the interaction of primaries with the interstellar medium. The solid lines are the total yields, while the dashed lines are components not due to π^0 decays (right).

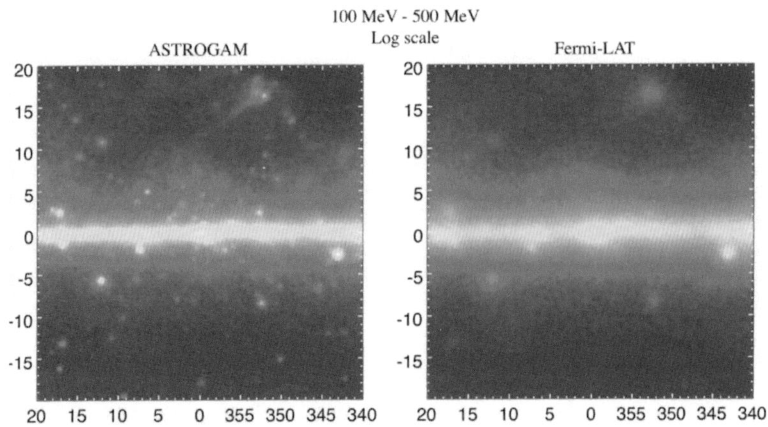

Figure 7: ASTROGAM view of the Galactic Center Region in the 100 MeV-500 MeV energy region compared with the Fermi view.

The Point Spread Function of ASTROGAM is shown in Figure 4, and the sensitivity is shown in figure 5 is for an effective exposure of 1 year of a high Galactic latitude source. Sensitivities above 30 MeV are given at the 5-sigma confidence level, whereas those below 10 MeV (30 MeV for COMPTEL) are at 3-sigma. The curves for Chandra/ACIS-S, Suzaku/HXD (PIN, GSO), INTEGRAL/IBIS and ASTRO-H (HXI, SGD) are given for an observing time T_{obs} = 100 ks. The COMPTEL and EGRET sensitivities are given for the observing time accumulated during the whole duration of the CGRO mission ($T_{obs} \sim 9$ years). The Fermi-LAT sensitivity is for a high Galactic latitude source and T_{obs} = 1 year. For MAGIC, H.E.S.S. and CTA the sensitivities are given for T_{obs} = 50 hours.

The importance of ASTROGAM for Dark Matter searches can be seen in figure 6 where the differential γ-ray energy spectra per annihilation of WIMP are plotted [26]. As one can see the bulk of the emission even for high WIMP masses is in the energy range 5 MeV - 100 MeV.

The very good angular resolution at low energies will help to resolve sources in the galactic center region and to desentangle a possible dark matter contribution (see Figure 7).

Conclusions

Detection of gamma rays from the annihilation or decay of dark matter particles is a promising method for identifying dark matter, understanding its intrinsic properties, and mapping its distribution in the universe (in synergy with the experiments at the LHC and in the underground laboratories). In the future it would be extremely important to extend the energy range of experiments at lower energies (compared to the Fermi energies) (ASTROGAM) and higher energies (HAWC, Dampe, HERD, Gamma-400, CTA, LHAASO)

References

[1] M. Ackermann *et al.* [Fermi-LAT Collaboration], Phys. Rev. Lett. **107** (2011) 241302 [arXiv:1108.3546 [astro-ph.HE]].
[2] A. A. Abdo *et al.* [Fermi-LAT Collaboration], Astrophys. J. **712** (2010) 147 [arXiv:1001.4531 [astro-ph.CO]].
[3] A. Geringer-Sameth and S.M. Koushiappas, Phys. Rev. Lett. **107** (2011) 241303 [arXiv:1108.2914 [astro-ph.CO]].
[4] W. B. Atwood *et al.* [Fermi-LAT Collaboration], Astrophys. J. **697** (2009) 1071 [arXiv:0902.1089 [astro-ph.IM]].
[5] F. Prada, A. Klypin, J. Flix Molina, M. Martinez and E. Simonneau, Phys. Rev. Lett. **93** (2004) 241301 [astro-ph/0401512].

[6] Ya. B. Zeldovich, A. A. Klypin, M. Yu. Khlopov and V. M. Chechetkin, Sov. J. Nucl. Phys. **31** (1980) 664.
[7] Y. Mambrini, C. Muñoz, E. Nezri and F. Prada, JCAP **01** (2006) 010 [hep-ph/0506204].
[8] G.R. Blumenthal, S.M. Faber, R. Flores and J.R. Primack, Astrophys. J. **301** (1986) 27.
[9] O. Y. Gnedin, A. V. Kravtsov, A. A. Klypin and D. Nagai, Astrophys. J. **616** (2004) 16 [astro-ph/0406247 [astro-ph]].
[10] M. Gustafsson, M. Fairbairn and J. Sommer-Larsen, Phys. Rev. D **74** (2006) 123522 [astro-ph/0608634 [astro-ph]].
[11] P. Colín, O. Valenzuela and A. A. Klypin, Astrophys. J. **644** (2006) 687 [astro-ph/0506627 [astro-ph]].
[12] P. B. Tissera, S. D. M. White, S. Pedrosa and C. Scannapieco, MNRAS **406** (2010) 922 [arXiv:0911.2316 [astro-ph.CO]].
[13] O. Y. Gnedin, D. Ceverino, N. Y. Gnedin, A. A. Klypin, A. V. Kravtsov, R. Levine, D. Nagai and G. Yepes, arXiv:1108.5736 [astro-ph.CO].
[14] M. Zemp, O. Y. Gnedin, N. Y. Gnedin and A. V. Kravtsov, Astrophys. J. **748** (2012) 54 [arXiv:1108.5384 [astro-ph.GA]].
[15] J. Sommer-Larsen and M. Limousin, MNRAS **408** (2010) 1998 [arXiv:0906.0573 [astro-ph.CO]].
[16] A. Abramowski et al. (H.E.S.S. Collaboration), Phys. Rev. Lett. 106, 161301 (2011) [arXiv:1103.3266]
[17] J. Aleksić et al. (MAGIC Collaboration), J. Cosmol. Astropart. Phys. 1402, 008 (2014) [arXiv:1312.1535]
[18] S. Adrían-Martínez et al. [Antares Coll.] arXiv:1505.04866
[19] M.G. Aartsen et al. [IceCube Collaboration] arXiv:1309.7007
[20] Lisa Goodenough, Dan Hooper arXiv:0910.2998
[21] V. Vitale and A. Morselli for the Fermi/LAT Collaboration, 2009 Fermi Symposium, eConf Proceedings C091122 [arXiv:0912.3828]
[22] A. Morselli, B.Cañadas, V.Vitale, Il Nuovo Cimento **34** C, N. 3 (2011) [arXiv:1012.2292]
[23] M. Ajello et al.[Fermi-LAT Coll.] Apj accepted [arXiv:1511.02938]
[24] F.Calore et al. arXiv:1409.0042
[25] T. Daylan et al., 2014, arXiv:1402.6703
[26] A.Cesarini, F.Fucito, A.Lionetto, A.Morselli, P. Ullio, Astropart. Phys. 21 (2004) 267 [astro-ph/0305075]
[27] J. Petrovic et al., 2014, JCAP, 10, 052
[28] E.Carlson , S.Profumo , 2014, Phys. Rev. D, 90, 023015
[29] Lee et al. arXiv:1506.05124
[30] G.A.Gomez-Vargas *et al.* JCAP 10 (2013) 029 [arXiv:1308.3515]
[31] M. Ackermann *et al.* [Fermi LAT Collaboration], Astrophys. J. **761** (2012) 91 [arXiv:1205.6474 [astro-ph.CO]].

[32] A. Birkedal, K. T. Matchev, M. Perelstein and A. Spray, hep-ph/0507194.
[33] G. Jungman, M. Kamionkowski and K. Griest, Phys. Rep. **267** (1996) 195 [arXiv:hep-ph/9506380].
[34] M. Ackermann *et al.* [Fermi LAT Collaboration], Phys. Rev. D. 91, 122002 (2015) [arXiv:1506.00013]
[35] M. Ackermann *et al.* [Fermi LAT Collaboration], Phys. Rev. Lett. 115, 231301 8 pg. (2015) [arXiv:1503.02641]
[36] A.Morselli et al., Nuclear Physics B - 239240 (2013) 193-198 [arXiv:1406.1071]
[37] http://astrogam.iaps.inaf.it
[38] X.Wu et al. Proceedings of the XI Multifrequency Behaviour of High Energy Cosmic Sources Workshop, 25-30 May 2015, Palermo, Italy

NATURE OF THE COSMIC RAY POWER LAW EXPONENTS

Allan Widom [a] and John.Swain [b]
Department of Physics, Northeastern University, Boston, Mass, USA
Yogendra Srivastava [c]
Department of Physics, University of Perugia, Perugia, Italy

Abstract. Two major unresolved issues with cosmic ray particle energy spectrum concern the isotropy & stable critical indices. Our basic hypothesis for isotropy is thermal particle production and thus the basic theoretical tool for critical indices is thermal field theory. Particles are supposed to evaporate from a hot source as in Landau-Fermi liquid theory. We compute both classical and quantum (Bosonic and Fermionic) critical indices. A quantum phase transition is found that explains the onset and the development of the "knee". The AMS data for $(e^- e^+)$ and cosmic ray data on nuclei are considered in detail and our theoretical predictions agree very well with experimental data. We have also made some energy loss estimates and show that for electrons and positrons of energy below about 1 TeV the energy losses are miniscule and thus the predicted critical indices remain robust and of fundamental importance. We conclude by highlighting our results and discuss further work in progress

Presented by YS at the 17th Lomonosov Conference on Elementary Particles, 21st August 2015, Moscow State University, Moscow, Russia

1 Introduction

Ever since Fermi [1] proposed his acceleration mechanism for high energy cosmic rays by *wandering* magnetic fields, it has become well nigh the "standard" explanation despite its well known deficiencies:

- First order Fermi acceleration mechanism does produce a power law behavior but with index = 1.
- Second order Fermi produces an incorrect index 2.
- Fermi was aware of the difficulty in explaining why heavy nuclei possessing the same indices as the proton are present in the cosmic radiation.
- Since multiple sources contribute to what is seen, one has to explain:
  A: why the observed spectrum is as smooth as it is (i.e. why each source has, within observed errors, the same exponent)
  B: why various sources should be distributed in such a way and with such intensities that there would, even with the same slopes, be no jumps in the spectrum.

[a]E-mail: allan.widom@hotmail.com
[b]E-mail: John.swain@cern.ch
[c]E-mail: Yogendra.srivastava@gmail.com

In a series of recent papers [2–6], we have tackled the above problems and found some remarkably simple power law spectra from thermal field theory that basically provide answers to the above issues and agree very well with their measured values. The reason for this success is that at high energies, the dynamics becomes simple thanks to the parton model and essentially a free thermal field theory results that is amenable to theoretical analysis.

2 Cosmic ray particle spectrum: Isotropy & power law indices

The differential energy distribution of nuclei ($1GeV < E < 1PeV$) in cosmic rays is (i) smooth, shows (ii) isotropy and has (iii) a stable power-law spectrum:

$$(\frac{dN}{dAdtd\Omega dE}) \approx (\frac{1.8\ Nucleons}{cm^2\ sec.\ ster\ GeV})(\frac{1\ GeV}{E})^{\alpha}, \tag{1}$$

where

$$\alpha(\text{experiment}) \approx\ 2.70 \tag{2}$$

Our theoretical result from thermal field theory [2, 3] gives for this index

$$\alpha(\text{theory}) = 3\frac{\zeta(4)}{\zeta(3)} \approx\ 2.7012, \tag{3}$$

in very good agreement with Eq.(2). In the above, $\zeta(s)$ is the Riemann zeta function of index s.

3 Statistical thermodynamic model

We apply entropy computations in a form originally due to Landau for the Landau-Fermi liquid drop model of a heavy nucleus. The energy distribution of some of the decay products are thought to be evaporating nucleons from the bulk liquid drops excited by a heavy nuclear collision. We propose as the sources of cosmic rays, the evaporating stellar winds from (say) neutron star surfaces

4 Cosmic ray evaporation & the thermal Boson condensate formalism

The structure of neutron stars consists of a big nuclear droplet facing a very dilute gas, i.e. the vacuum. Neutron stars differ from being simply very large nuclei in that most of their binding is gravitational rather than nuclear, But, the droplet model of the nucleus should still offer a good description of nuclear matter near the surface where it can evaporate. The evaporation is from the effective fields in the form of

$$\text{Scalar nuclei}: \ {}^4He \equiv \ \alpha \tag{4}$$

$$\text{Vector deuterons}: \ {}^2H \equiv \ d \tag{5}$$

- What evaporates from (say a) neutron star via scalar and vector fields are then energetic protons and alpha particles along with other nuclei to a much lesser extent.
- Deuterons are only weakly bound and would be expected to photodissociate on background photons present throughout space rapidly to produce protons and neutrons which in turn would produce protons when they decay.
- The energy distribution comes directly from the entropy of scalar (spin zero) and vector (spin 1) combinations of baryons.

Note that from neutron stars we will be at high temperatures where the heat capacity goes to a constant.

5 Power law index: Bosons & Fermions

An ultra relativistic particle of energy ϵ has a density of state $\propto \ \epsilon^2$. The mean energy per boson

$$E = \frac{\int_o^\infty (d\epsilon) f(\epsilon)\epsilon^3}{\int_o^\infty (d\epsilon) f(\epsilon)\epsilon^2} \tag{6}$$

For the Planck distribution

$$f(\epsilon) = \frac{1}{e^{\epsilon/k_B T} - 1}, \tag{7}$$

gives

$$E = cT; \ c = k_B \left[\frac{\Gamma(4)\zeta(4)}{\Gamma(3)\zeta(3)}\right] \ \approx \ (2.7012) k_B \tag{8}$$

The heat capacity per evaporated boson

$$c = \alpha k_B = T\frac{ds}{dT} = \frac{dE}{dT} \tag{9}$$

The entropy

$$\Delta s(E) = \alpha k_B \ ln(E/E_o) \tag{10}$$

The heat of vaporization

$$q = (\Delta s)T \tag{11}$$

Energy distribution of evaporated bosons given by the activation probability:

$$P(E) = e^{-(\Delta s)/k_B} = \Big(\frac{E_o}{E}\Big)^{\alpha}; \ \alpha = 2.701178 \tag{12}$$

Similarly, for fermions, we had predicted that

$$\alpha_{Fermi}(theory) = 3.151374, \tag{13}$$

which agrees very favourably with recent AMS data [7]

$$\alpha(experiment) = 3.170 \pm 0.008 \text{ (stat + syst.) } \pm 0.008 \text{ (energy scale)} \tag{14}$$
$$\text{for } e^{\mp} \text{ in the energy interval } (30\ GeV < E < 1\ TeV). \tag{15}$$

In general, the mean energy per particle for an ideal ultra-relativistic gas following any statistics

$$E = \frac{\int_o^{\infty}(d\epsilon)\frac{\epsilon^3}{[e^{\epsilon/k_BT}-\eta]}}{\int_o^{\infty}(d\epsilon)\frac{\epsilon^2}{[e^{\epsilon/k_BT}-\eta]}} = \alpha_\eta(k_BT);$$
$$\text{where } \eta = 1 \text{ [Bose] ; } \eta = 0 \text{ [Boltzmann] ; } \eta = -1 \text{ [Fermi]} \tag{16}$$

The thermal regulators are

$$\alpha_o = \frac{\Gamma(4)}{\Gamma(3)} = 3; \text{ [Classical Index];}$$
$$\alpha_1 = \frac{\Gamma(4)\zeta(4)}{\Gamma(3)\zeta(3)} = 2.701178; \text{ [Boson Index];}$$
$$\alpha_{-1} = \frac{1-(1/8)}{1-(1/4)}\alpha_1 = 3.151374; \text{ [Fermi Index]} \tag{17}$$

Incidentally, if electrons/positrons were classical particles, their index would have been 3. This is a remarkable confirmation of our theory as well as to the fact that the measured indices are telling us something fundamental about the system.

6 Cosmic quantum phase transition: The knee

There is a break in the cosmic particle spectrum at an energy $E_c \sim 1\ PeV$ [The Knee]. For energies of nuclei lower than the knee, the index $\alpha = 2.7$ (Boson value). For energies of nuclei higher than the knee, the index $\alpha = 3.1$

(Fermion value).The energy distribution depends on the heat and thereby entropy of evaporation from the cosmic ray source. There exists a quantum phase transition of cosmic proportions at the crossing energy. It is evidently a quantum phase transition since the order parameter involves the difference between Bose and Fermi statistical phases. In virtue of the experimental continuity of the energy distribution and thereby the entropy, the phase transition is higher than first order.

The boson phase arises from those evaporation nuclei described by the conventional collective Boson models of nuclear physics. There exist pairing correlations in odd-odd nuclei made up of deuterons. Correlations between two spin one deuterons lead to spin zero alpha particles and so forth all within the pairing condensate. The condensate resides near the surface of evaporating high baryon number $A >> 1$ nuclei. For example, a neutron star itself is merely a nucleus of extremely high baryon number ($A >> .. >> 1$) with superfluidity (and superconductivity) in the neighborhood of the nuclear surface. The "partons" from the pairing condensate are evidently the nucleons. The partons then turn into fermions for the phase that exists above the crossing energy. For energies above the knee, the cosmic rays must be composed mainly of protons. In order to comprehend the phase transition, it is necessary to understand how photons can photo-disintegrate the compound boson odd-odd nuclei ultimately into its nucleon parts

We can make an estimate of the cross-over energy using a simple kinetics model for photo disintegration. Consider an initial compound nucleus I that is hit by a photon leading to a final state F: $\gamma + I \to \ F$. The cross-over energy may be estimated as

$$E_c = \frac{< (M_F c^2)^2 - (M_I c^2)^2 >}{(6 k_b T_\gamma)}, \tag{18}$$

where T_γ is the noise temperature. Numerically

$$E_c \approx \frac{(100\ MeV)^2}{10 eV} \sim\ 1 PeV. \tag{19}$$

7 Index changes due to energy losses

In [6], we have investigated conditions under which energy losses due to EM radiation would not change the observed energy power law spectrum from the critical indices at the source. Our analysis is of particular relevance for the AMS electron/positron data as energy losses for these are (potentially) the largest. We find that for $E < 1\ TeV$, the measured index remains the same and reflects the index at the source. Hence, AMS data are quite robust and of fundamental importance.

We have computed energy losses for ultra-relativistic electrons and/or positrons due to classical electro-dynamic bremsstrahlung. The energy losses considered are (i) due to Thompson scattering from fluctuating electromagnetic fields in the background cosmic thermal black body radiation and (ii) due to the synchrotron radiation losses from quasi-static domains of cosmic magnetic fields. For distances to sources of galactic length proportions, the lepton cosmic ray energy must be less than about a TeV for negligible losses.

8 Conclusions and work in progress

Thermal origin (from neutron stars, supernoave,..) can explain the observed isotropy and smoothness in the cosmic ray energy spectrum. Using quantum relativistic thermal field theory, we have computed power law indices for bosons and fermions in very good agreement with data. A second order phase transition is shown to occur leading to the observed knee in the spectrum and the estimate of the cross-over energy is satisfactory. Our prediction of the critical index for electron/fermion index is in excellent agreement with the AMS data. Our computations of the EM energy losses largest for electrons and positrons-confirm that the AMS data for E ¡ 1 TeV, reflect the critical index at the source.

Ultra-relativistic cosmic ray particles should have tiny deviations from our computed critical indices due to asymptotic freedom that anchors the quasi-free parton model, leading to an ideal gas of bosons and fermions (IBM). We hope to present albeit small- one-loop corrections soon. Quantum EM energy losses need to be performed to confirm our classical energy loss estimates. Particularly relevant would be computation of changes in the critical indices for electrons/positrons with energies $> 1\ TeV$.

For cosmic energies far beyond the knee, we expect a dramatic change in the composition of the nuclei so much so that at asymptotic energies we expect to observe essentially only protons. This statement needs to be quantified and details of the spectrum need to be computed. An important problem that concerns future work is into the nature of the very high energy spectrum of the photon and the neutrino.

References

[1] E. Fermi, *Phys. Rev* **75** (1949) 1169.

[2] *Entropy and the Cosmic Ray Particle Energy Distribution Power Law Exponent*:
A. Widom, J. Swain, Y. Srivastava, arXiv:1410.1766 v1 [hep-ph] (2014)

[3] *Concerning the Nature of the Cosmic Ray Power Law Exponents*:
A. Widom, J. Swain, Y. Srivastava, arXiv:1410.6498 [hep-ph] (2014)

[4] *A Quantum Phase Transition in the Cosmic Ray Energy Distribution*: A. Widom, J. Swain, Y. Srivastava, arXiv:1501.07809v2[hep-ph](2015)
[5] *The Theoretical Power Law Exponent for Electron and Positron Cosmic Rays, A Comment on the Recent Letter of the AMS Collaboration*: A. Widom, J. Swain, Y. Srivastava, arXiv:1501.07810v2 [hep-ph] (2015)
[6] *Bremsstrahlung Energy Losses for Cosmic Ray Electrons and Positrons*: A. Widom, J. Swain, Y. Srivastava, arXiv:1509.08365v1 [hep-ph] (2015)
[7] M. Aguilar, *et. al.*, AMS Collaboration, *Phys. Rev. Lett.* **113** (2014) 221102.

INDIRECT DARK MATTER SEARCHES IN THE LIGHT OF THE RECENT AMS-02 OBSERVATIONS

Pierre Salati [a]
LAPTh, Université de Savoie Mont Blanc, CNRS;
9 Chemin de Bellevue, B.P.110, Annecy-le-Vieux F-74941, France

Abstract. If the astronomical dark matter is made of weakly interacting, massive and stable species, it should annihilate on itself into particles. This process should produce rare antimatter cosmic rays and lead to distortions in their energy distributions. The AMS-02 spectrometer has been measuring them with unprecedented accuracy. It is timely to investigate if anomalies have been found in the positron and antiproton spectra and if so, if they indirectly point toward the presence of DM particles annihilating inside the Milky Way.

1 Cosmic rays as an indirect probe for dark matter

The universe contains a substantial fraction of its mass under the form of the so-called astronomical dark matter (DM), a pressureless component found inside galaxies [1, 2], clusters of galaxies [3] and on cosmological scales. The recent observations of the Planck satellite [4] have confirmed the picture of a flat universe filled with dark energy (68.3%), dark matter (26.8%) and baryons (4.9%). According to this standard lore, the astronomical dark matter cannot be made of baryons and its nature is still unknown. Many solutions have been proposed for the last three decades. Among the numerous possibilities, a particular candidate under the form of a weakly interacting massive particle dubbed WIMP has attracted much attention. This species is naturally present in most extensions of the standard model of particle physics. It is stable by conservation of a quantum number, such as R-parity in supersymmetry or the momentum along the extra-dimensions in Kaluza-Klein inspired models. It interacts with its surroundings and annihilates on itself through typically weak interactions. The crucial consequence, that makes WIMPs so interesting, is that they are produced during the Big Bang with a relic abundance close to the Planck value of $\Omega_{\rm DM} h^2 = 0.1196 \pm 0.0031$. For this to happen, the annihilation cross section $\langle \sigma v \rangle$ should be close to the canonical value of 3×10^{-26} cm^3 s^{-1}.

The searches for WIMP-like DM have developed along three directions. First, WIMPs could be produced at colliders such as the LHC and appear in missing energy events. An abnormally large rate of gluon monojets or single gauge boson events could be explained by the fusion of quark-antiquark pairs into pairs of DM particles. A second line of research, called direct detection, is based on the potential collisions of WIMPs on a terrestrial instrument. As an impinging DM species collides on a nucleus, a kinetic energy of a few tens of keV is transferred to the later. Current techniques are sensitive to tiny recoil energies and

[a] E-mail: pierre.salati@lapth.cnrs.fr

the background is now strongely suppressed, making that technique particularly promising. Finally, WIMPs could also be indirectly detected through the particles which they would produce by annihilating inside the Milky Way. Although DM annihilation is a marginal process today, it is a potential source of high-energy photons, neutrinos and rare antimatter particles such as positrons e^+, antiprotons $\bar{p}$ or even antideuterons $\bar{D}$ through the set of reactions

$$\chi + \chi \; \rightarrow \; q + \bar{q} \, , W^+ + W^- \, , \, \ldots \; \rightarrow \; \bar{p}, \bar{D}, e^+, \gamma \;\&\; \nu. \tag{1}$$

Antimatter cosmic rays are already manufactured by conventional astrophysical processes. The dominant mechanism is the spallation of primary high-energy protons and helium nuclei on the gas of the Galactic plane. Positrons could also be accelerated by highly-magnetized neutron stars called pulsars. The messengers of DM annihilation are expected to generate distortions in the signals detected at the Earth or to appear in the γ-ray sky as hot spots with no optical counterpart – see the review [5] for more details.

In this presentation, I will be concerned with DM indirect searches in the light of the latest AMS-02 observations. The AMS-02 spectrometer has been measuring the fluxes of charged cosmic rays with unprecedented accuracy. Should there be WIMPs in the Galaxy, their annihilation products could leave imprints in the energy spectra of antimatter cosmic rays. Detecting an excess requires to know the background though. Modeling the transport of charged cosmic rays inside the Milky Way is of paramount importance insofar as the searched signals come out actually as deviations from the conventional astrophysical spectra. These need to be determined as precisely as possible. The propagation of charged particles inside the Galactic magnetic field is understood as a mere diffusion process where the cosmic ray species collide on the turbulent knots of the field. The magnetic halo can be seen as a slab a few kiloparsecs thick in the middle of which the Galactic disk lies, with its stars and gas. The particles can also be driven apart from the disk through Galactic convection. Above a few GeV, diffusion and energy losses are the dominant mechanisms. Primary cosmic rays are interstellar nuclei and electrons of the disk that are accelerated to very high energies by supernova driven shock waves. As they diffuse throughout the magnetic halo, they can interact on the disk to produce secondary species, such as positrons and antiprotons, generating a background to the exotic DM signals.

2 The cosmic ray positron excess – Dark matter versus pulsars

As a matter of fact, a positron excess was observed in 2008 above 10 GeV by the PAMELA satellite [6]. More recently, the AMS-02 measurements [7] confirmed this anomaly up to 500 GeV. As featured in Fig. 1, the signal well exceeds the astrophysical background of secondary species. There must be a

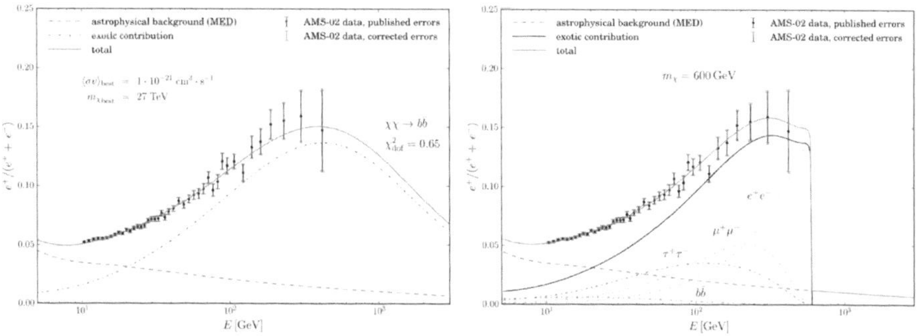

Figure 1: The WIMP annihilation signal and the astrophysical background contribute to the positron fraction, plotted here as a function of energy and compared with AMS data [7]. In the left panel, the case of a pure $b\bar{b}$ annihilation channel is featured. The best-fit values of $\langle\sigma v\rangle$ and DM mass m_χ are indicated. They correspond to a reduced chi-square of 0.65. In the right panel, the possibility of a WIMP annihilating into lepton and $b\bar{b}$ pairs is featured to investigate how mixed annihilation channels modify the best-fit DM parameters. The DM mass m_χ has been set equal to 600 GeV. The cross section $\langle\sigma v\rangle$ and branching ratios are left free to vary until they best fit the positron fraction data. The contribution of each channel to the signal is indicated. The branching ratio into $\tau^+\tau^-$ amounts to 50% whereas the quark contribution is 20%, with $\langle\sigma v\rangle = 1.11 \times 10^{-23}$ cm^3 s^{-1}. These values yield an excellent fit with $\chi^2_{\mathrm{dof}} = 0.5$. In both panels, the propagation parameters correspond to the MED model. Figure borrowed from [8].

source of additional positrons inside the Milky Way. An exciting possibility, which has been triggering a lot of activity and enthusiasm in the particle physics community, is that this extra component is made of primary positrons produced by WIMP annihilation. The fact that the anomaly is observed at high energy, where DM particles with a mass $\sim$ TeV are actually expected to contribute, is a strong incentive in favor of the DM hypothesis.

High-energy positrons lose rapidly their energy while they spiral inside the magnetic fields through which they propagate. Those detected by AMS-02 must have been produced in the vicinity of the solar system in order to make it to the Earth. The DM density in the solar neighborhood is ~ 0.3 GeV cm^{-3}, so that the production term for primary positrons from WIMP annihilation can be estimated. The only ingredient to be ajusted in the DM fit to the signal is the annihilation cross section $\langle\sigma v\rangle$. In the Boudaud *et al.* analysis [8], the DM mass m_χ and cross section $\langle\sigma v\rangle$ are fitted to the AMS-02 positron fraction. An agnostic scan over the various possible annihilation channels is performed. For the first time, mixed channels are also considered as displayed in the right panel of Fig. 1. This analysis confirms that the annihilation cross section must be orders of magnitude larger than the Big Bang canonical value of 3×10^{-26} cm^3 s^{-1}. In the left panel of Fig. 1 for instance, $\langle\sigma v\rangle$ needs to be boosted by 3×10^4. The leptonic channels are also disfavored by the data. Annihilation into tau pairs leads to a good fit, but muon pairs need to be produced via a light mediator while electrons are always excluded. All the other channels have P-values in excess of 99%. An additional problem lies in the fact that for precisely those channels, DM particles are coupled to quarks as well as to gauge and Higgs bosons, and eventually produce antiprotons through the hadronization of the final state species. Such couplings are severely constrained under the penalty of overproducing antiprotons as shown by Cirelli *et al.* [9] and confirmed by Donato *et al.* [10]. DM particles must be leptophilic on the one hand, but the AMS-02 data are not quite in favor of that possibility on the other hand. Another important piece of this puzzle comes from the observations of the gamma ray emission from dwarf spheroidal (dSph) satellite galaxies of the Milky Way. These systems are known to be dominated by DM and would shine in the gamma ray sky should DM strongly annihilate into high-energy photons. Recently, a comprehensive analysis by López *et al.* [11] indicates that the only viable channel that survives the Fermi/LAT constraint from dSph's and still produces a good fit to the AMS-02 positron fraction is DM annihilation via a mediator to 4 muons, or mainly to 4 muons in the case of multichannel combinations.

The excitement about the positron excess has by now receded. A more plausible explanation of the positron anomaly is based on pulsars. These highly magnetized neutron stars do exist and some of them even lie in our vicinity. Their rotating magnetic field can accelerate electron-positron pairs which are then released in the interstellar medium. The magnitude, spectral index and energy cut-off of the positron spectrum produced by each pulsar can be separately adjusted to the AMS-02 data. The fits are so good that even a single object is enough to explain the positron fraction, as demonstrated by [12,13,8]. Of course, nothing precludes DM to contribute to the positron excess. But this anomaly is no longer considered as a smoking gun signature of the presence of WIMPs inside the Milky Way.

3 Antiprotons – Has dark matter been discovered ?

A much more convincing signature for WIMPs would be an excess of antiprotons at high energy. No astrophysical objet is known to produce antiprotons and to release them in the interstellar medium like pulsars do for positrons. In that context, the preliminary measurement of the antiproton-to-proton ratio which the AMS-02 collaboration presented in April 2015 [14] aroused a renewed interest in WIMPs. As is clear in the left panel of Fig. 2, the gap between the astrophysical background of secondary antiprotons and the data is significant. It could leave room for a primary component produced by DM annihilation.

But before hastily reaching the conclusion that a new anomaly has been brought to light, the astrophysical background needs to be thoroughly investigated. As for positrons, it is produced by the collisions of high-energy protons

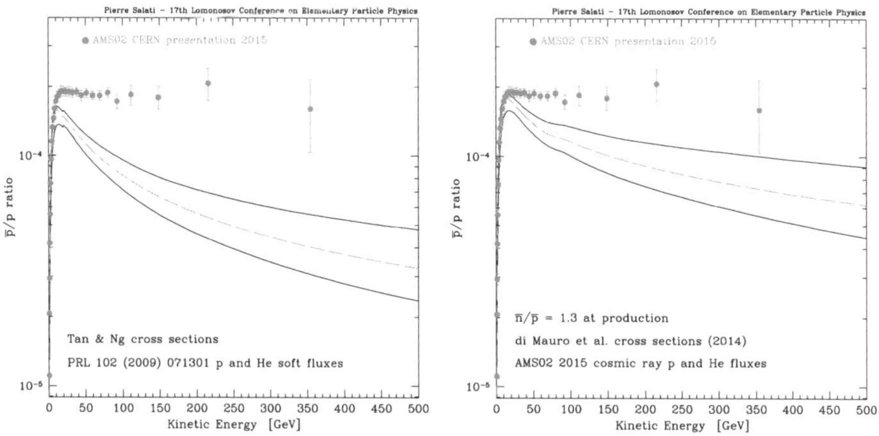

Figure 2: The antiproton-to-proton ratio $\bar{\mathrm{p}}/\mathrm{p}$ is plotted as a function of antiproton kinetic energy. The theoretical prediction for the secondary antiproton background is featured by the red long-dashed curve, whereas the preliminary results presented by the AMS-02 collaboration in April 2015 correspond to the red dots [14]. The yellow band encompasses the fluxes derived with cosmic ray propagation parameters compatible with the B/C ratio [21]. It illustrates the uncertainty arising from cosmic ray transport in the Milky Way. In the left panel, the background as calculated in 2008 by [10] is presented. The gap between the predictions and the data is obvious and may lead to the conclusion that an exotic antiproton component is necessary. The tension between the expected background and the data is less severe in the right panel where new calculations are featured [15]. See text for details.

and helium nuclei on interstellar gas. The left panel of Fig. 2 displays the result obtained in 2008 by Donato *et al.* [10] whereas the conspicuously larger background presented in the right panel is based on a recent analysis by Giesen *et al.* [15]. It leaves little room now for an exotic signal. As a matter of fact, three ingredients essential to the production of secondary antiprotons have changed since 2008. To commence, a secondary antiproton of energy E is produced on average by a proton of energy ηE impinging on an interstellar hydrogen atom at rest, with $\eta \sim 10$. The antiproton flux $\Phi_{\bar{\mathrm{p}}}(E)$ at energy E is proportional to the proton flux $\Phi_{\mathrm{p}}(\eta E)$ at energy ηE. As the latter is a power law with spectral index α, the $\bar{\mathrm{p}}/\mathrm{p}$ ratio scales like $\eta^{-\alpha}$. The spectral index α has been decreasing for the past years and hardenings in the proton and helium fluxes have recently been reported around 300 GeV [16,17]. Both observations lead to an increase of the $\bar{\mathrm{p}}/\mathrm{p}$ ratio. Then, the antiproton production cross section in pp collisions $d\sigma_{\mathrm{pH}\to\bar{\mathrm{p}}}/dE_{\bar{\mathrm{p}}}$ has been recently reinvestigated. A new parameterization based on data from the BRAHMS and NA49 experiments is now available [18]. It also induces a (modest) increase of the $\bar{\mathrm{p}}/\mathrm{p}$ ratio. Last but not least, pp interactions could yield more antineutrons than antiprotons as suggested by [19]. The cross sections of the reaction $\mathrm{pp} \to \bar{\mathrm{n}} + \mathrm{X}$ and its flipped counterpart $\mathrm{np} \to \bar{\mathrm{p}} + \mathrm{X}$ are related by isospin symmetry. The NA49

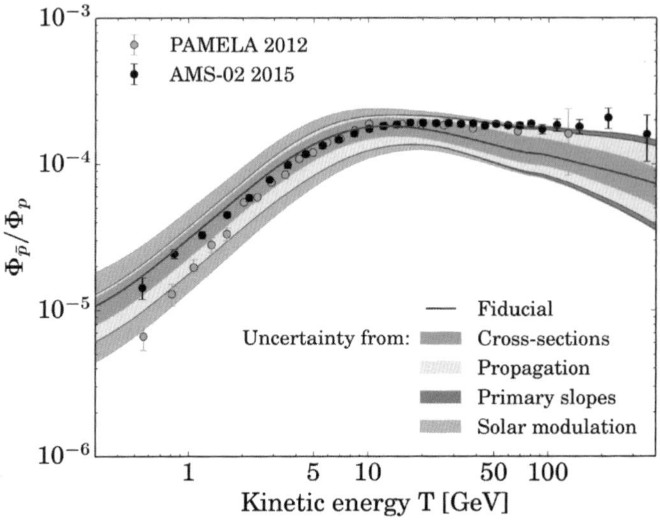

Figure 3: The combined total uncertainty on the predicted secondary $\bar{p}/p$ ratio, superimposed to the older PAMELA data [22] and the new AMS-02 data [14]. The curve labelled 'fiducial' assumes the reference values for the different contributions to the uncertainties: best fit proton and helium fluxes, central values for the cross sections, MED propagation and central value for the Fisk potential. We stress however that the whole uncertainty band can be spanned within the errors. Figure borrowed from [15].

experiment has measured a larger antiproton multiplicity in np compared to pp collisions [20]. This observation can only be understood if pp interactions yield more antineutrons than antiprotons, by a factor as large as 50%.

The calculations carried out by Giesen *et al.* [15] are presented in Fig. 3 where various sources of uncertainty are plotted as colored bands. The data are still compatible with the astrophysical background, although they lie close to the upper edge of the allowed region. Notice however that the fiducial prediction is based on the MED cosmic ray propagation model which has been derived from old B/C data. Forthcoming measurements of the B/C ratio may well induce a shift of the background region upward, as the analysis by [23] seems to indicate.

4 Closing thoughts

The cosmic ray positron anomaly has been confirmed by the AMS-02 collaboration. It is difficult to explain this excess solely by DM anihilation. The regions in the WIMP mass and cross section parameter space have moved because measurements are now so much precise. But very few of them survive after the Fermi/LAT constraints from dSph's are applied. The most plausible explanation has to be found in nearby pulsars. As regards antiprotons, the preliminary AMS-02 $\bar{p}/p$ ratio is compatible with a pure secondary component, although the data are close to the upper edge of the expected background. To decide whether a DM signal is hidden, cosmic ray propagation needs to be better constrained and the antiproton production cross sections in pp and NN collisions should be more accurately measured.

Acknowledgments

P.S. would like to thank the organizers of the 17th Lomonosov Conference on Elementary Particle Physics, in particular Alexander Studenikin, for their warm hospitality and the friendly and inspiring atmosphere of the meeting. This work has been supported by Institut universitaire de France.

References

[1] V.C. Rubin, W.K.J. Ford, N. Thonnard, Astrophys. J. **238**, 471 (1980).
[2] A. Bosma, P.C. van der Kruit, Astron. & Astrophys. **79**, 281 (1979).
[3] F. Zwicky, Helvetica Physica Acta **6**, 110 (1933).
[4] P. Ade *et al.*, Astron. & Astrophys. **571**, A16 (2014).
[5] J. Lavalle, P. Salati, Comptes Rendus Physique **13**, 740 (2012).
[6] O. Adriani *et al.*, Nature **458**, 607 (2009).
[7] L. Accardo *et al.*, Phys. Rev. Lett. **113**, 121101 (2014).
[8] M. Boudaud *et al.*, Astron. & Astrophys. **575**, A67 (2015).
[9] M. Cirelli *et al.*, Nucl. Phys. **B813**, 1 (2009).
[10] F. Donato *et al.*, Phys. Rev. Lett. **102**, 071301 (2009).
[11] A. Lopez *et al.*, JCAP **1603**, 033 (2016).
[12] T. Linden, S. Profumo, Astrophys. J. **772**, 18 (2013).
[13] I. Cholis, D. Hooper, Phys. Rev. **D88**, 023013 (2013).
[14] AMS-02 Collaboration, Talks at 'AMS Days at CERN', 15-17 april 2015.
[15] G. Giesen *et al.*, JCAP **1509**, 023 (2015).
[16] M. Aguilar *et al.*, Phys. Rev. Lett. **114**, 171103 (2015).
[17] M. Aguilar *et al.*, Phys. Rev. Lett. **115**, 211101 (2015).
[18] M. di Mauro *et al.*, Phys. Rev. **D90**, 085017 (2014).
[19] R. Kappl, M.W. Winkler, JCAP **1409**, 051 (2014).
[20] H.G. Fischer, Heavy Ion Phys. **17**, 369 (2003).
[21] D. Maurin *et al.*, Astrophys. J. **555**, 585 (2001).
[22] O. Adriani *et al.*, JETP Lett. **96**, 621 (2013) [Pisma Zh. Eksp. Teor. Fiz. **96**, 693 (2012)].
[23] R. Kappl, A. Reinert, M.W. Winkler, JCAP **1510**, 034 (2015).

EXTRAGALACTIC BACKGROUND STUDIES WITH FERMI-LAT AND IMAGING CHERENKOV TELESCOPES

Daniel Mazin [a]

ICRR, University of Tokyo, 277-8582 Chiba, Japan and MPI for Physics, 80805 Munich, Germany

Abstract. Very high energy (VHE, E > 30 GeV) gamma rays are absorbed via interaction with low-energy photons from the extragalactic background light (EBL) if the involved photon energies are above the threshold for electron-positron pair creation. The VHE gamma-ray absorption, which is energy dependent and increases strongly with redshift, distorts the VHE energy spectra observed from distant objects. The observed energy spectra of the AGNs carry, therefore, an imprint of the EBL. The detection of hard VHE gamma-ray spectra of distant sources (z = 0.11 - 0.94) by H.E.S.S., MAGIC and VERITAS enabled to set strong limits on the EBL density, using certain basic assumptions about blazar physics. Recently, Fermi/LAT and H.E.S.S. data led to a first clear detection of the EBL. Here, we critically review the assumptions for the constraints and discuss perspectives for the EBL and cosmology studies with the next generation observatory for ground-based gamma-ray observation CTA.

1 Extragalactic background light

During the star and galaxy formation history a diffuse extragalactic radiation field has been accumulated in the ultraviolet to far infrared wavelength regimes. This radiation field, commonly referred to as the extragalactic background light (EBL), is the second largest, in terms of the contained energy, background after the Cosmic Microwave Background of 2.7 K (CMB). While the CMB conserves the structure of the universe at the moment of the decoupling of matter and radiation following the Big Bang (at redshift $z \approx 1000$), the EBL is a calorimetric measure of the entire radiant energy released by processes of structure formation that have occurred since the decoupling.

A closer look to the UV – infrared backgrounds is given in Figure 1, left plot. From left to right, the spectral energy distributions of the three major components are shown: the EBL consisting of redshifted star light and redshifted emission by the dust in the galaxies, and the cosmic microwave background (CMB). Other contributions, like emission from AGN and quasars, are expected to produce no more than 5 to 20% of the total EBL density in the mid infrared (see, e.g., [1] and references therein).

The EBL is difficult to measure directly due to strong foregrounds from our solar system and the Galaxy. The observation of distant sources of VHE γ rays using Imaging Air Cherenkov Telescopes (IACT, such as H.E.S.S., MAGIC or VERITAS) provides a unique indirect measurement of the EBL. On the way from the source to the observer, VHE γ rays can suffer absorption losses by

[a]E-mail: mazin@icrr.u-tokyo.ac.jp

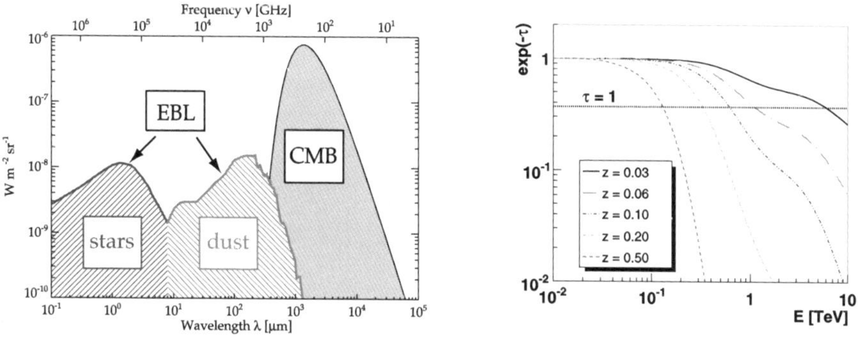

Figure 1: *Left:* Schematic Spectral Energy Distributions (SED) of the most important (by intensity) backgrounds in the universe. From left to right: the two EBL components, the one from the stars and the one from the dust emission in galaxies, as well as the cosmic microwave background (CMB). Plot adopted from [2]. *Right:* Attenuation coefficients $\exp(-\tau)$ as a function of γ-ray energy. The "best fit" model from [3] is used for the EBL density. The curves represent the expected attenuation of the VHE γ-ray spectra for different source distances. The horizontal dotted line corresponds to the optical depth $\tau = 1$.

interaction with the low energy photons of the EBL. The corresponding reaction is the pair-production of an electron-positron pair and is well understood. The resulting EBL imprint in the observed VHE γ-ray spectra depends on the distance to the source and the EBL density, see the expected attenuation in Fig. 1, right plot.

2 Status of the EBL and cosmology limits

Several constraints on the EBL were set in the recent years using data from the IACTs and Fermi/LAT. The first constraints were made using individual sources only, which led to criticism that the limits are source specific and if the selected sources are some what "special" the derived EBL limits would be model-dependent because the underlying assumption for the constraints is that the VHE sources are "typical".

An alternative is to use several avilable AGN spectra simultaneously to derive a consistent EBL strength for all of them. Strong convincing limits were obtained by, e.g., [4] and [5] using this method. Once Fermi/LAT collected sufficient number of photons, it was also used to set EBL constraints. In the work by [6] Fermi/LAT data was used to construct a global likelihood function for many AGNs, binned in three redshift bins. In the method, the intrinsic AGN spectral shapes are obtained first by fitting the available data in the unabsorbed energy regime (EBL independent, typically below 1 GeV). Then, a global likelihood function is maximized with the data in the absorbed energy range by utilizing the obtained expected spectral shapes, while the normalization of the

EBL level is the only free parameter.

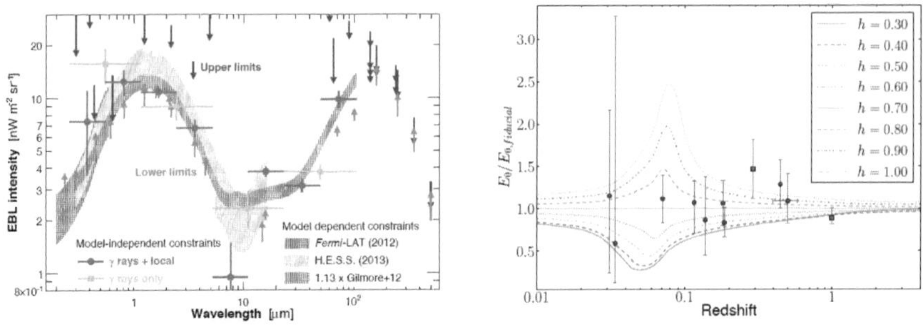

Figure 2: *Left:* EBL limits from Fermi/LAT and imaging Cherenkov telescopes. *Right:* First constraints on the Hubble constant using imaging Cherenkov telescope data.

In the work by [7], the state of the limits based on many AGN data (Fermi/ LAT and IACTs) are shown (Fig. 2, left plot). As one can see, the EBL is not only constrained from above, it is also well constrained from below. The upper limits on EBL derived with this technique are usually considered as more robust since they rely on the assumption that the AGN spectra do not have pile-ups at highest energy measured and that the intrinsic spectral shape is concave. The lower limits rely on the assumption that there are no wiggles or stronger than exponential breaks in the intrinsic spectrum, which is considered less well proven. Overall, the EBL limits are well in agreement with the recent EBL model predictions and can be understood as that there are not many unresolved sources contributing to the EBL.

Assuming that the EBL is resolved, one can use the EBL imprint in the AGN spectra to make tests on cosmological models. In particular, one can determine the distance to the AGN independently, thus setting constraints to the Hubble constant. A result of one such studies is shown in Fig. 2, right plot. See [8] for details.

3 Prospects for future measurements

Cherenkov Telescope Array (CTA) is the next generation IACT array, with ten-times better sensitivity than the current IACTs and an extended energy coverage between 30GeV and 300 TeV, see the sensitivity curves in Fig. 3, left plot [9]. The telescopes are now under design and in the prototyping phase with an aim to start scientific observations in both hemispheres by 2021. Several studies were performed to investigate the power of CTA to the cosmology and EBL, see, e.g., [10]. The power to resolve the EBL density as a function of redshift with CTA, is illustrated in Fig. 3, right plot by the red points. One

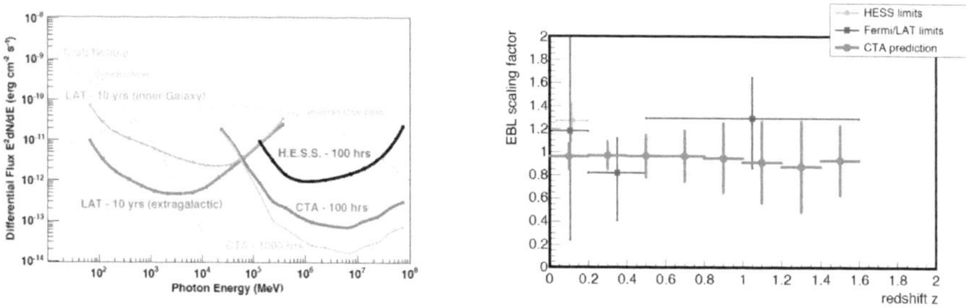

Figure 3: *Left:* Sensitivity of CTA (in prototyping phase compared to the existing telescopes and Fermi/LAT. *Right:* Simulated redshift resolved EBL measurements with CTA after 5 years of data taking.

can see that CTA measurements will be a factor of several superior to the existing ones. The statistical limits from such EBL measurement can become even tighter if the sources will be found at a higher flux level increasing the photon statistics in the EBL affected regime. However, systematic uncertainties will dominate the measurement when trying to resolve the EBL at a level better than 5-10% uncertainty. Still, CTA promises to become a great instrument to study EBL density and EBL evolution.

References

[1] I. Matute, F. La Franca, F. Pozzi, et al., *Astronomy & Astrophysics* **451**, 443–456 (2006).

[2] H. Dole, G. Lagache, Puget, et al., *Astronomy & Astrophysics* **451**, 417–429 (2006).

[3] T. M. Kneiske, K. Mannheim & D. H. Hartmann, *Astronomy & Astrophysics* **386**, 1 (2002).

[4] M. Meyer, M. Raue, D. Mazin & D. Horns *Astronomy & Astrophysics,* **542**, A59 (2012).

[5] H.E.S.S. Collaboration: A. Abramowski, F. Acero, F. Aharonian et al., *Astronomy & Astrophysics*, **550**, A4 (2013).

[6] Fermi/LAT Collaboration: M. Ackermann, M. Ajello, A. Allafort et al., *Science*, **338**, 6111 (2012).

[7] J. Biteau & D. Williams, *The Astrophysical Journal*, **812**, 60 (2015).

[8] A. Domínguez, & F. Prada, *The Astrophysical Journal, Letters*, **771**, L34 (2013).

[9] CTA Consortium: B. S. Acharya, M. Actis, T. Aghajani et al., *Astroparticle Physics*, **43**, 3 (2013).

[10] D. Mazin, M. Raue, M. Behera et al., *Astroparticle Physics*, **43**, 241 (2013).

GRAVITATIONAL INSTABILITY IN NON-STATIONARY EXTERNAL FIELDS

Elena Arbuzova [a]
Department of Higher Mathematics, Dubna State University, 141980, Dubna, Russia
Novosibirsk State University, 630090, Novosibirsk, Russia

Abstract. Gravitational instability in classical Jeans theory, General Relativity, and modified gravity is considered. It is shown that density perturbations in the background with an increasing density rise faster than in the case of time independent background. Evolution of density and metric perturbations in the background of high frequency oscillations of curvature in $F(R)$ gravity is studied. In addition to the usual Jeans-like instability new effects of amplification of perturbations, associated with parametric resonance and antifriction phenomena, are found.

The instability of self-gravitating systems was first investigated by Jeans [1] in non-relativistic Newtonian gravity. It was extended to General Relativity (GR) by Lifshitz [2] and nowadays it is widely used in cosmology to study the rise of density perturbations in the expanding universe [3, 4]. The comparison between theoretical calculations and astronomical data is a powerful tool for testing the Standard Model of cosmology.

At first let us consider a spherically symmetric cloud of particles with initially vanishing pressure and velocities, and study the classical non-relativistic Jeans problem in Newtonian gravity, as it was done in paper [5]. The essential equations are the well known Poisson, Euler, and continuity equations. In classical Jeans theory zero order approximation does not satisfy equations of motion. In our approach this problem is solved.

We consider the time dependent problem taking as initial value a homogeneous distribution $\rho = const$ inside a sphere with radius r_m, while outside this sphere $\rho = 0$. The potential Φ is supposed to be a solution of the Poisson equation:

$$\Phi_0(r > r_m) = -MG/r, \;\; \Phi_0(r < r_m) = 2\pi G\rho_0 r^2/3 + C_0, \tag{1}$$

where $C_0 = -2\pi G\rho_0 r_m^2$, and $M = 4\pi\rho_0 r_m^3/3$ is the total mass of the gravitating sphere. In what follows we will be interested in the internal solution for $r < r_m$.

Now we can find how the background quantities ρ, v, and P evolve with time at small t. In particular we find from the continuity equation:

$$\rho_b(t,r) = \rho_0 + \rho_1 = \rho_0\left(1 + \frac{2\pi}{3}G\rho_0 t^2\right). \tag{2}$$

The evolution of perturbations over this time-dependent background can be found as usually writing $\rho = \rho_b(r,t) + \delta\rho$, $\Phi = \Phi_b(r,t) + \delta\Phi$, $v = v_1(r,t) + \delta v$, and

[a] E-mail: arbuzova@uni-dubna.ru

$\delta P = c_s^2 \delta\rho$, where c_s is the speed of sound. All δ-quantities are infinitesimal, so we find the linear system of equations for them. Making the Fourier transform, we obtain the eigenvalue equation:

$$k^2(\gamma^2 - c_s^2 k^2 + 4\pi G\rho_0) = 0\,. \tag{3}$$

For small k we find the usual exponential Jeans instability:

$$\frac{\delta\rho_{J1}}{\rho_0} \sim \exp\left[t\left(4\pi G\rho_0 - c_s^2 k^2\right)^{1/2}\right]. \tag{4}$$

We can estimate the impact of the rising background energy density on the rise of perturbations making an adiabatic approximation, that is replacing the exponent in Eq. (4) with the integral:

$$\frac{\delta\rho_{J2}}{\rho_0} \sim \exp\left\{\int_0^t dt\,\left[4\pi G\rho_b(t,r) - k^2 c_s^2\right]^{1/2}\right\}. \tag{5}$$

where $\rho_b(t,r)$ is given by Eq. (2).

Estimating the above integral for small k we find that the enhancement factor $\delta\rho_{J2}/\delta\rho_{J1}$ grows with time [5]. We can interpret this fact as an indication that the rise of fluctuations is indeed faster then in the usual Jeans scenario.

The treatment of the Jeans instability in General Relativity starts from the Einstein equations:

$$G_{\mu\nu} \equiv R_{\mu\nu} - \frac{1}{2}\,g_{\mu\nu}R = 8\pi G\,T_{\mu\nu} \equiv \tilde{T}_{\mu\nu}\,, \tag{6}$$

which automatically include the equations of motion of matter.

As previously, we consider a spherically symmetric cloud of matter with initially constant energy density inside the limit radius $r = r_m$. We choose Schwarzschild-like isotropic coordinates in which the metric takes the form:

$$ds^2 = Adt^2 - B\,\delta_{ij}\,dx^i dx^j\,, \tag{7}$$

where the functions A and B may depend upon r and t.

We assume that the background metric changes slowly as a function of space and time and study small fluctuations around the background quantities: $\rho = \rho_b + \delta\rho$, $\delta P = c_s^2\delta\rho$, $\mathbf{v} = \delta\mathbf{v}$, $A = A_b + \delta A$, $B = B_b + \delta B$.

We look for the solution of the corresponding linear equations for infinitesimal perturbations in the form $\sim \exp[-i\gamma t + i\mathbf{k}\cdot\mathbf{x}]$ and obtain the following expressions for the frequency eigenvalues:

$$\gamma^2 = \frac{c_s^2 k^2 - \tilde{\rho}_b/2}{1 + 3\tilde{\rho}_b/(2k^2)}\,, \tag{8}$$

where $\tilde{\rho}_b = 8\pi G\rho_b$. This result almost coincides with the Newtonian one (3). An extra term in the denominator is induced by the volume variation, and it is small for $k \sim k_J = \sqrt{4\pi G\rho_0}/c_s$.

In modified gravity a nonlinear function of curvature $F(R)$ is added to the standard Einstein-Hilbert Lagrangian, so the modified Einstein equations have the form:

$$(1+F_{,R})\,R_{\mu\nu}-\frac{1}{2}\,(R+F)\,g_{\mu\nu}+(g_{\mu\nu}D_\alpha D^\alpha-D_\mu D_\nu)\,F_{,R}=\tilde{T}_{\mu\nu}\,, \tag{9}$$

where $F_{,R}=dF/dR$. Some examples of $F(R)$-theories which realistically describe the cosmological acceleration can be found in Refs. [7,8].

In these models, the function $F(R)$ takes very different values for $|R|\ll|R_c|$, $|R|\sim|R_c|$, and $|R|\gg|R_c|$, where R_c is the present average cosmological curvature. We consider the case $|R|\gg|R_c|$, which is realized in astronomical systems with the energy density grossly exceeding the cosmological one.

As usual the metric and the curvature tensors are expanded around their background values at first order in infinitesimal perturbations, i.e.

$$A=A_b+\delta A\,,\quad B=B_b+\delta B\,,\quad R=R_b+\delta R\,, \tag{10}$$

where A_b and R_b are quickly oscillating functions of time with possibly large amplitude ("spikes") which were found in our previous works [9].

Using technique described in papers [5,6] we derive the forth order equation for the single function $\delta B\equiv z$, which in terms of dimensionless quantities takes a form convenient for the analysis:

$$z''''+\alpha y'z'''+(\Omega^2+2\alpha y'')z''+\alpha(y'''+bcy')z'+\mu z=0\,. \tag{11}$$

We assume, that $y(\tau)$ is a periodic function describing curvature oscillations ($R_b=-\tilde{\rho}_b y$) and consider two possible forms of them: purely harmonic ones and spiky solutions found in Refs. [9]. If $y(\tau)$ is non-negligible, quite interesting new effects erise. It can induce an analogue of the parametric resonance instability resulting in a very fast rise of perturbations at a certain set of frequencies. Another new effect can be called "antifriction". It appears at sufficiently large amplitudes of oscillations of $y(\tau)$ such that the coefficients in front of the odd derivative terms become periodically negative. This phenomenon leads to an explosive rise of z in a wide range of frequencies.

Both effects do not exist in the standard General Relativity and, if discovered, would be a proof of modified gravity. On the contrary, the non-observation of these effects would allow to put restrictions on the parameters of $F(R)$-theories.

We solved Eq. (11) numerically for different values of the amplitude of oscillations, y_0, and the frequencies, $\Omega_{1,2}$. The parametric resonance is rather sensitive to variations of y_0 and, of course, of $\Omega_{1,2}$. In fact, the exponential growth is slower as we depart from the resonance frequency.

The antifriction amplification is observed at the frequencies away from the resonance values, if y_0 exceeds a threshold value, y_{th}. The farther away the frequency is from the resonance, the larger is the threshold. For example for harmonic oscillations with $\Omega_1/\Omega = 3.2$ the threshold value is $y_{th} = 169$, while for $\Omega_1/\Omega = 4.4$ the threshold is $y_{th} = 267$. The evolution of $z(\tau)$ for harmonic and spike-like oscillations is depicted in Fig. 1.

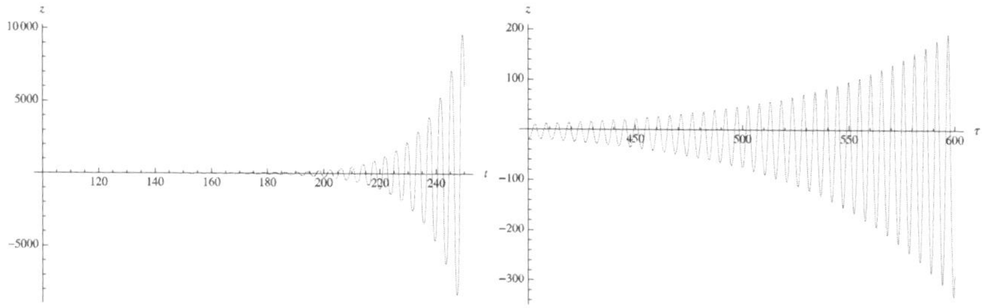

Figure 1: Antifriction effect in evolution of $z(\tau)$ for harmonic curvature oscillations with $y_0 = 169$, $\Omega_1/\Omega = 3.2$ *(left)* and for spike-type oscillations with $y_0 = 400$, $\Omega_2/\Omega = 0.6$*(right)*.

Acknowledgments

EA acknowledge the support of the grant of the Russian Federation government 11.G34.31.0047.

References

[1] J.H. Jeans, Philosophical Transactions of the Royal Society **A199**, 1 (1902).

[2] E.M. Lifshitz, ZhETF, **16**, 587 (1946).

[3] Ya.B. Zeldovich, I.D. Novikov *"Relativistic Astrophysics. Vol. 2. The Structure And Evolution Of The Universe"* (Chicago, Usa: Chicago Univ., 1983).

[4] S. Weinberg, *"Cosmology"* (Oxford University Press, 2008); V. Mukhanov, *"Physical Foundations of Cosmology"* (Cambridge University Press, 2005); D.S. Gorbunov, V.A. Rubakov, *"Introduction to the theory of the early universe: Cosmological perturbations and inflationary theory"* (Hackensack, USA: World Scientific, 2011).

[5] E.V. Arbuzova, A.D. Dolgov, L. Reverberi, Phys.Lett. **B 739**, 279 (2014).

[6] E. Arbuzova, A. Dolgov, L. Reverberi, Phys.Rev. **D 92**, 064041 (2015).

[7] A.A. Starobinsky, JETP Lett. **86**, 157 (2007); W. Hu, I. Sawicki, Phys. Rev. **D 76**, 064004 (2007); A. Appleby, R. Battye, Phys. Lett. **B 654**, 7 (2007).

[8] S.A. Appleby, R.A. Battye, A.A. Starobinsky, *JCAP* **1006**, 005 (2010).

[9] E.V. Arbuzova, A.D. Dolgov, L. Reverberi, Eur. Phys. J. **C 72**, 2247 (2012); Phys. Rev. **D 88** 024035 (2013); Astropart. Phys. **54**, 44 (2014).

ASTROPHYSICAL COMPONENTS FROM PLANCK MAPS

Carlo Burigana[1,2,3 a], Tiziana Trombetti[1,2 b], Daniela Paoletti[1,3 c], Nazzareno Mandolesi[1,4,2 d] and Paolo Natoli[2,5 e] (on behalf of *Planck* Collaboration)

[1]*INAF–IASF[f] Bologna, Via Piero Gobetti 101, I-40129 Bologna, Italy*

[2]*Dipartimento di Fisica e Scienze della Terra, Università degli Studi di Ferrara, Via Giuseppe Saragat 1, I-44122 Ferrara, Italy*

[3]*INFN[g], Sezione di Bologna, Via Irnerio 46, I-40126, Bologna, Italy*

[4]*Agenzia Spaziale Italiana, Viale Liegi 26, Roma, Italy*

[5]*Agenzia Spaziale Italiana Science Data Center, Via del Politecnico snc, 00133, Roma, Italy*

Abstract. The *Planck* Collaboration has recently released maps of the microwave sky in both temperature and polarization. Diffuse astrophysical components (including Galactic emissions, cosmic far infrared (IR) background, y-maps of the thermal Sunyaev-Zeldovich (SZ) effect) and catalogs of many thousands of Galactic and extragalactic radio and far-IR sources, and galaxy clusters detected through the SZ effect are the main astrophysical products of the mission. A concise overview of these results and of astrophysical studies based on *Planck* data is presented.

1 Introduction

The analysis of the microwave to sub-mm sky plays a crucial role not only for the cosmological information contained in the cosmic microwave background (CMB) but also for the wealth of astrophysical information it provides. The *Planck* Collaboration has recently released maps in temperature (resp. polarization) at nine (resp. seven) frequency bands between 30 and 857 GHz (resp. 353 GHz) reprocessing the data from the Low (up to 70 GHz) and High (from 100 GHz) Frequency Instruments (LFI and HFI). To extract relatively bright variable sources, the diffuse Zodiacal Light Emission and solar system bodies, a comparison between data taken at different times or maps derived from different surveys is performed, since signal dependence on time and/or on the relative observer-source or -interplanetary dust cloud positions. The separation of all the other astrophysical emissions and of CMB anisotropies is carried out on maps averaged over advantageous sets of surveys applying, respectively for point-like sources and diffuse emissions, dedicated filters, possibly complemented by external catalogs, and algorithms exploiting frequency dependence and angular correlation, possibly in combination with templates at other frequency bands.

[a]E-mail: `burigana@iasfbo.inaf.it`
[b]E-mail: `trombetti@iasfbo.inaf.it`
[c]E-mail: `paoletti@iasfbo.inaf.it`
[d]E-mail: `mandolesi@iasfbo.inaf.it`
[e]E-mail: `paolo.natoli@gmail.com`
[f]Istituto Nazionale di Astrofisica – Istituto di Astrofisica Spaziale e Fisica Cosmica
[g]Istituto Nazionale di Fisica Nucleare

2 Catalogs of sources and clusters of galaxies

The Second *Planck* Catalogue of Compact Sources (PCCS2), both Galactic and extragalactic, consists of (many tens of thousands) sources detected in single-frequency maps, combined from all *Planck* mission data over the entire sky, through the Mexican Hat Wavelet 2 algorithm (see Fig. 1). Four different flux-density estimates have been validated by simulations (internal validation) and comparison with other astrophysical data (external validation). Compact sources detected at lower (resp. higher) frequencies are assigned to the PCCS2 (resp. to the PCCS2 or PCCS2E sub-catalogues, depending on their sky positions). The PCCS2 covers most of the sky and allows to extract subsamples at higher reliabilities than the target 80% integral reliability of the catalogue. The PCCS2E contains sources whose detection reliability is affected by diffuse emission. Both catalogs include polarized flux densities, or upper limits, and orientation angles at frequencies 30 GHz $\leq \nu \leq$ 353 GHz for many hundreds sources. The spectral index proves the source nature: the low (resp. high) frequency channels are dominated by synchrotron (resp. dusty galaxies) sources. A significant steepening at $\nu \sim$ 50 GHz is found in blazar spectra and the bright tail of extragalactic source counts at high frequency (at least for $\nu \leq$ 217 GHz) is dominated by synchrotron emitters, not by dusty galaxies.

The Second *Planck* catalog of SZ galaxy clusters (PSZ2), with 1653 detections, is the largest SZ-selected sample and the deepest all-sky catalogue of galaxy clusters. They have been analyzed with observations in other wavebands (radio to IR, optical and X-rays) and 1203 of them have counterparts in datasets external to *Planck*. A population of low-redshift X-ray under-luminous clusters has been revealed by SZ selection. They appear in optical and SZ surveys with consistent mass properties, but are almost absent in ROSAT X-ray selected samples. Fig. 1 compares PSZ2 with deeper SZ catalogs obtained with South Pole and Atacama Cosmology Telescopes (SPT and ACT) on selected areas. Microwave and X-ray cluster data have been jointly used to independently constrain cosmological parameters. In particular, the evolution of galaxy cluster abundance with mass and redshift set constraints on the matter density fluctuation normalization, the mean matter density, the dark energy density and equation of state, and on extensions of the minimal cosmological model, including e.g. massive neutrinos and modified gravity.

Integrated information on cluster physics and cosmological evolution comes from the Comptonization parameter, y, fluctuations over all-sky *Planck* maps [3] other than observing it as SZ effect towards specific clusters. The analysis of sub-millimeter/far-IR background fluctuations, carried out mainly in selected *Planck* areas, proves high-z galaxy evolution and star formation history [4]. These themes are also probed by large scale CMB polarization, sensitive to cosmic reionization history [5,6] related to structure and star formation/evolution and their production of ionizing photons in the intergalactic medium.

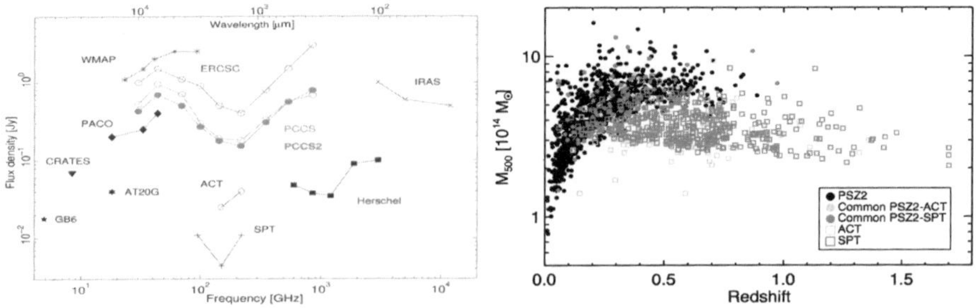

Figure 1: Left panel: sensitivity (the flux density at 90% completeness) of the PCCS2, compared with PCCS, ERCSC, WMAP and others. For the LFI channels, the sensitivities refer to the full sky. For the HFI channels, the 90% completeness limits plotted for the PCCS were evaluated in a selected extragalactic zone; the regions of sky to which the 90% completeness limits apply are similar but not identical to those of the PCCS2. From Ref. [1]. Right panel: distribution of PSZ2 clusters with associated redshift in the mass-redshift (M500 $-$ z) plane compared to SPT and ACT catalogues. Black circles: PSZ2 clusters; red (green) filled circles: common SPT/PSZ2 (ACT/PSZ2) clusters. Red (green) empty squares: the remaining SPT (ACT) clusters not detected by *Planck*. See also the text. From Ref. [2].

3 Galactic diffuse components

Astrophysical diffuse components and CMB anisotropies have been separated from *Planck* multifrequency maps by means of four algorithms. SEVEM and COMMANDER work in real space, NILC in real and harmonic (needlet) space and SMICA in harmonic or needlet domain. The CMB maps extracted with these methods are in remarkable agreement, implying that the overall foreground emission is analogously well set. COMMANDER, based on a-priori knowledge of foreground components characterized by parameters to be reconstructed with Bayesian methods in real space independently for each resolution element, has been found particularly suitable for characterizing astrophysical emissions varying across the sky. For temperature analysis, *Planck* data have been combined with WMAP [7] and 408 MHz [8] maps to jointly extract CMB, synchrotron, free-free, spinning dust, CO, line emission in the 94 and 100 GHz channels, and thermal dust emission. Fig. 2 shows the Stokes Q and U parameter maps for the synchrotron and thermal dust emissions, the only diffuse astrophysical components faithfully extracted in polarization. Another important emission clearly appears close to the Galactic center in *Planck* low frequency maps: the so-called "haze" component, whose full comprehension – related to astrophysics and/or dark matter phenomena – is still controversial. From the various source maps it is possible to quantify the contribution of each component to sky fluctuations, displayed in Fig. 2. A remarkable result of *Planck* is the assessment of the microwave sky complexity, calling for even more frequency channels for future CMB projects aimed at precise characterization of all components and of tiny CMB polarization features, such as primordial

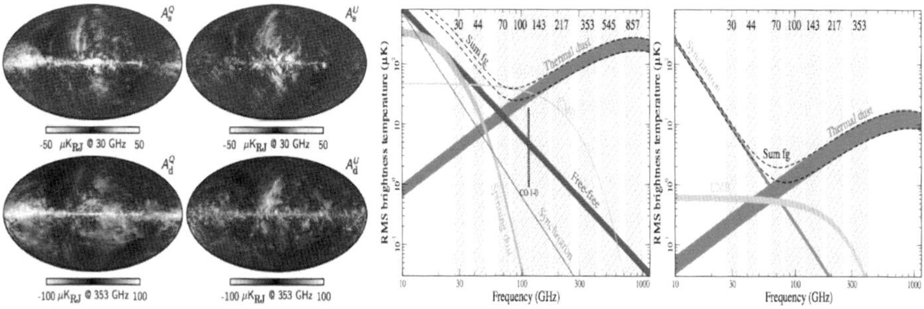

Figure 2: Maps: Stokes Q and U parameters for the synchrotron (s; at 30 GHz) and thermal dust (d; at 353 GHz) polarization. Panels: brightness temperature rms as a function of frequency and component for temperature (left), where each component is smoothed to an angular resolution of 1° FWHM and the lower and upper edges of each line are defined by masks covering respectively 81% and 93% of the sky, and polarization (right), where the corresponding smoothing scale is 40′ and the sky fractions are 73% and 93%. From Ref. [3].

B-modes. Note also the change of paradigm with respect to WMAP results in the relative weights of the various components. In temperature, synchrotron emission fluctuations are less than those of both free-free and spinning dust emissions around 30 GHz. The angular power spectrum (APS) of synchrotron and dust polarized emissions has been properly characterized on various wide sky areas [9] and the synchrotron frequency behavior has been studied as function of the Galactic latitude [10]. Polarized emission from Galactic dust, the main foreground above 70 GHz, is important (and sometimes underestimated in previous analyses) essentially everywhere in the sky. *Planck* maps provided new insights into interstellar dust physics and a precise determination of the contamination level in the CMB APS [11] for polarization experiments, allowing for a proper analysis of BICEP2 and *Keck* array observations [12].

Acknowledgments

We acknowledge the use of the NASA Legacy Archive for Microwave Background Data Analysis and of the ESA *Planck* Legacy Archive. The ASI/INAF Agreement 2014-024-R.0 for the *Planck* LFI Activity of Phase E2 is acknowledged. Some results of this paper have been derived using HEALPix [13] package.

References

[1] *Planck* Collaboration, A&A, forthcoming, arXiv:1507.02058 (2015).
[2] *Planck* Collaboration, A&A, forthcoming, arXiv:1502.01598 (2015).
[3] *Planck* Collaboration, A&A, submitted, arXiv:1502.01596 (2015).
[4] *Planck* Collaboration, A&A **571**, A30 (2014).
[5] T. Trombetti and C. Burigana, JMP **3**, 1918 (2012).

[6] *Planck* Collaboration, A&A, accepted, arXiv:1502.01582 (2015).
[7] C.L. Bennett, D. Larson, J.L. Weiland, et.al., ApJS **208**, 20B (2013).
[8] C. Haslam, H. Stoffel, C.J. Salter, W.E. Wilson, A&ASS **47**, 1 (1982).
[9] *Planck* Collaboration, A&A, submitted, arXiv:1502.01588 (2015).
[10] *Planck* Collaboration, A&A, submitted, arXiv:1506.06660 (2015).
[11] *Planck* Collaboration, A&A **586**, A133 (2016).
[12] BICEP2/*Keck* and *Planck* Collaborations, PRL **114**, 101301 (2015).
[13] K.M. Gorski, E. Hivon, A.J. Banday, et al., ApJ **622**, 759 (2005).

THE CHERENKOV TELESCOPE ARRAY OBSERVATORY AND THE ASTRI MINI-ARRAY PRECURSOR

Aleksandr Burtovoi[a]
Department of Physics and Astronomy, University of Padova, vicolo dell Osservatorio 3, I-35122 Padova, Italy
INAF - Astronomical Observatory of Padova, vicolo dell Osservatorio 5, I-35122 Padova, Italy
for the ASTRI Collaboration
http://www.brera.inaf.it/astri/
and the CTA Consortium
http://www.cta-observatory.org/

Abstract. Very High Energy (VHE) γ-ray astronomy studies the extremely energetic processes at the TeV regime in- and outside our Galaxy. Significant progress on the investigation of VHE γ-ray phenomena has been achieved in recent years due to the operation of ground-based Imaging Atmospheric Cherenkov Telescopes (IACTs) such as MAGIC, VERITAS and H.E.S.S. The Cherenkov Telescope Array (CTA) is an international initiative to build the next generation ground-based VHE γ-ray observatory. Full sky coverage will be assured by two arrays, with one located in the northern and one in the southern hemisphere. Three main classes of IACTs (large, medium, small) will cover a wide energy range, from tens of GeV up to hundreds of TeV. CTA will have a 10 times better sensitivity and angular resolution compared to the present Cherenkov telescope installations. CTA plans to install, at its southern site, about 70 small size telescopes (SSTs) dedicated to study the sky from a few TeV up to hundreds of TeV. The implementation foresees intermediate steps with the development of telescope precursor mini-arrays, like ASTRI led by the Italian National Institute of Astrophysics (INAF) in synergy with the Universidade de Sao Paulo (Brazil) and the North-West University (South Africa). The ASTRI mini-array will be composed of nine dual-mirror SSTs whose end-to-end prototype, developed by INAF during the first phase of the ASTRI program, has been recently inaugurated in Italy. In addition to the assessment of a number of technological aspects related to CTA, the ASTRI mini-array will allow us to perform scientific investigations with unprecedented flux sensitivity in the 5-100 TeV energy range.

1 Very High Energy Gamma-ray Astronomy

Very high energy (VHE) γ-ray astronomy investigates astrophysical sources that emit photons with energies above 100 GeV. It is a relatively new branch of high-energy astronomy. The first VHE source – the Crab Nebula – was detected in 1989 [1]. Significant progress on the investigation of the VHE γ-ray phenomena has been made in recent years thanks to the operation of ground-based Imaging Atmospheric Cherenkov Telescopes (IACTs) such as MAGIC, VERITAS and H.E.S.S. Nowadays, about 150 Galactic and extragalactic sources are known at VHE.

[a]E-mail: `aleksandr.burtovoi@studenti.unipd.it`

In general, VHE γ-ray emission can be associated with the acceleration of relativistic particles (leptons and hadrons). There is a wide range of astrophysical regions inside our Galaxy and beyond it, where acceleration processes occur with different intensities. It is very important to investigate the TeV energy domain because it provides information concerning the nature of particle acceleration in various environments across the Universe. By detecting energetic γ-ray photons from different astrophysical objects, we can study physical processes under particular conditions, unachievable in the laboratory, and constrain theoretical models of both astrophysics and fundamental physics.

The VHE γ-ray observations are carried out from the ground through the IACT technique. This technique corresponds to the indirect investigation of γ-ray photons by detecting short (~10 ns) flashes of blue Cherenkov light from energetic showers, produced after an interaction of primary γ-rays with the upper layers of the atmosphere. A typical shower produces a Cherenkov light cone with a diameter of ~250 m on the ground. By means of the analysis of the Cherenkov radiation, it is possible to reconstruct the energy and direction of the primary γ-ray. The effective collecting area of a single Cherenkov telescope can be roughly approximated by the area of the Cherenkov light cone on the ground ($\sim 5 \times 10^4$ m^2). In this respect, the IACT technique provides a significant improvement in the sensitivity of Cherenkov instruments as compared to space γ-ray telescopes (e.g. AGILE, *Fermi*[b]). Stereoscopic observations with several IACTs can lead to a larger collecting area of the IACT installation, higher angular and energy resolution (better reconstruction of the shower) and also to the more efficient discrimination of background events.

2 The Cherenkov Telescope Array Project

The Cherenkov Telescope Array (CTA) is a modern project of the next-generation ground-based VHE γ-ray facility [2]. CTA is designed as an array of a large number of telescopes of 3 different types: Large Size Telescopes (LSTs, 23-meter diameter), Medium Size Telescopes (MSTs with diameters of 10–12 meters) and Small Size Telescopes (SSTs, 4-meter diameter). This will allow us to perform observations in a wide energy range from a few tens of GeV to more than 100 TeV and also to reach a 10 times better sensitivity and an improved angular resolution, compared to the present Cherenkov telescope installations [3].

Sub-arrays of telescopes featuring different mirror sizes (large, medium, small) are designed for observations in different parts of the CTA spectral range. A few LSTs placed at the center of the array will effectively detect faint Cherenkov light from low-energy showers, produced by photons with energies >20 GeV. Several dozens of MSTs, which will cover an area of 1 km^2, will be most sensitive

[b]The effective area of the *Fermi*-LAT is about 1 m^2.

at 100 GeV – 10 TeV energies. Finally, a large number (∼70) of SSTs will be spread over an area of 4 km^2 in order to provide a sufficient level of area for the detection of rare energetic photons with energies up to 300 TeV.

The CTA observatory will comprise two arrays, one in each hemisphere for full coverage of the sky. According to preliminary Monte Carlo simulations[c], the northern CTA installation (CTA-North) will contain 4 LSTs and 15 MSTs, whereas the southern (CTA-South) – 4 LSTs, 24 MSTs and 72 SSTs. CTA-South can be possibly extended by 24 Schwarzschild-Couder Telescopes (SCTs), which represent a class of medium size telescopes (10-meter diameter) with a dual-mirror optical design[d]. This array is more extended than CTA-North, because the southern hemisphere observations will be focused on Galactic objects, including extended sources, which require a larger network of Cherenkov telescopes to be observed with better angular resolution, and also extremely energetic emitters of photons at energies >100 TeV (so-called PeVatrons), which will be studied mainly with an array of SSTs. CTA-North is designed mainly for observations of point-like extragalactic objects, whose VHE emission is significantly attenuated above 10 TeV through interactions with the extragalactic background light.

The main scientific objectives of the CTA Project are (i) to understand the origin and acceleration mechanism of relativistic cosmic-ray particles; (ii) to probe extreme environments and physical processes in close vicinity to compact objects (neutron stars, black holes) and also within relativistic jets/winds/ shocks/outflows; (iii) to investigate questions of fundamental physics, such as the nature and distribution of dark matter, the quantum gravity effects on photon propagation (Lorentz invariance violation), the existence of axion-like particles [3]. The high sensitivity of CTA will allow us to perform more detailed analysis of already investigated sources and also to detect new objects of potentially unknown classes. Detailed morphological studies of extended sources (e.g. pulsar wind nebulae (PWNe), supernova remnants (SNRs), star-forming regions) will be possible due to high angular resolution of this IACT (about 1-3 arc minutes). Covering at least two decades in energy, CTA will perform both broadband studies and investigations at the high-/low- energy limit, such as a discovery of PeVatrons (at energies >100 TeV) and observations of gamma-ray bursts and active galactic nuclei (AGNs) at cosmological distances z=5 (at energies <100 GeV). The large field of view (∼8°) of both the northern and southern installations makes CTA a powerful machine for both Galactic and Extragalactic Surveys.

As a part of the CTA project, a dual-mirror end-to-end prototype of the Small Size Telescope (SST-2M) is under undergoing commissioning activities within the framework of the ASTRI flagship project of the Italian Ministry of

[c]for more details see *https://portal.cta-observatory.org/Pages/CTA-Performance.aspx*
[d]in contrast to the single-mirror optical design of the 12-meter MSTs.

Research and Education led by INAF [4]. The Schwarzschild-Couder optical solution [5] implemented for the ASTRI telescopes allows us to obtain a smaller plate scale, to reduce the size and the weight of the Cherenkov camera, and to adopt silicon photo-multipliers as light detectors. The first prototype was recently inaugurated in Italy (Mount Etna, Sicily) [6]. In cooperation with the Universidade de Sao Paulo (Brazil) and the North-West University (South Africa), this project also foresees the construction of a mini-array of 9 ASTRI SST-2M units with a relative distance of about 300 m (ASTRI mini-array). The ASTRI mini-array will possibly be one of the precursors for the CTA southern array. This array will perform early scientific investigations of prominent VHE sources in the 5-100 TeV energy range. An improved sensitivity above 10 TeV with respect to the current IACTs, together with an angular resolution of few arc minutes and energy resolution of 10–15% will make the ASTRI mini-array a very promising facility for observations of bright known AGNs, PWNe, SNRs and also for scanning the Galactic center region at energies of hundreds of TeV [7].

Acknowledgments

This work was partially supported by the ASTRI “Flagship Project” financed by the Italian Ministry of Education, University, and Research (MIUR) and led by the Italian National Institute of Astrophysics (INAF). We acknowledge partial support by the CTA MIUR 2014 and TeChe.it 2014 Special Grants. We also acknowledge support from the Brazilian Funding Agency FAPESP (Grant 2013/10559-5) and from the South African Department of Science and Technology through Funding Agreement 0227/2014 for the South African Gamma-Ray Astronomy Programme. We gratefully acknowledge support from the agencies and organizations listed under Funding Agencies at *http://www.cta-observatory.org/* and from the University of Padova.

References

[1] T. C. Weekes, M. F. Cawley, D. J. Fegan, et al., ApJ, **342**, 379 (1989).
[2] M. Actis, G. Agnetta, F. Aharonian, et al., Experimental Astronomy, **32**, 193 (2011).
[3] B. S. Acharya, M. Actis, T. Aghajani, et al., Astroparticle Physics, **43**, 3 (2013).
[4] N. La Palombara, P. Caraveo, M. Fiorini, et al., in *“Astroparticle, Particle, Space Physics and Detectors for Physics Applications”* (Proceedings of the 14th ICATPP Conference), Ed. by S. Giani et al., 754–758 (2014).
[5] V. Vassiliev, S. Fegan, and P. Brousseau, Astroparticle Physics, **28**, 10 (2007).

[6] S. Vercellone for the ASTRI Collaboration and CTA Consortium, *[arXiv:1508.00799]* (2015).
[7] S. Vercellone, G. Agnetta, L. A. Antonelli, et al., *[arXiv:1307.5671]* (2013).

THE MOST PROMISING ASTROPHYSICAL SOURCES OF ELECTROMAGNETIC AND GRAVITATIONAL RADIATION

Giulia Stratta [a]
University of Urbino, Italy

Abstract. Gravitational wave (GW) astronomy will start soon with the advent of the second generation ground-based GW detectors operating in the 1 Hz - 10 kHz frequency range. The electromagnetic counterpart is very important to assess and characterize their astrophysical origin of the first GW signals. Among the astrophysical sources expected to be detected in the next years both by GW detectors and ground- and space-based telescopes, the most promising ones are coalescing binary systems of compact objects as two neutron stars (BNS) or a neutron star and a stellar-mass black hole (NSBH). Indeed, a number of indirect empirical evidence identify these systems with the progenitors of *short* Gamma Ray Burst (sGRB) and their multi-wavelength afterglow emission. The simultaneous detection of a sGRB and a GW signal from a BNS or NSBH will definitively confirm this scenario, and at the same time it will increase the confidence in the astrophysical origin of the detected GW signal. The status of the art of the observational properties of sGRB electromagnetic radiation as well as the observational strategies to simultaneously detect these objects in both the gravitational and electromagnetic waves are briefly summarized in this manuscript.

1 Introduction

In the framework of the second generation ground-based gravitational wave (GW) detector era (e.g. [1]), we discuss on the electromagnetic (EM) observational properties of the most promising astrophysical sources of both gravitational and EM radiation. This issue is fundamental for the multi-wavelength observational campaigns that astronomers will perform with ground and space based telescopes to follow-up a GW detection. Detecting a GW source also in the EM spectrum is of interest for many aspects. Beside the confidence increase in assessing an astrophysical origin to a GW signal, a number of useful information can be obtained from joint GW and EM data analysis (e.g. [2,3]). For example, source properties obtained from EM observations, as its cosmological distance, can provide useful priors for the GW data analysis and parameter estimation processes. The GW source sky localization, that is of the order of $10-100$ deg^2 [4], can be refined down to the arcsecond level, thus allowing for instance to the host galaxy identification, if any. Host galaxy and/or source optical counterpart spectroscopic studies can provide information on the source environment properties, as its density and/or chemical composition, useful to ultimately infer the source formation history.

The second generation laser interferometers Advanced LIGO (aLIGO, [5]) and Advanced Virgo (AdV, [6]) are starting in these years to take scientific

[a]E-mail: giulia.stratta@uniurb.it

data with fair chances of a GW detection [7]. Among the possible astrophysical sources that are known to emit EM radiation and GWs in the sensitive frequency range of aLIGO and AdV (1 Hz - 10 kHz), *short* Gamma-Ray Bursts (sGRBs) play a special role. GRBs are classified according to their burst bimodal distribution and their spectral properties as *long* GRBs and sGRBs [8]. This classification likely reflects two distinct progenitors. However, while we have direct evidence that *long* GRBs are produced by the gravitational collapse of peculiar massive stars, only indirect proofs associate sGRBs to binary neutron stars (BNS) or a neutron star and a stellar mass black hole (NSBH) in their coalescing phase (e.g. [9, 10]). The smoking gun of such association will be provided by a GW detection simultaneous with a sGRB. Indeed, BNS and NSBH systems are expected to release a large amount of GW energy output[b], and their well known GW morphology facilitates their imprint identification in the acquired data, contrary for example to the case of the unmodeled gravitational waveforms expected from collapsing massive stars (e.g. [1]). For these reasons, systems as BNS and NSBH are among the best candidates for a simultaneous detection in both gravitational and EM radiation in the aLIGO and AdV era.

2 Short GRB electromagnetic emission

In the paradigm of the *internal-external* shock model [12], gamma-ray photons of sGRBs are thought to originate from shocks occurring inside the relativistically expanding jets produced during the compact object merging phase. The jet impact with the surrounding interstellar medium eventually originate the afterglow radiation in the X-ray/optical/radio wavebands. The challenging estimates of the jet opening angles range so far from $\sim$ 3 to >30 degrees [13].

On-axis emission - sGRBs observed so far are only those discovered in gamma-rays, that is, with their jet axis oriented towards the Earth (on-axis). sGRB afterglows show on average rapidly power-law fading emission detected mainly in the X-rays and optical bands up to day-week after the burst onset, and for a small fraction of them also in the radio wavelengths after week-month from the burst (e.g. [14, 15]).

Orphan afterglow - An EM emission is expected also if sGRBs are beamed away from the Earth direction (off-axis sGRBs). Analytic modeling [16] and numerical simulations [17] predict that the decelerating relativistic jets experience also a lateral spread. Thus, an observer that is outside the initial jet cone, can in principle detect the late afterglow radiation from the spreading jet that is slowly entering in the line of sight. This emission is often labelled as *orphan afterglow* since it is not preceded by the high-energy burst. The lacking

[b]up to several order of magnitudes more energetic than the one expected from the collapse of massive stars [11]

gamma-ray counterpart makes their detection is extremely challenging from blind sky surveys without knowing the sGRB onset time (e.g. [18]). Despite the lack of evidence of any *orphan afterglow* so far, future all-sky search strategies could be tailored on the expected light curves [17] by knowing the burst onset time through the associated GW signal.

Kilonova emission - BNS models predict that a neutron rich small remnant of $\leq 0.1 - 0.01$ $M_{\odot}$ is isotropically ejected during the mergering phase [19, 20]. This remnant ultimately radiates optical/NIR thermal emission, and it possibly interact with the newlyborn NS magnetic field radiating X-rays and radio [19, 21]. So far, there is only one claim of evidence for a *kilonova* detection for the case of the short GRB 130603B [22]. Despite the predicted faintness of this EM component at the typical distances up to which a BNS could be detected by aLIGO and AdV ($\sim$ 200 Mpc, [7]), the expected lack of any collimation, makes the *kilonova* the most promising EM counterpart of a GW signal from BNS systems.

3 Observational strategies for joint GW and EM detection

The best possible scenario for a joint GW and EM observation is an independent detection of a high-energy burst from a sGRB and a simultaneous GW signal. However, the rate of on-axis sGRBs whithin the aLIGO and AdV distance range is very low due to the expected gamma-ray photon's collimation: for example, at $\sim$ 300 Mpc (an averaged[c] distance up to which a face-on BNS can be detected by aLIGO and AdV with nominal sensitivity [23]) $\leq 1 - 3$ on-axis sGRBs per year detected with the *Fermi* or *Swift* high-energy satellites are expected [23]. A marginally rate increase can be obtained considering also sub-threshold GW signals generated for example by sources farther than 300 Mpc, the detection of which in the acquired data in this case can be driven by the time and sky localization of the associated sGRBs [24].

In the case of a GW detection from a BNS or NSBH system with no associated sGRB detection from high energy satellites, two possible observational strategies can be carried on. One strategy looks for sub-threshold GRBs signals in the high-energy satellite data temporally coincident with the GW trigger: indeed, a significant fraction of faint bursts may be missed by the present high-energy satellites onboard detection algorithms. Another strategy deals with the possibility to have a sGRB beamed away from the Earth direction. In this case, the *orphan afterglow* and/or *kilonova* components are searched in the large sky error boxes obtained from the GW detector network through extensive follow-up campaigns with X-ray, optical, and radio telescopes. Hundreds of plausible transient source candidates will be identified with large field of

[c]averaged over all the possible source positions in the sky

view facilities (e.g. [25]), and then monitored and characterized through more sensitive telescopes in order to identify the true one (e.g. [2]).

4 Conclusions

Today, 74 astronomer teams have signed a Memoranum Of Understanding with the LIGO and Virgo Collaborations with the aim to detect simultaneous EM emission from GW sources. This issue is challenging even for the most promising astrophysical sources as BNS or NSBH due to a number of reasons as the poor sky localization from the GW detector network, the expected low rate of on-axis sGRB, and the rather speculative nature of the *orphan afterglow* and *kilonova* off-axis counterparts, but the expected outstanding scientific return widely support such efforts.

Acknowledgments

G. Stratta is supported by the Italian Ministry of Education, University and Research via grant FIRB 2012- RBFR12PM1F and aknowledge M. Branchesi for useful discussions.

References

[1] A. Viceré, Astrop. and Spac. Sci. Lib., SIPS, **404**, 21 (2014)
[2] M. Branchesi et al., JPhCS, **375**, 062004 (2014)
[3] E. Howell et al, PASA, **32**, 24, (2015)
[4] S. Fairhurst, Journal of Physics: Conference Series, **484**, 012007 (2014)
[5] J. Abbott, et al., Rep. Prog. Phys. **72**, 076901 (2009).
[6] T. Accadia, et al., Classical Quantum Gravity,**28**, 114002 (2011).
[7] J. Abadie et al., Classical and Quantum Gravity, **27**, 17 (2010).
[8] C. Kouveliotou et al., ApJ, **413**, 10 (1993)
[9] N. Gehrels and P. Meszaros, Science, **337**, 932 (2012)
[10] E. Berger, ARA&A, **52**, 43 (2014)
[11] C. D. Ott, Classical and Quantum Gravity,**26**, 6 (2009)
[12] T. Piran and R. Sari, AIP Conference Proceedings,**428**,662 (1998).
[13] W. Fong et al., ApJ, **780**, 118 (2014).
[14] W. Fong et al., ApJ, **815**, 102 (2015).
[15] A.D. Kann et al., ApJ **720**, 1513 (2010).
[16] J. Granot et al., Apj **570**, 61 (2002).
[17] H. van Eerten et al., Apj **722**, 235 (2010).
[18] G. Ghirlanda et al., A&A, **578**, 71 (2015).
[19] B.D. Metzger and E. Berger, E., ApJ, **746**, 48 (2012).
[20] D. Kasen et al., ApJ, **774**, 25 (2013)

[21] B.M.S. Hansen and M. Lyutikov, MNRAS, **322**,695 (2001)
[22] N. Tanvir et al., Nature, **500**, 547 (2013)
[23] J. Clark et al., ApJ, **809**, 53 (2015).
[24] J. Abadie et al., ApJ, **715**,1453 (2010).
[25] J. Aasi et al., ApJSS, **211**, 7 (2014).

WEIGHING OF DARK MATTER IN THE GALACTIC CENTER

Vyacheslav Dokuchaev[1,2, a] and Yury Eroshenko[1, b]
[1]*Institute for Nuclear Research of the Russian Academy of Sciences 60-letiya Oktyabrya prospekt 7a, Moscow, 117312, Russia*
[2]*National Research Nuclear University MEPhI (Moscow Engineering Physics Institute), Kashirskoe highway 31, Moscow, 115409, Russia*

Abstract. We describe a promising method for measuring the total dark matter mass near a supermassive black hole at the Galactic center based on observations of nonrelativistic precession of the orbits of fast S0 stars. An analytical expression for the precession angle has been obtained under the assumption of a power-law profile of the dark matter density. The awaited weighing of the dark matter at the Galactic center provides the strong constraints on the annihilation signal from the SUSY neuralino dark matter particle candidate.

In recent years significant advances have been achieved in the observations of stars gravitationally bound with the supermassive black hole SgrA* at the center of the Galaxy. Several so-called S0 stars, which move at very high velocities ($> 10^3$ km/s) in almost elliptic orbits around a very compact supermassive object, are observed in the infrared range [1–4]. According to the measured parameters of the Kepler orbits of S0 stars, the mass of supermassive black hole SgrA* is $M_{\rm BH} = (4.1 \pm 0.4) \times 10^6 M_\odot$ [1–4]. Independent and currently most accurate values of the mass $M_{\rm BH}$ and spin (Kerr parameter) a of the SgrA* black hole are determined from the observations of quasi-periodic oscillations with the average periods of 11.5 and 19 min [6,7]. They are $M_{\rm BH} = (4.2 \pm 0.2) \times 10^6 M_\odot$ and $a = 0.65 \pm 0.05$ [8,9].

At the center of the Galaxy, in addition to the supermassive black hole SgrA*, there are invisible sources of mass such as compact gas clouds, dim stars and their remnants, and a distributed mass in the form of the diffuse dark matter. This additional mass would result in the deviation of the total Newtonian gravitational potential from the point mass potential of the black hole $U = -GM_{\rm BH}/r$. As a result, the orbits of S0 stars gravitationally bound with the black hole would be unclosed and precess. The openness of the orbit of the most studied S0-2 star will be measured in the next one or two years. Thus, the total mass of the dark matter within the orbit of this star with a characteristic radius of 0.005 pc will be determined. The existence of fast S0 stars provides a unique possibility of reconstructing the gravitational potential and measuring the mass distribution at the center of the Galaxy by fitting their orbits. We discuss and develop a method for studying the distribution of the dark matter at the center of the Galaxy by measuring the precession angle of orbits of S0 stars. It is noteworthy that the expected measurement of the nonrelativistic precession of the orbit of the S0-2 star will allow either improving the indicated

[a]E-mail: dokuchaev@inr.ac.ru
[b]E-mail: eroshenko@inr.ac.ru

bound on the distributed dark mass by two or three orders of magnitude or determining this dark mass. In particular, we determined the dependence of the additional mass both on the profile of the central peak of the dark matter density and on the annihilation cross section of SUSY neutralino dark matter particles, necessary for the explanation of the excess of gamma-radiation from the center of the Galaxy detected recently by the Fermi-LAT space gamma-telescope [10, 11] and taking into account the Sommerfeld enhancement effect (see Fig. 1).

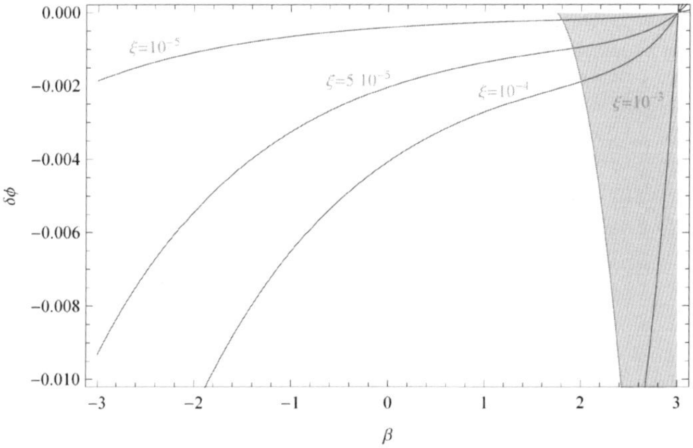

Figure 1: Shift angle of the apsis of the orbit of the star in one turn $\delta\phi$ versus the exponent of the power-law spectrum of the dark matter for realistic values of the mass fraction of the dark matter ξ within the orbit of the S0-2 star. The indicated region is excluded by the constraints caused by the annihilation of dark matter particles if the dark matter makes the main contribution to the mass fraction of the dark matter ξ.

We consider the power-law density profile of matter responsible for the correction δU to the potential of the black hole $\rho(r) = \rho_h(r/r_h)^{-\beta}$, where ρ_h, r_h, and β are the parameters. The corresponding total mass of the dark matter within the sphere with the radius r is

$$M_{\rm DM}(r) = \frac{4\pi\rho_h r_h^{\beta}}{3-\beta}\left[r^{3-\beta} - R_{\rm min}^{3-\beta}\right], \qquad (1)$$

where $R_{\rm min}$ is the minimum radius to which the density profile expands. We define the mass fraction of the dark matter within the orbit of the S0 star $\xi = [M_{\rm DM}(r_a) - M_{\rm DM}(r_p)]/M_{\rm BH}$, which is significant for the subsequent analysis.

In the presence of a small correction δU to the Newtonian potential of the black hole, the precession angle of the orbit of a probe particle (S0-2 star) in

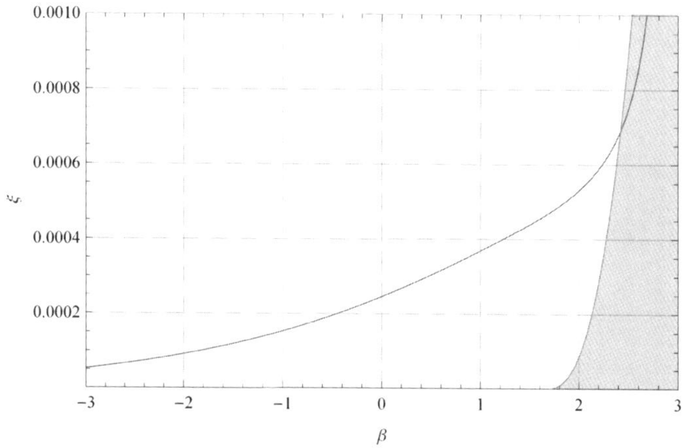

Figure 2: Mass fraction of the dark matter ξ versus the exponent β in the density profile at the precession angle $\delta\phi = 0.01$. The indicated region is excluded by the constraints caused by the annihilation of dark matter particles.

one turn is (see [12], section 15, Problem 3)

$$\delta\phi = \frac{\partial}{\partial L}\left(\frac{2m}{L}\int_0^{\pi} r^2(\phi)\delta U d\phi\right). \tag{2}$$

Here, integration is performed with the trajectory of the particle in the form of an unperturbed elliptic orbit $r(\phi) = p(1+e\cos\phi)^{-1}$, where e is the eccentricity of the ellipse, $p = L^2/(GM_{\mathrm{BH}}m) = a(1-e^2)$ is the parameter of the orbit, a is the major semiaxis, and L is the conserved angular momentum of the star with the mass m. The observed parameters of the Kepler orbit of the S0-2 star: the eccentricity $e = 0.898 \pm 0.0034$, the radius of the pericenter $r_p = a(1-e) = 0.585$ mpc, and the radius of the apocenter $r_a = a(1+e) = 9.42$ mpc. We note that, in the case of relativistic precession, the orbit would rotate in the direction of the rotation of the star, but Newtonian precession (2) occurs in the opposite direction, i.e., $\delta\phi < 0$. The calculation [13] of the precession angle of the orbit of the star in the time of one turn around the black hole $\delta\phi$ gives an expression with one hypergeometric function ${}_2F_1(a, b; c; z)$:

$$\delta\phi = -\frac{4\pi^2\rho_h r_h^{\beta} p^{3-\beta}}{(1-e)^{4-\beta}M_{\mathrm{BH}}}\,{}_2F_1\left(4-\beta, \frac{3}{2}; 3; -\frac{2e}{1-e}\right), \tag{3}$$

where $p = L^2/(GM_{\mathrm{BH}}m) = a(1-e^2)$ is the parameter of the elliptic orbit, e is the eccentricity of the orbital ellipse of the S0-2 star, a is the major semiaxis, and L is the conserved angular momentum of the star with the mass m.

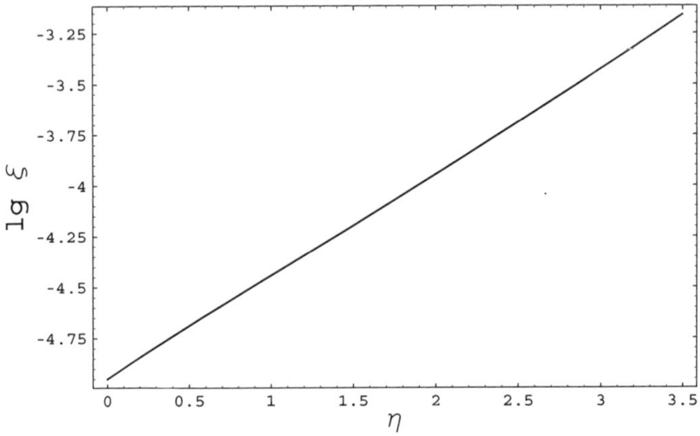

Figure 3: Mass fraction of the dark matter ξ required by the observed excess of gamma radiation in the case of the annihilation of the dark matter particles with the cross section due to the Sommerfeld enhancement effect $\langle\sigma v\rangle \propto v^{-\eta}$ versus an index η.

There is a high probability of reaching in the near future the accuracy required either for the measurement of the precession angle or for the determination of a strong bound, which in turn will make it possible to impose stringent dynamic constraints on the additional dark mass.

Acknowledgments

This work was supported in part by the Russian Foundation for Basic Research grant 15-02-05038.

References

[1] A. M. Ghez *et al.* Astrophys. J. **689**, 1044 (2008).
[2] S. Gillessen *et al.* Astrophys. J. **692**, 1075 (2009).
[3] S. Gillessen, *et al.* Astrophys. J. **707**, L114 (2009).
[4] L. Meyer, *et al.* Science. **338**, 84 (2012).
[5] S. L. Shapiro, S. A. Teukolsky, Astrophys. J. **292**, L41 (1985).
[6] B. Aschenbach *et al.* Astron. Astrophys. **417**, 71 (2004).
[7] R. Genzel *et al.* Nature. **425**, 934 (2003).
[8] V. I. Dokuchaev, Gen. Relativ. Gravit. **46**, 1832 (2014).
[9] V. I. Dokuchaev, Yu. N. Eroshenko, PhysicsUspekhi. **58**, 772 (2015).
[10] T. Daylan, *et al.* arXiv:1402.6703 [astro-ph.HE].
[11] B. D. Fields, *et al.* Phys. Rev. Lett. **113**, 151302 (2014).

[12] L. D. Landau, E. M. Lifshitz, *"Course of Theoretical Physics"*, Vol. 1: Mechanics (New York: Pergamon Press, 1988), Chapter 3.
[13] V. I. Dokuchaev, Yu. N. Eroshenko, JETP Lett. **101**, 777 (2015).

DIRECT DARK MATTER SEARCH WITH EDELWEISS-III: RECENT RESULTS AND FUTURE PROSPECTS

Valentin Kozlov [a] for the EDELWEISS collaboration
Karlsruhe Institute of Technology, Institut für Kernphysik,
Postfach 3640, 76021 Karlsruhe, Germany

Abstract. EDELWEISS experiment performs direct dark matter search by means of Ge bolometers. It is located in the underground laboratory of Modane (LSM, France). The third phase of the experiment is currently accumulating data using 20-kg array of detectors with improved resolution and rejection performance relative to the results of EDELWEISS-II. A part of the data has been analyzed and are presented, in particularly showing good performance for low-mass WIMPs. Prospects for the next-generation project, EURECA, are mentioned as well.

1 The EDELWEISS experiment

A large number of cosmological observations requires existence of dark matter. However, the nature of dark matter remains unknown. A promising class of dark matter candidates are Weakly Interacting Massive Particles (WIMPs). EDELWEISS is a direct dark matter search experiment looking for WIMPs by means of nuclear recoils resulting from a WIMP-nucleon elastic scattering. Germanium bolometers serve as the detection medium. Common challenges in direct searches are low count rate, i.e. well below 1 event/kg of target material/year, and small energy of nuclear recoils, e.g. $E_R < 75$ keV for $M_\chi \leq 1$ TeV in case of Ge target. This implies that a special care is taken against various sources of background: EDELWEISS experimental setup is located in a deep underground laboratory (Laboratoire Souterrain de Modane, France, overburden of 4800 m.w.e.), comprises a set of passive and active shields (Fig. 1), and dedicated material screening [1,2]. In the center of the setup are Ge detectors. The detector section is surrounded by a 10-cm polyethylene layer to attenuate neutrons. The lead layer of 20 cm follows it to reduce γ-ray background. Next, another 50-cm polyethylene layer is installed to suppress the external neutron background. To tag remaining muons an active muon-veto with a geometrical coverage of 98% completes the EDELWEISS setup (Fig. 1). The deradonized air flow purged inside the shields reduces the Rn level down to 30 mBq/m^3. Once Ge detectors are cooled down to 18 mK, one can simultaneously measure ionization and phonon signals, the ratio of which allows to distinguish between electron recoils, usually caused by natural $\gamma-$ and $\beta-$sources, and nuclear recoils expected from the elastic scattering of WIMPs. New FID800 detectors (Fig. 2) used in the current EDELWEISS-III phase of the experiment have γ-rejection factor below $5.8{\cdot}10^{-6}$ (90% C.L.) which is about 5 times

[a]E-mail: valentin.kozlov@kit.edu

better compared to EDELWEISS-II detectors [3]. The new detectors also provide better rejection power against surface background [3]: Events close to the detector surface suffer from incomplete charge collection, meaning that some electron recoils can be misidentified as nuclear recoils. The rejection of such events is realized through *fully interleaved design* of electrodes (Fig. 2), which implies that an alternating voltage is applied on neighboring electrodes of the same surface. Two general groups of electrodes are therefore distinguished: the most positive (negative) ones are called *collection* electrodes while those of reduced value and opposite polarity are called *veto* electrodes. The charge collection for bulk events takes place on *collection* electrodes of different surfaces while in case of surface events charges are collected on both *veto* and *collection* electrodes of the same surface. The surface-event rejection factor was measured by using ^{210}Pb source and was found to be $4{\cdot}10^{-5}$ above 15 keV. The fiducial volume is estimated to 75% [3].

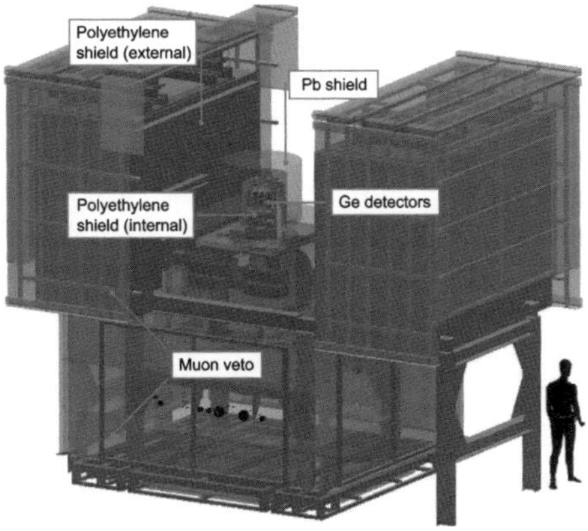

Figure 1: EDELWEISS-III experimental setup (as implemented in Geant4 simulations): location of Ge detectors, passive shielding and the muon-veto are indicated.

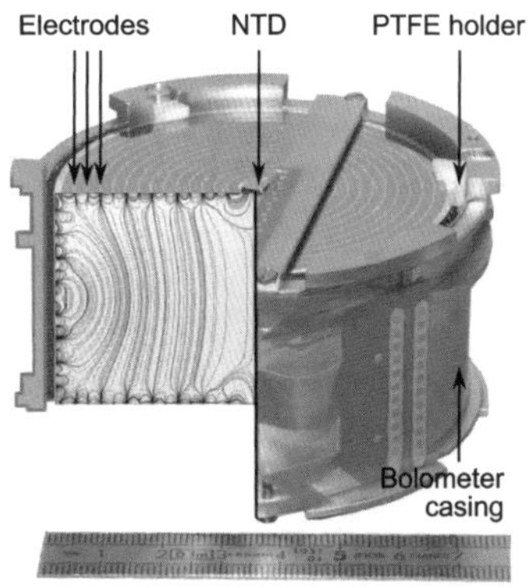

Figure 2: 870-g Ge bolometer with the fully interleaved electrode design (FID800): electrodes, NTD thermistor, and the electrical field configuration are shown.

2 Recent results on low-mass WIMP search

The recent data were acquired during period from July 2014 to April 2015 with 24 FID800 detectors: each of about 870 g total mass (Fig. 2). The data analysis presented here was optimized for low-mass WIMP search, i.e. for $M_\chi < 30$ GeV. As the sensitivity to low-mass WIMPs significantly depends on the energy threshold, only measurement periods with the online threshold (heat signal) below 1.5 keV$_{ee}$ were considered. This reduced the data sample down to 8 detectors, among which 4 detectors had online threshold of 1 keV$_{ee}$, which is an equivalent to 2.4 keV Ge nuclear recoil energy. A number of general cuts was implemented for each detector in terms of baseline resolutions, fiducial volume selection, reconstruction quality of ionization and heat signals. Multi-scattered events were excluded as well. The resulting 582 kg.d fiducial exposure was analyzed by means of Boosted Decision Tree (BDT) technique [4]. The WIMP-expected region was blinded, and a high statistics simulation of backgrounds and the WIMP signal were performed for BDT training. Six parameters are combined in one discriminating variable to decide on a signal or background origin of each event: 4 ionization signals (2 veto + 2 collection electrodes), one combined heat signal and a variable accounting for a time-dependent background. The WIMP signal was generated using the standard formula with standard halo parameters and taking into account the nuclear quenching of ionization yield $Y = 0.16 \cdot E_{rec}^{0.18}$. Background models are data driven and include: bulk gammas, surface events, neutrons, heat-only events. Regions outside of expected WIMP signal are considered for background model construction, while calibration data are used for cross-checks. For example, all γ-ray backgrounds of different ionization topology are modeled by a Compton-like flat component and a list of cosmogenic lines. The spectra are adjusted in the [3-15] keV recoil energy range, and the models are then extended down to 0 keV by extrapolating the flat component and including L-shell cosmogenic lines with an intensity derived from the K-shell intensity. Distribution of heat-only events, which are probably due to a mechanical stress in PTFE clamps of FID800 detectors (Fig. 2), is symmetric with respect to the ionization signal. We therefore use the negative part of the distribution to model the heat-only background. In order to optimize the BDT algorithm, a 'WIMP box' is applied to both simulated and measured events: signals on veto electrodes have to be less than 5σ of respective baseline resolutions, $E_{heat} < 15$ keV and $0 < E_{ion} < 8$ keV. A BDT is constructed for each detector and WIMP masses of 1, 4, 5, 6, 7, 10, 15, 20 and 30 GeV. BDT cuts are defined before unblinding, the Poisson statistics is then applied to derive limit on WIMP-induced events, which is translated into WIMP cross-section limit (Fig. 3). The derived limit shows significant improvement compared to Edelweiss-II [5] and excludes a number of potential dark-matter-detection claims.

3 Status and outlook

The upgrade of the setup resulted in a smaller heat threshold and better ionization resolution. This allowed us to reach significantly better sensitivity compared to EDELWEISS-II [5] at $M_\chi < 30$ GeV: e.g. by a factor 40 at 7 GeV. Ongoing R&D program includes elimination of heat-only events, further improvement of ionization baseline resolution down to 100 eV by means of HEMT electronics, optimization of the heat resolution in order to achieve

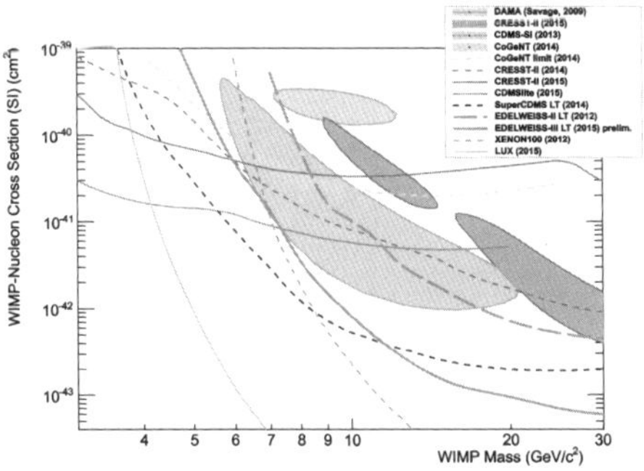

Figure 3: The Edelweiss-III low-mass upper limit (solid red line) on the WIMP-nucleon spin-independent cross-section as a function of WIMP mass. As a dashed red line the low-mass limit of Edelweiss-II is shown.

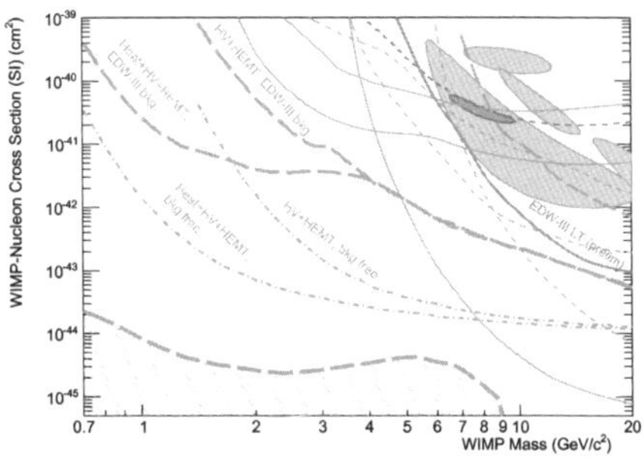

Figure 4: Sensitivities derived for different R&D activities of EDELWEISS (350 kg.d exposure): ionization readout based on HEMTs, using Neganov-Luke effect (HV), improving resolution on heat channel down to 100 eV.

100 eV, studies of Neganov-Luke voltage assisted amplification. Once goals of the R&D programs are fulfilled, EDELWEISS will be able to probe even lower WIMP masses (Fig. 4). To reach a sensitivity beyond the currently planned measurements in LSM will need to install upgraded detectors in much cleaner environment of an experimental setup being built in SNOLAB in the frame of EURECA – SuperCDMS collaboration. This would allow to go even further in sensitivity towards the neutrino floor.

Acknowledgments

The help of the technical staff of LSM and the participant laboratories is gratefully acknowledged. The EDELWEISS project is supported in part by the German ministry of science and education (BMBF Verbundforschung ATP Proj.-Nr. 05A14VKA), by the Helmholtz Alliance for Astroparticle Phyics (HAP), by the French Agence Nationale pour la Recherche (ANR) and the LabEx Lyon Institute of Origins (ANR-10-LABX-0066) of the Université de Lyon within the program "Investissements d'Avenir" (ANR-11-IDEX-00007), by the P2IO LabEx (ANR-10-LABX-0038) in the framework "Investissements d'Avenir" (ANR-11-IDEX-0003-01) managed by the ANR (France), by Science and Technology Facilities Council (UK), and the Russian Foundation for Basic Research (grant No. 07-02-00355-a).

References

[1] E. Armengaud *et al.*, Astropart. Phys. **47**, 1 (2013).
[2] S. Scorza *et al.* in *"Proceedings of Low Radioactivity Techniques"* (18–20 March 2015, Seattle, USA), Ed. by John L. Orrell, AIP Conf. Proc. 1672, 010001, 2015.
[3] J. Gascon *et al.*, J. Low. Temp. Phys. **176**, 870 (2014).

[4] S. Scorza *et al.* in *"Proceedings of Topics in Astroparticle and Underground Physics (TAUP)"* (7–11 September 2015, Toulon, Italy), to be published in Journal of Physics: Conference Series, ed. by IoP.
[5] E. Armengaud *et al.* Phys. Rev. **D 86(5)**, 051701 (2012).

MAGNETIC FIELD EVOLUTION OF NEUTRON STARS: LINKING MAGNETARS AND ANTI-MAGNETARS

Sergei B. Popov [a]
Sternberg Astronomical Institute, Lomonosov Moscow State University, 119991 Moscow, Russia
Alexander A. Kaurov
Department of Astronomy and Astrophysics, The University of Chicago, Chicago, IL 60637 USA
Alexander D. Kaminer
Ioffe Physical-Technical Institute, Politekhnicheskaya 26, Saint Petersburg, 194021, Russia

Abstract. We propose that observations of "hidden" magnetars in central compact objects can be used to probe crustal activity of neutron stars with large internal magnetic fields. Our calculations suggest that central compact objects, which are proposed to be "hidden" magnetars, must demonstrate flux variations on the time scale of months-years. We consider the source 1E161348-5055 in RCW103 as a "hidden" magnetar candidate. Employing a simple 2D-modeling we argue that properties of the source can be explained by the crustal activity of the magnetar type. Thus, this object may be supplemented for the known candidates for the "hidden" magnetars among central compact objects discussed in literature.

1 Introduction

Young isolated neutron stars (NSs) are observed as sources of many different types. Among them magnetars occupy a special place thanks to spectacular energy outputs demonstrated by these sources (see a review in [1]).

During last few years many attempts were made to describe diversity of young NSs in one evolutionary framework (see a brief review in [2]). One of the most important ingredients of such models is evolution of the magnetic field.

It is mostly believed that in not-so-young NSs the field is just decaying. However, in some of young ($< 10^4$ yrs) objects the external field can grow. This phenomenon is related to the fall-back accretion just after a supernova explosion, and screening of the magnetic field by the fallen matter (see, for example, [3]). Then the field slowly diffuses out, and we observe increasing external field (of course, in the crust the field is not growing, and can even slightly decay).

This process can work also for magnetars. Such "magnetars under cover" have been called "hidden" magnetars by [4]. Recently, it was proposed that a NS in the supernova remnant Kes 79 can belong to this class [5].

In this note we present our recent results [6] on "hidden" magnetars, where we also propose a new candidate.

[a]E-mail: polar@sai.msu.ru

2 Modeling and results

Our model is described in details in [6]. It is based on the scenario presented in [7]. These authors numerically modeled surface and neutrino luminosity of a NS as a result of a rather long (tens of kyrs) energy release event in the crust. The key parameter is the heat rate H [erg cm^{-3} s^{-1}].

1E161348-5055 (hereafter 1E) is the central compact X-ray source in the supernova remnant RCW103. The age of the remnant is about 2 kyrs. 1E is characterized by variable X-ray emission in the range of luminosities $\sim 10^{33} - 10^{35}$ erg s^{-1}. This variability is long term (months-years) and irregular.

We assume that the observational outbursts are powered by the energy of the magnetar's magnetic fields hidden in the bulk (deep in the crust) of the star. The magnetic energy is supposed to transform in the heat inside an internal region(s) of the NS crust. We investigate qualitatively possible parameters of the magnetar heater which are needed to provide observable energy release events with $\sim$ two orders enhancement of the observable flux and about ten years of relaxation.

Similar to [7] we introduce an internal phenomenological heat source located in a layer at $\rho_1 \leq \rho \leq \rho_2$. The basic difference is only in the scale of temporal behavior of the heater. The heating rate H [erg cm^{-3} s^{-1}] is taken in the form:

$$H = H(\rho, t) = H_0\, \Theta(\rho, \theta, t), \tag{1}$$

where H_0 is the characteristic heat intensity and $\Theta(\rho, \theta, t)$ is the step function, which equals unity when $(\rho_1 < \rho < \rho_2)$ & $(\theta < \theta_0)$ & $(t_0 < t < t_0 + \tau)$, and vanishes outside these spatial and temporal regions. The heater looks like a hot wide axisymmetric slab limited by the angle θ_0 and by densities ρ_1 and ρ_2 along radial axis. The solid angular size of the heater can be expressed as $\Delta\Omega_0/4\pi = (1 - \cos\theta_0)/2$. The temporal parameters are fixed in our model: the moment of the energy-input onset, t_0, is 2 kyr, the duration of energy input, τ, is chosen to be 120 days in order to match the interval between the first two observations.

Fig. 1 shows results of simplified simulations employing 2D-code (see [7]) of luminosity-time dependencies, $L(t)$, imposed on the X-ray data on long outburst of 1E in the time interval from September 1999 to July 2007. The X-ray fluxes $F(t)$ measured in different moments by different missions (all presented in [8]) have been normalized to the first observation in this series, when the object is presumably in a low state $(F/F_0 = L/L_0)$. The calculated curves are normalized to the luminosity $L_0 = 1.37 \times 10^{33}$ erg s^{-1} corresponding to a standard cooling of a NS (without heating) with $M = 1.4\, M_\odot$ at the age 2 kyr.

As it is seen in the Figure, results obtained in the framework of our model fits well the observational data on 1E, and so support the hypothesis of a "hidden" magnetar inside this source. More details on the calculations can be found in [6].

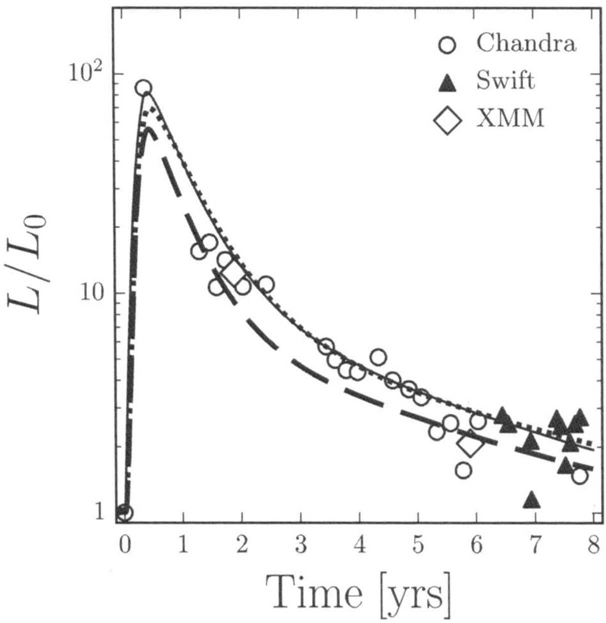

Figure 1: Simulated thermal X-ray luminosities L as a function of time t versus observational X-ray data on 1E 161348-5055 in the time interval from September 1999 to July 2007 [8]. Time zero corresponds to the first observation of the outburst. Observational fluxes and calculated luminosities are normalized to the flux F_0 and luminosity L_0 at the zero moment. Three light curves correspond to different heaters in the NS crust located in layers $\rho_1 \leq \rho \leq \rho_2$ within solid angles $\Delta\Omega_0$ around a symmetry axis. All heaters start energy release at the same moment and continue for 120 days. The age of the NS at this moment is $t_0 = 2\,\text{kyr}$. Chandra data is shown with empty circles, XMM-Newton data — with diamonds, and Swift data — with filled triangles (error bars are disregarded).

3 Discussion and conclusions

"Hidden" magnetars are useful to study the origin of magnetar activity. This is due to the absence of huge external magnetospheres in these objects. Then all magnetospheric processes are not important, and evolution is driven by processes in the crust. Observations of "hidden" magnetars of different ages can help to understand crustal activity of highly magnetizes NSs without complications of magnetospheric processes which can trigger energy releases.

Fall-back can be a natural consequence of a supernova explosion. It can be estimated that sources with screened fields (central compact objects in supernova remnants, so-called antimagnetars) can form up to 1/3 of all young NSs. If magnetars represent about 10% of newborn NSs, then we can expect that the fraction of "hidden" magnetars is about few percent, if fall-back and appearence of large magnetic field are uncorrelated.

It is tempting to propose a hypothesis that after the SN1987A a "hidden" magnetar was formed. Indeed, the progenitor of this supernova evolved in a binary system in which two massive stars coalesced. This situation potentially can result in a rapidly rotating core of pre-SN. Then, an effective dynamo-mechanism can operate in a newborn NS enhancing the magnetic field up to the magnetar value [9]. However, as we do not see any activity of the central object, even more than 25 years after the explosion, it is possible to assume that the field is screened due to fall-back. Future observations might help to probe this hypothesis.

Up to now we have several central compact objects in supernova remnants which are supposed to be NSs with screened fields. Some of them can be "hidden" magnetars, which we assume to be typically differ by peculiar properties of observed light curves, e.g. a high pulsed fraction or powerful outbursts. So we proposed that in addition to the source in the remnant Kes 79, the NS in RCW103 also can be a "hidden" magnetar. Its X-ray variability is well explained in the model of energy release in the crust. Note, however, that there are unusually warm compact objects like the NS in HESS J1731-347 (e.g. [10]) with more regular thermal radiation, which can be explained by the cooling theory without special energy release. Reliable difference of two (or more) branches of the central compact objects is a matter of future investigations.

Acknowledgments

SBP acknowledges support from the RFBR grant 14-02-00657. SBP is the Dynasty foundation fellow. ADK was supported by the RFBR grant 14-02-00868.

References

[1] R. Turolla, S. Zane, A. Watts, Rep. Prog. Phys. **78**, 116901 (2015).
[2] A. P. Igoshev, S. B. Popov, R. Turolla, Astron. Nach. **335**, 262 (2014).
[3] C. G. Bernal, D. Page, W. H. Lee, Astrophys. J. **770**, 106 (2013).
[4] U. Geppert, D. Page, T. Zannias, Astron. Astrophys. **345**, 847 (1999).
[5] N. Shabaltas, D. Lai, Astrophys. J. **748**, 148 (2012).
[6] S.B. Popov, A.A. Kaurov, A.D. Kaminker, Pub. Astron. Soc. Aus. **32**, 18 (2015).
[7] A.D. Kaminker, A.A. Kaurov, A.Y. Potekhin, D.G. Yakovlev, Mon. Not. R. Astron. Soc. **442**, 3483 (2014).

[8] A. De Luca et al., Astrophys. J. **682**, 1185 (2008).
[9] S.B. Popov, M.E. Prokhorov, Mon. Not. R. Astron. Soc. **367**, 732 (2006).
[10] D. Klochkov, V. Suleimanov, G. Pühlhofer, D.G. Yakovlev, A. Santangelo, K. Werner, Astron. Astrophys. **573**, A53 (2015)

IS THERE AN ORDINARY SUPERMASSIVE BLACK HOLE AT THE GALACTIC CENTER?

Alexander F. Zakharov[1,2,3,4]

[1] *Institute of Theoretical and Experimental Physics, B. Cheremushkinskaya 25, 117218 Moscow, Russia*

[2] *Bogoliubov Laboratory of Theoretical Physics, Joint Institute for Nuclear Research, 141980 Dubna, Russia*

[3] *National Research Nuclear University MEPhI (Moscow Engineering Physics Institute), 115409, Moscow, Russia*

[4] *North Carolina Central University, Durham, NC 27707, USA*

Abstract. Now there are two basic observational techniques to investigate a gravitational potential at the Galactic Center, namely, a) monitoring the orbits of bright stars near the Galactic Center to reconstruct a gravitational potential; b) measuring a size and a shape of shadows around black hole giving an alternative possibility to evaluate black hole parameters in mm-band with VLBI-technique. At the moment one can use a small relativistic correction approach for stellar orbit analysis (however, in the future the approximation will not be not precise enough due to enormous progress of observational facilities) while now for smallest structure analysis in VLBI observations one really needs a strong gravitational field approximation. We discuss results of observations, their conventional interpretations, tensions between observations and models and possible hints for a new physics from the observational data and tensions between observations and interpretations. We will discuss an opportunity to use a Schwarzschild metric for data interpretation or we have to use more exotic models such as Yukawa potential, Reissner – Nordström or Schwarzschild – de-Sitter metrics for better fits.

To evaluate the gravitational potential at the Galactic Center two teams of astronomers have observed trajectories of bright stars in the IR band for several years. One group led by A. Ghez (UCLA, USA) uses the twin 10-meter optical/infrared telescopes on Mauna Kea (Hawaii), and according to the Keck Strategic Mission, the first important goal is *high angular resolution astrophysics*[a] and it practically promises to be a world leader in the field, see their results on the observations of bright stars. [1, 2] Another group led by R. Genzel (ESO, MPE) uses four 8.2-meter VLT telescopes in Paranal (Chile). This European group got very important results [3–5] which are consistent with results presented by the US team. The important case of a G2 gas cloud is a very useful tracer of the gravitational potential at the Galactic Center [5] (later, the object has been called the Dusty S-cluster Object (DSO/G2) since further detailed observations were not completely consistent with the gas cloud model). It occurs that very likely this may be another example of a close peribothron[b]

---

[a]`http://www.keckobservatory.org/`

[b]The term "peribothron" has been introduced by Frank and Rees [6] following W.R. Stoeger's suggestion. Greek bothros means pit or hole. So, the peribothron means the point of least distance of an object orbiting a black hole.

passage of a dust-enshrouded star [7–9]. The analysis showed that the DSO/G2 is a rather young star than a coreless gas and dust cloud [10].

According to the list of the Top 10 ESO science discoveries, observations of stars orbiting the Milky Way black hole led to discovery # 1 "Several of ESO's flagship telescopes were used in a 16-year long study to obtain the most detailed view ever of the surroundings of the monster lurking at the heart of our galaxy– a supermassive black hole."[c] In 2012 the Royal Swedish Academy of Sciences has decided to award the Crafoord Prize in Astronomy to Reinhard Genzel[d] and Andrea Ghez, "for their observations of the stars orbiting the Galactic Centre, indicating the presence of a supermassive black hole".[e] The ESO and MPE formed a team to construct GRAVITY, the Very Large Telescope Interferometer for precise astrometry and interferometric imaging. The interferometer will provide precise astrometry of order 10 micro-arcseconds [11,12]. The GRAVITY equipment should be shipped to the VLT-observatory in Chile in 2015 and commissioning was planned to start in October 2015 [12].

A presence of dark matter and dark energy in the conventional cosmological approach may be treated as an anomaly. As it was note in paper [13], Le Verrier not only discovered the Mercury pericenter anomaly and explained 93% of the observed value, but a supplementary advance of 38 arcseconds/century was without explanation (later on, the value was corrected to 43 arcseconds/century), but also in 1859 Le Verrier analyzed the following options in order to explain the anomaly, namely, Le Verrier had analyzed the following options to explain an anomaly namely, a) The gravitational field of an invisible matter (planet, asteroids near Sun). This means an introduction of new objects; b) A deviation from the Newtonian law. It means a change of a law of Nature; c) There is a lack in precision, implying that the model has to be clarified. It was a rather general approach to resolve any anomaly. Le Verrier was very successful when in 1846 he suggested that the Uranus anomaly could be explained by unknown planet, and following his suggestion Neptune has been discovered. Le Verrier suggested to explain the Mercury anomaly with a presence of additional planet, he called it Vulcan, however, now we know that it was necessary to introduce a new fundamental gravity law as it was done by Einstein in 1915. Similarly to the Mercury anomaly, dark matter and dark energy problems may be explained by a change of the fundamental gravity law, for instance, one could change Einstein – Hilbert Lagrangian R with a function $f(R)$ and we have $f(R) = R$ for the standard GR [14]. For instance, in the framework of $f(R) = R^n$ (we have GR limit for $n = 1$), one could explain acceleration of Universe without dark energy and rotation curves

---

[c] `http://www.eso.org/public/science/top10/`

[d] In 2008 R. Genzel has been awarded with the Shaw prize for "recognition of his outstanding contributions in demonstrating that the Milky Way contains a supermassive black hole at its Center.

[e] `http://www.crafoordprize.se`

for spiral galaxies without dark matter, however, we have $n \approx 1$ from Solar system data [15, 16]. One could arrive at the same conclusion from S2 orbit data [17] (see also, paper [18], where an extended distribution near the Galactic Center has been taken into account).

The authors [19] constrained parameters of Yukawa gravity from the S2 star trajectory, see [20, 21], where it was shown that the Yukawa potential may be a weak gravitational field limit for a wide class of alternative gravitational theories

$$\Phi(r) = -\frac{GM}{(1+\delta)r}\left[1+\delta e^{-\left(\frac{r}{\Lambda}\right)}\right], \tag{1}$$

where Λ is a parameter with a length dimension and δ is a dimensionless constant. The authors [19] obtained the most probable Λ in the case of S2 star, is around 5000 – 7000 AU and that from the current observations it is very hard to obtain the reliable constraints on the universal constant δ. The authors [19] found also the same universal constant $\delta = 1/3$ which was successfully applied to clusters of galaxies [20] and rotation curves of spiral galaxies [21] also gives a good agreement in the case of observations of S2 star orbits.

A version of a Lorentz invariant massive gravity has been introduced by Fierz and Pauli [22] (see also a popular review on the subject [23]), however, later people found a number of problems with such theories such as existence of ghosts, vDVZ discontinuity and some other technical problems, however, some of them may be overcame [24]. Assuming a natural modification of the Newtonian potential corresponding to a massive graviton theory [23, 25]

$$V(r) = \frac{GM}{r}\exp(-r/\lambda_{\rm g}). \tag{2}$$

An opportunity to evaluate a graviton mass analyzing a time delay in electromagnetic waves such as supernova or gamma-ray bursts has been analyzed [25], moreover, earlier an opportunity was demonstrated to constrain a graviton mass from from a gravitational wave signal alone [26] ($\lambda_{\rm g} > 2.8 \times 10^{12}$ km).

Pulsar timing may be used to evaluate a graviton mass, [27] where the authors used the idea to find gravitational waves analyzing pulsar timing. In the paper it was concluded that, with 90% probability, massless gravitons can be distinguished from gravitons heavier than 3×10^{-22} eV (Compton wavelength $\lambda_{\rm g} = 4.1 \times 10^{12}$ km). A preliminary analysis of S2 orbit data from VLT and Keck telescopes showed that $\lambda_{\rm g} > 2350$ AU$= 3.5 \times 10^{11}$ km or $m_g < 3.5 \times 10^{-21}$ eV.

One can conclude that there are tensions between a size of the smallest spot at the Galactic Center and an expected shadow size, therefore one should use Reissner – Nordström or/and the Kottler (Schwarzschild – de-Sitter) metrics or there are systematic effects. Concerning the best fits for trajectories of S2 like stars with alternative theories of gravity, one concludes that R^n practically has to be ruled out with the observational dat. There are hints for a Yukawa potential for these data, because Yukawa potential provides a slightly better fit in comparison with the Newtonian fit. One needs more precise observations (such as VLBI in the mm band, GRAVITY interferometer or/and forthcoming large telescopes like E-ELT and TMT) for more definite claims on these issues. One can find a more detailed description of our results in Refs. [13, 15, 28–32].

References

[1] A.M. Ghez, A. Morris, E.E. Becklin et al. Nature, **407**, 349 (2000).
[2] M.R. Morris, L. Meyer, A.M. Ghez, RAA **12**, 995 (2012).
[3] R. Schödel, T. Ott, R. Genzel et al., Nature **419**, 694 (2002).
[4] S. Gillessen, F. Eisenhauer, S. Trippe et al., ApJ **692**, 1075 (2009).
[5] S. Gillessen, R. Genzel, T.K. Fritz et al., Nature **481**, 51 (2012).
[6] J. Frank, M.J. Rees, MNRAS **176**, 633 (1976).
[7] K. Phifer, K. Do, L. Meyer et al., ApJ **773**, L13 (2013).
[8] M. Zajaček, V. Karas, A. Eckart, A & A **565**, A17 (2014).
[9] M. Zajaček et al., arXiv:1507.00237v1[astro-ph.GA].
[10] M. Valencia-S., ApJ, **800**, 125 (2015).
[11] F. Eisenhauer et al. 2011, *The Messenger* **143**, 16.
[12] N. Blind et al., arXiv:1503.07303 [astro-ph.IM].
[13] A. F. Zakharov et al., Space Sci. Rev. **48**, 301 (2009).
[14] S. Capozziello, Int. J. Mod. Phys. D **11**, 483 (2002).
[15] A.F. Zakharov et al., Phys. Rev. **D 74**, 107101 (2006).
[16] Y. Maruccia et al., Nuovo Cim. **122 B**, 607 (2007).
[17] D. Borka et al., *Phys. Rev.* **D 85**, 124004 (2012).
[18] A.F. Zakharov et al., Adv. Space Res. **54**, 1108 (2014).
[19] D. Borka et al., JCAP **11** 050 (2013).
[20] S. Capozziello, V.F. Cardone, A. Troisi, Phys. Rev. **D 76**, 104019 (2007).
[21] V.F. Cardone, S. Capozziello, MNRAS **414**, 1301 (2011).
[22] M. Fierz, W. Pauli, Proc. Royal Soc. London **A173**, 211 (1939).
[23] M. Visser, Gen. Rel. Grav. **30**, 1717 (1998).
[24] V.A. Rubakov, P.G. Tinyakov, Phys. Usp. **51**, 759 (2008).
[25] C. Will, Living Reviews in Relativity **17**, 4 (2014).
[26] C. Will, Phys. Rev. **D 57**, 2061 (1998).
[27] K. J. Lee et al., ApJ **722**, 1589 (2010).
[28] A.F. Zakharov et al., Phys. Rev. **D 76**, 062001 (2007).
[29] A.F. Zakharov et al., *Phys. At. Nucl.* **73**, 1870 (2010).
[30] A.F. Zakharov, Phys. Rev. **D 90**, 062007 (2014).

[31] A.F. Zakharov, in *New Results and Actual Problems in Particle & Astroparticle Physics and Cosmology* Proceedings of The XXIXth International Workshop on High Energy Physics, Ed. by V.A. Petrov and R. Ryutin, (World Scientific, Singapore, 2014) p.141.

[32] A.F. Zakharov, J. Astrophys. Astr. **36**, 539 (2015).

THE DARKSIDE-50 EXPERIMENT: A LIQUID ARGON TARGET FOR DARK MATTER PARTICLES

P. Agnes[1], L. Agostino[2], I. F. M. Albuquerque[3,4], T. Alexander[5,6], A. K. Alton[7], K. Arisaka[8], H. O. Back[3,9], B. Baldin[6], K. Biery[6], G. Bonfini[10], M. Bossa[11,10], B. Bottino[12,13], A. Brigatti[14], J. Brodsky[3], F. Budano[15,16], S. Bussino[15,16], M. Cadeddu[17,18], L. Cadonati[5], M. Cadoni[17,18], F. Calaprice[3], N. Canci[19,10a], A. Candela[10], H. Cao[3], M. Cariello[13], M. Carlini[10], S. Catalanotti[20,21], P. Cavalcante[22,10], A. Chepurnov[23], A. G. Cocco[21], G. Covone[20,21], L. Crippa[24,14], D. D'Angelo[24,14], M. D'Incecco[10], S. Davini[11,10], S. De Cecco[2], M. De Deo[10], M. De Vincenzi[15,16], A. Derbin[25], A. Devoto[17,18], F. Di Eusanio[3], G. Di Pietro[10,14], E. Edkins[26], A. Empl[19], A. Fan[8], G. Fiorillo[20,21], K. Fomenko[27], G. Forster[5,6], D. Franco[1], F. Gabriele[10], C. Galbiati[3,10], C. Giganti[2], A. M. Goretti[10], F. Granato[20,28], L. Grandi[29], M. Gromov[23], M. Guan[30], Y. Guardincerri[6], B. R. Hackett[26], K. Herner[6], E. V. Hungerford[19], Al. Ianni[31,10], An. Ianni[3,10], I. James[15,16], C. Jollet[32], K. Keeter[33], C. L. Kendziora[6], V. Kobychev[34], G. Koh[3], D. Korablev[27], G. Korga[19,10], A. Kubankin[35], X. Li[3], M. Lissia[18], P. Lombardi[14], S. Luitz[36], Y. Ma[30], I. N. Machulin[37,38], A. Mandarano[11,10], S. M. Mari[15,16], J. Maricic[26], L. Marini[12,13], C J. Martoff[28], A. Meregaglia[32], P. D. Meyers[3], T. Miletic[28], R. Milincic[26], D. Montanari[6], A. Monte[5], M. Montuschi[10], M. Monzani[36], P. Mosteiro[3], B. J. Mount[33], V. N. Muratova[25], P. Musico[13], J. Napolitano[28], A. Nelson[3], S. Odrowski[10], M. Orsini[10], F. Ortica[39,40], L. Pagani[12,13], M. Pallavicini[12,13], E. Pantic[4], S. Parmeggiano[14], K. Pelczar[42], N. Pelliccia[39,40], S. Perasso[1], A. Pocar[5,3], S. Pordes[6], D. A. Pugachev[37,38], H. Qian[3], K. Randle[5], G. Ranucci[14], A. Razeto[10,3], B. Reinhold[26], A. L. Renshaw[8,19], A. Romani[39,40], B. Rossi[21,3], N. Rossi[10], D. Rountree[22], D. Sablone[10], P. Saggese[14], R. Saldanha[29], W. Sands[3], S. Sangiorgio[43], C. Savarese[11,10], E. Segreto[44], D. A. Semenov[25], E. Shields[3], P. N. Singh[19], M. D. Skorokhvatov[37,38], O. Smirnov[27], A. Sotnikov[27], C. Stanford[3], Y. Suvorov[8,10,37], R. Tartaglia[10], J. Tatarowicz[28], G. Testera[13], A. Tonazzo[1], P. Trinchese[20], E. V. Unzhakov[25], A. Vishneva[27], B. Vogelaar[22], M. Wada[3], S. Walker[20,21], H. Wang[8], Y. Wang[30,8,45], A. W. Watson[28], S. Westerdale[3], J. Wilhelmi[28], M. M. Wojcik[42], X. Xiang[3], J. Xu[3], C. Yang[30], J. Yoo[6], S. Zavatarelli[13], A. Zec[5], W. Zhong[30], C. Zhu[3], and G. Zuzel[42]

[1]*APC, Université Paris Diderot, CNRS/IN2P3, CEA/Irfu, Obs. de Paris, Sorbonne Paris Citée, Paris 75205, France;* [2] *LPNHE Paris, Université Pierre et Marie Curie, Université Paris Diderot, CNRS/IN2P3, Paris 75252, France;* [3]*Department of Physics, Princeton University, Princeton, NJ 08544, USA;* [4]*Instituto de Física, Universidade de São Paulo, São Paulo 05508-090, Brazil;* [5]*Amherst Center for Fundamental Interactions and Dept. of Physics, University of Massachusetts, Amherst, MA 01003, USA;* [6]*Fermi National Accelerator Laboratory, Batavia, IL 60510, USA;* [7]*Department of Physics, Augustana University, Sioux Falls, SD 57197, USA;* [8]*Department of Physics and Astronomy, University of California, Los Angeles, CA 90095, USA;* [9]*Pacific Northwest National Laboratory, Richland, WA 99354, USA;* [10]*Laboratori Nazionali del Gran Sasso, Assergi AQ 67010, Italy;* [11]*Gran Sasso Science Institute,*

L'Aquila AQ 67100, Italy; [12]*Department of Physics, Università degli Studi, Genova 16146, Italy;* [13]*Istituto Nazionale di Fisica Nucleare, Sezione di Genova, Genova 16146, Italy;* [14]*Istituto Nazionale di Fisica Nucleare, Sezione di Milano, Milano 20133, Italy;* [15]*Istituto Nazionale di Fisica Nucleare, Sezione di Roma Tre, Roma 00146, Italy;* [16]*Department of Physics and Mathematics, Università degli Studi Roma Tre, Roma 00146, Italy;* [17]*Department of Physics, Università degli Studi, Cagliari 09042, Italy;* [18]*Istituto Nazionale di Fisica Nucleare, Sezione di Cagliari, Cagliari 09042, Italy;* [19]*Department of Physics, University of Houston, Houston, TX 77204, USA;* [20]*Department of Physics, Università degli Studi Federico II, Napoli 80126, Italy;* [21]*Istituto Nazionale di Fisica Nucleare, Sezione di Napoli, Napoli 80126, Italy;* [22]*Department of Physics, Virginia Tech, Blacksburg, VA 24061, USA;* [23]*Skobeltsyn Institute of Nuclear Physics, Lomonosov Moscow State University, Moscow 119991, Russia;* [24]*Department of Physics, Università degli Studi, Milano 20133, Italy;* [25]*St. Petersburg Nuclear Physics Institute NRC Kurchatov Institute, Gatchina 188350, Russia;* [26]*Department of Physics and Astronomy, University of Hawai'i, Honolulu, HI 96822, USA;* [27]*Joint Institute for Nuclear Research, Dubna 14 1980, Russia;* [28]*Department of Physics, Temple University, Philadelphia, PA 19122, USA;* [29]*Kavli Institute, Enrico Fermi Institute, and Dept. of Physics, University of Chicago, Chicago, IL 60637, USA;* [30]*Institute of High Energy Physics, Beijing 100049, China;* [31]*Laboratorio Subterráneo de Canfranc, Canfranc Esta- ción 22880, Spain;* [32]*IPHC, Université de Strasbourg, CNRS/IN2P3, Strasbourg 67037, France;* [33]*School of Natural Sciences, Black Hills State University, Spearfish, SD 57799, USA;* [34]*Institute for Nuclear Research, National Academy of Sciences of Ukraine, Kiev 03680, Ukraine;* [35]*Radiation Physics Laboratory, Belgorod National Research University, Belgorod 308007, Russia ;* [36]*SLAC National Accelerator Laboratory, Menlo Park, CA 94025, USA;* [37]*National Research Centre Kurchatov Institute, Moscow 123182, Russia;* [38]*National Research Nuclear University MEPhI, Moscow 115409, Russia;* [39]*Department of Chemistry, Biology and Biotechnology, Università degli Studi, Perugia 06123, Italy;* [40]*Istituto Nazionale di Fisica Nucleare, Sezione di Perugia, Perugia 06123, Italy;* [41]*Department of Physics, University of California, Davis, CA 95616, USA;* [42]*Smoluchowski Institute of Physics, Jagiellonian University, Kra- kow 30348, Poland;* [43]*Lawrence Livermore National Laboratory, Livermore, CA 94550, USA;* [44]*Institute of Physics Gleb Wataghin, Universidade Estadual de Campinas, São Paulo 13083-859, Brazil;* [45]*School of Physics, University of Chinese Academy of Sciences, Beijing 100049, China.*

Abstract. The DarkSide-50 experiment, located at the "Laboratori Nazionali del Gran Sasso (INFN)", is based on low-radioactivity argon double phase time projection chamber, surrounded by an active liquid scintillator veto, designed for the zero background achievement. The liquid argon features sufficient self shielding and easy scalability to multi-tons scale. The impressive reduction of the ^{39}Ar isotope (compared to the atmospheric argon), along with the excellent pulse shape discrimination, make this technology a possible candidate for the forthcoming generation of multi-ton Dark Matter experiments.

Introduction

Astrophysical and Cosmological observations suggest the existence of a relevant amount of dark matter in the Universe [1]. According to some theoretical models, such a matter consists of weakly interactive massive particle (WIMP) [2] and interacts with target nuclei of experiments releasing energies of order few tens of keV. The DarkSide-50 experiment, after the promising results of its predecessor DarkSide-10 [3], searches for such nuclear recoils (NR) using a liquid argon double-phase Time Projection Chamber (LAr-TPC), surrounded by a liquid scintillator veto (LSV) located in the middle of a water Cherenkov veto (WCV), both used for rejecting the coincidences in the LAr-TPC induced by cosmic and material radiation (technical details are reported in [4–7]). Two arrays of 19 3" PMTs, facing from the top and the bottom the liquid argon active volume ($\sim$ 46.4 kg), detect the primary scintillation (S1) and the gas scintillation from drifted ionization electrons (S2). Thanks to S1 and S2 it is possible to reconstruct the full 3D position of the events. LAr intrinsic scintillation features allow us to reject electron recoils (β and γ events from background) at the level of 1.5×10^7 or even better [6]. The pulse shape discrimination between ER and NR is based on the f_{90} parameter, i.e. the fraction of S1 detected in the first 90 ns.

The DarkSide-50 experiment exploited two important data taking campaigns: first, the atmosheric argon (AAr) campaign, in which the main features of the detector have been tested [6]; second, the underground depleted argon (UAr) campaign in which the predicted characteristics have been confirmed and the impressive reduction of the ^{39}Ar has been proven [5]. The main features of both LAr-TPC and LSV detectors have been understood also by means of intense calibration campaigns with internal (^{83m}Kr) and external (γ, AmBe) sources.

1 The Underground Argon

UAr was extracted and purified with an effort lasted many years [8]. Fig. 1 shows a comparison between AAr and UAr spectrum in terms of S1 in a reduced volume (34 kg). The activity of ^{39}Ar and ^{85}Kr is determined through a simultaneous Monte Carlo (MC) fit of S1 spectrum (with field on and off) and z-position distribution of the corresponding events. The ^{39}Ar activity of

UAr is a factor $(1.4 \pm 0.2) \times 10^3$ lower than the AAr one, corresponding to (0.73 ± 0.11) mBq/kg.

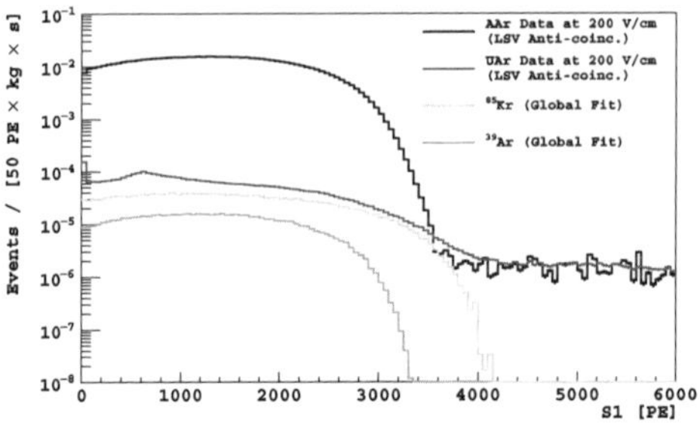

Figure 1: Livetime normalized S1 pulse integral comparison between AAr (black) and UAr (blue) events. For UAr events ^{85}Kr (green) and ^{39}Ar (orange) are shown separately [5].

The vetoing efficiency of the LSV, necessary to exclude NR events due to radiogenic or cosmogenic neutrons, during the UAr campaign is estimated to be $> 99.1\%$, using the AmBe calibration data and MC simulations.

Fig. 2a shows all UAr events in the dark matter region in the S1-f_{90} plane after applying all selection criteria (see for details [5]). NR acceptance is based on the median of f_{90} measured in SCENE experiment [9] combined with a suitable statistical model [10]. Finally, Fig. 2b shows the spin-independent WIMP-nucleon cross section 90% CL exclusion plots for DarkSide-50 AAr and UAr (independently and combined), compared with the most popular dark matter limits [11].

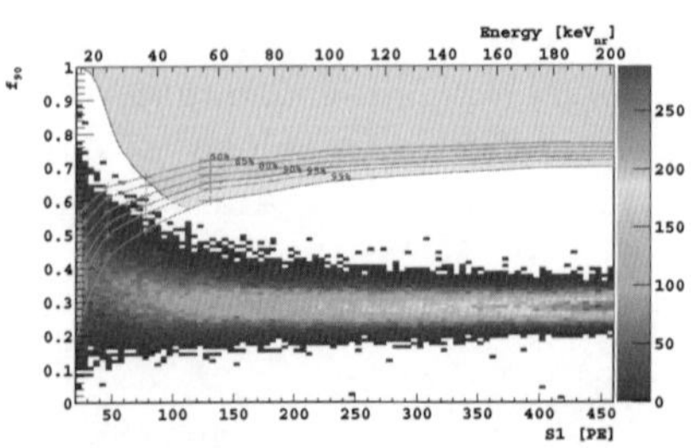

(a) Event distribution in the f_{90}-S1 plane. NR accenpance bands are derived from SCENE experiment. Shaded area represents the WIMP search region [5].

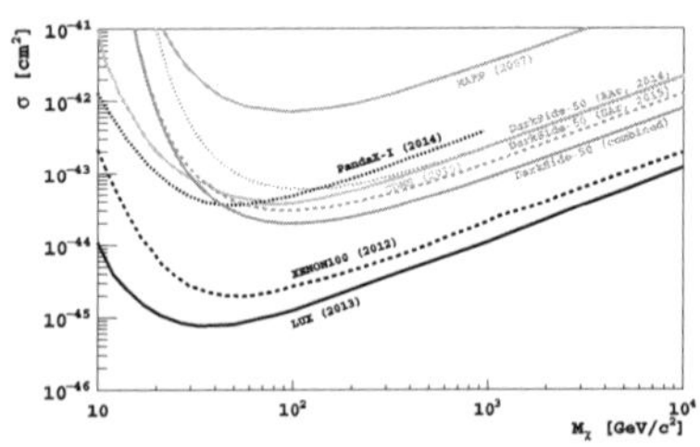

(b) Comparison between the DarkSide Spin-Independent WIMP-nucleon cross section 90% CL exclusion plot and the most popular experimental limits [5].

2 Future developments

LAr-TPC based on UAr is an excellent candidate for the next generation of dark matter search experiments, thanks especially to: low background, powerful pulse shape discrimination, sufficient self-shielding and good light yield. The DarkSide collaboration is proposing, in the same underground laboratory, the DarkSide-20k with an active mass of ~20 t. Yet this scaled version will be surrounded by a LSV inside a cylindrical WCV. Thanks to ultra-low background, this detector will be able to reach the sensitivity of $\sim 10^{-47}$ cm^2 for WIMP of 1 TeV/c^2 mass. SiPM light sensors will take the place of the usual PMTs, with a consequent increasing of the light collection. In parallel the "Urania" project, an improved extraction plant in Colorado, will guarantee an UAr production rate of ~ 100 kg/day. Finally the "Aria" project, a cryogenic distillation facility for isotopic separation of argon at the Seruci mine in Sardinia will further reduce the ^{39}Ar content and strongly suppress the ^{85}Kr presence.

References

[1] S. M. Faber and J. Gallagher, Annu. Rev. Astro. Astrophys. **17**, 135 (1979); D. N. Spergel, Phys. Rev. D **37**, 1353 (1988); D. Clowe et al., Ap. J. **648**, L109 (2006)

[2] J. L. Feng, Annu. Rev. Astro. Astrophys. **48**, 495 (2010).

[3] T. Alexander *et al.* [DarkSide Collaboration], Astropart. Phys. **49** (2013) 44 doi:10.1016/j.astropartphys.2013.08.004 [arXiv:1204.6218 [astro-ph.IM]].

[4] P. Agnes *et al.* [DarkSide Collaboration], JINST **11** (2016) no.03, P03016 doi:10.1088/1748-0221/11/03/P03016 [arXiv:1512.07896 [physics.ins-det]].

[5] P. Agnes *et al.* [DarkSide Collaboration], arXiv:1510.00702 [astro-ph.CO].

[6] P. Agnes *et al.* [DarkSide Collaboration], Phys. Lett. B **743** (2015) 456 doi:10.1016/j.physletb.2015.03.012 [arXiv:1410.0653 [astro-ph.CO]].

[7] P. Agnes *et al.* [DarkSide Collaboration], arXiv:1412.2969 [astro-ph.IM].

[8] D. Acosta-Kane et al., Nucl. Inst. Meth. A **587**, 46 (2008); H. O. Back et al., arXiv:1204.6024v2 (2012); H. O. Back et al., arXiv:1204.6061v2 (2012); J. Xu et al., Astropart. Phys. **66**, 53 (2015).

[9] T. Alexander et al. (The SCENE Collaboration), Phys. Rev. D **88**, 092006 (2013); H. Cao et al. (The SCENE Collaboration), Phys. Rev. D 91, 092007 (2015).

[10] D. V. Hinkley, Biometrika **56**, 635 (1969); M. G. Boulay and A. Hime, Astropart. Phys. **25**, 179 (2006).

[11] D. S. Akerib et al. (The LUX Collaboration), arXiv:1512.03506v1 (2015); E. Aprile et al. (The XENON100 Collaboration), Phys. Rev. Lett. 109,

181301 (2012); X. Xiao et al., Phys. Rev. D 92, 052004 (2015); R. Agnese et al. (The SuperCDMS Collaboration), Phys. Rev. D 92, 072003 (2015); C. Amole et al. (The PICO Collaboration), Phys.Rev. Lett. 114, 231302 (2015); D. Y. Akimov et al., Phys. Lett. B 709, 14 (2012); P. Benetti et al. (The WArP Collaboration), Astropart. Phys. 28, 495 (2008).

MAGNETIC FIELDS IN MAGNETARS: FROM WHERE AND HOW STRONG?

Victor Semikoz [a] and Maxim Dvornikov [b]
Pushkov Institute of Terrestrial Magnetism, Ionosphere and Radio Wave Propagation of the Russian Academy of Sciences (IZMIRAN), 142190, Troitsk, Moscow, Russia

Abstract. We suggest a new model for generation of strongest magnetic fields in magnetars based on the parity violation for electron-nucleon (eN) interaction in the Standard Model of particle physics.

The most plausible explanation of radiation of soft gamma repeaters [1] and anomalous X-ray pulsars [2] is the presence of strong magnetic fields $B \geq 10^{15}$ G in a neutron star (NS). Such highly magnetized NSs are called magnetars. Various models, explaining the origin of such strong astrophysical magnetic fields, were reviewed in Ref. [3]. Nevertheless, the issue of the magnetic fields generation in magnetars still remains open.

Recently in Refs. [4, 5] we proposed the new model for the generation of strong magnetic fields in magnetars based on the instability of magnetic fields in dense degenerate matter composed of nonrelativistic neutrons and ultra-relativistic electrons interacting by parity violating electroweak forces. The idea that electroweak interaction can induce the magnetic field instability was put forward first in Ref. [6]. Our model is based on the parity violating electroweak electron-nucleon interaction (the eN interaction). We shall take that the background nuclear matter consists of neutrons and protons. This matter is supposed to be unpolarized and nonmoving macroscopically. In Ref. [4] we derived the averaged effective Lagrangian of the eN interaction in the Fermi approximation as

$$\begin{aligned}
\mathcal{L}_{\text{int}} &= -\bar{\psi}_e \gamma^0 \left(V_{\text{L}} P_{\text{L}} + V_{\text{R}} P_{\text{R}} \right) \psi_e, \\
V_{\text{L}} &= \frac{G_{\text{F}}}{\sqrt{2}} \left[n_n - n_p (1 - 4\xi) \right] (1 - 2\xi), \\
V_{\text{R}} &= -\frac{G_{\text{F}}}{\sqrt{2}} \left[n_n - n_p (1 - 4\xi) \right] 2\xi,
\end{aligned} \tag{1}$$

where $G_{\text{F}} \approx 1.17 \times 10^{-5} \text{GeV}^{-2}$ is the Fermi constant, $n_{n,p}$ are the constant and uniform densities of neutrons and protons, ψ_e is the bispinor electron wave function, $\xi = \sin^2 \theta_{\text{W}} \approx 0.23$ is the Weinberg parameter, $P_{\text{L,R}} = (1 \mp \gamma^5)/2$ are the chiral projectors, $\gamma^5 = \mathrm{i}\gamma^0\gamma^1\gamma^2\gamma^3$, and $\gamma^\mu = (\gamma^0, \vec{\gamma})$ are the Dirac matrices. Now let us consider the interaction of ultrarelativistic electrons with background matter, described by Eq. (1), and an external magnetic field $\mathbf{B} =$

[a] E-mail: semikoz@yandex.ru
[b] E-mail: maxdvo@izmiran.ru

$(0, 0, B)$. The total Lagrangian has the form, $\mathcal{L} = \mathcal{L}_{\text{em}} + \mathcal{L}_{\text{int}}$, where $\mathcal{L}_{\text{em}} = \bar{\psi}_e \gamma^\mu (\mathrm{i}\partial_\mu + eA_\mu) \psi_e$ is the Lagrangian for the interaction of an ultrarelativistic electron with the electromagnetic field $A^\mu = (0, 0, Bx, 0)$, and $e > 0$ is the absolute value of the electron charge. Using the solution of Dirac equation generated by $\mathcal{L}$, exactly accounting for both the matter interaction and the magnetic field, one can compute the induced electric current along the magnetic field $J_z = -e\langle\bar{\psi}_e \gamma^3 \psi_e\rangle +$ *positron contribution*, averaged using the Fermi-Dirac distribution. This current, which is additive to the ohmic current $\mathbf{J}_{\text{ohm}} = \sigma_{\text{cond}} \mathbf{E}$, where σ_{cond} is the matter conductivity and $\mathbf{E}$ is the electric field, turns out to be nonzero. If we restore the vector notations, one gets for the induced electric current $\mathbf{J} = \Pi \mathbf{B}$ directed along the magnetic field. The parameter Π reads

$$\Pi = \frac{2\alpha_{\text{em}}}{\pi} (\mu_5 + V_5),$$
$$\mu_5 = \frac{1}{2} (\mu_{\text{R}} - \mu_{\text{L}}),$$
$$V_5 = \frac{1}{2} (V_{\text{L}} - V_{\text{R}}) \approx \frac{G_{\text{F}}}{2\sqrt{2}} n_n, \tag{2}$$

where $\alpha_{\text{em}} = e^2/4\pi \approx 7.3 \times 10^{-3}$ is the fine structure constant. Note that, since we consider ultrarelativistic electrons, we can assume that right and left chiral projections of the electron-positron field behave independently and possess different chemical potentials: μ_{R} and μ_{L}. To obtain Eq. (2) we assume that $n_n \gg n_p$ inside NS.

Using Eq. (2), in Ref. [5] we derived the system of kinetic equations for the spectra of the magnetic helicity density $h(k, t)$ and magnetic energy density $\rho_{\text{B}}(k, t)$ as well as the chiral imbalance $\mu_5(t)$ in the form,

$$\frac{\partial h(k,t)}{\partial t} = -\frac{2k^2}{\sigma_{\text{cond}}} h(k,t) + \left(\frac{4\Pi}{\sigma_{\text{cond}}}\right) \rho_{\text{B}}(k,t),$$
$$\frac{\partial \rho_{\text{B}}(k,t)}{\partial t} = -\frac{2k^2}{\sigma_{\text{cond}}} \rho_{\text{B}}(k,t) + \left(\frac{\Pi}{\sigma_{\text{cond}}}\right) k^2 h(k,t),$$
$$\frac{\mathrm{d}\mu_5(t)}{\mathrm{d}t} = \frac{\pi \alpha_{\text{em}}}{\mu^2 \sigma_{\text{cond}}} \int \mathrm{d}k \left[k^2 h(k,t) - 2\Pi \rho_{\text{B}}(k,t)\right] - \Gamma_f \mu_5, \tag{3}$$

where μ is the chemical potential of electrons in NS, $\Gamma_f \sim m_e^2$ is the chirality flip rate in the electron-proton (ep) collisions, and m_e is the electron mass. Note that the chirality flipping term in Eq. (3) should contain μ_5 since the equilibrium in the system of right and left electrons is achieved when $\mu_{\text{R}} = \mu_{\text{L}}$ (the explicit form of Γ_f derived via QFT method see in [7]). The total magnetic helicity H and the magnetic field strength B can be found on the basis of $h(k, t)$

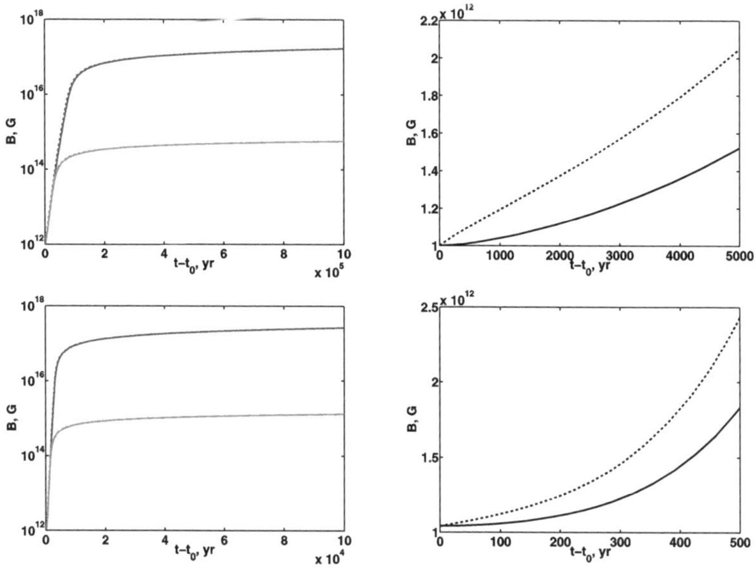

Figure 1: The growth of quenched and unquenched magnetic fields in magnetars versus $t-t_0$ for different $k_{\rm max}$, corresponding to different length scales $\Lambda_{\rm B}$. Red and blue lines in left panels are the solutions of Eq. (3) for quenched and unquenched Π in Eq. (6) respectively. Solid lines correspond to initially nonhelical fields and dashed lines to maximally helical fields. Left top: the magnetic field evolution for $k_{\rm max} = 2\times 10^{-10}$ eV or $\Lambda_{\rm B}^{(\rm min)} = 1$ km. Right top: the behavior of the quenched magnetic field with the parameters as in left top panel for shorter evolution time $t_0 < t < 5\times 10^3$ yr. Left bottom: the magnetic field growth for $k_{\rm max} = 2\times 10^{-9}$ eV or $\Lambda_{\rm B}^{(\rm min)} = 100$ m. Right bottom: the behavior of the quenched magnetic field with the parameters as in left bottom panel for shorter evolution time $t_0 < t < 5\times 10^2$ yr.

and $\rho_{\rm B}(k,t)$ as

$$H(t) = \int {\rm d}^3x(\mathbf{A}\cdot\mathbf{B}) = V\int h(k,t){\rm d}k,$$
$$\rho_B(t) = \frac{1}{2}B^2(t) = \int {\rm d}k\rho_{\rm B}(k,t), \tag{4}$$

where V is the normalization volume and the integration is over all the range of the wave number k variation. It should be mentioned that in Eqs. (3) and (4) we assume the isotropic spectra.

The numerical solution of non-linear self-consistent equations (3) for the initial Kolmogorov's spectrum of the magnetic energy density $\rho_B(k,t_0) = Ck^{-5/3}$, an arbitrary initial disbalance $\mu_5(t_0)$ and magnetic helicity $h(k,t_0)$ is illustrated in Fig. 1. The most important question , where the energy feeding the growth of magnetic field is coming from, was answered in our paper [8]. The energy source for that is the thermal part of the kinetic energy of degenerate non-relativistic nucleons and ultra-relativistic electrons,

$$\rho_T = \frac{B_{\mathrm{eq}}^2}{2} = \delta\rho_e + \delta\rho_p + \delta\rho_n = \left[\frac{M_n p_{F_n} + M_p p_{F_p}}{2} + \mu^2\right]\frac{T^2}{2}. \qquad (5)$$

We can consider the quantity $\rho_{\mathrm{T}} = B_{\mathrm{eq}}^2/2$ in Eq. (5) as the inexhaustible energy source requiring that $B_{\mathrm{eq}}^2 \gg B^2$. Thus, we do not violate the total energy conservation for the extended system which includes the background matter and the magnetic field. From the energy conservation law,

$$\frac{\mathrm{d}}{\mathrm{dt}}\left[\rho_T(t) + \rho_B(t)\right] = 0,$$

and the NS cooling due to neutrino emission [4, 10],

$$T(t) = T_0\left(\frac{t}{t_0}\right)^{-1/6},$$

one can easily find the growth of magnetic field more than two orders of magnitude starting at $t_0 = 100$ yr from a typical value of magnetic field in pulsars $B_0(t_0) = 10^{12}$ G. This is in a good agreement with Fig. 1.

To avoid a back reaction on matter from a strong magnetic field, or to obey the criterion $\rho_B \ll \rho_T$, we should modify Eq. (3). As known from the solar dynamo theory [9], one can avoid the infinite growth of the magnetic field by quenching of the dynamo α-parameter. Thus we can introduce the quenching of the parameter Π, given in Eq. (2), as

$$\Pi \to \frac{\Pi}{1 + B^2/B_{\mathrm{eq}}^2}, \qquad (6)$$

where B^2 and B_{eq}^2 can be found in Eqs. (4) and (5).

References

[1] E. P. Mazets, S. V. Golenetskij, and Y. A. Guryan, Sov. Astr. Lett. **5**, 343 (1979).

[2] G. G. Fahlman and P. C. Gregory, Nature **293**, 202 (1981).

[3] L. Ferrario, A. Melatos, and J. Zrake, Space Sci. Rev. (2015), doi:10.1007/s11214-015-0138-y; arXiv:1504.08074.

[4] M. Dvornikov and V. B. Semikoz, Phys. Rev. D **91**, 061301 (2015); arXiv:1410.6676.

[5] M. Dvornikov and V. B. Semikoz, J. Cosmol. Astropart. Phys. 05 (2015) 032; arXiv:1503.04162.

[6] A. Boyarsky, O. Ruchayskiy, and M. Shaposhnikov, Phys. Rev. Lett. **109**, 111602 (2012); arXiv:1204.3604.
[7] M. Dvornikov, arXive:1510.06228.
[8] M. Dvornikov and V. B. Semikoz, Phys. Rev. D **92**, 083007 (2015); arXiv:1507.03948.
[9] P. Charbonneau, Annu. Rev. Astron. Astrophys. **52**, 251 (2014).
[10] C. J. Pethick, Rev. Mod. Phys. **64**, (1992) 1133.

STATUS AND PROSPECT OF CHINA DARK MATTER EXPERIMNENT AND CHINA JINPING UNDERGROUND LABORATORY

Shukui Liu [a] and Qian Yue

Department of Engineering Physics, Tsinghua University, Beijing, 100084, China

Abstract. The China Dark Matter Experiment (CDEX) pursues direct searches of light Weakly Interacting Massive Particles (WIMPs) using point contact germanium (PCGe) detector at the China Jinping Underground Laboratory (CJPL). The results from CDEX-1A and CDEX-0 are reported. CDEX-10 with the PCGe array of 10 kg target mass range is being constructed and tested. The multi-purpose CDEX-1T experiment and the development of CJPL-II will also be discussed.

1 Introduction

Compelling evidence from astroparticle physics and cosmology indicates that about 85 % of the matter is composed of dark matter [1, 2]. The essence of dark matter is still unknown and WIMPs is one of the most popular cold dark matter candidates. The direct detection approach which studies the recoil nuclei scattered off by an incident WIMPs have been attempted with different detector technologies. Especially at the low energy region, some experiments observe anomalous excess of unidentified events which has been interpreted as signatures of WIMPs [4–8], however they are in conflict with the null results from other experiments [9–17].

The CDEX established in 2009 aims at the direct detection of low mass WIMPs with a tonne-scale point contact germanium detector (PCGe) array. PCGe with low energy threshold and good stability were identified as effective ways to probe the light WIMPs regions. The CDEX experiments were performed at CJPL with a rock overburden of more than 2400 m which is the deepest operating laboratory for astroparticle search in the world. The cosmic-ray rate at CJPL has been measured to be 61.7 $y^{-1}\cdot m^{-2}$ [18] which provides an ideal location for dark matter experiment where the contribution of muon can be negligible.

2 The status of CDEX

CDEX is now in the first stage called CDEX-1, with the objectives of the performance study and background understanding based on 1 kg-scale-mass PCGe detector. In this stage, both the passive and active shielding systems are used, as shown in Fig. 1. The passive shielding system consists of 1 m of polyethylene, 20 cm of lead, 20 cm of borated polyethylene, and 20 cm oxygen-free high conductivity cooper. A NaI(Tl) scintillator detector enclosing the

[a]E-mail: abc6081@163.com

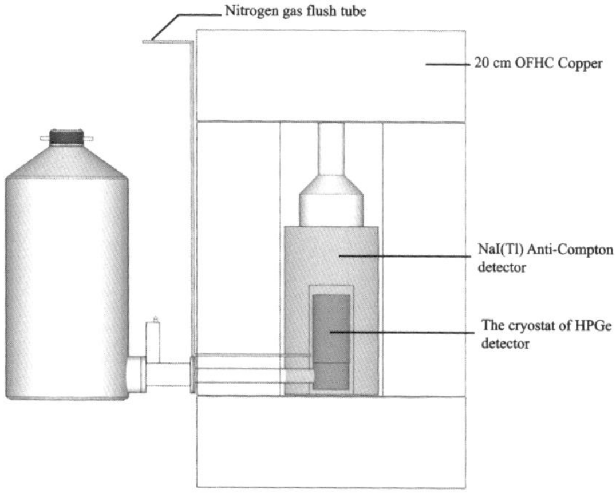

Figure 1: Schematic diagram of the CDEX-1 experiment, including PCGe and NaI(Tl) detectors and copper shielding.

germanium cryostat served as the active anti-Compton detector. In addition, the nitrogen gas was purged into the inner space of the copper to reduce the influence of the radioactive radon gas.

Several kinds of the detectors have been used in CDEX-1. The most recent experiment results have been published [17], based on the 53.9 kg-days of data from CDEX-1A detector which is the first designed one single module 1kg-scale p-type PCGe. This experiment has two important improvements including the presence of anti-Compton detector and surface events suppression, compared to the previous experiment [16] with the identical detector. The p-type PCGe has about 1 mm dead layer on the surface of the Ge crystal where the deposit energy can not be totally collected. The surface events, which have anomalous rise-time distribution, mainly come from gamma, electron or alpha particles and hardly come from the WIMPs. We can get rid off those events by the rise-time distribution. The event rate and spectrum above the analysis threshold of 475 eVee are consistent with the understood background model and no dark matter events have been observed. As shown in Fig.2 [17], an order of magnitude improvement has been achieved over our pervious results. With the identical detector technique, our result strongly disfavors the CoGeNT allowed region that it claimed the excess events were induced by dark matter [6].

The results from another experiment called CDEX-0 which is the first completed program of the pilot experiment in CDEX have also been published [15]. The detector is a 20 g germanium detector array which consist of four 5 g crystals. Each crystal is a semiplanar n-type Ge detector. It has been achieved very low energy threshold of 177 eVee at 50 % signal efficiency with 0.784 kg-days of data. It show a good direction to lower energy threshold further.

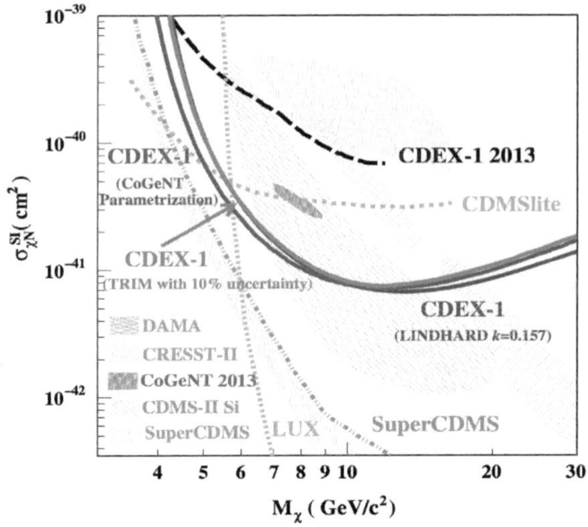

Figure 2: CDEX-1A exclusion plot of spin-independent WIMP-nucleon coupling at 90% confidence level, superimposed with the previous result.

Another improved 1 kg-mass p-type PCGe detector CDEX-1B is accumulating data. It is expected that CDEX-1B has a comparable energy threshold to CDEX-0 and has the similar background level as CDEX-1A.

3 Plan and prospects of CDEX

With the data accumulation of CDEX-1A and CDEX-1B, we focus more studies on the background understanding and the dark matter annual modulation. At the same time, the construction and test of CDEX-10, which is an important stage toward tonne-scale Ge experiment, employing the techniques of the liquid argon/liquid nitrogen shielding system and kg-scale germanium detector arrays, are ongoing. Each Ge detector string has three 1 kg-scale PCGe crystals. And all the Ge strings submerge into a tonne-scale liquid argon/liquid nitrogen served as the shielding system and the cooling medium. Through CDEX-10 studies, clearly understanding the performance of the Ge array detector and the feasibility of liquid argon or liquid nitrogen shielding system could provide the stable fundamental to CDEX-1T.

CDEX-1T aims at the direct searches of light WIMPs and double beta decay, featuring tonne-scale mass germanium detector arrays submerged in liquid argon or liquid nitrogen for cooling and shielding. To achieve this target, some of the key technologies must be broken through, including Ge purification, Ge crystal growth, Ge detector fabrication, the production of electro-form copper

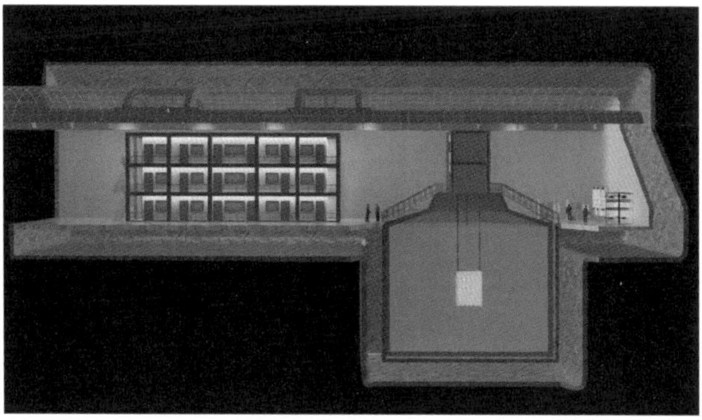

Figure 3: Conceptual drawing of the CDEX new space in CJPL-II and the CDEX-1T

with ultra low radioactivity in CJPL for CDEX, the development of the electronics with the low noise low radioactivity and the production of the large volume cooling tank. These relevant R&D have been carried on.

4 CJPL-II development

Now the CJPL has been fully used, which can not satisfy the future large experiment. At the end of 2014, CJPL-II located next to the CJPL started to construct and it will be finished in 2016. CJPL-II will have four tunnels with the dimension of 14(W)×14(H)×130(L) m and each tunnel will be divided into two individual laboratories. As shown in Fig.3, CDEX-1T experiment will be performed in one of those laboratories, and the new space for CDEX-1T detector system with the size of $\phi 18 \times 30$ m will be ready in 2016.

References

[1] J. Beringer *et al., Phys. Rev. D.* **86**, 010001 (2012), and references therein.
[2] P.A.R. Ade *et al., arXiv:1303.5076.*
[3] C. Kelso, D. Hooper, and M. R. Buckley, *Phys. Rev. D* **85**, 043515 (2012), and references therein.
[4] P. Belli, R. Bernabei, A. Bottino, F. Cappella, R. Cerulli, N. Fornengo, and S. Scopel *Phys. Rev. D.* **84**, 055014 (2011).
[5] R. Bernabei *et al., Eur. Phys. J. C* **67**, 39 (2010).
[6] C. E. Aalseth *et al., Phys. Rev. D.* **88**, 012002 (2013).
[7] G. Angloher *et al., Eur. Phys. J. C* **72**, 1971 (2012).
[8] R. Agnese *et al., Phys. Rev. Lett.* **111**, 251301 (2013).
[9] E. Aprile *et al., Phys. Rev. Lett.* **109**, 181301 (2012).
[10] H. B. Li *et al., Phys. Rev. Lett.* **110**, 261301 (2013).

[11] H. B. Li *et al.*, *Astropart. Phys.* **56**, 1 (2014).
[12] R. Agnese *et al.*, *Phys. Rev. Lett.* **112**, 041302 (2014).
[13] D. S. Akerib *et al.*, *Phys. Rev. Lett.* **112**, 091303 (2014).
[14] R. Agnese *et al.*, *Phys. Rev. Lett.* **112**, 241302 (2014).
[15] S.K. Liu *et al.*, *Phys. Rev. D.* **90**, 032003 (2014).
[16] W. Zhao *et al.*, *Phys. Rev. D.* **88**, 052004 (2013).
[17] Q. Yue *et al.*, *Phys. Rev. D.* **90**, 091701 (2014).
[18] Q. Yue *et al.*, *Chin. Phys. C* **37**, 086001 (2013).

SOME REMARKS ON THE EXPERIMENTS FOR DARK MATTER SEARCH

O.G. Ryazhskaya[a]

Institute for Nuclear Research of the Russian Academy of Sciences, 117312 Moscow, Russia

Abstract. The DAMA collaboration experiment gave the results similar to seasonal winter-summer cosmic ray muons modulations. If the DAMA effect is connected to just this mechanism the phase shift of the curves describing it in the North and South hemispheres will be half a year. If in the DAMA experiments mainly WIMPs are detected the maximum distribution position of the counting rate will not depend on the hemisphere.

In 1932 F. Zvikky [1] while observing galaxy motion in the Berenice's Hair Constellation found out that the situation can only be understood assuming that there exists a mass 10 times higher than the visible mass which has an impact on these galaxies. This fact can be considered as an experimental proof of dark matter existence at the macrolevel. Later on, the idea about the dark matter at the macrolevel came up as weakly interacting massive elementary particles, WIMPS, having mass more than 80 MeV. These particles travel at a speed of 200 km/sec. relative to the Earth due to the Sun movement inside the Galaxy. During their interaction with the ordinary matter WIMPs lose their energy in small portions, at the rate of about several keV, and can be registered by NaI crystals and other sensitive detectors.

DAMA detector designed for WIMP search is located in the underground laboratory Grand Sasso (LNGS) and has been operating in the low-background environment for over 15 years [2]. It registered a temporary variation of the annual rate of count. This variation could be associated with the annual Earth revolution around the Sun. The result obtained in the DAMA experiment is quite similar to the measurements of the well-known seasonal winter-summer modulations of cosmic ray muons.

Fig. 1 shows the results of muon count rate measurements obtained in 2001 2009 by LVD detector in LNGS [3]. In this work published in 2011 the authors draw attention to taking into account temperature muon modulations in experiments of dark matter search. If the effect of the DAMA experiment is connected with the temperature mechanism the position of modulation curve maximum values obtained in the Northern and Southern hemispheres will be shifted in relation to each other for half a year. If the effect is connected with WIMP registration the curve will be the same for experiments in both hemispheres. And this will serve an important indication to the nature of this effect.

[a]E-mail: Ryazhskaya@lvd.ras.ru

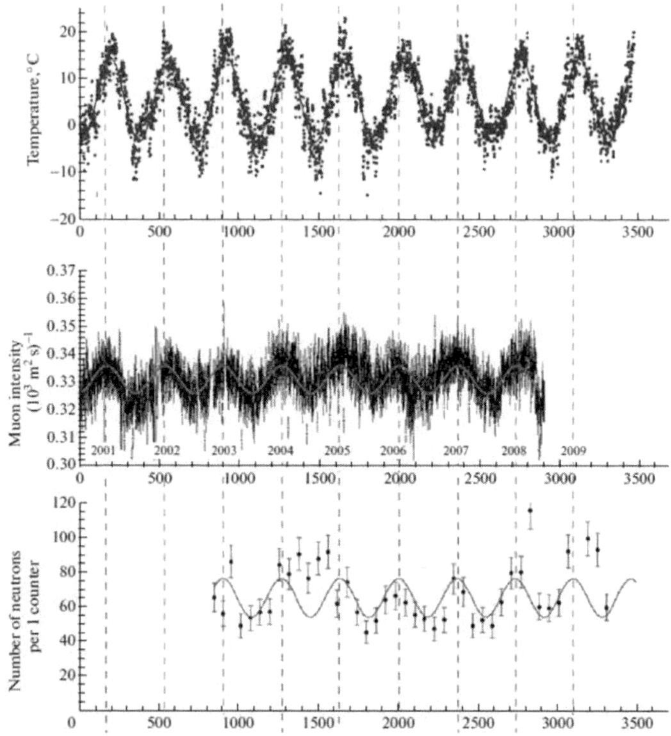

Figure 1: Upper panel: Temperature variations from January 2001 to June 2010. Middle panel: muon intensity over eight years of LVD operation; the curve is a fit to the data according to the formula $I_\mu = I_\mu^0 + dI_\mu cos(2\pi(t - t_0)/T)$. Lower panel: The number of neutrons produced by muons, per one counter, from April 2003 to April 2010. It is clear from the graphic that there is a correlation between the intensity of muons and neutrons produced by muons.

Acknowledgments

This work was supported by the Russian Foundation for Basic Research (project No. 15-02-01056 a) and was also funded within the program for support of leading scientific schools (grant No. 3110.2014.2) and the program of investigations of the Presidium of Russian Academy of Sciences Fundamental Properties of Matter and Astrophysics.

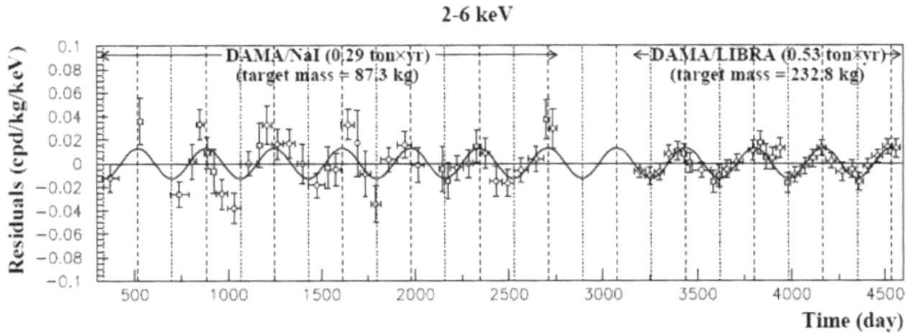

Figure 2: Model-independent residual rate of the single-hit scintillation events, measured by the new DAMA/LIBRA experiment in the (2 - 6) keV energy intervals as a function of the time in [2]

References

[1] F. Zwicky, Die Rotverschiebung von extragalaktischen Nebeln. Helvetica Physica Acta 6, 110-127 (1933)

[2] R. Bernabei et. al., (DAMA Coll.) // arXiv:0804.2741, Eur. Phys. J. **C 56**, 333-355 (2008)

[3] N.Yu. Agafonova et al. (LVD Collaboration), Bulletin of the Russian Academy of Sciences. Physics, Vol. **75**, No. 3, 427430, (2011)

MEASUREMENT OF THE MUON-INDUCED NEUTRON YIELD IN Fe AND Pb USING THE LARGE VOLUME DETECTOR AT LNGS

N. Yu. Agafonova [a] on behalf of the LVD Collaboration
Institute for Nuclear Research of the Russian Academy of Sciences, 117312 Moscow, Russia

Abstract. The experiment to measure the number of neutrons produced in iron (Fe) and lead (Pb) by muons at the mean muon energy 280 GeV using LVD is carried out. The preliminary results of the measurements are presented.

1 Introduction

After the article by G.T. Zatsepin and O.G. Ryazhskaya in 1965 [1] the study of muon-induced neutrons have become important as one of sources of background in underground low-background experiments. These experiments remain relevant today due to intense search for dark matter particles and neutrino investigations.

The motivation of this work is study of the muon-induce neutron production processes in differ material at muon energy about 280 GeV. There are few experimental results in this muon energy range. In the same time data of calculations using Geant4 and FLUKA are a large spread and do not correspond to each other.

2 Experimental set-up

The experiment to measure the number of neutrons generated by muons (neutron yield) is carry out using special construction with the matter under study which is fixed between two counters of LVD detector [2]. A muon, passing through the matter being studied and two vertical counters, produces an additional number of neutrons (dN). This value we normalize to one muon and refer to the length of the muon track (in g/cm^2) in the material under investigation.

Such direct method allows to determine the neutron production value for different materials, without calculating the proportion of neutrons produced in the mixed substance of the LVD set-up. Using this method the neutron yield in materials with high atomic number, such as Fe, Pb, Cu, Ca, is most easily measured.

In May 2014 a measurement of the neutron yield in iron [3] has been started. Since October 2015 the neutron yield in the lead is measuring. Total weight of additional iron mounted on four counters is 4×470 kg $= 1880$ kg, the total weight of lead is $4 \times 510 = 2040$ kg.

[a]E-mail: `agafonova@lngs.infn.it`

3 The results

Neutrons are detected by means of gamma quanta emitted as a result of the radiactive captures of thermal neutrons by protons of the scintillator and iron nuclei in the LVD structure. The number of neutrons is determined from the time distribution of low-energy pulses (1 - 12 MeV) in the time window 50 - 550 μs after the passage of muon. To approximate the time distribution we use the law $N_n(t) = N_0 \times exp(t/\tau) + B$, where $\tau = 145 \mu s$ is the exponent of thermalized neutron capture in the scintillator, B is a constant that depends on background conditions of a counter.

The LVD muon data from 2005 to 2013 without additional iron have been analyzed. The specific number of neutrons per muon 0.036 was obtained. Duration of data taking at additional iron is 22 months (from 06.2013 to 04.2015). The specific number of neutrons per muon 0.049 is obtained.

When determining of the neutron yield in lead the integral temporal distribution of neutron-like pulses was used. At the statistics during 7 months from 11.2014 till 04.2015 the number 0.096 neutrons per muon was obtained.

Considering the mean path for muon 35.6 g/cm^2 in additional iron and 38.7 g/cm^2 in lead and at the same time the neutron detection efficiency (24.4% and 28.0% correspondingly) the values of neutron production in iron $Y_{Fe} = (15 \pm 2) \times 10^{-4} n/muon/(g/cm^2)$ and in lead $Y_{Pb} = (50 \pm 20) \times 10^{-4} n/muon/(g/cm^2)$ were obtained.

The planned increase in the data statistic will provide better accuracy in determining of neutron production in lead.

Acknowledgments

This work was supported by the Russian Foundation for Basic Research (project No. 15-02-01056 a) and was also funded within the program for support of leading scientific schools (grant No. 3110.2014.2) and the program of investigations of the Presidium of Russian Academy of Sciences Fundamental Properties of Matter and Astrophysics.

References

[1] O. Ryazhskaya, G. Zatsepin, in (Proceedings of the 9 International Conference of Cosmic Ray, London, UK, 1965, Vol. 2 (Ed. A Stichland) (London: Institute of Physics and the Physical Society, 1966) p. 987)

[2] G. Bari et al., Nucl. Instrum. Methods Phys. Res., **A 277**, 11 (1989)

[3] N. Yu. Agafonova, V. V. Ashikhmin, V. L. Dadykin, E. A. Dobrynina, R. I. Enikeev, A. S. Malgin, V. V. Manukovsky, O. G. Ryazhskaya, V. G. Ryasny, I. R. Shakiryanova, A. V. Yudin, V. F. Yakushev, and the LVD Collaboration, Bulletin of the Russian Academy of Sciences. Physics, Vol. **79**, No. 3, pp. 401403, (2015)

THE SPATIAL STRUCTURE OF THE WMAP/PLANCK HAZE

Yana Zhezher [a]
Institute for Nuclear Research, Russian Academy of Sciences, 117312 Moscow, Russia
Faculty of Physics, Moscow State University, 119991 Moscow, Russia

Abstract. It was recently proposed that the two phenomena, WMAP/Planck haze and Fermi bubbles may have a common origin. In the present paper we analyze the spatial structure of the haze using the Planck 2015 data. It is found that the spatial dimensions and locations of WMAP/Planck haze and Fermi bubbles are compatible within experimental uncertainties. Both extended emissions possess north-south symmetry. No substructures similar to Fermi bubbles cocoon are identified in Planck data.

1 Introduction

The Wilkinson Microwave Anisotropy Probe spacecraft (or WMAP) was launched in 2001. The main objective of the mission was precise measurement the of cosmic microwave background anisotropy. The product of the spacecraft are so-called full-sky maps, which may also be used for the galactic matter emission study. The mechanisms of galactic emission are known, and described theoretically, so a full-sky map can be built for every emission mechanism. One may expect that if all relevant mechanisms are included, the full-sky map in the specific wave band is equal to the sum of all galactic emission components, CMB radiation and radiation of the bright sources.

A component separation procedure was conducted in 2003 by D. Finkbeiner [1]. As a result, an excess of emission in the Galactic center region was found, which was called WMAP haze (after the confirmation by the Planck mission - WMAP/Planck haze). The radiation is assumed to be radially symmetric, extending to more than 20°.

The Fermi satellite with Large Area Telescope (LAT) apparatus was launched in 2008 in order to study full sky in gamma-ray band. A similar analysis may be conducted using the gamma-ray sky maps produced using Fermi-LAT event data. It was made in 2010 by M. Su et al. [2], and an excess of radiation was found. It was later called the Fermi bubbles – a symmetric radiation extending to 50° above and below the Galactic Plane.

For now, there is no single explanation for the WMAP/Planck haze phenomenon. One class of theories assumes that several emission phenomena observed in the Galactic Center region have the same origin: WMAP/Planck haze, Fermi bubbles, soft X-ray emission and 2.3 GHz radio-emission. This scenario has a characteristic property, which is a coincidence of spatial structures of all emission maps. The purpose of this paper is the examination of this hypothesis.

[a]E-mail: zhezher.yana@physics.msu.ru

2 Method

Template fitting is a basic minimising scheme which was used in original WMAP analysis.

The initial map may be represented as:

$$d_i = a_j P_{ji} + x_i, \tag{1}$$

where a_j is an amplitude vector, P_{ji} is template matrix and x_i is an unknown component.

White noise makes a main contribution to noise maps of full-sky maps. It is statistically uncorrelated and it's covariation matrix is diagonal:

$$\langle d_i d_j \rangle = N_{\underline{i}} \delta_{\underline{ij}}, \tag{2}$$

where N_i is a vector with length N_{pix}.

The minimum of chi-squared criterion is obtained:

$$a_{\alpha'} = \left(P_{\alpha' i} N_i^{-1} P_{\alpha i}\right)^{-1} \left(P_{\alpha i} N_i^{-1} d_i\right). \tag{3}$$

Three interstellar medium emission mechanisms were considered:

- free-free emission – emission due to interaction between electrons and protons at the regions of partially ionised hydrogen (H_{II}). H_α emission (Lyman line transition of ionized hydrogen) is used as a template for this mechanism [3].

- synchrotron emission – an emission of electromagnetic waves by charged particles which propagate in a magnetic field with relativistic speed [4], [5].

- dust emission – emission of particles, including among their components silicon and carbon. An important dust property which allows to detect it is an ability to absorb and scatter light, it is called extinction [6].

One should notice, that the template for an unknown component should be also included in the template matrix P_{ji}, otherwise other components may compensate the part of studied emission. A traditional template for the WMAP/Planck haze is 2-D Gaussian ellipse with characteristic lengths $\sigma_l = 15°$ and $\sigma_b = 20°$.

Bright objects and the Galactic Plane are covered with a mask.

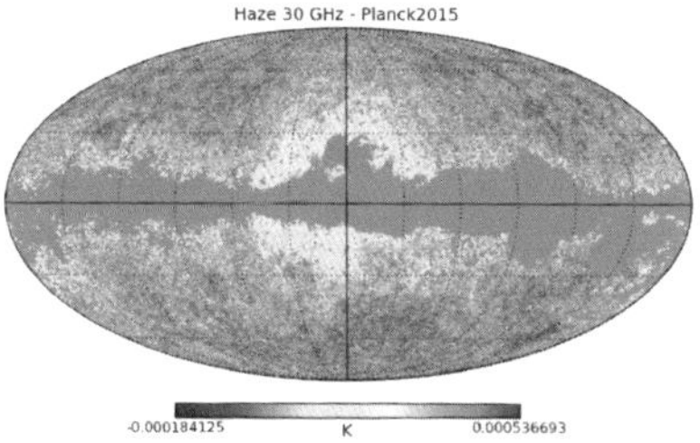

Figure 1: Extracted WMAP/Planck haze emission on 30 GHz.

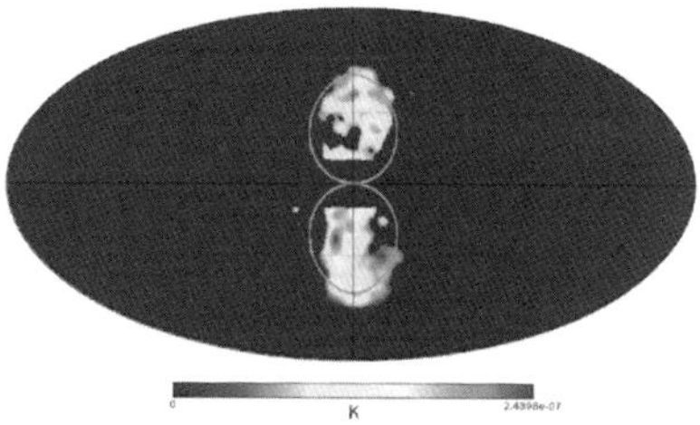

Figure 2: Comparison of the WMAP/Planck haze and Fermi bubbles sizes.

3 Results

Planck 2015 Low Frequency Instrument (LFI) full-sky maps were used as an initial data [7]. LFI creates maps on three different frequencies: 30, 44 and 70 GHz. The original WMAP analysis showed that the WMAP/Planck haze is the most intensive in at the frequencies 20-50 GHz, so 30 and 44 GHz separated emission were used for the extended emission analysis.

According to extracted data,new component of extended emission is more intensive in northern hemisphere. But considerable part is covered of it is covered with the mask and exact conclusions should not be drawn.

The latest Fermi collaboration research [8] have shown the existence of substructures in the Fermi bubbles. The Planck data investigation doesn't indicate the existence of such substructures in the WMAP/Planck haze.

The size of the WMAP/Planck haze are approximated by two ellipses symmetric below and above the Galactic Plane. The derived values are $\sigma_l = \sigma_b = b_0 = 23.7^\circ \pm 0.2^\circ$ in good agreement with the Fermi bubbles sizes.

4 Conclusion

The indication on the similarity of spatial sizes of the WMAP/Planck haze and the Fermi bubbles is obtained. For the future work, it is proposed to improve the techniques used, include more detailed theoretical analysis of Galactic matter and plane emission and extend the mask.

References

[1] D.P. Finkbeiner, "Microwave ISM Emission Observed by WMAP", ApJ **614** (2004) 186-193, astro-ph/0311547.

[2] M. Su, T. R. Slatyer, D. P. Finkbeiner, "Giant Gamma-ray Bubbles from Fermi-LAT: AGN Activity or Bipolar Galactic Wind?", Astrophys.J. **724** 1044-1082 (2010), astro-ph/1005.5480.

[3] D. P. Finkbeiner, "A Full-Sky H-alpha Template for Microwave Foreground Prediction", Astrophys.J.Suppl **146** 407 (2003), astro-ph/0301558.

[4] C. G. T. Haslam et al., "A 408 MHz all-sky continuum survey. I - Observations at southern declinations and for the North Polar region", Astronomy and Astrophysics **100** 2 209-219 (1981).

[5] C. G. T. Haslam et al., "A 408 MHz all-sky continuum survey. II - The atlas of contour maps", Astronomy and Astrophysics Supplement Series **47** 1, 2, 4-51, 53-142 (1982).

[6] D. P. Finkbeiner, M. Davis, D. J. Schegel, "Extrapolation of Galactic Dust Emission at 100 Microns to CMBR Frequencies Using FIRAS", Astrophys.J. **524** 867-886 (1999), astro-ph/9905128.

[7] J. A. Tauber et al., "Planck pre-launch status: The Planck mission", Astronomy&Astrophysics **520** A1 (2010).

[8] Fermi-LAT Collaboration, "The Spectrum and Morphology of the Fermi Bubbles", Astrophys.J. **793** 1 34 (2014), astro-ph/1407.7905.

CP Violation and Rare Decays

RECENT RESULTS FROM BELLE

Denis Epifanov [a] [b]
The University of Tokyo, 7-3-1 Hongo Bunkyo-ku, Tokyo 113-0033, Japan

Abstract. We report about recent studies of the $\Upsilon(10860)$ and $\Upsilon(11020)$ resonances, charged bottomonium-like states $Z_b(10610)$ and $Z_b(10650)$ and their decays at Belle. Study of the $\bar{B} \to D^{(*)}\tau^-\bar{\nu}_\tau$ with the whole Belle data sample and search for dark photon and dark Higgs boson are also briefly discussed.

1 Introduction

The Belle detector at the asymmetric energy e^+e^- collider KEKB (so called B-factory) collected the world largest data sample of about one inverse attobarn in the region of $\Upsilon(4S)$ resonance and around it [1,2]. Initially, it was designed to search for CP symmetry violation (CPV) in the B meson decays, hence the largest luminosity integral was recorded on or near the $\Upsilon(4S)$ resonance to produce B mesons via $\Upsilon(4S) \to B\bar{B}$ process. Finally, large CPV in B meson system was discovered by Belle and BaBar [3], in good agreement with Cabibbo-Kabayashi-Maskawa mechanism of CPV in the Standard Model [4]. However, Belle also recorded unique data sets at the $\Upsilon(1S)$, $\Upsilon(2S)$, $\Upsilon(3S)$, $\Upsilon(10860)$ and $\Upsilon(11020)$ resonances. The data sample in the region of $\Upsilon(10860)$ and $\Upsilon(11020)$ resonances is of particular interest in the bottomonium spectroscopy due to the exotic charged bottomonium-like states $Z_b^\pm$ recently found in the transitions $\Upsilon(10860, 11020) \to b\bar{b}\pi^+\pi^-$ and $\Upsilon(10860) \to B^{(*)}\bar{B}^{(*)}\pi$.

2 Study of bottomonia and charged bottomonium-like states $Z_b^\pm$

Recently Belle observed anomalously large rates (by about two orders of magnitude larger than that of $\Upsilon(2,3,4\text{S}) \to \Upsilon(1\text{S})\pi^+\pi^-$) of the events $e^+e^- \to \Upsilon(n\text{S})\pi^+\pi^-$ ($n = 1,\ 2,\ 3$) at the $\Upsilon(10860)$ resonance [5]. The rates of $e^+e^- \to h_b(m\text{P})\pi^+\pi^-$ ($m = 1,\ 2$) were found to be of the same order of magnitude, although the $\Upsilon(10860) \to h_b(m\text{P})\pi^+\pi^-$ decay should be suppressed due to a b-quark spin-flip [6]. In the study of the dynamics of $\Upsilon(10860) \to \Upsilon(n\text{S})\pi^+\pi^-$ and $\Upsilon(10860) \to h_b(m\text{P})\pi^+\pi^-$ processes Belle discovered two new charged bottomonium-like resonances, $Z_b^\pm(10610)$ and $Z_b^\pm(10650)$, decaying to $\Upsilon(n\text{S})\pi^\pm$ and $h_b(m\text{P})\pi^\pm$ final states [7]. The measured masses of $Z_b^\pm(10610)$ and $Z_b^\pm(10650)$ are close to the thresholds of the open beauty channels $B\bar{B}^*$ and $B^*\bar{B}^*$, and their quantum numbers were unveiled to be $J^P = 1^+$ [8]. The quark structure of the decay products indicates that $Z_b^\pm$ could be interpreted as four-quark state [9–11]. Also the properties of the $Z_b^\pm$ can be naturally explained

[a] On behalf of the Belle collaboration
[b] E-mail: `epifanov@inp.nsk.su`

by the "molecular" structure of these states [12]. Several years ago Belle found $Z_b^{\pm}(10610)$ and $Z_b^{\pm}(10650)$ states in the analysis of the three-body decay $\Upsilon(10860) \to (B^{(*)}\bar{B}^*)^{\mp}\pi^{\pm}$ [13]. The dominating decay of $Z_b^{\pm}$ into the $B^{(*)}\bar{B}^*$ final state strongly supports the molecular hypothesis. The neutral partner of the charged $Z_b^{\pm}(10610)$, $Z_b^0(10610)$, has been also observed in the $\Upsilon(10860) \to \Upsilon(2S, 3S)\pi^0\pi^0$ decay, while the statistics was not enough to resolve $Z_b^0(10650)$ [14].

Although $\Upsilon(10860)$ is usually considered as a radial exitation state of bottomonium, $\Upsilon(5S)$, with the quantum numbers $J^{PC} = 1^{--}$, its decay to the charged exotic states $Z_b^{\pm}$ and recent results of the energy scan of the $\sigma(e^+e^- \to \Upsilon(nS)\pi^+\pi^-)$ cross section [15] raised questions about the nature of the $\Upsilon(10860)$ state [16]. Conventionally, further we call $\Upsilon(10860)$ and $\Upsilon(11020)$ as $\Upsilon(5S)$ and $\Upsilon(6S)$, respectively.

2.1 $\sigma(e^+e^- \to \Upsilon(1S, 2S, 3S)\pi^+\pi^-)$ and $\sigma(e^+e^- \to b\bar{b})$ in the region of $\Upsilon(5S)$ and $\Upsilon(6S)$ resonances

Recent Belle result on the ratios, $R_{\Upsilon\pi\pi} = \sigma(e^+e^- \to \Upsilon(nS)\pi^+\pi^-)/\sigma^0_{\mu^+\mu^-}$ and $R_b = \sigma(e^+e^- \to b\bar{b})/\sigma^0_{\mu^+\mu^-}$ ($\sigma^0_{\mu^+\mu^-}$ is Born cross section of the $e^+e^- \to \mu^+\mu^-$ process), at six energy points near the $\Upsilon(5S)$ and $\Upsilon(6S)$ resonances showed inconsistency in the $\Upsilon(5S)$ mass of (9 ± 4) MeV/c^2 [15]. To clarify this anomaly Belle repeated this study with the whole available data sample at the center-of-mass (CM) energies, $\sqrt{s}$, between 10.60 and 11.05 GeV in the region of the $\Upsilon(5S)$ and $\Upsilon(6S)$ resonances [17]. Data sample of about 19 fb^{-1} was used to measure R_b. Hadronic $b\bar{b}$ events were required to have at least five good tracks, more than 1 cluster in the electromagnetic calorimeter (ECL) with the energy threshold of 0.1 GeV, ECL total energy in the range $(0.1 \div 0.8)\sqrt{s}$, total energy of the tracks and photons more than $0.5 \times \sqrt{s}$ and the reconstructed event vertex within 1.5 and 3.5 cm of the interaction point in the transversal and longitudinal directions, respectively. The $b\bar{b}$ detection efficiency $\epsilon_{b\bar{b}}$, estimated via Monte Carlo (MC) simulation, increases almost linearly from about 70% up to 74%. Selected sample comprises events of $b\bar{b}$, $q\bar{q}$ continuum, and bottomonia produced via e^+e^- annihilation with initial state radiation (ISR): $e^+e^- \to \gamma_{\rm ISR}\Upsilon(1S, 2S, 3S)$. The number of selected events at the $i^{\rm th}$ CM energy point is $N_i = \mathcal{L}_i(\sigma_{b\bar{b},i}\epsilon_{b\bar{b},i} + \sigma_{q\bar{q},i}\epsilon_{q\bar{q},i} + \sum_{\Upsilon(1,2,3S)} \sigma_{{\rm ISR},i}\epsilon_{{\rm ISR},i})$ ($\mathcal{L}_i$ - integrated luminosity; detection efficiencies, $\epsilon_{q\bar{q},i}$ and $\epsilon_{{\rm ISR},i}$, were evaluated with MC simulation; $\sigma_{q\bar{q},i}$ was estimated from the experimental data; $\sigma_{{\rm ISR},i}$ was calculated). As a result the ratios, $R'_{b,i} = \sigma_{b\bar{b},i}/\sigma^0_{\mu^+\mu^-,i}$ and $R_{b,i} = R'_{b,i} + \sum_{\Upsilon(1,2,3S)} \sigma_{{\rm ISR},i}/\sigma^0_{\mu^+\mu^-,i}$, were extracted and fitted by the function $\mathcal{F}(s) = |A_{\rm nr}|^2 + |A_{\rm r} + A_{5S}e^{i\phi_{5S}}f_{5S} + A_{6S}e^{i\phi_{6S}}f_{6S}|^2$, where $f_{nS} = M_{nS}\Gamma_{nS}/[(s - M^2_{nS}) + iM_{nS}\Gamma_{nS}]$ and $A_{\rm r}$ and $A_{\rm nr}$ are coherent and incoherent continuum amplitudes, respectively. Results of the fits of R'_b and R_b are shown in Table 1.

Table 1: $\Upsilon(5S)$ and $\Upsilon(6S)$ masses and widths extracted from the fits (the first error is statistical and the second is systematic one).

	M_{5S} (MeV/c^2)	Γ_{5S} (MeV)	M_{6S} (MeV/c^2)	Γ_{6S} (MeV)
R_b	$10881.9 \pm 1.0 \pm 1.2$	$49.8 \pm 1.9^{+2.1}_{-2.8}$	$11002.9 \pm 1.1^{+0.8}_{-0.9}$	$38.5^{+1.6}_{-1.5}{}^{+1.3}_{-2.4}$
R_b'	$10881.8^{+1.0}_{-1.1} \pm 1.2$	$48.5^{+1.9}_{-1.8}{}^{+2.0}_{-2.8}$	$11003.0 \pm 1.1^{+0.9}_{-1.0}$	$39.3^{+1.7}_{-1.6}{}^{+1.3}_{-2.4}$
$R_{\Upsilon\pi\pi}$	$10891.1 \pm 3.2^{+0.6}_{-1.5}$	$53.7^{+7.1}_{-5.6}{}^{+0.9}_{-5.4}$	$10987.5^{+6.4}_{-2.5}{}^{+9.0}_{-2.1}$	$61^{+9}_{-19}{}^{+2}_{-20}$

$\Upsilon(5S)$ and $\Upsilon(6S)$ masses and widths from the R_b fit are consistent with those measured by Belle [15] and BABAR [18]. To measure $R_{\Upsilon\pi\pi}$, events of the $e^+e^- \to \Upsilon(nS)[\to \mu^+\mu^-]\pi^+\pi^-$ ($n = 1, 2, 3$) process were fully reconstructed. Four-track events were selected, with two oppositely-charged tracks identified as muons with the $\mu^+\mu^-$ invariant mass $M(\mu^+\mu^-) > 8$ GeV/c^2 and the remaining oppositely-charged tracks identified as pions. Signal candidates were selected by requiring $|M(\mu^+\mu^-\pi^+\pi^-) - \sqrt{s}_i/c^2| < 200$ MeV/c^2 and $\delta\Delta M \equiv |\Delta M - (\sqrt{s_i}/c^2 - m_\Upsilon)| < 25$ MeV/c^2 ($\Delta M = M(\mu^+\mu^-\pi^+\pi^-) - M(\mu^+\mu^-)$), sideband events with 50 MeV/c^2 $< |\delta\Delta M| <$ 100 MeV/c^2 were used to evaluate background. Detection efficiencies were estimated from MC simulation, they vary from about 15% up to 45% over the $\sqrt{s}$ range. The distribution of $R_{\Upsilon\pi\pi,i} = N_{\Upsilon\pi\pi,i}/(\mathcal{L}_i\mathcal{B}(\Upsilon(nS) \to \mu^+\mu^-)\sigma^0_{\mu^+\mu^-}(\sqrt{s}_i))$ was fitted by the function $\mathcal{F}(s)$. A_{nr} and A_r were found to be consistent with zero for all channels; obtained $\Upsilon(5S)$ and $\Upsilon(6S)$ masses and widths are shown in Table 1. Figure 1 shows the $R_{\Upsilon\pi\pi}$ data and fit result. The masses and widths of the $\Upsilon(5S)$ and

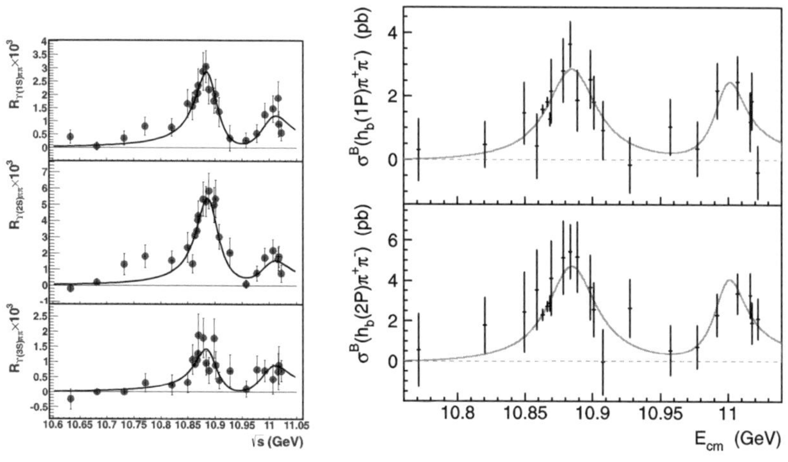

Figure 1: Left: $R_{\Upsilon\pi\pi}$ data for $\Upsilon(1S, 2S, 3S)$ (from top to bottom) with results of fit. Right: $\sigma(e^+e^- \to h_b(1P, 2P)\pi^+\pi^-)$ data (points with error bars) and result of the fit (solid curves).

Υ(6S) from the R'_b and $R_{\Upsilon\pi\pi}$ fits are in agreement. Large interference of the continuum, A_r (with unknown shape), with the resonance amplitudes results in the unknown systematic errors of the masses and widths, hence $R_{\Upsilon\pi\pi}$ is more suitable to measure Υ(5S) and Υ(6S) properties.

2.2 Energy scan of the $\sigma(e^+e^- \to h_b(1P,2P)\pi^+\pi^-)$ and evidence for the $\Upsilon(11020) \to Z_b^{\pm}\pi^{\mp}$ decay

Recently the energy dependence of the $e^+e^- \to h_b(mP)\pi^+\pi^-$ $(m = 1, 2)$ cross section in the region of Υ(5S) and Υ(6S) resonances was measured for the first time by Belle [19]. Data sample of about 142 fb^{-1} at the CM energies between 10.77 and 11.02 GeV was analyzed. Signal events were reconstructed inclusively fitting the missing mass of the $\pi^+\pi^-$ pairs, $M_{\rm miss}(\pi^+\pi^-)$. The detection efficiency, ϵ, for the $h_b(1P)\pi^+\pi^-$ ($h_b(2P)\pi^+\pi^-$) process varies with CM energy in the range $(40 \div 55)\%$ ($(35 \div 50)\%$). Born cross section is $\sigma^B(e^+e^- \to h_b(mP)\pi^+\pi^-) = N/(L\epsilon|1 - \Pi|^2)$, where N is the ISR corrected number of signal events, L is the integrated luminosity at the particular energy point, and $|1 - \Pi|^2$ is the vacuum polarization correction. The obtained cross sections, shown in Fig. 1, were fitted simultaneously by the function $\mathcal{F}(s) = A_{h_b(mP)} f(s)|BW(s, M_{5S}, \Gamma_{5S}) + ae^{i\phi}BW(s, M_{6S}, \Gamma_{6S}) + be^{i\delta}|^2$, where $f(s)$ is the phase space factor, $BW(s, M, \Gamma) = M\Gamma/(s - M^2 + iM\Gamma)$, free parameters in the fit are: $A_{h_b(1P)}$, $A_{h_b(2P)}$, M_{5S}, Γ_{5S}, M_{6S}, Γ_{6S}, a, ϕ and (optionally) b, δ. The non-resonant continuum contribution, $be^{i\delta}$, was found to be insignificant and was set to zero in the fits. The optimal masses and widths were found to be $M_{5S} = (10884.7^{+3.2+8.6}_{-2.9-0.6})$ MeV/c^2, $\Gamma_{5S} = (44.2^{+11.9+2.2}_{-7.8\ -15.8})$ MeV, $M_{6S} = (10998.6 \pm 6.1^{+16.1}_{-1.1})$ MeV/c^2, $\Gamma_{6S} = (29^{+20+2}_{-12-7})$ MeV, they agree with the values measured in the $e^+e^- \to \Upsilon(1S, 2S, 3S)\pi^+\pi^-$ processes. The $\Upsilon(6S) \to h_b(mP)\pi^+\pi^-$ transitions were studied with the data collected at the six highest CM energy points. The first evidence of the $e^+e^- \to h_b(1P)\pi^+\pi^-$ and observation of the $e^+e^- \to h_b(2P)\pi^+\pi^-$ at Υ(6S) were found. By fitting $M_{\rm miss}(\pi^+\pi^-)$ distribution in bins of $M_{\rm miss}(\pi)$ the dynamics of $\Upsilon(6S) \to h_b(mP)\pi^+\pi^-$ was investigated. An evidence of the $\Upsilon(6S) \to Z_b^{\pm}\pi^{\mp} \to h_b(mP)\pi^+\pi^-$ mechanism was found, but the statistics was not enough resolve between $Z_b^{\pm}(10610)$ and $Z_b^{\pm}(10650)$ states.

2.3 Study of $e^+e^- \to B^{(*)}\bar{B}^{(*)}\pi$ at the Υ(5S) resonance

To clarify the nature of the $Z_b^{\pm}$ states an analysis of the three-body $e^+e^- \to B\bar{B}\pi$, $e^+e^- \to B\bar{B}^*\pi$ and $e^+e^- \to B^*\bar{B}^*\pi$ processes has been performed recently at Belle [20]. Data sample with an integrated luminosity of 121.4 fb^{-1} collected at the Υ(5S) resonance was used. B mesons were reconstructed in the eighteen decay modes: $B^+ \to J/\psi K^{(*)+}$, $B^+ \to \bar{D}^{(*)0}\pi^+$, $B^0 \to J/\psi K^{(*)0}$, $B^0 \to D^{(*)-}\pi^+$. The number of B candidates, $N_B = 12263 \pm 168$, was determined from the fit of the reconstructed invariant mass ($M(B)$) distribution. After that B^+ or $\bar{B}^0$ were combined with the π^- candidates (right-sign combinations) and the modified missing mass to $B\pi$ system, $M^*_{\rm miss}(B\pi) = M_{\rm miss}(B\pi) + M(B) - m_B$, was analyzed. Wrong-sign combinaions with π^+ were used to estimate the combinatorial background. The $M^*_{\rm miss}$ distribution was approximated and numbers of signal events were extracted: $N_{B\bar{B}\pi} = 13 \pm 25$, $N_{B\bar{B}^*\pi} = 357 \pm 30$ (9.3σ) and $N_{B^*\bar{B}^*\pi} = 161 \pm 21$ (8.1σ). To study dynamics of the $B^{(*)}\bar{B}^*\pi$ production, events satisfying $|M^*_{\rm miss} - m_{B^*}| < 15$ MeV/c^2 ($B\bar{B}^*\pi$) and $|M^*_{\rm miss} - (2m_{B^*} - m_B)| < 12$ MeV/c^2 ($B^*\bar{B}^*\pi$) criteria were selected, and the recoil mass against primary pion, $M_{\rm miss}(\pi)$, was analyzed, see Fig. 2. The $m \equiv M_{\rm miss}(\pi)$ distribution was approximated by the function $F(m) =$

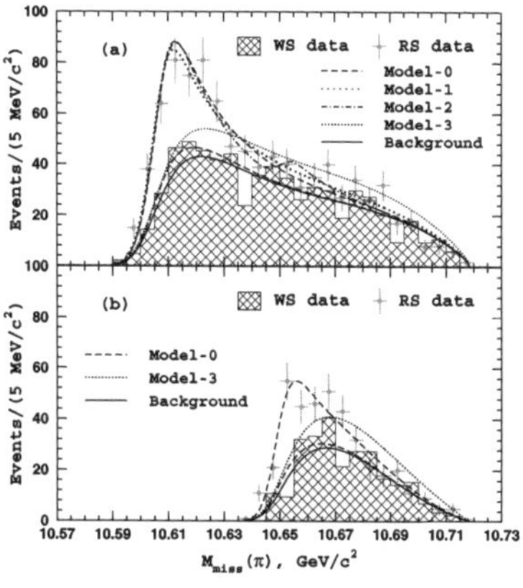

Figure 2: The $M_{\rm miss}(\pi)$ distribution for the (a) $B\bar{B}^*\pi$ and (b) $B^*\bar{B}^*\pi$ candidate events.

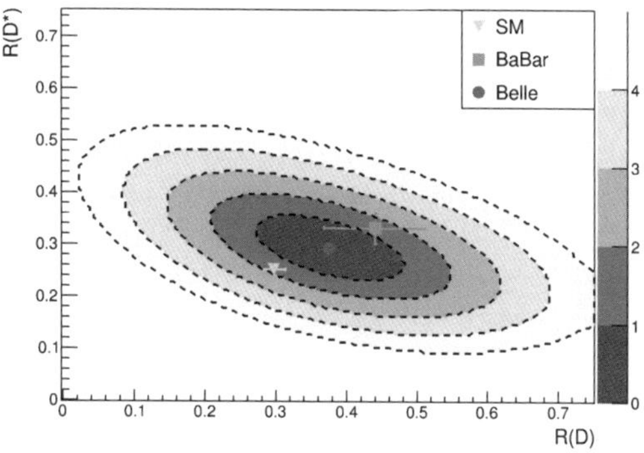

Figure 3: Exclusion levels of the $R(D)$ vs. $R(D^*)$ values in the units of the standard deviations.

$[S(m) + B(m)]\epsilon(m) F_{\rm PHSP}(m)$, where $S(m)$ and $B(m)$ are the signal and background probability density functions, $\epsilon(m)$ is detection efficiency, $F_{\rm PHSP}(m)$ is the phase space factor. $S(m) = |\mathcal{A}_{Z_b(10610)} + \mathcal{A}_{Z_b(10650)} + a_{\rm nr}e^{i\phi_{\rm nr}}|^2$, where $\mathcal{A}_{Z_b} = a_Z e^{i\phi_Z}/(m^2 - m_Z^2 - i\Gamma_Z m_Z)$, masses and widths of the Z_b's are fixed at the values obtained in [7, 8]. Results of the fit of the $B\bar{B}^*\pi(B^*\bar{B}^*\pi)$ events by the model including only $Z_b(10610)(Z_b(10650))$ contribution are shown in Fig. 2. The non-resonant contribution, $a_{\rm nr}e^{i\phi_{\rm nr}}$, was found to be insignificant, the production of the $B\bar{B}^*\pi$ and $B^*\bar{B}^*\pi$ are dominated by the intermediate $Z_b(10610)^\pm\pi^\mp$ and $Z_b(10650)^\pm\pi^\mp$, respectively. Visible cross sections were measured to be $\sigma_{\rm vis}(e^+e^- \to B\bar{B}^*\pi) = (11.2 \pm 1.0 \pm 1.2)$ pb, $\sigma_{\rm vis}(e^+e^- \to B^*\bar{B}^*\pi) = (5.61 \pm 0.73 \pm 0.66)$ pb, $\sigma_{\rm vis}(e^+e^- \to B\bar{B}\pi) < 2.1$ pb (90% C.L.).

3 Search for new Physics in $\bar{B} \to D^{(*)}\tau^- \bar{\nu}_\tau$

New measurement of the branching fraction ratios $R(D^{(*)}) = \mathcal{B}(\bar{B} \to D^{(*)}\tau^- \bar{\nu}_\tau)/\mathcal{B}(\bar{B} \to D^{(*)}\ell^- \bar{\nu}_\ell)$ (where $\mathcal{B}(\bar{B} \to D^{(*)}\ell^- \bar{\nu}_\ell) = [\mathcal{B}(\bar{B} \to D^{(*)}e^- \bar{\nu}_e)+\mathcal{B}(\bar{B} \to D^{(*)}\mu^- \bar{\nu}_\mu)]/2)$ has been recently published by Belle [21]. The whole data sample of $772 \times 10^6 B\bar{B}$ pairs collected at the $\Upsilon(4S)$ resonance was analyzed. One of the B mesons, $B_{\rm tag}$, was identified using hadronic full reconstruction algorithm, which includes 1149 B final states. The τ is reconstructed in the leptonic decays $\tau^- \to \ell^- \bar{\nu}_\ell \nu_\tau$ $(\ell = e, \mu)$, so in the events with $B_{\rm tag}$ the $D^{(*)}\ell$ combination $(D^+\ell^-,\ D^0\ell^-,\ \ D^{*+}\ell^-,\ D^{*0}\ell^-)$ was reconstructed among the remaining charged and neutral particles. The signal mode (with three neutrinos) was distinguished from the normalization mode (with one neutrino) with help of the missing mass squared $M^2_{\rm miss} = (p_{\rm CM} - p_{\rm tag} - p_{D^{(*)}} - p_\ell)^2/c^2$ (where $p_{\rm CM}$, $p_{\rm tag}$, $p_{D^{(*)}}$, and p_ℓ are the four-momenta of the initial e^+e^- system, the $B_{\rm tag}$ candidate, and the signal-B daughters, respectively), which peaks at (above) zero for the normalization (signal) events. The $M^2_{\rm miss}$ range was split into two regions: $M^2_{\rm miss} < 0.85$ GeV$^2/c^4$, populated mostly by normalization $\bar{B} \to D^{(*)}\ell^- \bar{\nu}_\ell$ events; $M^2_{\rm miss} > 0.85$ GeV$^2/c^4$, enriched by signal $\bar{B} \to D^{(*)}\tau^- \bar{\nu}_\tau$ events, both subsamples were approximated simultaneously. In the low-$M^2_{\rm miss}$ region, the $M^2_{\rm miss}$ distribution was fitted to evaluate $\mathcal{B}(\bar{B} \to D^{(*)}\ell^- \bar{\nu}_\ell)$, while in the high-$M^2_{\rm miss}$ region, contaminated by background with multiple missing particles, the neural-network output, $O'_{\rm NB}$, was approximated to extract $\mathcal{B}(\bar{B} \to D^{(*)}\tau^- \bar{\nu}_\tau)$. The measured ratios $R(D) = 0.375 \pm 0.064(\text{stat.}) \pm 0.026(\text{syst.})$ and $R(D^*) = 0.293 \pm 0.038(\text{stat.}) \pm 0.015(\text{syst.})$ are shown in Fig. 3. The results agree with the previous measurements as well as with the Standard Model expectation. They are also compatible with the 2HDM(type II) model in the region around $\tan\beta/m_{H^+} = 0.5\ c^2/\text{GeV}$ [22].

4 Search for dark photon and dark Higgs boson

The dark photon, A', and the dark Higgs boson, h', are introduced in the Dark Sector Models [23]. Recently Belle performed a search for their production in the Higgs-strahlung channel, $e^+e^- \to A'h'$ $(h' \to A'A')$ using full data sample of 977 fb^{-1} [24]. Ten exclusive final-states with $A' \to e^+e^-$, $\mu^+\mu^-$, or $\pi^+\pi^-$, and three inclusive final-states, $2(e^+e^-)X$, $2(\mu^+\mu^-)X$, and $(e^+e^-)(\mu^+\mu^-)X$ (where X denotes A' candidate detected via missing mass) were investigated. No significant signal was observed. Upper limits on the dark photon coupling to the dark Higgs boson (α_D) times the kinetic mixing (ϵ) between the Standard Model photon and the dark photon squared, $\alpha_D \times \epsilon^2$ were obtained. These limits improve upon results of the previous measurements, $3(\pi^+\pi^-)$ and $2(e^+e^-)X$ final states were studied for the first time.

References

[1] A. Abashian *et al.* [Belle Collaboration], Nucl. Instrum. Meth. Phys. Res. Sect. A **479**, 117 (2002); also see detector section in: J. Brodzicka *et al.* [Belle Collaboration], Prog. Theor. Exp. Phys. **2012**, 04D001 (2012) [arXiv:1212.5342 [hep-ex]].
[2] S. Kurokawa and E. Kikutani, Nucl. Instrum. Methods Phys. Res. Sect. A **499**, 1 (2003), and other papers included in this Volume; T.Abe *et al.*, Prog. Theor. Exp. Phys. **2013**, 03A001 (2013), and following articles up to 03A011.
[3] A. J. Bevan *et al.* [BaBar and Belle Collaborations], Eur. Phys. J. C **74**, 3026 (2014).
[4] M. Kobayashi and T. Maskawa, Prog. Theor. Phys. **49**, 652 (1973).
[5] K.-F. Chen *et al.* [Belle Collaboration], Phys. Rev. Lett. **100**, 112001 (2008).
[6] I. Adachi *et al.* [Belle Collaboration], Phys. Rev. Lett. **108**, 032001 (2012).
[7] A. Bondar *et al.* [Belle Collaboration], Phys. Rev. Lett. **108**, 122001 (2012).
[8] A. Garmash *et al.* [Belle Collaboration], Phys. Rev. D **91**, no. 7, 072003 (2015).
[9] M. Karliner and H. J. Lipkin, arXiv:0802.0649 [hep-ph].
[10] D. Bugg, Europhys. Lett. **96**, 11002 (2011).
[11] I. V. Danilkin, V. D. Orlovsky and Yu. A. Simonov, Phys. Rev. **D 85**, 034012 (2012).
[12] A.E. Bondar, A. Garmash, A.I. Milstein, R. Mizuk and M.B. Voloshin, Phys. Rev. **D 84** , 054010 (2011).
[13] I. Adachi *et al.* (Belle Collaboration), arXiv:1209.6450 [hep-ex].
[14] P. Krokovny *et al.* [Belle Collaboration], Phys. Rev. D **88**, no. 5, 052016 (2013).
[15] K.-F. Chen *et al.* [Belle Collaboration], Phys. Rev. D **82**, 091106(R) (2010).
[16] N. Brambilla *et al.*, Eur. Phys. J. C **74**, 2981 (2014).
[17] D. Santel *et al.* [Belle Collaboration], arXiv:1501.01137 [hep-ex].
[18] B. Aubert *et al.* [BABAR Collaboration], Phys. Rev. Lett. **102**, 012001 (2009).
[19] A. Abdesselam *et al.* [Belle Collaboration], arXiv:1508.06562 [hep-ex].
[20] A. Garmash *et al.* [Belle Collaboration], arXiv:1512.07419 [hep-ex].
[21] M. Huschle *et al.* [Belle Collaboration], Phys. Rev. D **92**, no. 7, 072014 (2015).
[22] M. Tanaka, Z. Phys. C **67** (1995) 321.
[23] B. Batell, M. Pospelov, and A. Ritz, Phys. Rev. D **79**, 115008 (2009).
[24] I. Jaegle [Belle Collaboration], Phys. Rev. Lett. **114**, no. 21, 211801 (2015).

STUDY OF χ_b MESON SPECTROSCOPY AND PRODUCTION IN pp COLLISIONS AT $\sqrt{S} = 7$ AND 8 TeV AT LHCb

Alexander Mazurov [a], on behalf of the LHCb collaboration
School of Physics and Astronomy, University of Birmingham

Abstract. A study of χ_b meson spectroscopy and production at LHCb is performed on proton-proton collision data, corresponding to $3.0\,\text{fb}^{-1}$ of integrated luminosity collected at centre-of-mass energies $\sqrt{s} = 7$ and 8 TeV. The fraction of $\Upsilon(\text{nS})$ mesons originating from χ_b decays is measured as a function of the Υ transverse momentum in the rapidity range $2.0 < y^{\Upsilon} < 4.5$. The radiative transition of the $\chi_b(3\text{P})$ meson to $\Upsilon(3\text{S})$ is observed for the first time. The $\chi_{b1}(3\text{P})$ mass is determined to be

$$m_{\chi_{b1}(3\text{P})} = 10515.7^{+2.2}_{-3.9}\,(\text{stat})^{+1.5}_{-2.1}\,(\text{syst})\,\text{MeV}/c^2$$

where the first uncertainty is statistical and the second is systematic.

1 Introduction

The comparison of experimental data for prompt production of S-wave quarkonia, *e.g.* J/ψ or $\Upsilon(1\text{S})$ mesons, with theory predictions requires knowledge of feed-down contributions from P-wave quarkonia states, *e.g.* radiative $\chi_b \to \Upsilon\gamma$ decays. This contribution could significantly influence the interpretation of the measured polarization of S-wave vector quarkonia. In addition, measurements of the relative production rates of P-wave to S-wave quarkonia, as well as the tensor-to-vector ratios, provide valuable information on colour-octet matrix elements.

Depending on the relative orientation of the quark spins, the χ_b states can be either scalar, vector or tensor mesons, denoted by χ_{bJ} with total angular momentum $\text{J} = 0, 1, 2$. The fractions of $\Upsilon(\text{nS})$ decays originating from $\chi_b(\text{mP})$ decays, where n and m are radial quantum numbers of the bound states, are defined as

$$\mathcal{R}^{\chi_b(\text{mP})}_{\Upsilon(\text{nS})} \equiv \frac{\sigma(\text{pp} \to \chi_{b1}(\text{mP})\text{X})}{\sigma(\text{pp} \to \Upsilon(\text{nS})\text{X})} \times \mathcal{B}_1 + \frac{\sigma(\text{pp} \to \chi_{b2}(\text{mP})\text{X})}{\sigma(\text{pp} \to \Upsilon(\text{nS})\text{X})} \times \mathcal{B}_2, \qquad (1)$$

where $\mathcal{B}_{1(2)}$ denotes the branching fraction for the decay $\chi_{b1(2)}(\text{mP}) \to \Upsilon(\text{nS})\gamma$. Possible contributions from $\chi_{b0}(\text{mP}) \to \Upsilon(\text{nS})\gamma$ decays are neglected because of the small branching fraction for the corresponding radiative decays.

A measurement of the fractions $\mathcal{R}^{\chi_b(\text{mP})}_{\Upsilon(\text{nS})}$ was recently performed by LHCb using calorimetry photons [1]. This measurements supersede earlier LHCb measurements [2,3] by increasing the statistical precision and exploiting more decay modes in the rapidity range $2.0 < y < 4.5$.

[a]E-mail: alexander.mazurov@cern.ch

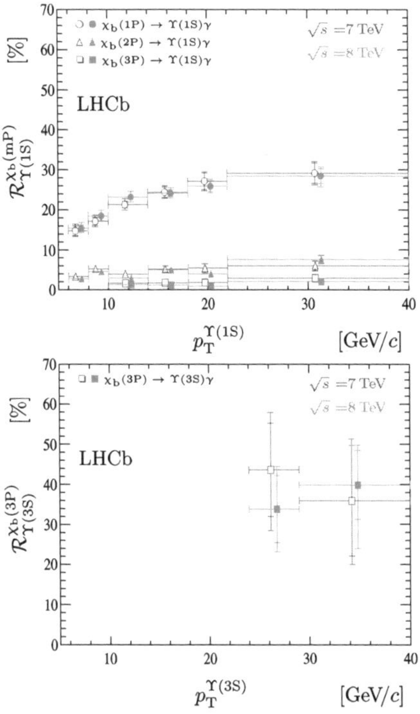

Figure 1: Fractions $\mathcal{R}^{\chi_b(mP)}_{\Upsilon(nS)}$ as functions of p_T^{Υ} [1]. Points with blue open (red solid) symbols correspond to data collected at $\sqrt{s} = 7(8)$ TeV, respectively. For better visualization the data points are slightly displaced from the bin centres. The inner error bars represent statistical uncertainties, while the outer error bars indicate statistical and systematic uncertainties added in quadrature.

2 Results

The measured fractions $\mathcal{R}^{\chi_b(mP)}_{\Upsilon(nS)}$ [1] are presented in Fig. 1. The results are dominated by the statistical uncertainties, and show no dependence on the pp collision energy. A measurement of the $\mathcal{R}^{\chi_b(3P)}_{\Upsilon(3S)}$ fraction is performed for the first time. The large value of this fraction impacts the interpretation of experimental data on Υ production and polarization. When data on Υ production and polarization are compared with theory predictions, as well as when different theory predictions are compared among themselves, it is often implicitly assumed that the fraction of $\Upsilon(3S)$ mesons produced by feed down from higher states is small. The large measured value of $\mathcal{R}^{\chi_b(3P)}_{\Upsilon(3S)}$ indicates that these assumptions need to be revisited.

The recent results supersede earlier LHCb measurements [2, 3] by increasing the statistical precision and exploiting more decay modes and higher transverse momentum regions. In particular, the full data sample collected by LHCb at $\sqrt{s} = 7$ and 8 TeV has been used and the measured fractions $\mathcal{R}^{\chi_b(mP)}_{\Upsilon(nS)}$ are reported for all six kinematically allowed transitions: $\chi_b(1P) \to \Upsilon(1S)\gamma$, $\chi_b(2P) \to \Upsilon(1S)\gamma$, $\chi_b(2P) \to \Upsilon(2S)\gamma$, $\chi_b(3P) \to \Upsilon(1S)\gamma$, $\chi_b(3P) \to \Upsilon(2S)\gamma$ and $\chi_b(3P) \to \Upsilon(3S)\gamma$ in bins of transverse momentum of the Υ mesons in the rapidity range $2.0 < y < 4.5$. LHCb also measured the fractions $\mathcal{R}^{\chi_b(mP)}_{\Upsilon(nS)}$ with the radiative photon reconstruted from a e^+e^- pair $\mathcal{R}^{\chi_b(mP)}_{\Upsilon(nS)}$ [4]. The results are consistent with those obtained in [1].

Using photons converted into electron pairs, and assuming the mass splitting $m_{\chi_{b2}(3P)} - m_{\chi_{b1}(3P)} = 10.5\,\mathrm{MeV}/c^2$, the mass of the $\chi_{b1}(3P)$ state is measured to be

$$m_{\chi_{b1}(3P)} = 10515.7^{+2.2}_{-3.9}\,(\mathrm{stat})^{+1.5}_{-2.1}\,(\mathrm{syst})\,\mathrm{MeV}/c^2,$$

This result [4] is compatible and significantly more precise than the event yield average mass of $\chi_{b1}(3P)$ and $\chi_{b2}(3P)$ states, $10\,530 \pm 5 \pm 17\,\mathrm{MeV}/c^2$ and $10\,551 \pm 14 \pm 17\,\mathrm{MeV}/c^2$, reported by the ATLAS [5] and D0 [6] experiments, respectively.

References

[1] LHCb collaboration, R. Aaij et al., *Study of χ_b meson production in pp collisions at $\sqrt{s} = 7$ and 8TeV and observation of the decay $\chi_b(3P) \to$* $\Upsilon(3S)\gamma$, Eur. Phys. J. C 74,3092 (2014).

[2] LHCb collaboration, R. Aaij et al., *Measurement of the fraction of $\Upsilon(1S)$ originating from $\chi_b(1P)$ decays in pp collisions at $\sqrt{s} = 7$ TeV*, JHEP 11 (2012) 031

[3] LHCb collaboration, *Observation of $\chi_b(3P)$ state at LHCb in pp collisions at $\sqrt{s} = 7$* TeV, LHCb-CONF-2012-020

[4] LHCb collaboration, R. Aaij et al., *Measurement of the $\chi_b(3P)$ mass and of the relative rate of $\chi_{b1}(1P)$ and $\chi_{b2}(1P)$ production*, JHEP 14, 88 (2014).

[5] ATLAS collaboration, G. Aad et al., *Observation of a new χ_b state in radiative transitions to $\Upsilon(1S)$ and $\Upsilon(2S)$ at ATLAS*, Phys. Rev. Lett. 108 (2012) 152001

[6] D0 collaboration, V. M. Abazov et al *Observation of a narrow mass state decaying into $\Upsilon(1S) + \gamma$ in $p\bar{p}$ collisions at $\sqrt{s} = 1.96$ TeV*, Phys. Rev. D86 (2012) 031103.

RECENT RESULTS FROM NA48/2 EXPERIMENT

Milena Misheva[a][b] on behalf of the NA48/2 Collaboration[c]
Joint Institute for Nuclear Research,
Joliot-Curie 6, 141980 Dubna, Moscow region, Russia

Abstract. The first observation of about 2000 candidates, with a background contamination below 3%, of the rare decay $K^{\pm} \to \pi^{\pm}\pi^{0}e^{-}e^{+}$ is reported. The preliminary branching ratio in the full kinematic region is obtained to be: $\mathrm{BR}(K^{\pm} \to \pi^{\pm}\pi^{0}e^{-}e^{+}) = (4.06 \pm 0.17) \cdot 10^{6}$ by analyzing the data collected in 2003. A sample of 1.69×10^{7} $\pi^{0} \to e^{-}e^{+}\gamma$ decay candidates with a negligible background contamination collected in 2003–2004 data is analyzed to search for the dark photon (A') production in the $\pi^{0} \to \gamma A'$ decay followed by the prompt $A' \to e^{-}e^{+}$ decay. No signal is observed, and exclusion region in the plane of the dark photon mass ($m_{A'}$) and mixing parameter ε^{2} is established.

1 The NA48/2 experiment

The NA48/2 experiment at the CERN SPS collected a large sample of charged kaon ($K^{\pm}$) decays during its 2003-04 data taking period. Two simultaneous $K^{\pm}$ beams were produced by 400GeV/c protons from the CERN/SPS impinging on a beryllium target. Opposite charge particles with a central momentum of (60±3)GeV/c were selected and focused about 200m downstream at the first spectrometer chamber. A magnetic spectrometer (a dipole magnet surrounded by two sets of drift chambers) measured themomentum (p) of charged decay products with a relative precision of 1% for 10 GeV/c momentum tracks. It was followed by a scintillator hodoscope consisting of two planes segmented into horizontal and vertical strips achieving a very good 150 ps time resolution. A liquid krypton calorimeter (LKr), 27 radiation length thick, was used to measure electromagnetic deposits (E) and to identify electrons through their E/p ratio (the energy and transverse position resolutions were 1% and ~1.5

---

[a]also INRNE, BAS, Sofia, Bulgaria

[b]E-mail: `dyulendarova.milena@gmail.com`

[c] **NA48/2 Collaboration:** G. Anzivino, R. Arcidiacono, W. Baldini, S. Balev, J.R. Batley, M. Behler, S. Bifani, C. Biino, A. Bizzeti, B. Bloch-Devaux, G. Bocquet, N. Cabibbo, M. Calvetti, N. Cartiglia, A. Ceccucci, P. Cenci, C. Cerri, C. Cheshkov, J.B. Chze, M. Clemencic, G. Collazuol, F. Costantini, A. Cotta Ramusino, D. Cow- ard, D. Cundy, A. Dabrowski, P. Dalpiaz, C. Damiani, M. De Beer, J. Derr, H. Dibon, L. DiLella, N. Doble, K. Eppard, V. Falaleev, R. Fantechi, M. Fidecaro, L. Fiorini, M. Fiorini, T. Fonseca Martin, P.L. Frabetti, L. Gatignon, E. Gersabeck, A. Gianoli, S. Giudici, A. Go- nidec, E. Goudzovski, S. Goy Lopez, M. Holder, P. Hristov, E. Iacopini, E. Imbergamo, M. Jeitler, G. Kalmus, V. Kekelidze, K. Kleinknecht, V. Kozhuharov, W. Kubischta, G. Lamanna, C. Lazzeroni, M. Lenti, L. Litov, D. Madigozhin, A. Maier, I. Mannelli, F. Marchetto, G. Marel, M. Markytan, P. Marouelli, M. Martini, L. Masetti, E. Mazzucato, A. Michetti, I. Mikulec, N. Molokanova, E. Monnier, U. Moosbrugger, C. Morales Morales, D.J. Munday, A. Nappi, G. Neuhofer, A. Norton, M. Patel, M. Pepe, A. Peters, F. Petrucci, M.C. Petrucci, B. Peyaud, M. Piccini, G. Pierazzini, I. Polenkevich, Yu. Potrebenikov, M. Raggi, B. Renk, P. Rubin, G. Ruggiero, M. Savri, M. Scarpa, M. Shieh, M.W. Slater, M. Sozzi, S. Stoynev, E. Swallow, M. Szleper, M. Valdata- Nappi, B. Vallage, M. Velasco, M. Veltri, S. Venditti, M. Wache, H. Wahl, A. Walker, R. Wanke, L. Widhalm, A. Winhart, R. Win- ston, M.D. Wood, S.A. Wotton, A. Zinchenko, M. Ziolkowski.

mm (resp.) for 10 GeV showers). Detailed description of the NA48/2 detector can be found in Ref. [1].

2 First observation of $K^{\pm} \to \pi^{\pm}\pi^{0}e^{+}e^{-}$ decay and preliminary branching ratio measurement

The $K^{\pm} \to \pi^{\pm}\pi^{0}e^{+}e^{-}$ decay (denoted as $K_{\pi\pi ee}$ below) proceeds by one virtual photon exchange which undergoes internal conversion into electron-positron pair, i.e. $K^{\pm} \to \pi^{\pm}\pi^{0}\gamma^{*} \to \pi^{+}\pi^{0}e^{+}e^{-}$. Two possible mechanisms of this rare kaon channel are possible: Inner Bremsstrahlung (IB), where the γ^{*} is emitted by one of the charged mesons and Direct Emission (DE), when γ^{*} is radiated off at the weak vertex of the intermediate state of the process. There are a few theoretical articles related to $K_{\pi\pi ee}$ decay channel where detailed studies of the various contributions to its amplitude have been performed [2], [3], [4]. The selection criteria of $K_{\pi\pi ee}$ candidates, described in Ref. [5], are applied to the data samples collected in 3-month NA48/2 run during 2003. The number of the $K_{\pi\pi ee}$ events in the signal region is 1916 where the overall background is below 3%. All background contributions are estimated by generation of GEANT3-based [7] MC (MC) simulations for each kaon channel.
The $K_{\pi\pi ee}$ branching ratio (BR) is determined as: $BR(s) = ((N_s - N_b)/N_n) \cdot (A_n \cdot \epsilon_n/A_s \cdot \epsilon_s) \cdot BR(n)$ where N_s is the number of the signal candidates (1916), N_b is the number of the estimated background events (56), A_s and ϵ_s are the geometrical acceptance and the trigger efficiency of the signal. The index "$_n$" is used to denote the corresponding notations which are valid for the normalisation channel, particularly in our case it is chosen the $K^{\pm} \to \pi^{\pm}\pi^{0}$ with a subsequent $\pi^{0} \to e^{-}e^{+}\gamma$ decay ($K_{2\pi_D}$) decay. The normalization mode branching ratio $BR(n) = (2.425 \pm 0.076) \cdot 10^{3}$ is obtained from the PDG [6] world average. Both the signal and the normalisation kaon modes are selected concurrently by using the same trigger logic. A common event reconstruction is considered as much as possible aiming partial cancellation of systematic effects such as particle identification and trigger inefficiencies. The preliminary result of the $K_{\pi\pi^{0}ee}$ branching ratio is obtained: $BR(K^{\pm} \to \pi^{\pm}\pi^{0}e^{+}e^{-}) = (4.06 \pm 0.10_{stat.} \pm 0.06_{syst.} \pm 0.13_{ext.}) \times 10^{-6}$,
where systematic errors include uncertainties on acceptance, particle identification, trigger efficiencies and radiative corrections. External error originates from the normalization mode BR uncertainties and it is the dominant error in the present result. All major issues about the analysis of the BR($K_{\pi\pi^{0}ee}$) measurement can be found in Ref. [5].

3 Search for dark photon in π^0 decays

The large sample of π^0 mesons produced and decaying in vacuum collected by the NA48/2 experiment allows a high sensitivity search for the dark photon (A'), a new gauge boson introduced in hidden sector new physics models with an extra U(1) gauge symmetry. The simplest DP model is characterised by two free parameters, the DP mass $m_{A'}$ and the mixing parameter with the standard model ε . Its possible production in the π^0 decay and subsequent decay proceed via the following chain: $\pi^0 \to \gamma A'$, $A' \to e^+e$, producing a final state with three charged particles and a photon.
The NA48/2 experiment provides pure $\pi^0 \to e^-e^+\gamma$ (π^0_D) decay samples through the reconstruction of $K^\pm \to \pi^\pm\pi^0$ ($K_{2\pi}$) and $K^\pm \to \pi^0\mu^\pm\nu$ ($K_{\mu 3}$). Event selection of $K_{2\pi}$ and $K_{\mu 3}$ decays followed by the prompt $\pi^0 \to \gamma A'$, $A' \to e^-e^+$ decay chain are identical up to the momentum, invariant mass and particle identification conditions. All details about the selection criteria and MC simulations of the above mentioned decay modes are described in Ref. [9]. The procedure of scanning for a DP signal in the mass range 9 MeV/$c^2 \leq m_{A'} < 120$ MeV/c^2 is given in Ref [9]. The resolution on m_{ee} as a function of m_{ee} evaluated with MC simulation is parametrized $\sigma_m(m_{ee}) = 0.067$ MeV/$c^2 + 0.0105 \cdot m_{ee}$, and varies from 0.016 MeV/c^2 to 1.33 MeV/c^2 over the mass range of the scan. The mass step of the scan and the half-width of the DP search window are defined depending on the value of A' mass as $\sigma_m(m_{A'})/2$ and $\delta m = 1.5\sigma_m(m_{A'})$, respectively (and both are rounded to the nearest multiple of 0.02 MeV/c^2). The lower boundary of the mass range is determined by the limited accuracy of the π^0_D background simulation at low e^-e^+ mass. At high DP mass approaching the upper limit of the mass range, the sensitivity to the mixing parameter ε^2 is not competitive with the existing limits due to the kinetic suppression of the $\pi^0 \to \gamma A'$ decay [9].
A search for DP production in the $\pi^0 \to \gamma A'$ decay followed by the prompt $A' \to e^-e^+$ decay has been performed using 2003-2004 data sample. No DP signal is observed. The obtained upper limits on ε^2 represent an improvement over the existing data in the DP mass range 9–70 MeV/c^2. Under the assumption that the DP couples to the Standard Model particles, the NA48/2 result excludes the DP as an explanation for the muon (g-2) anomaly in the range 10-1000MeV/c^2.

References

[1] NA48: V. Fanti *et al.*, Nucl. Instrum. Methods **A 574**, 433 (2007)
[2] L. Cappiello *et al.*, Eur. Phys. J. **C 72**, 1872 (2012); Erratum-ibid. **C 72**, 2208 (2012)
[3] H. Pichl, Eur. Phys. J. **C 20**, 371 (2001)
[4] G. R. Gevorkyan and M. H. Misheva, Eur. Phys. J. **C** 74, no. 5, 2860 (2014)
[5] M. Raggi on behalf of the NA48/2 Collaboration, *"NA48/2 studies of rare decays"*, arXiv:1508.01307v1[hep-ex] (to be published in the La Thuile 2015 proceedings)
[6] K.A. Olive et al. (Particle Data Group), Chin. Phys. **C**38, 090001 (2014).

[7] GEANT Detector Description and Simulation Tool, CERN Program Library W5013 (1994)
[8] B. Holdom, Phys. Lett. B166, 196 (1986).
[9] NA48/2: J. R. Batley *et al.*, Physics Letters **B** 746, pages 178-185 (2015)

Hadron Physics

PROTON STRUCTURE IN THE LHC ERA

Katerina Lipka[a] for PROSA Collaboration
Deutsches Elektronen Synchrotron (DESY) Hamburg, Germany

Abstract. The precise understanding of the proton structure is one of major problems of the modern physics at hadron colliders, which must be addressed by both, experiment and theory. Experimentally, the proton structure is determined by using the data from deep-inelastic scattering experiments. In this work, the additional impact of a verity of processes, measured in the proton-proton collision at the LHC on the precision of the valence and the gluon distribution and on the details of the quark-sea composition in the proton is investigated and quantified.

1 Introduction

Besides the experimental errors and the uncertainties of theory predictions, the lack of knowledge of the details of the proton structure is the key factor limiting the interpretation of the LHC measurements. The international network Proton Structure Analyses in hadronic collisions (PROSA) addresses this issue in a common effort by the theorists and experimentalists with the goal to improve the understanding of the proton structure, expressed by the parton distribution functions (PDFs) and enhance the discovery potential of the LHC. The PDFs depend on the scale of the process and the fraction x of the proton momentum, carried by the struck parton. The x-dependence of the PDFs has to be constrained from the experimental measurements, with the major data set being obtained with the HERA collider. These data are however not essential to constrain e.g. the gluon distribution in the low-x and the high-x ranges, in particular interesting for searches for new physics. Additional important constraints are provided by the LHC measurements. Production of W bosons, jets and heavy quark pairs is important in this context. Here, the potential of HERA and the LHC measurements to tighten constraints on PDFs and to improve precision of the strong coupling constant and heavy quark masses is illustrated. Among other results, the role of top-quark pair production measurements at the LHC for determination of the fundamental QCD parameters and for improving precision of the gluon distribution at high x is discussed. The impact of the heavy-quark production measurements at the LHCb on the gluon distribution at low x is presented. The open-source QCD analysis framework HERAFitter [1] is used for the analyses presented here.

2 Impact of different LHC processes on PDF constraints

The Drell-Yan production in proton-proton collisions probes different bi-linear combination of light quarks in proton. For example, measurement of a charge

[a]E-mail: `katerina.lipka@desy.de`

asymmetry in $W^{\pm}$ production at the LHC provides important information about the u and d valence quarks. Measurements of the charge asymmetry in the muon channel of $W^{\pm}$ decays at 7 TeV and 8 TeV [2, 3] are included in a QCD analysis at NLO together with the combined [4] inclusive cross section measurements in DIS at HERA. By including the LHC measurements, the precision of the valence quark distributions is significantly improved, as shown in Fig. 1. Associated production of the W-bosons with charm quarks probes the

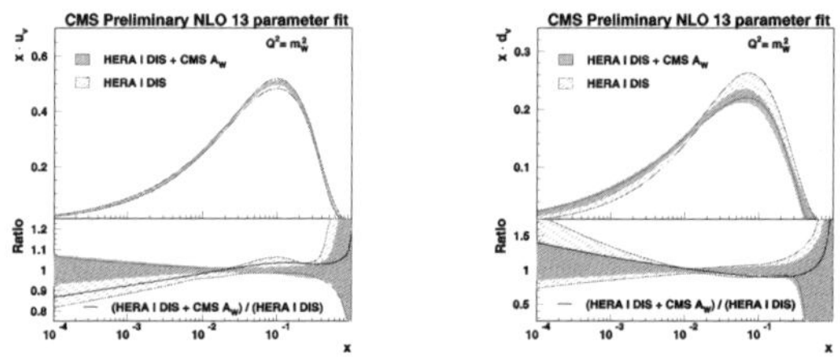

Figure 1: Distributions of u and d valence quarks as a function of x, shown at the scale of m_W and their relative uncertainties. The results obtained by using inclusive DIS measurements only (by adding LHC measurements) are shown by hatched (filled shaded) bands.

s-quark in the proton directly via the LO process $sg \to W^- c$. For the first time, the hadron collider data directly sensitive to the strange-quark distribution, are included into the QCD analysis [2] of the proton structure. Both ATLAS and CMS collaborations have performed similar analyses with the s-quark distribution being enhanced as determined by ATLAS, but consistent with earlier neutrino-experiment measurements as obtained by CMS. In a joined analysis [5] of ATLAS, CMS and neutrino-experiment data, the full consistency of the LHC measurements among each other was demonstrated and an improvement of precision on s-quark density in the proton is achieved, as demonstrated in Fig. 2. The processes in proton-proton collisions at the LHC, dominated by

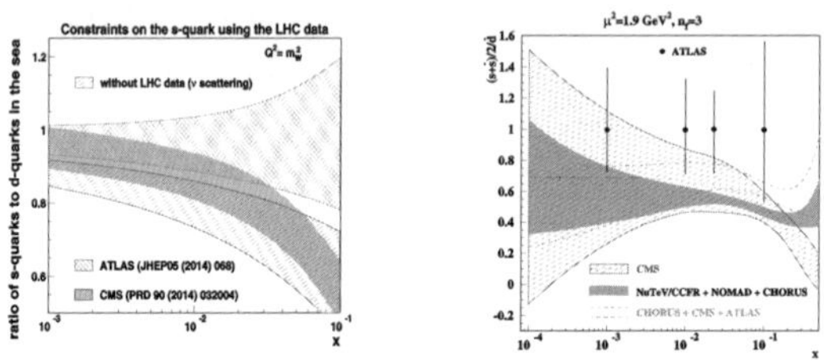

Figure 2: Distribution of the s-quark in the proton as a function of x, as obtained individually by ATLAS and CMS collaborations (left) and the result of the joined QCD analysis including the measurements at the LHC and in neutrino-scattering experiments (right).

gluon-gluon fusion, like production of (multi-) jets or top-quark pairs, probe the gluon distribution in the x- ranges, inaccessible by inclusive DIS measurements. The measurements of inclusive jet production at $\sqrt{s}$=7 TeV and 8 TeV

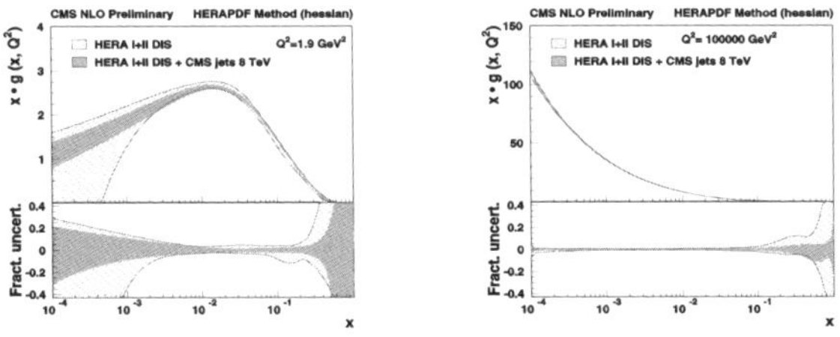

Figure 3: Distribution of the gluon and of its fractional uncertainty as a function of x, shown at low(left) and high (right) energy scales, as obtained by using the HERA data only (hatched bands) and by adding the CMS measurements of inclusive jet production at $\sqrt{s}$=8 TeV.

at the CMS experiment were included in the QCD analyses [6,7], together with the combined DIS cross-section measurements [4] and [8], respectively, and a significant improvement of the precision of the gluon distribution is observed, as illustrated in Fig. 3. Top-quark pair production in proton-proton collisions probes the gluon distribution, top-quark mass and the strong coupling α_S [9]. The impact of the top-quark pair production measurements at the LHC on the gluon distribution is investigated in details in [10], where for the first time the details of the phenomenology uncertainty on the prediction for differential cross sections of top-quark pair production are investigated. For this purpose, the novel calculation Difftop is developed. Therein, differential cross sections of heavy-quark pair production in proton- (anti)proton collisions are calculated in pQCD at approximate NNLO (α_s^4), by using methods of threshold resummation beyond the leading logarithmic accuracy. Predictions for the single-particle inclusive kinematics (top-quark p_T^t, y^t) provide effective description of the kinematic distributions capturing the main features of the full calculation, at the same time suitable for QCD analysis. In Fig. 4, the Difftop prediction is compared to the recent measurement by CMS experiment of the transverse momentum of the top-quark [11], where different contributions to the theory uncertainty are investigated. The dominant contribution to the full theory error arise from the uncertainty on the knowledge of parton distribution, in particular, of the gluon. By including the measurements of the top-quark pair production measurements at LHC and Tevatron into the QCD analysis at NNLO using Difftop for calculation of the top-quark kinematics, the constraint on the gluon distribution is improved in particular at high x, as shown in Fig. 5 (right), the kinematic range in particular important for searches for new heavy particles. The improvement of the precision of the gluon distribution is necessary also at very low x, a kinematic range in particular important for calculation of neutrino fluxes in the atmosphere. Forward production of c- and b- quark pairs at the LHCb experiment probes x down to $x \approx 10^{-5}$. For the first time, these data were included into a QCD analysis [12] and a significant improvement of the gluon distribution at low x is demonstrated, as

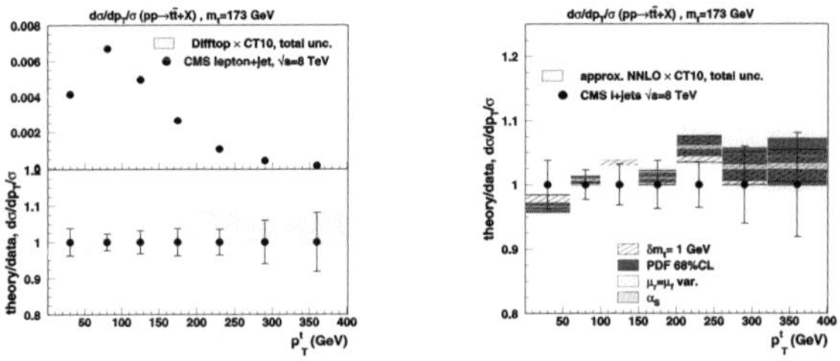

Figure 4: Normalised cross section of $t\bar{t}$ production measured by the CMS experiment at $\sqrt{s}$=8 TeV(closed symbols), compared to Difftop prediction with total theory uncertainty (left) and its breakdown into different contributions (right).

shown in Fig. 5 (left).

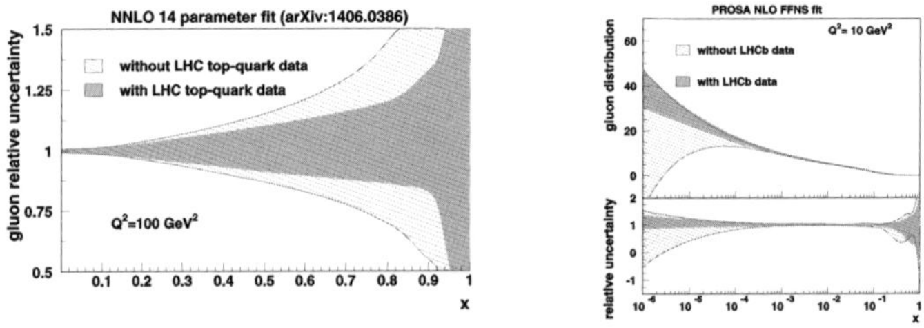

Figure 5: Uncertainty of $g(x)$, improved by using the differential cross sections of $t\bar{t}$ production (left) at the LHC. Distribution of $g(x)$ and of its uncertainty improved by using c and b-quark pairs production measurements at the LHC.

References

[1] S. Alekhin et al., Eur. Phys. J. **C75** (2015) 7, 304.
[2] [CMS Collaboration], Phys. Rev. **D90** (2014) 032004.
[3] [CMS Collaboration], CMS-PAS-SMP-14-022.
[4] [H1 and ZEUS Collaborations], JHEP 1001:109 (2010).
[5] S. Alekhin et al. Phys. Rev. **D91**, 094002 (2015).
[6] [CMS Collaboration], Eur. Phys. J. **C75** (2015) 288.
[7] [CMS Collaboration], CMS-PAS-SMP-14-001.
[8] [H1 and ZEUS Collaborations], arXiv:1506.06042.
[9] [CMS Collaboration], Phys. Lett. **B728** (2013) 496.

[10] M. Guzzi, K. Lipka, S.-O. Moch, JHEP 1501 (2015) 082.
[11] [CMS Collaboration], arXiv:1505.04480.
[12] O. Zenaiev et al. [PROSA Collaboration], Eur. Phys. J. **C75** (2015) 8, 396.

OVERVIEW OF RECENT ALICE RESULTS

Igor Lakomov (on behalf of the ALICE Collaboration) [a]
European Organization for Nuclear Research (CERN), CH-1211, Geneva 23, Switzerland

Abstract. ALICE is one of the four main experiments at the Large Hadron Collider. It is dedicated to the study of the characteristics of the hot medium produced at large energy densities in heavy-ion collisions, which is called Quark-Gluon Plasma. In addition, ALICE also participates in the pp and p–Pb collisions program at the LHC. The overview of the latest ALICE results is presented in these proceedings.

1 Introduction

ALICE is an experiment at the Large Hadron Collider (LHC) at CERN optimized for the study of ultra-relativistic heavy-ion collisions which create nuclear matter at extreme temperature and energy densities. At sufficiently high energy densities QCD predicts a transition from ordinary nuclear matter to a phase of de-confined quarks and gluons, called the Quark-Gluon Plasma (QGP). In order to quantify nuclear effects, the measurements in Pb–Pb collisions are compared to incoherent superpositions of pp collisions. Comparison to p–Pb collisions allow to disentangle hot (QGP-related) and cold nuclear matter (CNM) effects. In these proceedings an overview of the latest ALICE results from pp, p–Pb and Pb–Pb collision systems is given.

2 Jets

High-$p_{\rm T}$ partons are ideal probes of the medium produced in hard interactions at the early stage of heavy-ion collisions. They traverse the medium losing energy due to induced gluon radiation and elastic scattering (quenching). They eventually fragment into sprays of hadrons (jets) that can be observed experimentally. Jet quenching has been previously observed both at RHIC and the LHC. Jet production was found to be strongly modified in central heavy-ion collisions compared to the expectations from pp collisions. ALICE has measured the jet nuclear modification factor $R_{\rm AA}$ in Pb–Pb collisions for different centrality intervals [1] and $R_{\rm pPb}$ in p–Pb collisions integrated over centrality [2]. Both measurements are presented in Fig. 1. Given the absence of nuclear modifications in p–Pb collisions compared to pp collisions, it can be concluded that the strong jet suppression seen in central and semi-central Pb–Pb collisions can be fully attributed to hot nuclear matter effects while CNM effects do not affect jet production.

[a]E-mail: igor.lakomov@cern.ch

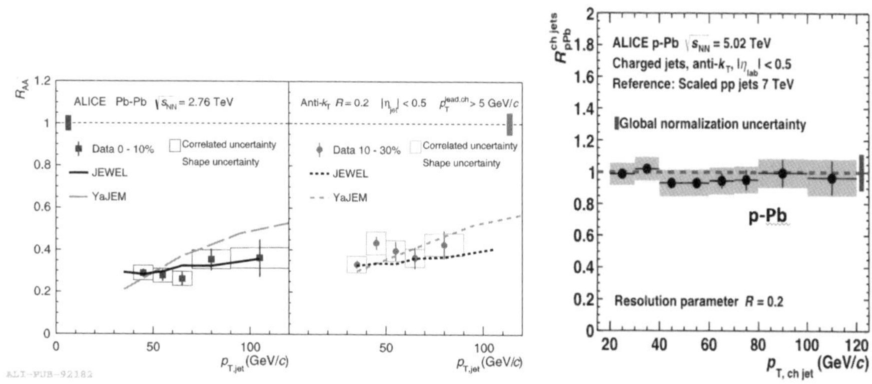

Figure 1: Nuclear modification factor of fully reconstructed jets, using the anti-k_T algorithm with $R = 0.2$, in Pb–Pb collisions at 2.76 TeV (left) and in p–Pb at 5.02 TeV. Jets are reconstructed in $|\eta| < 0.5$.

3 Two-particle correlations

Two-particle azimuthal correlations provide insight into the underlying mechanism of particle production in collisions of hadrons and nuclei at high energy. These studies involve the measurements of the differences in azimuthal angle, $\Delta\phi$, and in pseudorapidity, $\Delta\eta$, between two particles: trigger and associated particle. Similar long-range ridge structures ($\Delta\phi$ correlations elongated in $\Delta\eta$) have been measured in pp, p–Pb and Pb–Pb collisions by the LHC experiments. Since in Pb–Pb these patterns are attributed to hydrodynamic flow, their presence in small systems has raised great interest in the community. Further insight into the origin of these long-range correlation structures can be achieved by studying their η-dependence. ALICE has recently reported the measurement of the angular correlations between trigger particles (muons) in the pseudorapidity range $2.5 < \eta < 4$ and associated particles (charged particles) at mid-rapidity ($|\eta| < 1$) in p–Pb collisions at 5.02 TeV [3]. In order to suppress contributions from low-p_T jets the difference between correlations at high and low multiplicities is studied. From a Fourier decomposition of the difference distribution the second order coefficient for muons, $v_2^{\mu}\{2\text{PC,sub}\}$, is obtained. Figure 2 shows the $v_2^{\mu}\{2\text{PC, sub}\}$ for high-multiplicity events in the Pb-going (Pb–p) and p-going (p–Pb) directions (left) and their ratio (right). Results are compared to a multi-phase transport model (AMPT) [4]. A sizable muon v_2 is seen which is larger in the Pb-going direction. The AMPT model agrees with the p-going direction results at low p_T while at high p_T there is no agreement with the data. In the Pb-going direction the model agrees with the data only for one particular p_T interval while no agreement is seen for all the other intervals. The ratio presented in the right panel of Fig. 2 shows no p_T dependence within the quoted uncertainties. For $p_T > 2$ GeV/c the inclusive muon yield is dominated by heavy-flavour (HF) decays, and the data may support a finite value for the HF muons v_2, or a different composition of the parent distribution or their v_2 values in the AMPT compared to data.

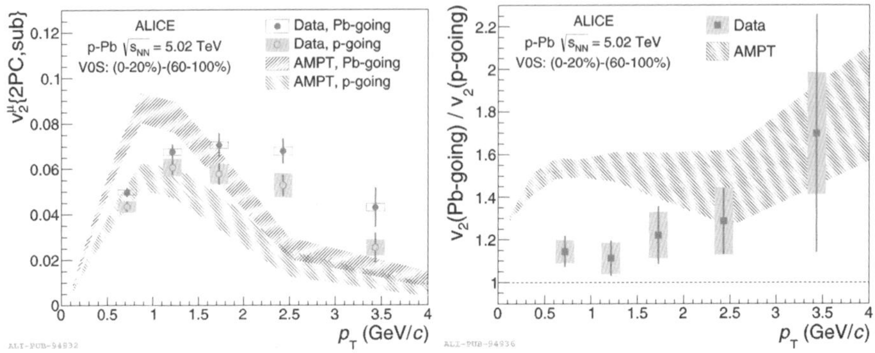

Figure 2: The $v_2^{\mu}\{2\text{PC,sub}\}$ coefficients from muon-tracklet correlations in p-going and Pb-going directions in p–Pb collisions (left) and their ratio (right). The data are compared to model calculations from AMPT. See text for details.

4 Quarkonium production

Production of quarkonia, bound state of quark and its own anti-quark pair, represents another powerful probe to study the properties of the QGP. It was predicted that at sufficiently high energy densities the color screening of the heavy-quarks potential in deconfined QCD matter will lead to a sequential suppression of the production of different quarkonium states [5]. Figure 3 shows one of the latest measurement of quarkonium production by the ALICE Collaboration. The left panel of the figure shows the nuclear modification factor R_{AA} of the J/ψ production in Pb–Pb collisions at 2.76 TeV as a function of the average number of participants, $\langle N_{part}\rangle$. The latter is related to the collision centrality: the higher $\langle N_{part}\rangle$ the more central is the collision. The measurement is performed in three p_T intervals. For $p_T > 0.3$ GeV/c, the data show a suppression of the J/ψ production in Pb–Pb collisions compared to pp collisions for all $\langle N_{part}\rangle$ which is expected to be a QGP effect, while for lower p_T, a significant excess of the J/ψ production is seen at $\langle N_{part}\rangle < 100$, in particular R_{AA} reaches up to 7 for the most peripheral collisions. Comparing the p_T distribution of the opposite-sign dimuons around the J/ψ mass for the most peripheral collisions (right panel of Fig. 3) with STARLIGHT calculations [6] for coherent photo-production of J/ψ, one can see a perfect agreement. It means that the observed excess of the low-p_T J/ψ production in peripheral events can be assigned to the photo-production effect.

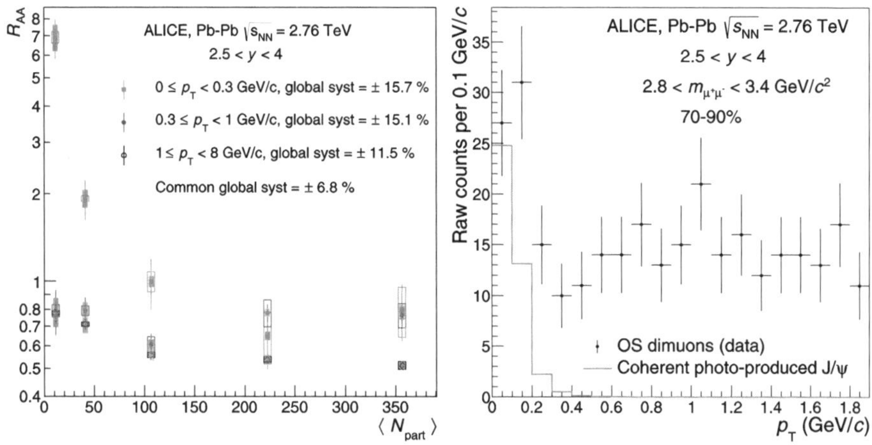

Figure 3: (Color online). Low-p_T excess of the J/ψ production in Pb–Pb collisions measured by ALICE. Left: the nuclear modification factor R_{AA} as a function of centrality for three p_T ranges. Right: the p_T-distribution of the opposite-sign dimuons for the most peripheral collisions. Red line represents the calculations from STARLIGHT Monte Carlo generator [6]. Both figures are taken from [7].

Quarkonium production has been also studied by ALICE in p–Pb collisions at 5.02 TeV. As mentioned above, such measurements allow to study the CNM effects in detail since no QGP-related effects are expected in these collisions. In addition to previously published measurements of J/ψ production in p–Pb collisions integrated over centrality [8, 9], ALICE recently reported the centrality dependence of the J/ψ production in p–Pb [10]. Figure 4 represents centrality dependence of the J/ψ nuclear modification factor Q_{pPb} measured by ALICE in three rapidity intervals: backward (left), mid- (central) and forward (right). Centrality dependence is presented as a function of the mean number of binary nucleon-nucleon collisions, $\langle N_{\text{coll}} \rangle$. Highest values of $\langle N_{\text{coll}} \rangle$ correspond to higher centrality. The J/ψ production is suppressed in p–Pb collisions compared to pp collisions in both mid- and forward-rapidity intervals, decreasing for higher $\langle N_{\text{coll}} \rangle$ at forward rapidity. At backward rapidities, Q_{pPb} is consistent with unity indicating zero CNM effects, with a hint of an excess of the J/ψ production at higher centrality. CNM models show a fair agreement with the data within the quoted uncertainties for all rapidity intervals.

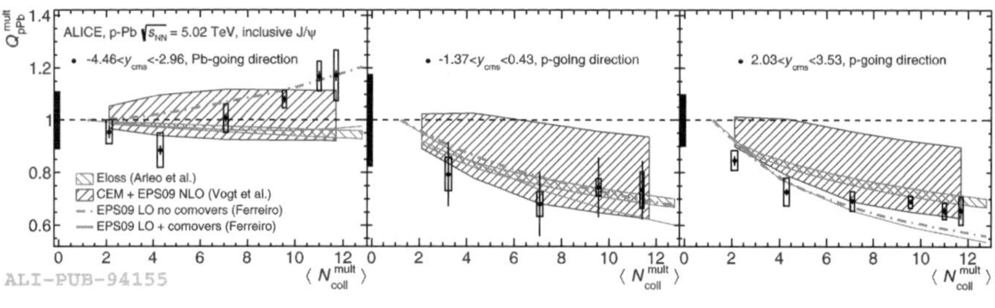

Figure 4: (Color online) Nuclear modification factor of J/ψ in p–Pb collisions as a function of centrality at backward (left), mid- (center) and forward (right) y intervals [10]. Bands are theoretical calculations from [11, 12, 13, 14].

5 Open charm production

The comparison of the HF hadron R_{AA} with that of pions allows the colour-charge dependence of parton energy loss to be tested. ALICE has performed such measurements for Pb–Pb collisions at 2.76 TeV [15]. They are shown in the left panel of Fig. 5 where R_{AA} of prompt D-mesons are compared to that of charged pions in the same p_{T} interval. The results for charged pions and for prompt D-mesons are compatible within uncertainties. A suppression is observed already for the centrality interval $50-80\%$, increasing towards higher centrality and reaching a factor of $5-6$ for the most central collisions. The right panel of Fig. 5 shows the comparison of the same measurement for the prompt D-mesons to the CMS results on the non-prompt J/ψ R_{AA}. The J/ψ suppression is smaller compared to that of D-mesons at all centralities. These results hint at a difference in the suppression of particles originating from c and b quarks (mass ordering). Finally, the results are compared to theoretical calculations in which in-medium parton energy loss increases with increasing colour charge factor and decreases with increasing quark mass [16]. The model provides a fair agreement with the data.

6 Summary

An overview of recent ALICE results is presented in these proceedings. ALICE measured jets, two-particle correlations, quarkonium and open-charm production in pp, p–Pb and Pb–Pb collisions. This overview is not exhaustive, many other ALICE measurements could not be covered in these proceedings, including for example an important measurement of the d-$\overline{d}$ and ^{3}He-$\overline{\mathrm{He}}$ mass-over-charge ratio difference performed by ALICE with an ultra-high precision confirming the CPT-invariance expectations [17].

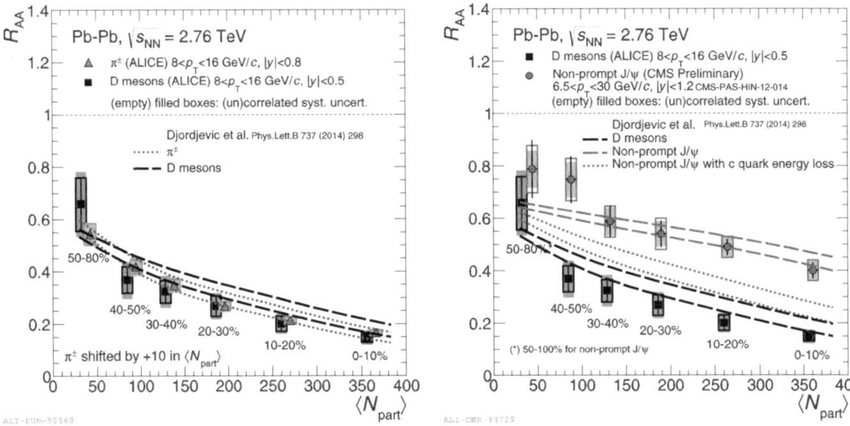

Figure 5: Left: Comparison of the D meson and pion $R_{\rm AA}$ measurements ($8 < p_{\rm T} < 16$ GeV/c) with the calculations by Djordjevic et al. including radiative and collisional energy loss. Lines of the same style enclose a band representing the theoretical uncertainty. Right: Comparison of the calculation by Djordjevic et al. with the D mesons in $8 < p_{\rm T} < 16$ GeV/c and preliminary non-prompt J/ψ mesons from CMS in $6.5 < p_{\rm T} < 30$ GeV/c. For the latter the model results for the case in which the b quark interactions are calculated using the c quark mass are shown as well.

References

[1] **ALICE** Collaboration, J. Adam *et al.*, "Measurement of jet suppression in central Pb-Pb collisions at $\sqrt{s_{\rm NN}} = 2.76$ TeV," *Phys. Lett.* **B746** (2015) 1–14, `arXiv:1502.01689 [nucl-ex]`.

[2] **ALICE** Collaboration, J. Adam *et al.*, "Measurement of charged jet production cross sections and nuclear modification in p-Pb collisions at $\sqrt{s_{\rm NN}} = 5.02$ TeV," *Phys. Lett.* **B749** (2015) 68–81, `arXiv:1503.00681 [nucl-ex]`.

[3] **ALICE** Collaboration, J. Adam *et al.*, "Forward-central two-particle correlations in p-Pb collisions at $\sqrt{s_{\rm NN}} = 5.02$ TeV," *Phys. Lett.* **B753** (2016) 126–139, `arXiv:1506.08032 [nucl-ex]`.

[4] Z.-W. Lin *et al.*, "A Multi-phase transport model for relativistic heavy ion collisions," *Phys. Rev.* **C72** (2005) 064901, `arXiv:nucl-th/0411110 [nucl-th]`.

[5] T. Matsui and H. Satz, "J/ψ Suppression by Quark-Gluon Plasma Formation," *Phys.Lett.* **B178** (1986) 416.

[6] STARLIGHT website. Http://starlight.hepforge.org/, 2013.

[7] **ALICE** Collaboration, J. Adam *et al.*, "Measurement of an excess in the yield of J/ψ at very low $p_{\rm T}$ in Pb-Pb collisions at $\sqrt{s_{\rm NN}} = 2.76$ TeV," arXiv:1509.08802 [nucl-ex].

[8] **ALICE** Collaboration, B. Abelev *et al.*, "J/ψ production and nuclear effects in p-Pb collisions at $\sqrt{s_{\rm NN}} = 5.02$ TeV," *JHEP* **1402** (2014) 073, arXiv:1308.6726 [nucl-ex].

[9] **ALICE** Collaboration, J. Adam *et al.*, "Rapidity and transverse-momentum dependence of the inclusive J/ψ nuclear modification factor in p-Pb collisions at $\sqrt{s_{\rm NN}} = 5.02$ TeV," *JHEP* **1506** (2015) 055, arXiv:1503.07179 [nucl-ex].

[10] **ALICE** Collaboration, J. Adam *et al.*, "Centrality dependence of inclusive J/ψ production in p-Pb collisions at $\sqrt{s_{\rm NN}} = 5.02$ TeV," *JHEP* **11** (2015) 127, arXiv:1506.08808 [nucl-ex].

[11] R. Vogt, "Cold Nuclear Matter Effects on J/ψ and Υ Production at the LHC," *Phys.Rev.* **C81** (2010) 044903, arXiv:1003.3497 [hep-ph].

[12] J. Albacete *et al.*, "Predictions for p-Pb Collisions at $\sqrt{s_{\rm NN}} = 5$ TeV," *Int.J.Mod.Phys.* **E22** (2013) 1330007, arXiv:1301.3395 [hep-ph].

[13] E. G. Ferreiro, "Excited charmonium suppression in proton-nucleus collisions as a consequence of comovers," *Phys. Lett.* **B749** (2015) 98–103, arXiv:1411.0549 [hep-ph].

[14] F. Arleo *et al.*, "Centrality and $p_{\rm T}$ dependence of J/ψ suppression in proton-nucleus collisions from parton energy loss," *JHEP* **1305** (2013) 155, arXiv:1304.0901 [hep-ph].

[15] **ALICE** Collaboration, J. Adam *et al.*, "Centrality dependence of high-p_T D meson suppression in Pb-Pb collisions at $\sqrt{s_{\rm NN}} = 2.76$ TeV," *JHEP* **11** (2015) 205, arXiv:1506.06604 [nucl-ex].

[16] M. Djordjevic *et al.*, "RHIC and LHC jet suppression in non-central collisions," *Phys. Lett.* **B737** (2014) 298–302, arXiv:1405.4250 [nucl-th].

[17] **ALICE** Collaboration, J. Adam *et al.*, "Precision measurement of the mass difference between light nuclei and anti-nuclei," *Nature Phys.* **11** no. 10, (2015) 811–814, arXiv:1508.03986 [nucl-ex].

HADRONIC CONTRIBUTIONS TO ELECTROWEAK OBSERVABLES IN THE FRAMEWORK OF DPT

A.V. Nesterenko [a]
BLTPh JINR, Dubna, 141980, Russian Federation

Abstract. The hadronic contributions to the muon anomalous magnetic moment and to the shift of the electromagnetic fine structure constant at the scale of Z boson mass are evaluated within dispersively improved perturbation theory (DPT). The latter merges the corresponding perturbative input with intrinsically nonperturbative constraints, which originate in the respective kinematic restrictions. The obtained results conform with recent assessments of the quantities on hand.

Certain nonperturbative information on the low–energy hadron dynamics is embodied within dispersion relations. In particular, the latter render the kinematic restrictions on the relevant physical processes into the mathematical form and impose nonperturbative constraints on such quantities as the hadronic vacuum polarization function $\Pi(q^2)$, R–ratio of $e^+e^- \to$ hadrons, and the Adler function $D(Q^2)$. These constraints have been merged with corresponding perturbative input within dispersive approach to QCD [1–3], which provides the unified integral representations for the functions on hand:

$$\Delta\Pi(q^2,\, q_0^2) = \Delta\Pi^{(0)}(q^2,\, q_0^2) + \int_{m^2}^{\infty} \rho(\sigma) \ln\left(\frac{\sigma - q^2}{\sigma - q_0^2}\frac{m^2 - q_0^2}{m^2 - q^2}\right)\frac{d\,\sigma}{\sigma}, \tag{1}$$

$$R(s) = R^{(0)}(s) + \theta(s - m^2)\int_{s}^{\infty} \rho(\sigma)\frac{d\,\sigma}{\sigma}, \tag{2}$$

$$D(Q^2) = D^{(0)}(Q^2) + \frac{Q^2}{Q^2 + m^2}\int_{m^2}^{\infty} \rho(\sigma)\frac{\sigma - m^2}{\sigma + Q^2}\frac{d\,\sigma}{\sigma}. \tag{3}$$

Here $\Delta\Pi(q^2,\, q_0^2) = \Pi(q^2) - \Pi(q_0^2)$, $m^2 = 4m_\pi^2$, $\theta(x)$ is the unit step–function [$\theta(x) = 1$ if $x \ge 0$ and $\theta(x) = 0$ otherwise], the leading–order terms read

$$\Delta\Pi^{(0)}(q^2,\, q_0^2) = 2(\varphi - \tan\varphi)/\tan^3\varphi - 2(\varphi_0 - \tan\varphi_0)/\tan^3\varphi_0, \tag{4}$$

$$R^{(0)}(s) = \theta(s - m^2)\big[1 - (m^2/s)\big]^{3/2}, \tag{5}$$

$$D^{(0)}(Q^2) = 1 + 3\big[1 - \sqrt{1+\xi^{-1}}\,\sinh^{-1}\big(\xi^{1/2}\big)\big]\xi^{-1}, \tag{6}$$

$\sin^2\varphi = q^2/m^2$, $\sin^2\varphi_0 = q_0^2/m^2$, $\xi = Q^2/m^2$, see Refs. [1–3] for the details. The perturbative part of the spectral density $\rho(\sigma)$ entering Eqs. (1)–(3) can be expressed in terms of the strong correction to the Adler function

$$\rho_{\mathrm{pert}}(\sigma) = \lim_{\varepsilon \to 0_+}\big[d_{\mathrm{pert}}(-\sigma - i\varepsilon) - d_{\mathrm{pert}}(-\sigma + i\varepsilon)\big]/(2\pi i), \tag{7}$$

as well as in terms of the strong corrections to the functions $\Pi(q^2)$ and $R(s)$. The integral representations (1)–(3) contain no unphysical singularities and

[a]E-mail: nesterav@theor.jinr.ru, nesterav@gmail.com

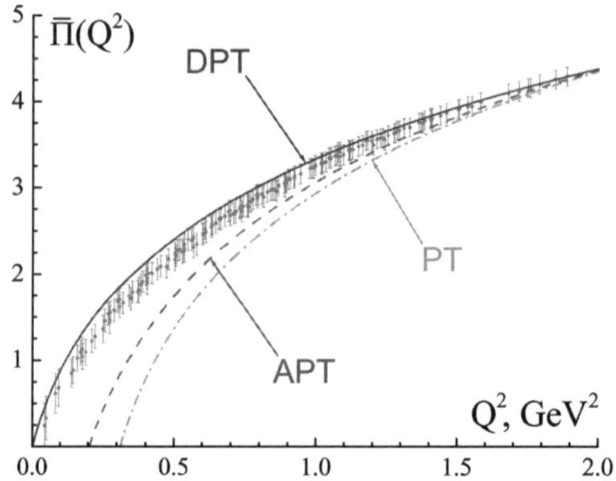

Figure 1: Hadronic vacuum polarization function in the framework of various approaches: DPT expression (solid curve), APT prediction (dashed curve), perturbative approximation (dot–dashed curve), and lattice data [13] (circles).

substantially extend the range of applicability of QCD perturbation theory towards the infrared domain. In particular, the dispersive approach to QCD enables one to describe experimental data [4] on inclusive τ lepton hadronic decay in vector and axial–vector channels in a self–consistent way [2,5,6]. The representations (1)–(3) conform with the results of Bethe–Salpeter calculations [7] as well as of lattice simulations [8]. The Adler function (3) agrees with its experimental prediction in the entire energy range [1,9,10].

The integral representations (1)–(3) along with respective perturbative input (7) constitute the "dispersively improved perturbation theory" (DPT) expressions for the functions on hand. At the one–loop level the spectral function (7) assumes a simple form, specifically, $\rho^{(1)}_{\rm pert}(\sigma) = (4/\beta_0)[\ln^2(\sigma/\Lambda^2) + \pi^2]^{-1}$ ($\beta_0 = 11 - 2n_{\rm f}/3$, $n_{\rm f}$ is the number of active flavors, and Λ is the QCD scale parameter), whereas at the higher loop levels Eq. (7) becomes rather cumbersome, see Refs. [11,12] for the details.

It is worth noting that in the massless limit ($m = 0$) for the case of perturbative spectral function (7) Eqs. (2) and (3) become identical to those of the "analytic perturbation theory" (APT) [14]. However, the massless limit ignores some of the nonperturbative constraints, which relevant dispersion relations impose on the functions on hand, that appears to be substantial for the studies of hadron dynamics at low energies, see Refs. [1–3,6,10].

For practical purposes it is convenient to deal with the subtracted at zero form of Eq. (1), namely

$$\bar{\Pi}(Q^2) = \Delta\Pi(0,-Q^2) = \Delta\Pi^{(0)}(0,-Q^2) + \int_{m^2}^{\infty} \rho(\sigma) \ln\left(\frac{1+Q^2/m^2}{1+Q^2/\sigma}\right)\frac{d\,\sigma}{\sigma}. \quad (8)$$

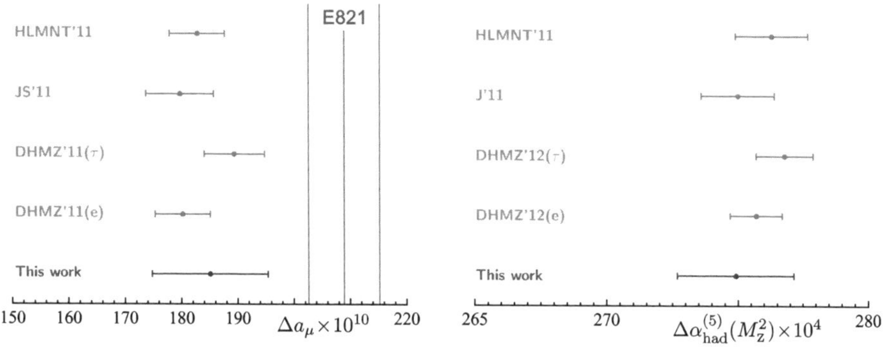

Figure 2: Left–hand plot: theoretical evaluations (circles) and experimental measurement (shaded band) of the subtracted muon anomalous magnetic moment ($\Delta a_\mu = a_\mu - a_0$, $a_0 = 11659 \times 10^{-7}$). Right–hand plot: theoretical evaluations of hadronic contribution to the shift of electromagnetic fine structure constant at the scale of Z boson mass.

As one can infer from Fig. 1, the DPT expression for the hadronic vacuum polarization function (8) contains no unphysical singularities and proves to be in a good agreement with lattice data [13]. The presented result corresponds to the four–loop level, $\Lambda = 419\,\text{MeV}$, and $n_f = 2$. Figure 1 also displays the one–loop Eq. (1) in the massless limit (which corresponds to APT) as well as the one–loop perturbative approximation of $\Pi(q^2)$. However, the latter is inapplicable at low energies due to unphysical singularities, whereas the APT prediction for $\Pi(q^2)$ diverges at $Q^2 \to 0$, that invalidates it in the infrared domain, too.

The DPT expression for the hadronic vacuum polarization function (8) is applicable in the entire energy range, that enables one to perform the assessment of the hadronic contributions to electroweak observables without involving experimental data on R–ratio. In particular, the four–loop DPT prediction of the value of the leading–order hadronic contribution to the muon anomalous magnetic moment [3]

$$a_\mu^{\text{HLO}} = \frac{1}{3}\left(\frac{\alpha}{\pi}\right)^2 \int_0^1 (1-x)\bar{\Pi}\left(m_\mu^2 \frac{x^2}{1-x}\right) dx = (696.1 \pm 9.5) \times 10^{-10}, \qquad (9)$$

appears to be in a good agreement with its recent estimations [15–17]. The complete muon anomalous magnetic moment includes the QED contribution, the electroweak contribution, as well as the higher–order and light–by–light hadronic contributions, that, together with a_μ^{HLO} (9) yields $a_\mu = (11659185.1 \pm 10.3) \times 10^{-10}$. The obtained a_μ corresponds to two standard deviations from the experimental measurement $a_\mu^{\text{exp}} = (11659208.9 \pm 6.3) \times 10^{-10}$ and, as one can infer from Fig. 2, conforms with its recent evaluations [15–17], see Ref. [3] for the details.

Another quantity of an apparent interest is the hadronic contribution to the electromagnetic running coupling

$$\Delta\alpha_{\text{had}}(q^2) = -\frac{\alpha}{3\pi}\, q^2\, \mathcal{P}\!\int_{m^2}^{\infty} \frac{R(s)}{s-q^2}\frac{d\,s}{s}. \tag{10}$$

The four–loop DPT prediction for the five–flavor hadronic contribution to the shift of the electromagnetic fine structure constant at the scale of Z boson mass

$$\Delta\alpha^{(5)}_{\text{had}}(M_{\text{Z}}^2) = (274.9 \pm 2.2) \times 10^{-4} \tag{11}$$

agrees with its recent evaluations [15, 17, 18], see Fig. 2. In turn, Eq. (11) together with leptonic and top quark contributions leads to $\alpha_{\text{em}}^{-1}(M_{\text{Z}}^2) = 128.962 \pm 0.030$, that also conforms with recent assessments of the quantity on hand [15, 17, 18], see Ref. [3] for the details.

References

[1] A.V. Nesterenko and J. Papavassiliou, J. Phys. G **32**, 1025 (2006).
[2] A.V. Nesterenko, Phys. Rev. D **88**, 056009 (2013).
[3] A.V. Nesterenko, J. Phys. G **42**, 085004 (2015); arXiv:1509.00197[hep-ph].
[4] D. Boito *et al.*, Phys. Rev. D **84**, 113006 (2011); **91**, 034003 (2015); M. Davier *et al.*, Eur. Phys. J. C **74**, 2803 (2014).
[5] A.V. Nesterenko, Nucl. Phys. B (Proc. Suppl.) **234**, 199 (2013); **258**, 177 (2015); SLAC eConf C1106064, 23 (2011); arXiv:1110.3415 [hep-ph]; AIP Conf. Proc. (in press); arXiv:1508.03705 [hep-ph].
[6] A.V. Nesterenko, PoS (Confinement X), 350 (2013); 1401.0620 [hep-ph].
[7] M. Baldicchi *et al.*, Phys. Rev. Lett. **99**, 242001 (2007); Phys. Rev. D **77**, 034013 (2008).
[8] B. Blossier *et al.*, Phys. Rev. D **85**, 034503 (2012); **89**, 014507 (2014); Nucl. Phys. B (Proc. Suppl.) **234**, 217 (2013).
[9] A.V. Nesterenko, SLAC eConf C0706044, 25 (2007).
[10] A.V. Nesterenko, Nucl. Phys. B (Proc. Suppl.) **186**, 207 (2009).
[11] A.V. Nesterenko, Int. J. Mod. Phys. A **18**, 5475 (2003); Nucl. Phys. B (Proc. Suppl.) **133**, 59 (2004).
[12] A.V. Nesterenko, C. Simolo, Comp.Phys.Comm. **181**, 1769 (2010); **182**, 2303 (2011); A.P. Bakulev, **184**, 183 (2013); G. Cvetic, **190**, 182 (2015).
[13] M. Morte, B. Jaeger, A. Juttner, and H. Wittig, JHEP **1203**, 055 (2012).
[14] D.V. Shirkov and I.L. Solovtsov, Phys. Rev. Lett. **79**, 1209 (1997); Theor. Math. Phys. **150**, 132 (2007).
[15] K. Hagiwara, R. Liao, A.D. Martin, D. Nomura, and T. Teubner, J. Phys. G **38**, 085003 (2011).

[16] F. Jegerlehner and R. Szafron, Eur. Phys. J. C **71**, 1632 (2011).
[17] M. Davier, A. Hocker, B. Malaescu, and Z. Zhang, Eur. Phys. J. C **71**, 1515 (2011); **72**, 1874(E) (2012).
[18] F. Jegerlehner, Nuovo Cim. C **034S1**, 31 (2011).

QCD MEASUREMENTS OF VECTOR-BOSON PRODUCTION WITH THE ATLAS DETECTOR

Nataliia Kondrashova[a] on behalf of the ATLAS Collaboration
DESY, Notkestrasse 85, 22607, Hamburg, Germany.

Abstract. Measurements of W and Z bosons production at the Large Hadron Collider with the ATLAS detector are presented. These measurements include total and differential cross-sections of inclusive vector-boson production as well as in association with jets. The results show sensitivity to the parton distribution functions, are used to test predictions from Quantum Chromodynamics and to tune Monte Carlo predictions.

1 Introduction

The measurement of vector-boson production at hadron colliders provides a benchmark of understanding the Quantum Chromodynamics (QCD) and electroweak processes and can be used as an input to constrain parton distribution function (PDF) uncertainties. The presented measurements include LHC Run I results based on the datasets recorded by the ATLAS [1] detector in 2010 and 2011 at the centre-of-mass energy of 7 TeV and early Run II results based on the dataset recorded in 2015 at the centre-of-mass energy of 13 TeV.

2 Results on vector-boson production measurements

The differential cross-section of the low mass Drell-Yan production is measured as a function of the dilepton invariant mass M_{ll} [2]. The measurement is performed in the range of $26 < M_{ll} < 66$ GeV in both the electron (e) and muon (μ) channels for the 2011 data. For the 2010 data, the dilepton mass range is extended to $12 < M_{ll} < 66$ GeV for the μ channel. The measured cross-section is compared to pure next-to-leading order (NLO), next-to-next-to-leading order (NNLO) calculations and NLO calculations matched to leading-logarithm (LL) resummed parton shower (PS) calculations. In the low M_{ll} region the NNLO QCD predictions provide a significantly better description of the results compared with the NLO QCD predictions, unless the latter are matched to the PS calculations.

The measurement of the Z boson transverse momentum p_T^Z [3] is sensitive to the initial state radiation and the intrinsic transverse momentum of the initial partons at low p_T^Z and perturbative QCD (pQCD) at high p_T^Z. The differential cross-section of the Z/γ^* boson production is measured near the Z pole as a function of p_T^Z and compared to pure NNLO QCD calculations, to NNLO QCD calculations in combination with soft-gluon resummation at next-to-next-to-leading-logarithmic accuracy (NNLL), as well as to the predictions from

[a]E-mail: nataliia.kondrashova@desy.de

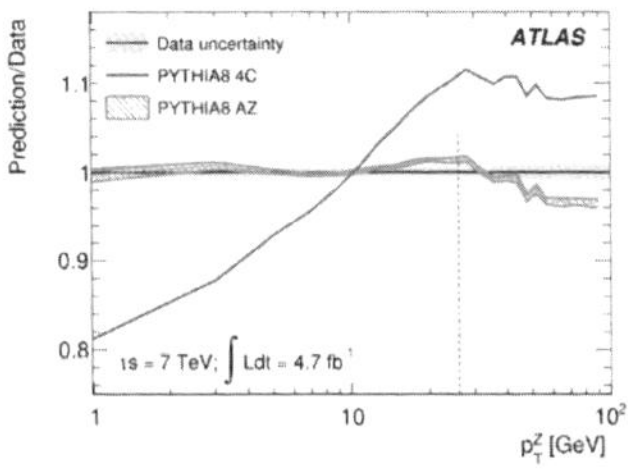

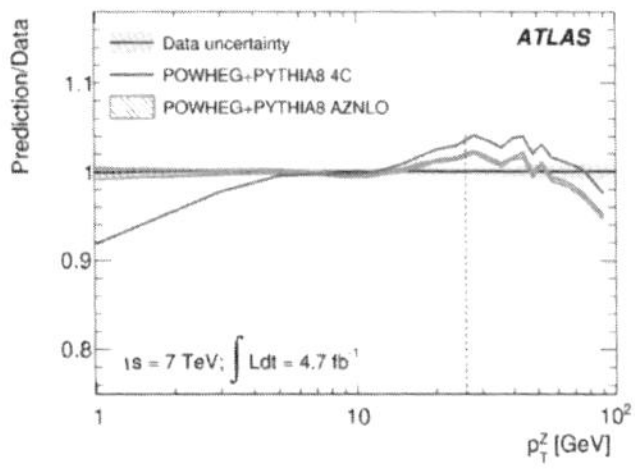

Figure 1: Comparison of tuned predictions to the muon-channel p_T^Z differential cross section [3]. The AZ tune for PYTHIA8 (left) and the AZNLO tune for POWHEG+PYTHIA8 (right) are used. The predictions with the base PYTHIA8 tune 4C are shown for the comparison. The vertical dashed lines show the upper limit of the tuning range.

different Monte Carlo (MC) generators. Disagreement between data and pure NNLO calculations is observed at low p_T^Z due to the lack of resummation. The new tunes AZ and AZNLO for MC generators PYTHIA8 [9] and POWHEG [8] showered with PYTHIA8, correspondingly, are obtained using the new data by varying the PS model components. Tuned predictions agree with the measured cross-section within 2 % over the tuning range, as shown in Figure 1.

The measurement of the production of vector bosons in association with jets provides an excellent test of pQCD calculations. Differential cross-sections of the production of a W boson in association with jets are studied for about 40 observables [4], including jet multiplicities for events with up to seven associated jets and jet transverse momentum of up to 1 TeV. Comparisons are made to a variety of theoretical predictions, including NLO pQCD calculations, resummation calculations and predictions from MC generators. The cross-section ratio of W+jet and Z+jet [5] is measured with significantly higher precision due to the cancellation of many systematic effects. In addition, this measurement probes the difference between kinematics of the jet system recoiling against W or Z bosons, in the context of searches for new particles or interactions beyond the Standard Model. Differential cross-sections are studied for about 20 observables and compared to predictions from NLO pQCD calculations and MC generators. Predictions agree well with data in most regions.

The production of Z boson in association with b-jets is an important background to ZH process with H decaying to $b\bar{b}$. In addition, Z production with at least one b-jet ($Z+\geq 1$ b-jet) allows to probe poorly known b-quark PDF. The cross-sections of $Z+\geq 1$ b-jet and $Z+\geq 2$ b-jets processes are measured differentially as a function of different observables [6]. The level of agreement between data and NLO predictions is different in the $Z+\geq 1$ b-jet and $Z+\geq 2$ b-jets cases. The $Z+\geq 1$ b-jet process is better described by the five-flavour number scheme prediction, which allows a b-quark density in the initial state, while the $Z+\geq 2$ b-jets is better described by the four-flavour number scheme prediction, which considers only parton density of gluons and first four quarks.

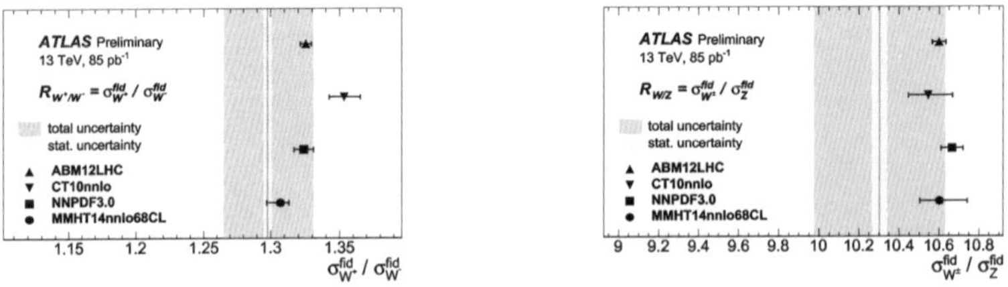

Figure 2: Ratio of W^+ to W^- (left) and ratio of $W^\pm$ to Z (right) production combined fiducial cross-sections compared to theory predictions based on different PDF sets [7].

The cross-sections of W and Z bosons production are measured in both e and μ channels using an early Run II dataset with the integrated luminosity of 85 pb^{-1} [7]. The obtained results are in a good agreement with theoretical calculations based on NNLO QCD. The measurements of cross-section ratios benefit from the cancellation of experimental uncertainties and are powerful tools to constrain PDF uncertainties. The cross-section ratio of W^+ to W^- is sensitive to the $u_v - d_v$ valence-quark distribution at low Bjorken-x, and the cross-section ratio of $W^\pm$ to Z constraints the strange quark distribution. The ratios of the fiducial cross-sections are shown in Figure 2. For the ratio of the W^+ to W^- fiducial cross-sections a significant scatter is observed corresponding to the different PDF predictions. Better agreement is seen for PDF sets that include LHC measurements from Run I. The measurement of the ratio of $W^\pm$ to Z agrees with all considered PDF predictions within uncertainties.

3 Summary

Measurements at hadron colliders of inclusive vector-boson production as well as in association with jets provide an excellent test of pQCD calculations and allow to constrain relevant PDFs. An overview of measurements based on ATLAS Run I dataset is presented, as well as early measurements of W and Z bosons production cross-sections using Run II data.

References

[1] ATLAS Collaboration, JINST **3** (2008) S08003.
[2] ATLAS Collaboration, JHEP **06** (2014) 112.
[3] ATLAS Collaboration, JHEP **09** (2014) 145.
[4] ATLAS Collaboration, Eur. Phys. J. C **75** (2015) 82.
[5] ATLAS Collaboration, Eur. Phys. J. C **74** (2014) 3168.
[6] ATLAS Collaboration, JHEP **10** (2014) 141.

[7] ATLAS Collaboration, ATLAS-CONF-2015-039 (2015).
[8] S. Frixione, P. Nason and C. Oleari, JHEP **0711** (2007) 070
[9] T. Sjostrand, S. Mrenna and P. Z. Skands, Comput. Phys. Commun. **178** (2008) 852

ANISOTROPY IN pPb AND PbPb COLLISIONS FROM CMS

Jovan Milosevic on behalf of the CMS Collaboration [a]
University of Belgrade and Vinca Institute of Nuclear Sciences, Belgrade, Serbia

Abstract. The recent CMS results on anisotropic particle emission will be presented. A multi-particle nature of the long-range correlations observed in pPb collisions is revealed through consistency between the results obtained using four-, six and eight-particle correlation as well as the Lee-Yang zero method. The magnitude of the elliptic and triangular flow of strange particles from both pPb and PbPb collisions have been extracted by correlating an identified strange hadron (K_S^0 or $\Lambda/\bar{\Lambda}$) with a charged particle separated by a large relative pseudorapidity. The results for strange, K_S^0 and $\Lambda/\bar{\Lambda}$, particles scaled by the number of constituent quarks plotted as a function of transverse kinetic energy per number of constituent quarks are in a rather good mutual agreement for both v_2 and v_3 over a wide range of particle transverse kinetic energy and event multiplicities. The initial-state fluctuations induce that the event-plane angle is not any more a global quantity but depends on both, transverse momenta (p_T) and pseudorapidity (η), which further induces the factorization breaking of the two-particle azimuthal anisotropy into a product of single-particle anisotropies. In the p_T direction, maximal effect of factorization breaking of about 20% is observed in ultra-central PbPb collisions. In the η direction, the effect is weakest for mid-central PbPb events and gets larger for more central or peripheral PbPb collisions as well as for high multiplicity pPb collisions. The experimental results are compared with recent hydrodynamic predictions which involve the factorization breakdown effect. The effect is sensitive to the initial-state conditions rather than the shear viscosity of the medium.

1 Introduction

The methods of the two-, four-, six- and eight-particle correlations [1] as well as the Lee-Yang Zero method (LYZ) [2,3] could be used to extract the magnitude of the azimuthal anisotropy which is characterized by Fourier coefficients, v_n. The long-range ($|\Delta\eta| > 2$) correlation known as the ridge [4], observed in high-multiplicity pPb collisions using the two-particle correlations, gave a hint that such a structure could have a hydrodynamic origin. In order to give a strong evidence of the multi-particle nature of the observed long-range effect, the correlations among four or more charged particles from pPb collisions at $\sqrt{s_{NN}}$ = 5.02 TeV are performed [5]. Beside charged particles, two-particle angular correlations are formed also between an identified strange hadron (K_S^0 or $\Lambda/\bar{\Lambda}$) and a charged particle [6] separated by a large relative pseudorapidity ($|\Delta\eta| >$ 2). These correlations revealed similar ridge structures. The extracted Fourier coefficients, v_2 and v_3, scaled to the number of constituent quarks showed that found hydrodynamic behavior happens on the partonic level. It is shown that even in the case when the hydrodynamic flow is the only source of the long-range correlations, initial-state fluctuations makes the event plane angle dependent

[a] E-mail: Jovan.Milosevic@cern.ch

on both, p_T and η. This then leads to factorization breaking of the two-particle azimuthal anisotropy into a product of single-particle anisotropies [7–9].

2 The CMS experiment and data used

A super-conducting solenoid surrounds the tracker detector of the CMS experiment [10]. Produced magnetic field of 3.8 T enabled precise measurements of p_T above 0.3 GeV/c. The data at the LHC energies of $\sqrt{s_{NN}} = 2.76$ TeV and 5.02 TeV in PbPb and pPb collisions with integrated luminosities of 160 μb^{-1} and 35 nb^{-1}, respectively, have been collected. The CMS detector has a wide pseudorapidity coverage ($|\eta| < 2.5$) which is excellently suited for studying the long-range correlations.

3 Results

3.1 Collectivity in pPb collisions

The magnitude of the elliptic flow, v_2, is measured using the two-, four-, six- and eight-particle correlation as well as using the LYZ method in both PbPb and pPb collisions [5]. In Fig. 1 is shown multiplicity dependence of the v_2 coefficient, averaged over $0.3 < p_T < 3.0$ GeV/c range and within $|\eta| < 2.4$.

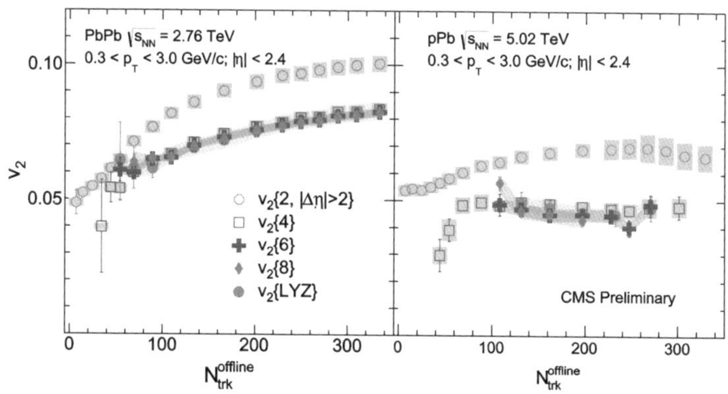

Figure 1: The v_2 values as a function of multiplicity obtained from two-, four-, six- and eight-particle cumulants, and LYZ methods in PbPb at $\sqrt{s_{NN}} = 2.76$ TeV (left) and pPb at $\sqrt{s_{NN}} = 5.02$ TeV (right) collisions. Statistical and systematic uncertainties are indicated by the error bars and shaded regions, respectively.

In order to suppress jet-related non-flow effects, the elliptic flow magnitudes from two-particle correlations, $v_2\{2\}$, are obtained applying the $|\Delta\eta| > 2$ cut. The obtained $v_2\{2\}$ results are consistently above the results obtained from the multi-particle correlation. Such feature appeared due to the event-by-event participant geometry fluctuations of the v_2 coefficient which affect the two- and multi-particle correlations differently. Also, the v_2 results from higher order cumulants and from LYZ method are in mutual agreement within 2% and 10% for PbPb and pPb collisions respectively. This mutual agreement between the v_2 results in high-multiplicity pPb collisions means that measured v_2 magnitude does not depend on the number of particles used in its reconstruction. This feature could be used as a strong evidence to support interpretation of the long-range correlation as a collective phenomenon not only in PbPb, but also in small systems formed in pPb collisions.

3.2 Collective flow of strange particles

Detailed description of the reconstruction technique for K_S^0 and $\Lambda/\bar{\Lambda}$ particles can be found in [6]. The magnitude of the single-particle flow coefficients, v_2 and v_3, is extracted from the two-dimensional $\Delta\phi - \Delta\eta$ distributions formed by correlating reconstructed strange particles candidates with charged particles. In order to remove short-range correlations such as jet fragmentation, a Fourier decomposition of the corresponding $\Delta\phi$ projection is applied after averaging over $|\Delta\eta| > 2$ region. The v_2 results for K_S^0 and $\Lambda/\bar{\Lambda}$ particles emitted from

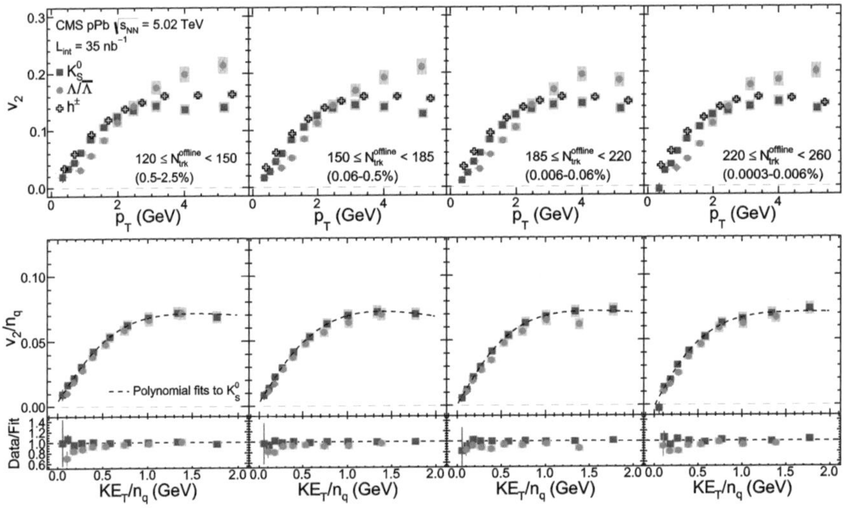

Figure 2: Top: the v_2 for K_S^0 (filled squares), $\Lambda/\bar{\Lambda}$ (filled circles), and charged particles (open crosses) as a function of p_T for four multiplicity ranges in high-multiplicity pPb events at $\sqrt{s_{NN}} = 5.02$ TeV. Middle: the v_2/n_q ratios for K_S^0 (filled squares) and $\Lambda/\bar{\Lambda}$ (filled circles) particles as a function of transverse kinetic energy per quark KE_T/n_q. Bottom: ratios of v_2/n_q for K_S^0 and $\Lambda/\bar{\Lambda}$ particles to a smooth fit function of v_2/n_q for K_S^0 particles *vs* KE_T/n_q. The error bars correspond to statistical uncertainties, while the shaded areas denote the systematic uncertainties.

high-multiplicity pPb collisions are shown in Fig. 2. The corresponding peripheral PbPb results, obtained within the same multiplicity ranges as in the case of pPb collisions, are shown in Fig. 3. The top raw of Fig. 2 and Fig. 3 shows the v_2 in function of p_T for four high-multiplicity bins in pPb collisions and for four peripheral PbPb bins respectively. In both figures are also shown v_2

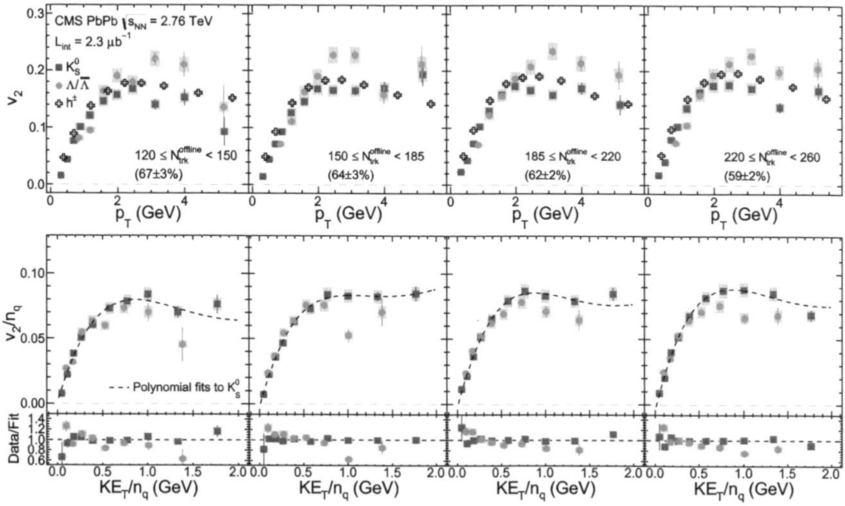

Figure 3: Top: the v_2 for K_S^0 (filled squares), $\Lambda/\bar{\Lambda}$ (filled circles), and charged particles (open crosses) as a function of p_T for peripheral PbPb events at $\sqrt{s_{NN}} = 2.76$ TeV. Middle: the v_2/n_q ratios for K_S^0 (filled squares) and $\Lambda/\bar{\Lambda}$ (filled circles) particles as a function of transverse kinetic energy per quark KE_T/n_q. Bottom: ratios of v_2/n_q for K_S^0 and $\Lambda/\bar{\Lambda}$ particles to a smooth fit function of v_2/n_q for K_S^0 particles *vs* KE_T/n_q. The error bars correspond to statistical uncertainties, while the shaded areas denote the systematic uncertainties.

results for charged particles. Huge majority of them are pions. The v_2 results exhibit a mass ordering effect: lighter particle species have a stronger azimuthal anisotropy at the given small p_T with respect to the heavier kind of particles. The effect is a bit less pronounced in PbPb collisions with respect to pPb ones. At high p_T appear two branches where the baryonic flow is greater than the mesonic v_2. In order to deeper investigate the seen effect, in the middle raw of Fig. 2 and Fig. 3 is shown the v_2 magnitude divided with the number of constituent quarks, n_q, and plotted as a function of transverse kinetic energy per quark, KE_T/n_q, with $KE_T = \sqrt{m^2 + p_T^2} - m$. This scaling makes that the v_2 of the shown particle species collapse onto a unique distribution. The bottom raw of Fig. 2 depicts the ratio between the data and a polynomial fit to the K_S^0 data which shows that the scaling is valid to better than 10% (25%) over most of the KE_T/n_q range in pPb (PbPb) collisions. For p_T values below 0.2 GeV/c the scaling breaks. The fact that the scaling behavior mainly holds could be related to the quark recombination model and suggests that the collective flow happens on the partonic level.

Beside the elliptic flow, the triangular flow, v_3, is also measured for strange K_S^0 and $\Lambda/\bar{\Lambda}$ and charged particles in both colliding systems. Their magnitudes are depicted in Fig. 4. Due to limited statistics, the triangular flow has been analyzed within a wide multiplicity range ($185 \leq N_{trk}^{offline} < 350$) only. As in the case of the elliptic flow, a mass ordering effect is observed also for the triangular flow in both colliding systems. The distributions of the magnitude of the triangular flow scaled to the number of the constituent quarks for K_S^0 and $\Lambda/\bar{\Lambda}$ particles are in a mutual agreement better than 20% over the whole KE_T/n_q range. According to our knowledge, no calculations on the triangular flow scaled to the n_q has been performed in parton recombination models.

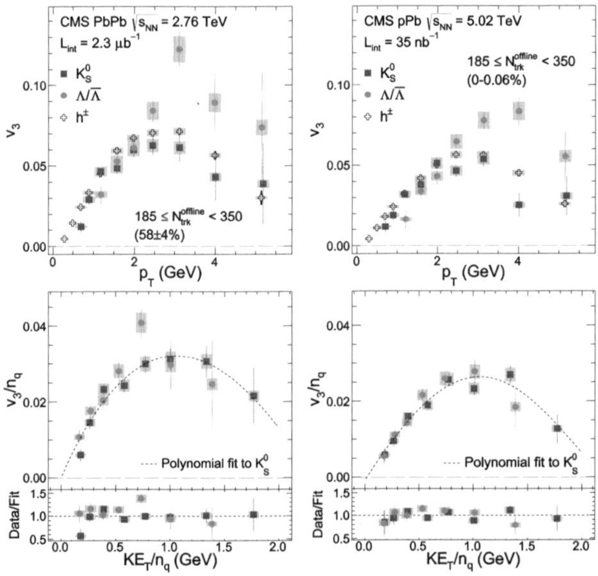

Figure 4: Top: the v_3 for K_S^0 (filled squares), $\Lambda/\bar{\Lambda}$ (filled circles), and charged particles (open crosses) as a function of p_T for the multiplicity range $185 \leq N_{trk}^{offline} < 350$ in PbPb collisions at $\sqrt{s_{NN}} = 2.76$ TeV (left) and in pPb collisions at $\sqrt{s_{NN}} = 5.02$ TeV (right). Middle: the n_q-scaled v_3 values of K_S^0 and $\Lambda/\bar{\Lambda}$ particles as a function of KE_T/n_q. Bottom: Ratios of v_n/n_q to a smooth fit function of v_n/n_q for K_S^0 *vs* KE_T/n_q. The error bars correspond to statistical, while the shaded areas to systematic uncertainties.

3.3 Initial-state fluctuations and factorization breaking

Due to the event-by-event fluctuations in positions of nucleons which form the colliding nuclei, higher-order Fourier harmonics (n $\geq$ 3), measured with respect to their corresponding global event plane angles, Ψ_n, appear. Recently, it was discovered [7,8] that not only magnitudes of Fourier harmonics, v_n, but also Ψ_n could depends on p_T due to initial-state fluctuations even if the hydrodynamic flow is the only source of the correlations. It has an important consequence: the factorization of the two-particle azimuthal anisotropy, $V_{n\Delta}(p_T^{(a)}, p_T^{(b)})$, into a product of single-particle anisotropies, $v_n(p_T^{(a)}) \times v_n(p_T^{(b)})$, does not hold any more precisely. A new observable, r_n, is introduced. It is formed from the two-particle Fourier coefficients

$$r_n(p_T^{(a)}, p_T^{(b)}) \equiv \frac{V_{n\Delta}(p_T^{(a)}, p_T^{(b)})}{\sqrt{V_{n\Delta}(p_T^{(a)}, p_T^{(a)}) V_{n\Delta}(p_T^{(b)}, p_T^{(b)})}}, \tag{1}$$

and it is approximately equal to $\cos[n(\Psi_n(p_T^{(a)}) - \Psi_n(p_T^{(b)}))]$ (more details in [8,9]). If the factorization breaks (holds), then the r_n achieves value smaller (equal) than 1. If the r_n has the value greater than 1 then it means that there are unremoved non-flow effects. A most pronounced effect has been found in ultra-central PbPb collisions at $\sqrt{s_{NN}}$ = 2.76 TeV [11]. In the left (right) panel of Fig. 5 are shown p_T- (η-)dependent magnitudes of the factorization

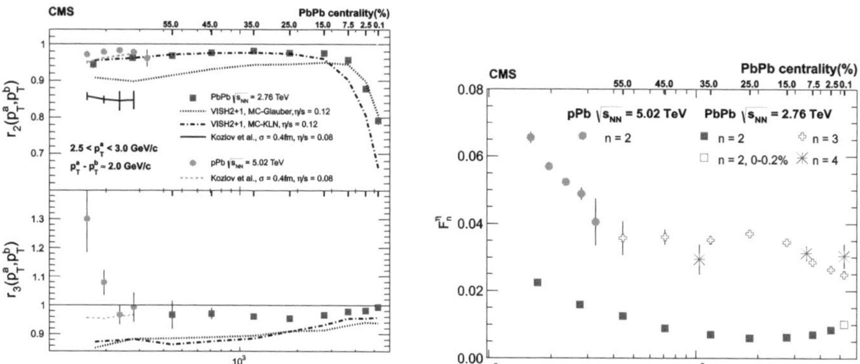

Figure 5: Left: The p_T-dependent r_2 and r_3 results as a function of multiplicity in pPb and PbPb collisions. The curves show the calculations for PbPb collisions from viscous hydrodynamics in Ref. [8] with MC-Glauber and MC-KLN initial condition models, and also hydrodynamic predictions for PbPb and pPb data in Ref. [7]. Right: The η-dependent F_n^η parameter as defined in Eq. (12) in [9] as a function of multiplicity in PbPb collisions for n = 2, 3 and 4 and pPb collisions for n = 2. The error bars correspond to statistical uncertainties, while systematic uncertainties are negligible.

breaking effect over a wide multiplicity range. The centrality axis, drawn at the top of the figures, is applicable only to the PbPb collisions. In the case of the 2nd harmonic, the smallest effect is observed for semi-central collisions, both for p_T- and η- dependencies. Going to more central or more peripheral collisions the size of the effect increases. At the same multiplicity, the size of the effect in the p_T- dependence is rather similar in the two colliding systems, while in the η- dependence the effect is much stronger in pPb than in PbPb collisions. In PbPb collisions, the size of the p_T-dependent r_3 is small and nearly independent of centrality, while in pPb collisions the r_3 approaches to 1 and then goes significantly above 1 at lower multiplicities. Oppositely to the p_T-, the η-dependence of the effect for the 3rd harmonic is stronger than for the 2nd one. The p_T-dependent data are qualitatively described by viscous hydrodynamic models [7,8] with fluctuating initial-state conditions, while they are mainly insensitive to the shear viscosity to entropy density ratio [9]. This promises of using the factorization data to disentangle contributions of the initial-state conditions and the mediums transport properties to the collective flow in the final state. The η-dependent factorization data show the influence of the initial-state fluctuations along the longitudinal direction. This could improve the three-dimensional modeling of the evolution of the strongly-coupled quark gluon medium.

Acknowledgments

We wish to acknowledge the support by the Ministry of Education, Science and Technological Development of the Republic of Serbia through the project 171019 which financed this contribution to the 17-th Lomonosov conference on particle physics. Also, we thank to our colleagues in the CERN accelerator departments for the excellent performance of the LHC and to other CMS institutes for their contributions in collecting the CMS data. In addition, we are thankful to the computing centers and personnel of the Worldwide LHC Computing Grid for the use of the computing infrastructure which was essential in the above presented analyses.

References

[1] N. Borghini, P. M. Dinh, and J.-Y. Ollitrault, Phys. Rev **C 64**, 054901 (2001).
[2] R. S. Bhalerao, N. Borghini, and J.-Y. Ollitrault, Nucl. Phys. **A 727**, 373 (2003).
[3] N. Borghini, R. S. Bhalerao, and J.-Y. Ollitrault, J. Phys. **G 30**, S1213 (2004).
[4] CMS Collaboration, Phys. Lett. **B 718**, 795 (2013).
[5] CMS Collaboration, Phys. Rev. Lett. **115**, 012301 (2015).
[6] CMS Collaboration, Phys. Lett. **B 742**, 200 (2015).

[7] F. G. Gardim *et al.*, Phys. Rev. **C 87**, 031901 (2013).
[8] U. Heinz *et al.*, Phys. Rev. **C 87**, 034913 (2013).
[9] CMS Collaboration, Phys. Rev. **C 92**, 034911 (2015).
[10] CMS Collaboration, JINST **3**, S08004 (2008).
[11] CMS Collaboration, JHEP **1402**, 088 (2014).

RECENT RESULTS on HARD DIFFRACTION at HERA

Irina A. Korzhavina [a] (for the H1 and ZEUS Collaborations)
Skobeltsyn Institute of Nuclear Physics,
Moscow State University, 119991 Moscow, Russia

Abstract. Recently measured differential cross sections of diffractive processes with dijets and prompt photons in the final state of *ep* collisions at the HERA collider are reviewed. Results are compared to QCD calculations based on different models of diffraction.

1 Introduction

The HERA collider experiments H1 and ZEUS recently presented new results on diffractive processes with dijets and prompt photons in the final state of *ep* collisions at $\sqrt{s} = 318\ GeV$ [1–4]. The measurements were performed in the range of photon (γ^*) virtualities running from the photoproduction region at $Q^2 < 1.0\ GeV^2$ to the deep inelastic scattering (DIS) region with $Q^2 \gtrsim 2.0\ GeV^2$. Diffractive events were identified either by a large rapidity gap (defind with pseudorapidity η_{max}) in the final state (LRG method) or by tagging the leading scattered proton (LP method). Both detectors were designed for measurements in a wide range of the kinematic variables of diffractive events such as $x_{I\!P}$, the fraction of the proton momentum carried by the Pomeron ($I\!P$) and fractions of the photon, x_γ, and of the Pomeron, $z_{I\!P}$, momenta, participating in the hard process. Hadronic and photon jets were reconstructed with the k_T algorithm [5]. Hard diffraction data were compared to pQCD calculations with the Resolved-Pomeron [6] and the Two-Gluon-Exchange [7] models of the diffraction mechanism.

2 Dijet Production in Diffractive *ep* Collisions

New H1 measurements (LRG and LP methods) [1, 2] of diffractive dijet cross sections in DIS and in photoproduction and also the ZEUS earlier measurements (LRG method) [8] were compared to the Resolved-Pomeron model [6]. The measurement ranges of both experiments mostly overlap, but they are not identical. Good agreement within experimental and theoretical uncertainties in both shape and normalization between measured distributions (Fig.1,a,b) and pQCD calculations based on HERA DPDFs [8–10] supports QCD factorization validity for diffractive dijet processes in DIS. In diffractive dijet photoproduction H1 and ZEUS observations are different. While the ZEUS cross sections are in agreement with the calculations, the new H1 distributions, as well as H1 earlier ones, are systematically below the calculations with a factor of $\sim 0.55 \pm 0.21$. The H1 data alone indicate factorisation violation for diffractive

[a] E-mail: irina@sinp.msu.ru, irina@mail.desy.de

dijet photoproduction. New proof of this indication was obtained with a quotient $\mathcal{R} = (DATA/NLO)_{\gamma^* p}/(DATA/NLO)_{DIS} = 0.51 \pm 0.09(exp) \pm 0.02(th)$ (double ratio of data to NLO) [2] where the errors are somewhat reduced due to the cancellation of some experimental systematic and theoretical uncertainties.

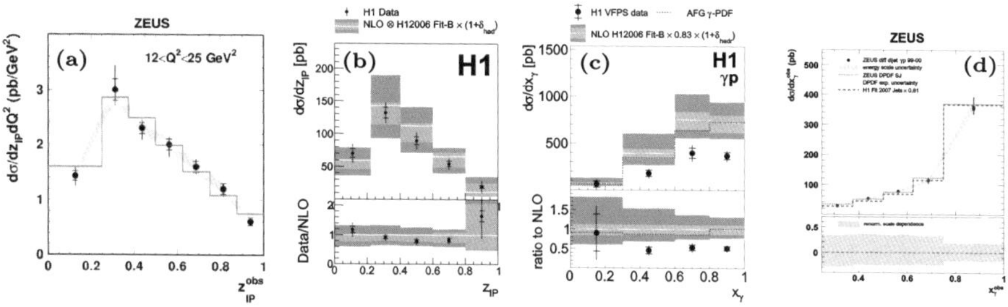

Figure 1: Differential cross-sections in $z_{I\!P}$ (a,b) for diffractive DIS production and in x_{γ^*} (c,d) for diffractive photoproduction of dijets compared to NLO QCD calculations with HERA DPDFs [8–10].

3 Exclusive Dijet Production in Diffractive DIS

The exclusive production of dijets, in diffractive deep inelastic $e^{\pm}p$ scattering was studied with the ZEUS detector using an integrated luminosity of 372 pb^{-1} for the first time [3]. The diffractive DIS range was restricted by $Q^2 > 25\ GeV^2$, $90 < W < 250\ GeV$, $x_{I\!P} < 0.01$, $\eta_{max} < 2$ and $M_X > 5\ GeV$(diffractive mass). Jets with $p_T^{jet} > 2\ GeV$ were selected in $ep \to e + jet_1 + jet_2 + p$ reaction using the k_T algorithm. The differential cross sections as a function of $\beta = x/x_{I\!P}$, where x is the Bjorken variable, and ϕ which is the angle between the jet and the lepton planes (details in Fig.2 of [3]) are shown in Fig.2. As is seen from Fig.2, both models, the Resolved-Pomeron model [6] (red histograms) based on DPDFs H1 2006 fits A and B [9] and the Two-Gluon-Exchange [7] model (blue histograms) with the GRV gluon density [11] of the proton, underestimate the data. As predicted by both models, the shapes of the measured ϕ distributions are well reproduced by function $(1 + A \cdot cos2\phi)$. As β increases, the parameter A flips sign (Fig.2, lower row) in accord with the Two-Gluon-Exchange model. Such behaviour is inconsistent with an almost constant, positive value of A in the whole β range as expected with the Resolved-Pomeron model.

4 Diffractive Photoproduction of Isolated Photons

Differential cross sections of isolated photon diffractive photoproduction were measured with the ZEUS detector at HERA collider using an integrated luminosity of 374 pb^{-1} inclusively and with at least one accompanying jet [4]. The measurement region was defined by $Q^2 < 1\ GeV^2$, $0.2 < y < 0.7$ and $x_{I\!P} < 0.03$, $\eta_{max} < 2.5$. Photon and hadronic jets were identified in the

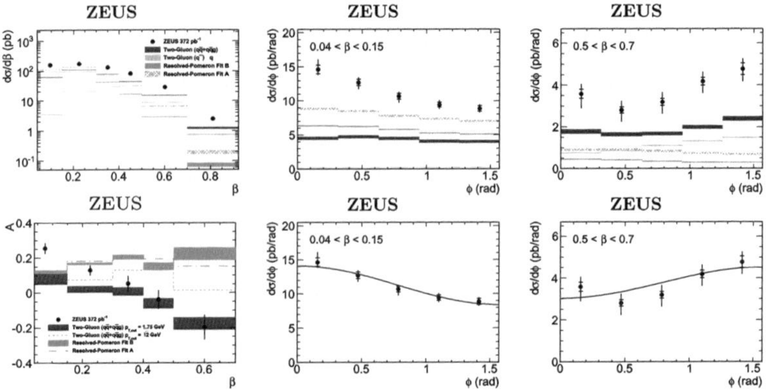

Figure 2: Differential cross-sections in β and ϕ for the diffractive DIS production of exclusive dijets compared to Resolved-Pomeron (red histograms), Two-Gluon-Exchange models (blue histograms) and function $(1 + A \cdot cos2\phi)$ (black lines).

ranges $5 < E_T^{\gamma} < 15\ GeV$, $-0.7 < \eta^{\gamma} < 0.9$ and $4 < E_T^{jet} < 35\ GeV$, $-1.5 < \eta^{jet} < 1.8$. Most photons ($\sim 80\%$) were found to be accompanied by a jet. A proton excitation admixture ($\sim 16\%$) was not subtracted. The shapes of the measured cross sections (Fig.3) are reasonably described by RAPGAP (with DPDF-p H1-2006-B [9], PDF-γ SAS-G 1D LO [12]) predictions normalized to data, except for a peak at $z_{I\!P} \sim 1$ (Fig.3f). In diffractive data, direct $\gamma^* p$ processes appeared to be strongly dominated over the resolved ones (Fig.3e): $\mathcal{R}_{diff} = direct/resolved \sim 4/1$ (in non-diffractive case: $\mathcal{R}_{non-diff} \sim 1.17/1$ [13]).

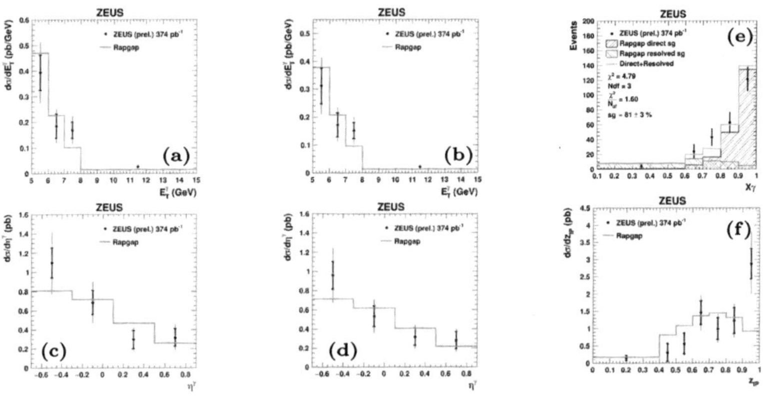

Figure 3: Differential cross sections of diffractive photoproduction of isolated photonin E_T^{γ}, η^{γ} without (a,b) and in E_T^{γ}, η^{γ}, $z_{I\!P}$, x_{γ} with (c,d,e,f) an accompanying jet compared to RAPGAP MC model.

5 Summary

The HERA collider experiments recently presented new results on diffractive processes with dijets and prompt photon in final states of ep collisions. The NLO pQCD calculations with HERA DPDFs agree with the new H1 data on diffractive dijet cross sections in DIS, whereas in photoproduction they overestimate the new H1 measurements, as well as H1 earlier, independent ones, indicating possible factorisation violation in these processes. However, earlier ZEUS diffractive dijet data are compatible with NLO pQCD calculations in photoproduction and DIS, suggesting that the QCD factorisation holds in both ranges. The first measured cross sections of diffractive production of exclusive dijets in DIS from the ZEUS detector are underestimated by QCD based models of diffraction. The Two-Gluon-Exchange model reproduces the shape of the ϕ distributions whereas the Resolved-Pomeron model does not. The diffractive photoproduction of isolated photons with and without an accompanying jet has been measured with the ZEUS detector. The RAPGAP model reasonably describes the shape of differential cross sections in the transverse energy and pseudorapidity of the photon and of the jet. A prominent peak near $z_{I\!P}^{meas} = 1$ is not described by this model and requires further study.

Acknowledgments

I would like to thank H1 and ZEUS colleagues for their efforts to produce physics results presented at the conference. I thank also the spokesman of ZEUS collaboration, Matthew Wing, for giving me an opportunity to report these results here.

References

[1] H1 Coll., V. Andreev et al., JHEP **03** (2015) 092.
[2] H1 Coll., V. Andreev et al., JHEP **05** (2015) 056.
[3] ZEUS Coll., H. Abramowicz et al., Eur. Phys. J. **C 76** (2016) 1.
[4] ZEUS Coll., ZEUS-prel-15-001.
[5] S. Catani, Y. L. Dokshitzer, M. H. Seymour, B. R. Webber, Nucl. Phys. **B 406** (1993) 187.
[6] G. Ingelman and P. E. Schlein, Phys. Lett. **B 152** (1985) 256.
[7] J. Bartels et al., Phys. Lett. **B 386** (1996) 389 and refs. therein V.M. Braun and D. Yu. Ivanov, Phys. Rev. **D 72** (2005) 034016.
[8] ZEUS Coll., S. Chekanov et al., Nucl. Phys. **B 831** (2010) 1.
[9] H1 Coll., A. Aktas et al., Eur. Phys. J. **C 48** (2006) 715.

[10] H1 Coll., A. Aktas et al., JHEP **10** (2007) 042.
[11] M. Glück, E. Reya and A. Vogt, Z. Phys. **C 67** (1995) 433.
[12] G. A. Schuler and T. Sjöstrand, Phys. Lett. **B 376** (1996) 193.
[13] ZEUS Coll., H. Abramowicz et al., JHEP **08** (2014) 023.

New Developments in Quantum Field Theory

60 YEARS TO NONABELIAN GAUGE FIELDS

A.A. Slavnov [a]
Steklov Mathematical Institute and Moscow State University, Moscow, Russia

Abstract. The development of nonabelian gauge fields for last 60 years is reviewed. The new method of quantization of gauge fields applicable beyond perturbation theory is proposed.

1 Introduction

Most people, presented here, hardly realize the situation which existed in high energy physics at 50-60-ties.

The Standard model and Weinberg- Salam model did not exist and even the question what are the particle which transfer the interaction was not clear. Classical gauge fields, introduced by Yang and Mills, gave the answer to this question.

It is postulated in the field theory that any particle is described by the field, which for any x belongs to representation of semi simple compact Lie group, that is if $\omega(x)$ is a function with values in the gauge group Ω

$$\psi(x) \to \psi(x)^{\omega} = \Gamma[\omega(x)]\psi(x) \tag{1}$$

The field configurations $\psi(x), A_\mu(x)$ and $\Gamma[\omega(x)]\psi(x), A_\mu^{\omega(x)}$ describe the same physical situation. This principle is called the relativity principle in the charge space. It allows uniquely introduce the interaction by changing in the free Lagrangian the usual derivatives by the covariant ones,

$$\partial_\mu \to D_\mu = \partial_\mu - g\Gamma(A_\mu) \tag{2}$$

Here

$$A_\mu = A_\mu^a T^a \tag{3}$$

In physical language $A_\mu(x)$ is called the gauge field. It describes the parallel transport in the charge space. From the point of view of mathematics the Yang-Mills field is the connection in the principle bundle. Under the gauge transformation it transforms as follows

$$A_\mu \to A_\mu^{\omega} = \omega(A_\mu)\omega^{-1} + \partial_\mu \omega \omega^{-1} \tag{4}$$

It is interesting to note that the notion of connection appeared in mathematics almost simultaneously with physics. The first mathematician who introduced this notion was Likhnerovich. In Russian translation his book appeared in1960, and it was written not long ago.

[a]E-mail: slavnov@mi.ras.ru

The free Lagrangian for the Yang-Mills field is the square of the curvature tensor

$$L = -\frac{1}{4}[F^a_{\mu\nu}F^a_{\mu\nu}] \tag{5}$$

$$F_{\mu\nu} = \partial_\mu A_\nu - \partial_\nu A_\mu + g[A_\mu, A_\nu] \tag{6}$$

This Lagrangian obviously describe the massless field. It prevented for a long time its using for description of weak interactions, which are known to be short range, that is correspond to the exchange by the massive particle. The gauge invariance of the Lagrangian plays the crucial role in proof of the renormalisability of the theory. A simple introduction of the mass term breaks the gauge invariance of the theory, and simultaneously breaks renormalisability.

In the works by P.Higgs, F.Englert and R.Broute it was shown how one can introduce the mass term for the vector field preserving the gauge invariance of the model. To do that one should introduce the gauge invariant interaction of scalar field with a special potential, generating the spontaneous breaking of the global symmetry of the model, but preserving the gauge invariance. The corresponding Lagrangian for the group $SU(2)$ looks as follows

$$L = (D_\mu\varphi(x))^*(D_\mu\varphi(x)) + m^2(\varphi(x))^*(\varphi(x)) - \lambda(\varphi(x)^*\varphi(x))^2 - \frac{1}{4}F^a_{\mu\nu}F^a_{\mu\nu} \tag{7}$$

The potential

$$-V = m^2\varphi^*(x)\varphi(x) - \lambda(\varphi^*(x)\varphi(x))^2 \tag{8}$$

where

$$\varphi^(x) = (\frac{iB_1 + B_2}{\sqrt{2}}; \frac{\sigma - iB_3}{\sqrt{2}}) \tag{9}$$

is unstable and generates the spontaneous breaking of the goobal symmetry responsible for the conservation of the charge of the field φ. From the formal point of view the fields φ acquire the imaginary mass. To develop the perturbation theory one should shift the fields σ by the constant $\varphi \to \varphi + C$. Here C is a constant doublet with zero first component and real second component. Clearly the shift of the fields by constant cannot break the gauge invariance of the theory, but the form of transformations change.

Before the change of variables the gauge transformation of the fields φ was a phase transformation but after the change it includes the shift by arbitrary function

$$\varphi(x) \to \exp\{ig\alpha(x)\}\varphi(x) - (C\exp\{ig\alpha(x)\} - 1) \tag{10}$$

The field $\varphi(x)$ under the gauge transformation is changed by an arbitrary function, and together with the field $A_\mu(x)$ plays the role of a gauge field. Under the shift of the field φ by a constant spinor the term $(D_\mu\varphi)^*(D_\mu\varphi)$

generates the mass term for the vector field. For the Higgs model the explicit form of the transformation of the field φ looks as follows

$$\delta\varphi^a = m\alpha^a + \frac{g}{2}\epsilon^{abc}\varphi^b\alpha^c + \frac{g}{2}\varphi^0\alpha^a \tag{11}$$

The resulting theory is obviously gauge invariant, and the gauge invariance allows to pass from the unitary gauge $B = 0$ to some explicitly renormalizable gauge, for example the Lorentz gauge $\partial_\mu A_\mu = 0$.

The simple introduction of the mass term breaks the gauge invariance of the theory. The gauge invariance of the theory may be reconstructed, if one introduces the mass term in the so-called Stueckelberg form

$$\frac{m^2}{2}[(A_\mu - \omega\partial_\mu\omega^{-1})^a]^2 \tag{12}$$

However it was shown by L.Faddeev and A.Slavnov that in one loop approximation this theory describes additional scalar particle and in the higher loops the limit $m \to 0$ does not exist.

The main merit of H.Higgs and F.Englert and R.Brout is the fact that they showed how to introduce the vector field mass without breaking the gauge invariance. One can notice that both papers appeared in 1964, and the correct quantization procedure for nonabelian gauge fields was constructed by L.Faddeev and V.Popov and independently by B. DeWitt three years later. For that reason strictly speaking H.Higgs and F.Englert and R.Brout solved the problem only in Abelian case. For nonAbelian fields the corresponding procedure was constructed later by T.Kibble, following the same ideas.

The gauge invariant action describes the constrained system, which does not allow to apply directly the methods based on the Hamiltonian character of the action.

$$A = \int dx[p\dot{q} - H(p,q) + \lambda^a C^a] \tag{13}$$

The first people who succeeded to construct the Lorentz invariant diagram technique for the Yang-Mills field were L.Faddeev and V.Popov and B.DeWitt. They reduced the constrained system, describing the gauge field, to the really Hamiltonian system, described by the Hamiltonian H^* , solving the constraints and applying the suitable gauge conditions. $H(p,q) \to H^*(p^*,q^*)$. These methods may be used as for the systems without spontaneous symmetry breaking and for the spontaneously broken systems. This method was applied by G.t'Hooft to Weinberg-Salam model.

Later the other methods of quantization of gauge invariant systems were developed. I have in mind the methods based on the BRST symmetry of the effective action. These methods do not appeal for their justification to the non relativistic Coulomb gauge, and allow to work always without breaking the manifest Lorentz invariance. They also are easily generalized to more complicated gauge invariant systems, such as strings. Moreover, as I am going to show these methods may be applied beyond perturbation theory as well. All the methods developed so far strictly speaking are applicable only in perturbation theory.

The construction of manifestly Lorentz invariant formulation of QED is impossible without introduction of the vector-potential A_μ, which describes not only physical excitations corresponding to three dimensionally transversal components, but also the longitudinal and time-like photons. At the same time the action of electromagnetic field is gauge invariant. Such situation is typical for the field theory- one introduces the unphysical degrees of freedom, and simultaneously introduces a new symmetry, which provides their decoupling. It is used for example in the BRST quantization of gauge systems.

We shall use this idea for the construction of the quantization procedure of nonabelian gauge fields, free of ambiguity beyond perturbation theory. The procedure of quantization of gauge fields includes the hypothesis, that if the the gauge condition is imposed, the Cauchi problem for the gauge field becomes uniquely solvable. It is true for QED, based on the Abelian group $U(1)$. Choosing the Coulomb gauge $\partial_i A_i = 0, \quad i = 1, 2, 3$ we see that any gauge transformed field must satisfy the equation

$$\partial_i A_i^\epsilon = 0 = \partial_i A_i + \triangle\epsilon = 0 \tag{14}$$

he condition $\partial_i A_i = 0$ is fulfilled therefore we must have $\triangle\epsilon = 0$, that is ϵ is a harmonic function. The harmonic function has an extremum at the boundary, therefore $\epsilon = 0$.

Let us pass to the nonabelian group $SU(N)$.

In the Coulomb gauge the field A_i must satisfy the equation $\partial_i A_i^\epsilon = 0$. For infinitesimal ϵ it follows

$$\triangle\epsilon_a + g t^{abc} A_i^b \partial_i \epsilon^c = 0 \tag{15}$$

It was noticed by V.Gribov, that this equation has nontrivial solutions vanishing at spatial infinity. Therefore the uniqueness of the solution of the Cauchi problem is broken.

It does not influence the perturbation theory. In this case the eq.(15) has only trivial solution.

If one ignores this problem the transition from one gauge to another becomes singular. The problem of quantization of nonabelian gauge fields beyond perturbation theory remains open.

Problems. Efficient nonperturbative methods.

1. Structure of a ground state in QCD.
2. Quark confinement
3. Mass spectrum

4. Phase transitions.
5. Quantitative theory of strong interactions.
6. Nonuniqueness of the gauge fixing does not allow to quantize nonabelian gauge fields beyond perturbation theory.

The only systematic way to deal with nonperturbative effects at present is given by the gauge theories on the lattice treated with the help of supercomputers.

The problem of Gribov ambiguity may be avoided using the alternative method of quantization, based on the equivalence theorems for spectrum changing transformation.

Here the technique of BRST quantization is heavily used.

We shall apply this technique to the Yang-Mills theory

$$L = -\frac{1}{4} F^a_{\mu\nu} F^a_{\mu\nu} \tag{16}$$

This theory is equivalent to the theory described by the Lagrangian

$$\begin{aligned}\tilde{L} = -\frac{1}{4} F^a_{\mu\nu} F^a_{\mu\nu} + (D_\mu \varphi)^*(D_\mu \varphi) + (D_\mu \chi)(D_\mu \chi)^* \\ -[(D_\mu b)^*(D_\mu e) + (D_\mu e)^*(D_\mu b)]\end{aligned} \tag{17}$$

Here the fields φ and χ commute and b, e are anti commuting elements of Grassman algebra. Indeed integrating over φ, χ, b, e in the sector which does not contain these fields, we get the usual Yang-Mills theory. Of course one should prove that the theory is unitary in this sector.

2 The spectrum

Ghost exitations: $\varphi_\pm, b, e$, longitudinal and temporal components of A^a_μ
Physical exitations: three dimensionally transversal components of the Yang-Mills field. Here

$$\varphi_\pm = \frac{\varphi \pm \chi}{\sqrt{2}} \tag{18}$$

The supersymmetry of the effective action generates a conserved nilpotent charge Q. Physical states are separated by the condition

$$Q|\psi>_{ph} = 0 \tag{19}$$

The states separated by this condition describe only three dimensionally transversal components of the Yang-Mills field.

The Lagrangian is invariant with respect to supertransformations

$$\delta\varphi_- = b\epsilon; \quad \delta e = \varphi_+ \epsilon \tag{20}$$

Here

$$\varphi_{\pm} = \frac{\varphi \pm \chi}{\sqrt{2}} \tag{21}$$

Obviously this invariance is preserved if we perform before the shift of the fields φ_-

$$\varphi_- \to \varphi_- + m; \quad \varphi_+ \to \varphi_+ \tag{22}$$

Let us demonstrate the equivalence of this theory to the standard Yang-Mills theory.

Equivalence theorems: canonical transformations $\to$ point transformations $\varphi = \varphi' + f(\varphi')$

More general transformations:

$$\varphi = \frac{\partial^n \varphi'}{\partial t^n} + f(\frac{\partial^{n-1}\varphi'}{\partial t^{n-1}}, \dots \frac{\partial \varphi'}{\partial t}) = \tilde{f}(\varphi') \tag{23}$$

The spectrum is changed. What about the unitarity?

Path integral representation for the scattering matrix

$$S = \int \exp\{i \int L(\varphi) dx\} d\mu(\varphi); \quad lim_{t \to \pm\infty} \varphi(x) = \varphi_{out,in}(x) \tag{24}$$

If the change (23) does not change the asymptotic conditions, then the only effect of such transformation is the appearance of a nontrivial jacobian

$$L(\varphi) \to \tilde{L}(\varphi') = L[\varphi(\varphi')] + \bar{c}^a \frac{\delta \varphi^a}{\delta \varphi'^b} c^b \tag{25}$$

For all new excitations one should take the vacuum boundary conditions. It is possible because the new Lagrangian is invariant with respect to supertransformations

$$\delta \varphi'_a = c_a \epsilon$$
$$\delta \bar{c}_a = \frac{\delta L}{\delta \varphi_a(\varphi')} \epsilon; \quad \delta c_a = 0 \tag{26}$$

On mass shell these transformations are nilpotent and generate a conserved charge Q. In this case there exists an invariant subspace of states annihilated by Q, which has a semidefinite norm (A.A.S.). For asymptotic space this condition reduces to

$$Q_0 |\phi>_{as} = 0 \tag{27}$$

The scattering matrix is unitary in the subspace which contains only excitations of the original theory. However the theories described by the L and the $\tilde{L}$ are different, and only expectation values of the gauge invariant operators coincide.

The structure of the vectors, annihilated by the operators Q may be studied in the following way. It is possible to introduce the operators K, satisfying the following conditions

$$[\hat{K}, \hat{Q}]_{+} = \hat{N} \tag{28}$$

where $\hat{N}$ is the number operator for the ghost particles. Any vector which contains at least one ghost particle may be presented in the form

$$|\chi> = \frac{[\hat{K}, \hat{Q}]_{+}}{N}|\chi> = \frac{\hat{K}\hat{Q}}{N}|\chi> + \frac{\hat{Q}\hat{K}}{N}|\chi> \tag{29}$$

The first term is zero for the state vectors annihilated by $\hat{Q}$ The second term has zero norm.

Any vector which satisfies the equation (27) has a form

$$|\phi>_{as} = |\phi>_{tr} + |N> \tag{30}$$

where $|\phi_{tr}>$ is the physical Yang-Mills vector, having only transversal polarizations, and $|N>$ is a zero norm vector,which is proportional to $Q|\lambda>$. Factorizing the space of the vectors, satisfying the equation (27), with respect to zero norm vectors we see that the asymptotic space of our model coincides with the space of the Yang-Mills theory.

In our case the supersymmetry leads to decoupling of excitations corresponding to the fields $\varphi_{\pm}, b, e$.

It allows also to use the gauge $\varphi_{-} = 0$. This gauge is algebraic and therefore does not produce the Gribov ambiguity, at the same time it is manifestly Lorentz invariant. This gauge does not lead to Gribov ambiguity, and it may be used beyond perturbation theory. It goes without saying that one should prove the decoupling of all unphysical degrees of freedom. It was done in the papers of A.A.S., Quadri and A.A.S.

3 Conclusion

1. The gauge invariant theory may be free of Gribov ambiguity beyond perturbation theory.

2. The method described above allows also to construct a manifestly gauge and Lorentz invariant infrared regularization of nonabelian gauge theories, which does not contain any additional fields.

3. Contrary to a general belief the classical finite energy localized solution (soliton) exists in the theory of the Yang-Mills, modified as described above.

4 Acknowledgements

This work was supported in part by RBRF and the program "Nonlinear dynamics".

References

[1] C.N.Yang, R.L.Mills, Phys.Rev. 96(1954)191.
[2] A.Lichneroviich, M.IL.1960
[3] P.W.Higgs, Phys.Lett.12(1964)132.
[4] R.Brout, F.Englert, Phys.Rev.Lett.13(1964)321.
[5] L.D.Faddeev, V.N.Popov, Phys.Lett. B25(1967)30.
[6] B.De Witt, Phys.Rev.160(1967)1113, 1195.
[7] T.W.B.Kibble, Phys.Rev.155(1967)1554.
[8] G.t'Hooft, Nucl.Phys.B35(1971)167.
[9] G.Cursi, R.Ferrari, Nuovo Cimento A Ser 11, 35(1976)273.
[10] T.Kugo, I.Ojima, Suppl.Progr.Theor.Phys.66(1979)
[11] V.N.Gribov, Nucl.Phys. B139 (1978)1.
[12] A.A.Slavnov, JHEP 8(2008)047
[13] A.A.Slavnov, Theor . Math. Phys.161(2009)1497.
[14] A.A.Slavnov, Proceedings of the Steklov Institute of Mathematics, 272(2011)235.
[15] A.A.Slavnov,Theor.Math.Phys. 183(2015)585
[16] A.Quadri, A.A.Slavnov, JHEP 1007 (2010).
[17] A.Quadri, A.A.Slavnov, Theor.Math.Phys. 166(2011)291.

HIGGS BOSON IN THE STANDARD MODEL AND ITS EXTENSIONS

Mikhail Dubinin [a] and Elena Petrova [b]

[a] *Skobeltsyn Institute of Nuclear Physics, Moscow State University, 119991 Moscow, Russia*

[b] *Physics Department, Moscow State University, 119991 Moscow, Russia*

Abstract. Precise experimental data on the Higgs boson properties at the LHC provide significant constraints on both the general model-independent extensions of the SM by effective operators of higher dimension and model-dependent extensions of the SM, among the latter Minimal Supersymmetric Standard Model (MSSM) seems to be one of the most attractive candidates. Production channels include $\gamma\gamma$, ZZ, W^+W^-, $\tau^+\tau^-$, $b\bar{b}$ and $\mu^+\mu^-$ in the data samples collected at the integrated luminosities 5.1 fb^{-1} at $\sqrt{s}$ =7 TeV and 19.7 fb^{-1} at 8 TeV. They give possibilities to probe deviations of the couplings of the Higgs boson from the Standard Model in many ways and to limit significantly the MSSM parameter space.

1 Introduction

Identification of the mechanism providing electroweak symmetry breaking (EWSB) is one of the most important objectives of the Large Hadron Collider (LHC) physics program. In the Standard Model (SM) EWSB is a consequence of Higgs potential which includes one scalar SU(2) isodoublet that leads to one physical Higgs boson [1]. Masses of gauge bosons $W^{\pm}$, Z are generated by the kinetic term of the potential and masses of SM fermions are ensured by Yukawa interaction with the Higgs field [2]. A new boson with mass 125 GeV was observed by ATLAS and CMS Collaborations at the LHC in 2012 [3]. In the following years the observation has been confirmed at a higher levels of statistical significance [4]. Extensive studies of production rates in various channels [5] and quantum numbers [6] demonstrated that properties of the new boson are very similar to those expected for the SM Higgs boson. Evidence for a signal consisent with ATLAS and CMS measurements was reported also by CDF and DO collaborations at the Tevatron [7].

Summary of a large number of calculations performed with the help of various theoretical tools can be found in [8]. The largest cross section (pp, $\sqrt{s}$ =7 TeV or 8 TeV) from the one-loop mediated gluon-gluon fusion $gg \to H$ is accompanied in descending order by vector boson fusion (VBF), associated WH and ZH production and associated $t\bar{t}$ production, the sum of all production modes yields approximately 20 pb at the energy $\sqrt{s}$ =7-8 TeV. The following Higgs boson decays in the channels $b\bar{b}$, W^+W^-, ZZ, $\tau^+\tau^-$, $\mu^+\mu^-$ and $\gamma\gamma$ as well as searches for invisible decays of the Higgs boson give a huge array of experimental data for a combination of various production modes with decay

[a] E-mail: `dubinin@theory.sinp.msu.ru`

[b] E-mail: `petrova@theory.sinp.msu.ru`

channels, which contains rich information about the couplings of the new boson. The best energy resoluton and high sensitivity (i.e. signal/irreducible background ratio) of the $H \to \gamma\gamma$ and $H \to ZZ \to 4l$ ($l = e, \mu$) channels is combined with good sensitivity of $H \to W^+W^- \to l\nu l\nu$ ($\nu = \nu_e, \nu_\mu$) channel having poor mass resolution. Moderate mass resolution of the channels $b\bar{b}$ and $\tau^+\tau^-$ which are spoiled by large backgrounds allows nevertheless to extract an evidence for the Higgs boson coupling to fermions [9].

	ggH	qqH	$W/Z\,H$	ttH	significance	resolution
$H \to \gamma\gamma$	•	•	•	•	5.6σ	1-2%
$H \to ZZ$ 4 leptons	•	•			6.5σ	1-2%
$H \to W^+W^-$ $2l{+}2\nu$	•	•	•	•	4.7σ	15%
$H \to \tau^+\tau^-$	•	•	•	•	3.8σ	15%
$H \to b\bar{b}$			•	•	2.0σ	10%
$H \to \mu^+\mu^-$	•	•			0.1σ	1-2%

Table 1: Higgs boson signal significance in various production channels calculated for m_H =125 GeV with rearrangement of final states (for example, $\tau^+\tau^-$ signal is a combination of two production channels $t\bar{t}H(\tau^+\tau^-) + H(W^+W^- \to \tau^+\tau^-)$.

In order to investigate couplings of the Higgs boson to vector bosons and fermions of the SM, a combined analysis of all channels has been performed. Combined analysis gives a possibility to extract 'pseudo observables' [10] which are dependent on the ratios of the experimentally observed production cross sections and decay widths to those of the SM. In such analyses it is assumed that properties of SM Higgs boson are modified by a new physics at the TeV scale Λ, when SM is an effective theory at the $W^\pm, Z$ scale. Either generic set of higher-dimension effective operators is introduced in addition to the SM Lagrangian (the same set of effective operators can be a remnant of different field theory models at a multiTeV energy scale), or consequences of a definite renormalizeable model are evaluated at the electroweak scale. In both types of extensions the question to what extent the observed boson is consistent with the SM Higgs boson seems most interesting to be addressed.

2 Model-independent extensions: dimension-six effective operators

Extensions by dimension-six effective operators have been considered recently in different bases (e.g. [11]). In the following we use [12] where a rather general framework of the SM extension by the dimension-six effective operators was analysed. At the electroweak energy scale the new physics degrees of freedom can be integrated out and the low-energy effective Lagrangian can be written as

$$L_{eff} = L_{SM} + \frac{1}{\Lambda^2} \sum_{k=V,F} c_{k\Phi} O_{k\Phi}$$

where the anomalous couplings $c_{k\Phi}$ modify the SM Higgs boson couplings to the vector bosons and to the fermions. A general set of the dimension-six operators from [12] in the Buchmueller-Wyler basis [13] modified by the subtraction of v.e.v.: $\Phi^\dagger\Phi \to \Phi^\dagger\Phi - v^2/2$ (see [14]) to avoid undesirable mixing in the gauge field kinetic terms, can be reduced to a restricted set of only five fermion-Higgs and vector boson-Higgs operators $O_{t\Phi}$, $O_{b\Phi}$, $O_{\tau\Phi}$ and $O_{\Phi}^{(1)}$, $O_{\Phi G}$, disposing the tensor structure of interaction vertices identical to the SM and dependent only on the two anomalous couplings c_V, which rescales the vector boson-Higgs vertices, and c_F, which rescales the fermion-antifermion-Higgs vertices. The anomalous couplings C_n, C_{mn} in front of the dimension-six operators O_n, O_{mn} are conformably redefined [20], for example

$$c_F = 1 + C_{t\Phi} \cdot \frac{v^2}{\Lambda^2}, \quad c_V = 1 + \frac{v^2}{2\Lambda^2} \cdot C_{\Phi}^{(1)}, \quad c_G = c_F + \frac{6\pi}{\alpha_s} \cdot C_{\Phi G} \cdot \frac{v^2}{\Lambda^2}, \quad \ldots$$

(v =246 GeV). The last equation here means that in the case under consideration the one-loop $H \to \gamma\gamma$ and $H \to gg$ vertices are "resolved" in the sense that the anomalous parameters c_V and c_F are included in the one-loop effective c_G and c_γ couplings which parametrise $H \to gg$ and $H \to \gamma\gamma$ vertices. This circumstance is different from 'kappa-framework'.

2.1 Statistical methodology for combination of channels

The method of exclusion contours reconstruction [16] in the anomalous parameter space has been introduced to CompHEP version 4 [17]. Theoretical signal strengths in the infinitely small width approximation (ISW or production × decay approximation) and for complete gauge-invariant sets of diagrams are defined as

$$\mu_i^{ISW} = \frac{[\sum_j \sigma_{j\to h} Br(h \to i)]_{SME}}{[\sum_j \sigma_{j\to h} Br(h \to i)]_{SM}}; \qquad \mu_i = \frac{\left[\sum_{j=1}^{N_{ch}} \sigma_{j\to H(off-shell)\to i}\right]_{SME}}{\left[\sum_{j=1}^{N_{ch}} \sigma_{j\to H(off-shell)\to i}\right]_{SM}}$$

where for ISW i is the number of Higgs boson decay channel and j is the number of Higgs production process for a given final state in the SM and the SM extension (SME). The signal strength $\hat{\mu}_i$ can be expressed using the observed cross section σ_{obs}, the background cross section σ_{backgr} and the SM signal cross section σ^{SM}_{signal}, then the global χ^2 for the signal strengths $\hat{\mu}_i$ is defined as

$$\hat{\mu}_i = \frac{\sigma_{obs,i} - \sigma_{backgr,i}}{\sigma^{SM}_{signal,i}}; \quad \chi^2(\mu_i) = \sum_i^{N_{ch}} \frac{(\mu_i - \hat{\mu}_i)^2}{\sigma_i^2}$$

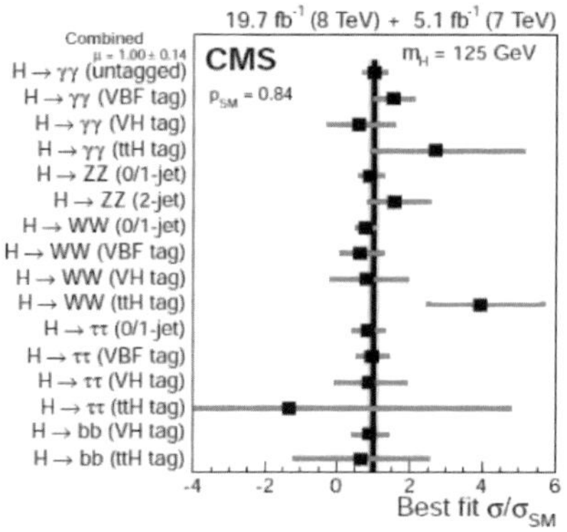

Figure 1: Signal strength and signal strength error for varions groups of production channels.

Value of the best fit σ/σ_{SM} for the overall combined analysis (vertical line) where combinations of channels grouped by production mode and specific kinematics in the final state taken from [15] are shown in Fig.1. The horisontal bars indicate $\pm 1\sigma$ uncertainties in the best-fit σ/σ_{SM} values for individual channels, both statistical and systematic uncertainties are included. Combinations of channels are grouped by dominant decay modes when more than one decay mode contributes to the same final state. Tags in brackets indicate a specific production mechanism.

for the number of production channels N_{ch}. Minimization of $\chi^2 \to \chi^2_{min}$ gives us 1σ, 2σ and 3σ regions $\chi^2 = \chi^2_{min} + \Delta\chi^2$ where $\Delta\chi^2$ is defined by cumulative distribution function. Assuming that the signal strengths of various channels have Gaussian distributions with the probability density functions (PDF's) having the expected values $\hat{\mu}_i$ and the dispersions σ_i normalized to one, combined PDF for a number of production channels can be found by multiplication of PDF's for individual channels. Combined probability density function is also Gaussian characterized by μ_c and σ_c; $1/\sigma_c^2 = \sum_i^{N_{ch}} \cdot 1/\sigma_i^2$. The combined PDF allows one to determine, for example, 95% CL exclusion upper μ_U and lower μ_L limits on the signal strength parameter integrating the combined pdf from $\hat{\mu}$ to μ_U and from μ_L to $\hat{\mu}$, respectively, then equating the result to 0.95/2. If the SM is fully adequate, the values of μ_i are as close to one as allowed by experimental errors. Calculation of the $\Delta\chi^2$, see Fig.2, for the best fit defines a *given number* $\times$ *CL* contours corresponding to the departure of the SM point $(1, 1)$ from the best fit point in the (c_V, c_F) parameter plane. Contours in Fig.3 correspond to 65%, 90% and 99% best fit CL regions with $\Delta\chi^2$ less than 2.10, 4.61 and 9.21, respectively.

2.2 Global fits

At the LHC calculations of complete gauge invariant sets of diagrams are more precise than production $\times$ decay approximation. They take into account (1) untrivial interference between the signal diagrams (2) untrivial interference between the signal and the irreducible background diagrams (3) correct kinematics of lepton and jet distributions. The four-lepton final state $l^+l^-l^+l^-$, for example, is produced through $H \to Z^*Z^*$ and $\gamma\gamma$ intermediate states with interferences not accounted for in the production $\times$ decay approximation. The $2l, 2\nu$ state is formed by a mixture of WW and ZZ production mechanisms. Data at the energies 7 TeV and 8 TeV from LC2013 (Hamburg, May 2013)

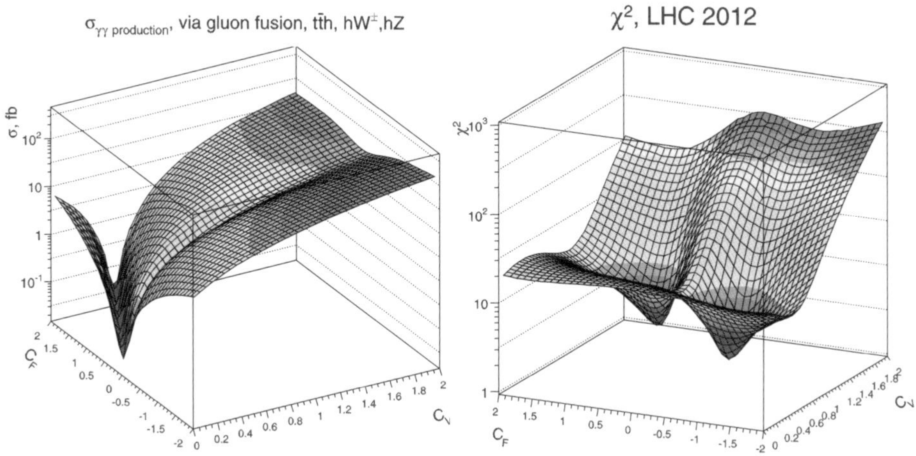

Figure 2: Three-dimensional plots in the (c_V,c_F) parameter space: (a) total cross section for $\gamma\gamma$ production processes at the LHC, $\sqrt{s}$ =8 TeV; (b) three-dimensional plot of χ^2 as a function of (c_V,c_F), LHC $\sqrt{s}$ =8 TeV - all channels

improve significantly the precision which has been achieved for $b\bar{b}$ and $\tau^+\tau^-$ channels in 2012. Contours generated with LC2013 data are shown in Fig.3, right panel. The area at negative c_V available with 2012 data, see Fig.2, left panel, disappeared almost completely while contours of the positive (c_V,c_F) quadrant demonstrate the consistency of the SM hypothesis with the data on the level of 2σ.

3 Model-dependent extension: Minimal Supersymmetric Standard Model

An absence of superparner signals at the LHC have not decreased an interest to the concept of supersymmetry. Supersymmetric extensions solve the gauge hierarchy problem, include candidates for dark matter, can ensure first order phase transition in the early Universe, important for generation of baryon

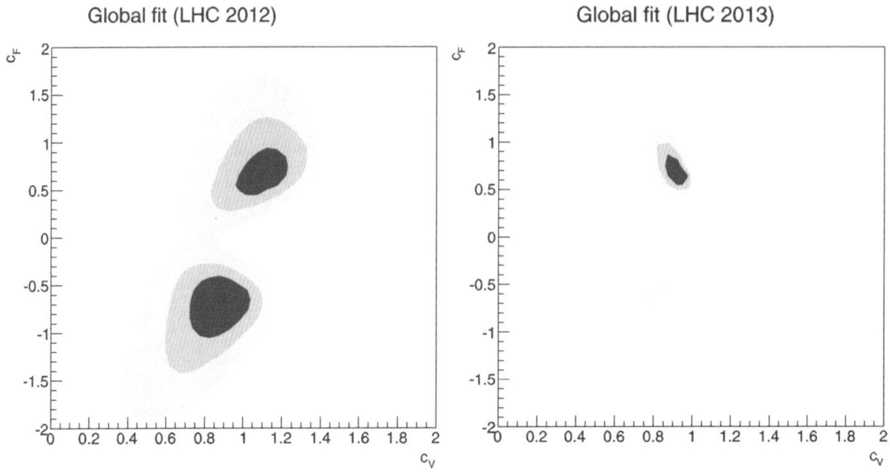

Figure 3: (b) - three exclusion contours for the combined χ^2 fit in the (c_V,c_F) plane based on 2012 LHC data, (c) - the same based on 2013 LHC data. Blue, green and yellow areas correspond to $\Delta\chi^2$ =2.10, 4.61 and 9.21 (CL of the fit is 65%, 90% and 99%), respectively.

asymmetry, and contain a candidate for inflaton. The MSSM two-doublet Higgs sector includes five physical states: two CP-even neutral Higgs bosons h and H, one CP-odd neutral Higgs boson A and two charged scalars $H^{\pm}$. Large radiative corrections from the third generation quarks and quark superpartners and other sources in a number of MSSM scenarios [18] strongly influence m_h and play an important role in generation of the effective two-doublet Higgs potential at the scale below the sparticle mass scale $M_{\mathbf{SUSY}}$. Precision data taken at the LHC have led to strong constraints on the Minimal Supersymmetric Standard Model (MSSM) parameter space. Natural assumption for the Higgs sector mass splittings is the 'decoupling limit' when m_h =125 GeV for the lightest CP-even Higgs boson mass and all other scalars are heavy with masses at the TeV scale. Decoupling limit is not a unique possibility [19]. In some aligned regions of parameter space h has SM-like couplings to fermions and gauge bosons, while other scalars are light. In both decoupling and alignment limits the constraint m_h =125 GeV has a significant impact on the allowed regions of the MSSM parameter space. For the case of 'heavy supersymmetry' in the limit close to decoupling, so that sparticle mass scale is of the order of TeV and the CP-odd scalar mass $m_A << M_{SUSY}$, the effective field theory model at the scale m_{top} is a two-Higgs doublet model which in the following is considered in the limit of CP-conserving two-doublet Higgs potential. Domains of the five-dimensional parameter space m_A, $\tan\beta$ (the ratio of vev's v_2/v_1), M_{SUSY}, $A_{t,b}$ (trilinear soft SUSY breaking terms parameter) and μ (Higgs superfield mass parame-

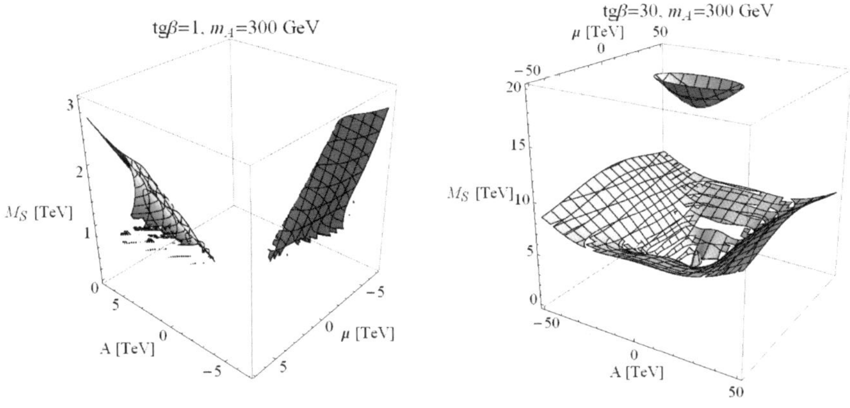

Figure 4: Surface of m_h = 125 GeV in the M_{SUSY},μ,A parameter space. $\tan\beta$ =1, m_A =300 GeV (left) and $\tan\beta$ = 30 (right). For illustrative purposes the region of very large μ,A greater than 10 is shown, where the effective two-doublet potential decomposition up to operators of dimension four only is not adequate.

ter) are delimited, see Fig.4 and Fig.5, in the scenario [20] when the dominant radiative corrections to the two-doublet Higgs potential come from the third generation quark superpartners. The scenario [20] is simplified in comparison with [21], where the four-step RGE evolution is performed from the top scale to the intermediate m_A and then to M_{SUSY} scale, where the threshold corrections are calculated, with the following evolution down to m_A and m_{top} mass scales. Surface of m_h =125 GeV in the tree-dimensional $M_{SUSY}, A_{t,b}, \mu$ parameter space at moderate m_A =300 GeV in combination with small $\tan\beta$ (of the order of 1) and large $\tan\beta$ (of the order of 10^1) is shown in Fig.4. In the first case the squark mass scale M_{SUSY} is bounded from above by a value of a few TeV, while in the second case there is an ambiguity, when both small and high superpartner mass scales are possible. Enhancement of the couplings of the heavy Higgs bosons at large $\tan\beta$ (for example, $pp \to t\bar{t} \to b\bar{b}H^+H^-$ is enhanced by $\tan^4\beta \sim 10^6$) lead to strong experimental constraints in the $(m_A, \tan\beta)$ plane [22]. In the representative characteristic scenarios [23] values of $\tan\beta$ greater than 15 are not allowed if m_A is less than 500 GeV, so the parameter plane in Fig.5 is expected to be further constrained by the LHC data. Lower $\tan\beta$ are not excluded yet leading to significant branchings of the heavy CP-even Higgs boson decay to W^+W^-, ZZ and hh [24].

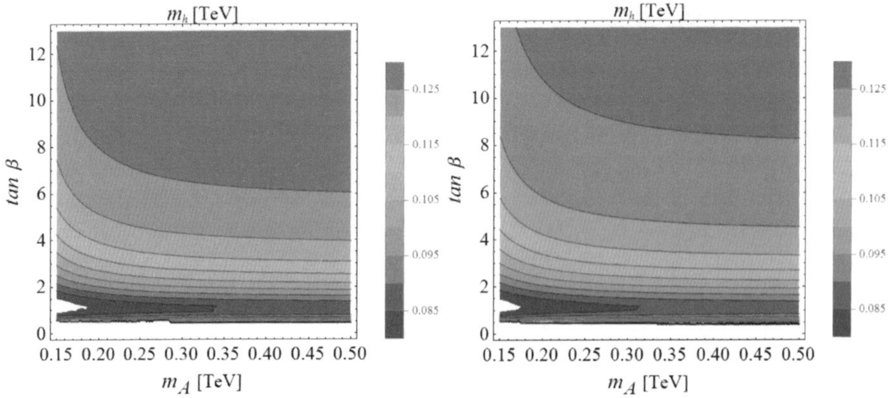

Figure 5: Domains of m_h in the μ,A plane for $A_{t,b}$ =2 TeV and μ =1.5 TeV. M_{SUSY} =1 TeV (left) and 1.25 TeV (right). Very narrow red domain m_h =125 GeV at small $\tan\beta \sim 1$ is not resolved, although would be observable in the figure with better resolution.

4 Summary

The event yields which are observed in different production channes at the LHC are consistent with those expected for the SM Higgs boson m_h =125.02 GeV, as measured by the ATLAS and the CMS experiments with statictical and systematical errors of approximately $\pm$0.5 GeV. Pseudo-observables c_V and c_F introduced in the SM extension by the dimension-six effective operators rescale in a definite way the SM interaction vertices thus introducing possible effects of 'new physics' at the energy scale Λ. The SM limiting case $(c_V, c_F) = (1,1)$, although is not the best fit of the data, confirms the SM at the level of statistical significance better than 2σ. While in the model-independent extensions of the SM by the dimension-six effective operators efforts were concentrated on the restriction of pseudo-observables c_F and c_V at some levels of statistical significance by means of global fits, in the MSSM one can restrict multidimensional parameter space of the model using explicit constraints imposed by the observed m_h value, which are consistent with explicit transformation to the mass basis of Higgs bosons and radiative corrections evaluated at m_{top} scale within a definite MSSM scenarios and characteristic parameter sets. MSSM parameter sets characterized by moderate m_A and small $\tan\beta$ are not excluded by the data on heavy Higgs bosons searches at the LHC and lead to interesting and untrivial phenomenological consequences.

Acknowledgments

The authors acknowledge partial support from NSh-3042.2014.2.

References

[1] F. Englert, A. Brout, Phys. Rev. Lett. **13**, 321 (1964); P.W. Higgs, Phys. Rev. Lett. **13**, 508 (1964); T.W.B. Kibble, Phys. Rev. **155**, 1554 (1967).

[2] Y. Nambu, G. Jona-Lasinio, Phys. Rev. **122**, 345 (1961); M. Gell-Mann, M. Levy, Nuovo Cim. **16**, 705 (1960).

[3] ATLAS Collaboration, Phys. Lett. **B 716**, 1 (2012); CMS Collaboraton, Phys. Lett. **B 716**, 30 (2012); JHEP **06**, 081 (2013).

[4] ATLAS Collaboration, Phys. Rev. **D 90**, 052004 (2014); CMS Collaboraton, Eur. Phys. J. **C 74**, 3076 (2014)

[5] CMS Collaboration, JHEP **05**, 145 (2013); Phys. Rev. **D 89**, 012003 (2014); JHEP **01**, 096 (2014); JHEP **05**, 104 (2014); Phys. Lett. **B 726**, 587 (2013); Phys. Lett. **B 736**, 64 (2014); Eur. Phys. J. **C 74**, 2980 (2014); JHEP **09**, 087 (2014); Phys. Lett. **B 738**, 68 (2014)
ATLAS Collaboration, Phys. Lett. **B 726**, 88 (2013); Phys. Lett. **B 732**, 8 (2014); Phys. Rev. Lett. **112**, 201802 (2014); Phys. Lett. **B 738**, 68 (2014); JHEP **09**, 112 (2014); Phys. Lett. **B 738**, 234 (2014); Phys. Rev. **D 91**, 012006 (2015); Phys. Rev. **D 90**, 112015 (2014); Phys. Lett. **B 740**, 222 (2015); JHEP **01**, 069 (2015).

[6] CMS Collaboration, Phys. Rev. Lett. **110**, 081803 (2013); Phys. Rev. **D 92**, 012004 (2015); ATLAS Collaboration, Phys. Lett. **B 726**, 120 (2013).

[7] CDF and D0 Collaborations, Phys. Rev. Lett. **109**, 071804 (2012); Phys. Rev. **D 88**, 052014 (2013).

[8] LHC Higgs Cross Section Working Group, *Handbook of LHC Higgs Cross Sections, (1) Inclusive Observables*, arXiv:1101.0593[hep-ph], 2011; *(2) Differential Distributions*, arXiv:1201.3084[hep-ph], 2012; *(3) Higgs Properties*, arXiv:1307.1347[hep-ph], 2013.

[9] CMS Collaboration, Nat. Phys. **10**, 557 (2014).

[10] M. Gonzalez-Alonso, A. Greljo, G. Isidori, D. Marzocca, Eur. Phys. J. **C 75**, 128 (2015)
G. Passarino, Nucl. Phys. **B 834**, 77 (2010)

[11] A. Falkowski, F. Riva, JHEP **1502**, 039 (2015)

[12] E. Boos, V. Bunichev, M. Dubinin, Y. Kurihara, Phys. Rev. **D89**, 035001 (2014); Phys. Lett. **739**, 410 (2014)

[13] W. Buchmuller, D. Wyler, Nucl.Phys. **B268**, 621 (1986)

[14] G. Passarino, Nucl. Phys. **B868**, 416 (2013)

[15] CMS Collaboration, CERN-PH-EP/2013-037; 2015/06/16; arXiv:1412.8662[hep-ph], 2015

[16] B. Dumont, S. Fichet, G. von Gersdorff JHEP **1307**, 065 (2013)
J.R. Espinosa, C. Grojean, M. Muhlleitner, M. Trott, JHEP **1212**, 045 (2012); JHEP **1209**, 126 (2012); JHEP **1205**, 097 (2012)

S. Kraml, B.C. Allanach, M. Mangano, H.B. Prosper, S. Sekmen, C. Balazs, A. Barr, P. Bechtle, G. Belanger, A. Belyaev et al., Eur.Phys.J. **C72**, 1976 (2012)
A. Azatov, R. Contino, J. Galloway, JHEP **1204**, 127 (2012) 127, Erratum-ibid. **1304**, 140 (2013)
[17] E. Boos *et al.*, Nucl. Instrum. Meth. **A534**, 250 (2004); A. Pukhov *et al.*, *CompHEP 3.3 users' manual*, hep-ph/9908288; see also `http://theory.npi.msu.su/comphep`
[18] P. Draper, G. Lee, C.E.M. Wagner, Phys. Rev. **D 89**, 055023 (2014)
B. Allanach, A. Djouadi, J. Kneur, W. Porod, P. Slavich, JHEP **0409**, 044 (2004)
G. Degrassi, S. Heinemeyer, W. Hollik, P. Slavich, G. Weiglein, Eur.Phys.J. **C 28**, 133 (2003)
J.F. Gunion, H.E. Haber, Phys. Rev. **D 67**, 075019 (2003)
M. Carena, H.E. Haber, S. Heinemeyer, W. Hollik, C.E.M. Wagner, G. Weiglein, Nucl. Phys. **B 580**, 29 (2000)
J.R. Espinosa, R.J. Zhang, JHEP **0003**, 26 (2000)
A. Pilaftsis, C. Wagner, Nucl. Phys. **553**, 3 (1999)
H.E. Haber, R. Hempfling, Phys. Rev. **D 48**, 4280 (1993)
[19] M. Carena, H. Haber, I. Low, N. Shah, C. Wagner, Phys.Rev. **D 91**, 035003 (2015)
[20] E. Akhmetzyanova, M. Dolgopolov, M.Dubinin, Phys.Rev. **D 71**, 075008 (2005); Phys. Part. Nuclei **37**(5), 677 (2006)
M. Dubinin, A. Semenov, Eur.J.Phys. **C 28**, 223 (2003)
[21] G. Lee, C.E.M. Wagner, Phys.Rev. **D 92**, 075032 (2015)
[22] CMS Collaboration, JHEP **1410**, 160 (2014); ATLAS Collaboration, JHEP **1411**, 056 (2014); JHEP **1503**, 088 (2015)
[23] M. Carena, S. Heinemeyer, O. Stal, C.E.M. Wagner, G. Weiglein, Eur.Phys.J. **C 73**, 2552 (2013)
[24] E. Bagnaschi et al, LHCHXSWG Note 2015-004, 2015
A. Djouadi, J. Quevillon, JHEP **1310**, 028 (2013)

THE SPIN-CHARGE-FAMILY THEORY OFFERS THE EXPLANATION FOR ALL THE ASSUMPTIONS OF THE STANDARD MODEL, FOR THE DARK MATTER, FOR THE MATTER-ANTIMATTER ASYMMETRY, FOR..., MAKING SEVERAL PREDICTIONS

Norma Susana Mankoč Borštnik [a]

Faculty of Mathematics and Physics, University of Ljubljana, 1000 Ljubljana, Slovenia

Abstract. The *spin-charge-family* theory, which is a kind of the Kaluza-Klein theories, but with two kinds of the spin connection fields - the gauge fields of the two kinds of spins [1, 2] - is offering the explanation for the appearance and properties of family members (quarks and leptons), of families, of vector gauge fields (weak, hyper, colour), of scalar higgs and Yukawa couplings. It offers also the explanation for the appearance of the dark matter and matter/anti-matter asymmetry. In this talk the achievements of this theory and its predictions are briefly presented. Not yet solved problems are discussed in the talk [3], the shorter version of which is the present paper.

1 Introduction

More than 40 years ago the *standard model* offered an elegant new step in understanding elementary fermion and boson fields. It postulated (without explanation):

i. The existence of the massless family members with the assumed charges in the fundamental representation of the colour, weak and hyper charge groups, the left handed quarks and leptons distinguishing from the right handed ones in the weak and hyper charges and correspondingly mass protected.

ii. The existence of massless families to each of the family member.

iii. The existence of the massless vector gauge fields (colour octet, weak triplet, hyper singlet) to the observed (colour, weak and hyper) charges of the family members, all in the adjoint representations with respect to the charges.

iv. The existence of a massive self interacting scalar field carrying the weak and the hyper charge ($\pm\frac{1}{2}$, $\mp\frac{1}{2}$), respectively (doublets like fermions), with the "nonzero vacuum expectation values", what breaks the weak and the hyper charge, breaking correspondingly the mass protection of fermions and of weak and hyper bosons.

v. The existence of the Yukawa couplings, which together with (the gauge fields and) the scalar higgs take care of the properties of the fermions.

The *standard model* assumptions have been confirmed without offering surprises. The last unobserved field, the higgs scalar, was detected in June 2012 and confirmed in March 2013.

[a]E-mail: norma.mankoc@fmf.uni-lj.si

There are several attempts in the literature, offering the extensions of the *standard model* by enlarging groups, but do not explain the *standard model* assumptions.

An elegant trustworthy step beyond the *standard model* must offer, to my opinion, the explanation:
o for the origin of the fermion charges, **o** for the origin of the families of fermions and their properties, **o** for the origin of the vector gauge fields and their properties, **o** for the origin of the scalar field, its properties and of the Yukawa couplings, **o** for the origin of the dark matter, **o** for the origin of the "ordinary" matter/anti-matter asymmetry, **o** possibly also for the other observed phenomena.

Inventing a next step, which covers only one of the open questions, can hardly be the right step.

The *spin-charge-family* theory [1,2,4–9] does offer the explanation for all the assumptions of the *standard model.* The more I am working (together with the collaborators) on the *spin-charge-family* theory, the more answers to the open questions of the elementary fermion and boson fields and cosmology the theory is offering.

I shall briefly present the achievements of the *spin-charge-family* theory. Some of still open questions and answers to some of the most often posed questions and criticisms are presented in Ref. [3].

2 The *Spin-charge-family* theory, action and assumptions

The *spin-charge-family* theory is built on a very simple action and on *assumptions*, presented in what follows.

A i. In the action [1,2] fermions ψ carry in $d = (13+1)$ as the *internal degrees of freedom only two kinds of spins* (no charges), which are determined by the two kinds of the Clifford algebra objects (there exist no additional Clifford algebra objects) - γ^a and $\tilde{\gamma}^a$ - and *interact correspondingly with the two kinds of the spin connection fields* - $\omega_{ab\alpha}$ and $\tilde{\omega}_{ab\alpha}$, the *gauge fields* of $S^{ab}(= \frac{i}{4} (\gamma^a\gamma^b - \gamma^b\gamma^a))$ (the generators of $SO(13,1)$) and $\tilde{S}^{ab}(= \frac{i}{4} (\tilde{\gamma}^a\tilde{\gamma}^b - \tilde{\gamma}^b\tilde{\gamma}^a))$ (the generators of $\widetilde{SO}(13,1)$) - and the *vielbeins* $f^{\alpha}{}_{a}$.

$$
\begin{aligned}
\mathcal{A} &= \int d^dx\, E\, \mathcal{L}_f + \int d^dx\, E\, (\alpha\, R + \tilde{\alpha}\, \tilde{R})\,, \\
\mathcal{L}_f &= \frac{1}{2}\, (\bar{\psi}\, \gamma^a p_{0a} \psi) + h.c., \\
p_{0a} &= f^{\alpha}{}_a p_{0\alpha} + \frac{1}{2E}\, \{p_\alpha, E f^{\alpha}{}_a\}_-, \quad p_{0\alpha} = p_\alpha - \frac{1}{2} S^{ab}\omega_{ab\alpha} - \frac{1}{2}\tilde{S}^{ab}\tilde{\omega}_{ab\alpha}, \\
R &= \frac{1}{2}\, \{f^{\alpha[a} f^{\beta b]}\, (\omega_{ab\alpha,\beta} - \omega_{ca\alpha}\, \omega^{c}{}_{b\beta})\} + h.c.\,,
\end{aligned}
$$

$$\tilde{R} = \frac{1}{2} \{ f^{\alpha[a} f^{\beta b]} \, (\tilde{\omega}_{ab\alpha,\beta} - \tilde{\omega}_{ca\alpha} \, \tilde{\omega}^{c}{}_{b\beta}) \} + h.c. \,. \tag{1}$$

Here [b] $f^{\alpha[a} f^{\beta b]} = f^{\alpha a} f^{\beta b} - f^{\alpha b} f^{\beta a}$. R and $\tilde{R}$ are the two scalars (R is a curvature).

This starting action *enables to represent* the *standard model* as an *effective low energy manifestation* of the *spin-charge-family* theory [1,2,4–9], *offering the explanation for all* the *standard model assumptions*: **a.** One representation of $SO(13,1)$ contains, if analyzed with respect to the *standard model* groups ($SO(3,1) \times SU(2) \times U(1)$ $\times SU(3)$) all the members of one family [2, Table 3], left and right handed, with the quantum numbers required by the *standard model.* **b.** The action explains the appearance of families due to the two kinds of generators of groups, the infinitesimal generators of one being S^{ab} (determining the charges), of the other $\tilde{S}^{ab}$ [2, Table 4] (determining the family quantum numbers). There are two decoupled groups of four families, remaining massless up to the electroweak break due to the "mass protection mechanism", that is due to the fact that the right handed members have no left handed partners with the same charges. **c.** The action explains the appearance of the gauge fields of the *standard model* [1,2]. [c] Before the electroweak break are all observable gauge fields massless: the gravity, the colour octet, the weak triplet and the hyper singlet vector gauge fields. All are the superposition of the $f^{\alpha}{}_{c} \, \omega_{ab\alpha}$ spinor gauge fields. **d.** The action explains the appearance of the scalar higgs and Yukawa couplings. There are scalar fields with the space index ($s = (7,8)$) and with respect to the space index with the weak and the hyper charge of the Higgs's scalar. They belong with respect to additional quantum numbers either to one of the two family groups of two triplets (expressible as superposition of $\tilde{\omega}_{abs}$), or they belong to three singlets, the scalar gauge fields of (Q, Q', Y') (expressible as superposition of $\omega_{tt's}$), which couple to the family members quantum numbers. Both kinds of the scalar fields determine the fermion masses [9], offering the explanation for the higgs, the Yukawa couplings and the heavy bosons masses. **e.** The starting action contains also additional $SU(2)_{II}$ vector gauge field (one of the components contributes to the hyper charge gauge fields) as well as the scalar fields with the space index $s \in (5,6)$ and $t \in (9, 10, \ldots, 14)$. All these fields gain masses of the scale of the condensate (Table 1), which they interact with. They all are expressible with the superposition of $f^{\mu}{}_{m} \, \omega_{ab\mu}$ or of $f^{\mu}{}_{m} \, \tilde{\omega}_{ab\mu}$ [2].

---

[b] $f^{\alpha}{}_{a}$ are inverted vielbeins to $e^{a}{}_{\alpha}$ with the properties $e^{a}{}_{\alpha} f^{\alpha}{}_{b} = \delta^{a}{}_{b}$, $e^{a}{}_{\alpha} f^{\beta}{}_{a} = \delta^{\beta}_{\alpha}$, $E = \det(e^{a}{}_{\alpha})$. Latin indices $a, b, .., m, n, .., s, t, ..$ denote a tangent space (a flat index), while Greek indices $\alpha, \beta, .., \mu, \nu, ..\sigma, \tau, ..$ denote an Einstein index (a curved index). Letters from the beginning of both the alphabets indicate a general index ($a, b, c, ..$ and $\alpha, \beta, \gamma, ..$), from the middle of both the alphabets the observed dimensions $0, 1, 2, 3$ ($m, n, ..$ and $\mu, \nu, ..$), indices from the bottom of the alphabets indicate the compactified dimensions ($s, t, ..$ and $\sigma, \tau, ..$). We assume the signature $\eta^{ab} = diag\{1, -1, -1, \cdots, -1\}$.

[c] In Ref [2] the proof is presented, that gauge fields can in the Kaluza-Klein theories be equivalently represented by either the vielbeins or the spin connection fields.

A ii. The manifold $M^{(13+1)}$ breaks first into $M^{(7+1)}$ times $M^{(6)}$ (manifesting as $SO(7,1) \times SU(3) \times U(1)$), affecting both internal degrees of freedom - the one represented by S^{ab} and the one represented by $\tilde{S}^{ab}$. Since the left handed (with respect to $M^{(7+1)}$) spinors couple differently to scalar (with respect to $M^{(7+1)}$) fields than the right handed ones, the break can leave massless and mass protected $2^{((7+1)/2-1)}$ families (which decouple into twice four families). The rest of families get heavy masses [10].

A iii. The manifold $M^{(7+1)}$ breaks further into $M^{(3+1)}\times M^{(4)}$.

A iv. The scalar condensate (Table 1) of two right handed neutrinos with the family quantum numbers of one of the two groups of four families, brings masses of the scale of the unification ($\propto 10^{16}$ GeV) to all the vector and scalar gauge fields, which interact with the condensate [1]. The condensate makes massive all the scalar gauge fields (with the space index (s=$(5,6,\ldots,14)$) and the vector gauge fields, manifesting nonzero τ^4, τ^{23}, $\tilde{\tau}^4$, $\tilde{\tau}^{23}$, $\tilde{N}_R^3$ [1,2]. Only the vector gauge fields of Y, $SU(3)$ and $SU(2)$ remain massless, since they do not interact with the condensate.

A v. There are nonzero vacuum expectation values of the scalar fields with the space index $s = (7,8)$, conserving the electromagnetic and colour charge, which cause the electroweak break and bring masses to all the fermions and to the heavy bosons.

All the rest scalar fields keep masses of the scale of the condensate and are correspondingly unobservable at the low energy regime.

Rewriting the fermion part of the action, Eq.(1), we can clearly see that the action does in the low energy regime manifest by the *standard model* required degrees of freedom of the fermions, as well as of the vector and scalar gauge fields [2, 4, 7–9].

$$\mathcal{L}_f = \bar{\psi}\gamma^m (p_m - \sum_{A,i} g^A \tau^{Ai} A^{Ai}_m)\psi + \sum_{s=7,8} \bar{\psi}\gamma^s p_{0s}\psi + \sum_{t=5,6,9,..,14} \bar{\psi}\gamma^t p_{0t}\psi\,, \quad (2)$$

where $p_{0s} = p_s - \frac{1}{2}S^{s's"}\omega_{s's"s} - \frac{1}{2}\tilde{S}^{ab}\tilde{\omega}_{abs}$, $p_{0t} = p_t - \frac{1}{2}S^{t't"}\omega_{t't"t} - \frac{1}{2}\tilde{S}^{ab}\tilde{\omega}_{abt}$, with $m \in (0,1,2,3)$, $s \in (7,8)$, $(s', s") \in (5,6,7,8)$, (a,b) (appearing in $\tilde{S}^{ab}$) run within either $(0,1,2,3)$ or $(5,6,7,8)$, t runs $\in (5,6,9,\ldots,14)$, $(t', t")$ run either $\in (5,6,7,8)$ or $\in (9,10,\ldots,14)$. The spinor function ψ represents all family members of all the $2^{\frac{7+1}{2}-1} = 8$ families.

The first two terms of Eq. (2) determine (in $d = (3+1)$) the kinematics and dynamics of spinor (fermion) fields, coupled to the vector gauge fields. The generators τ^{Ai} of the charge groups are expressible in terms of S^{ab} through the

state	τ^{23}	τ^{4}	Y	Q	$\tilde{\tau}^{23}$	$\tilde{\tau}^{4}$	$\tilde{N}_R^3$
$(\vert\nu_{\mathbf{1R}}^{\mathbf{VIII}}>_{\mathbf{1}}\ \vert\nu_{\mathbf{2R}}^{\mathbf{VIII}}>_{\mathbf{2}})$	1	−1	0	0	1	−1	1
$(\vert\nu_{1R}^{VIII}>_1\ \vert e_{2R}^{VIII}>_2)$	0	−1	−1	−1	1	−1	1
$(\vert e_{1R}^{VIII}>_1\ \vert e_{2R}^{VIII}>_2)$	−1	−1	−2	−2	1	−1	1

Table 1: The condensate of the two right handed neutrinos ν_R, with the VIIIth family quantum numbers ($\tilde{\tau}^{13} = 0 = \tilde{Y} = 0 = \tilde{Q} = 0 = \tilde{N}_L^3$), coupled to spin zero ($S^{03} = 0 = S^{12}$), the weak charge ($\tau^{13} = 0$) zero and belonging to a triplet with respect to the generators τ^{2i}, is presented, together with its two partners. The right handed neutrino has $Q = 0 = Y$. The triplet carries $\tau^4 = -1$, $\tilde{\tau}^{23} = 1$, $\tilde{\tau}^4 = -1$, $\tilde{N}_R^3 = 1$. The family quantum numbers are presented in [2, Table 4].

complex coefficients [d], ${c^{Ai}}_{ab}$ and fulfill the commutation relations

$$\tau^{Ai} = \sum_{a,b} {c^{Ai}}_{ab}\, S^{ab}\,, \quad \{\tau^{Ai}, \tau^{Bj}\}_- = i\delta^{AB} f^{Aijk} \tau^{Ak}\,. \tag{3}$$

They represent the colour, the weak and the hyper charge. The corresponding vector gauge fields A^{Ai}_m are expressible with the spin connection fields ω_{stm}, with (s,t) either $\in (5,6,7,8)$ or $\in (9,\dots,14)$, in agreement with the assumptions **A ii.** and **A iii.**

All the vector gauge fields, appearing in Eq. (2), except $A^{2\pm}_m$ and $A^{Y'}_m$, are massless before the electroweak break. $\vec{A}^3_m$ carries the colour charge, $\vec{A}^1_m$ carries the weak charge and A^Y_m carries the hyper charge Y. (See Ref. [2].)

Since spinors (fermions) carry besides the family members quantum numbers also the family quantum numbers (determined by $\tilde{S}^{ab}$), there are correspondingly $2^{(7+1)/2-1} = 8$ families [2], which split into two groups of $\widetilde{SU}(2)_{\widetilde{SO}(3,1)}$ $\times \widetilde{SU}(2)_{\widetilde{SO}(4)}$ families.

After the electroweak break - when the scalar fields with the space index $s = (7,8)$ (those with the family quantum numbers, which are the superposition of $\tilde{\omega}_{abs}$ and those with the family members quantum numbers (Q, Q', Y'), which are the superposition of $\omega_{s'ts}$) start to self interact and gain nonzero vacuum expectation values - only the charges $\vec{\tau}^3$ and $Q = Y + \tau^{13}$ are the conserved charges. No family quantum numbers are conserved, since all scalar fields with the family quantum numbers and the space index $s = (7,8)$ gain nonzero vacuum expectation values.

---

[d] $\vec{\tau}^1 := \frac{1}{2}(S^{58} - S^{67},\, S^{57} + S^{68},\, S^{56} - S^{78})\,, \vec{\tau}^2 := \frac{1}{2}(S^{58} + S^{67},\, S^{57} - S^{68},\, S^{56} + S^{78})$, $\vec{\tau}^3 := \frac{1}{2}\,\{S^{9\,12} - S^{10\,11}\,, S^{9\,11} + S^{10\,12},\, S^{9\,10} - S^{11\,12}\,, S^{9\,14} - S^{10\,13},\, S^{9\,13} + S^{10\,14}\,, S^{11\,14} - S^{12\,13}\,, S^{11\,13} + S^{12\,14},\, \frac{1}{\sqrt{3}}(S^{9\,10} + S^{11\,12} - 2S^{13\,14})\}\,, \tau^4 := -\frac{1}{3}(S^{9\,10} + S^{11\,12} + S^{13\,14})$. After the electroweak break the charges $Y := \tau^4 + \tau^{23}\,, Y' := -\tau^4 \tan^2\vartheta_2 + \tau^{23}\,, Q := \tau^{13} + Y\,, Q' := -Y\tan^2\vartheta_1 + \tau^{13}$ manifest. θ_1 is the electroweak angle, breaking $SU(2)_I$, θ_2 is the angle of the break of the $SU(2)_{II}$ from $SU(2)_I \times SU(2)_{II}$.

Below a common notation A^{Ai}_s is introduced for all the scalar fields with the space index $s = (7,8)$, since all are contributing to masses of fermions and weak bosons. $Ai = (Q, Q', Y')$, if A^{Ai}_s are superposition of ω_{abs}, while if A^{Ai}_s are superposition of $\tilde{\omega}_{\tilde{a}\tilde{b}s}$, they belong to $SU(2)$ triplets.

$$A^{Ai}_s = (A^{Q}_s, A^{Q'}_s, A^{Y'}_s, \vec{\tilde{A}}^{\tilde{1}}_s, \vec{\tilde{A}}^{\tilde{N}_{\tilde{L}}}_s, \vec{\tilde{A}}^{\tilde{2}}_s, \vec{\tilde{A}}^{\tilde{N}_{\tilde{R}}}_s),\ \tau^{Ai} = (Q, Q', Y', \vec{\tilde{\tau}}^{1}, \vec{\tilde{N}}_L, \vec{\tilde{\tau}}^{2}, \vec{\tilde{N}}_R). \quad (4)$$

3 Achievements and predictions of the *spin-charge-family* theory

Achievements and predictions of the spin-charge-family theory:
A. The theory does offer the explanation for all the assumptions of the *standard model* (Ref. [1,2]):
i. It explains all the properties of family members of one family - their spins and all the charges - clarifying the relationship between the spins and charges.
ii. It explains the properties of the vector fields, the gauge fields of the corresponding fermion charges.
iii. The scalar fields with the space index $s \in (7,8)$ belong to two doublets with respect to the space index s - carrying the weak and hyper charge equal to ($\pm\frac{1}{2}$, $\mp\frac{1}{2}$), respectively, although they are bosons (explaining the quantum numbers of the higgs) - while they belong with respect to additional quantum numbers either to three singlets or to twice two triplets of the family charges, explaining the appearance of the higgs and its quantum numbers and the Yukawa couplings.
iv. There are the nonzero vacuum expectation values of these scalar gauge fields, which, when starting to interact among themselves and gaining the nonzero vacuum expectation values, cause the break of the weak and the hyper charge symmetry.
B. The theory does offer the explanation for the dark matter. Neutral clusters of the members of the stable among the upper four families explain the appearance of the dark matter [8], *predicting a different "nuclear force" among the heavy nucleons of the upper four families.*
C. The scalar fields with the space index $s \in (9, \ldots, 14)$ belong with respect to the space index s to a triplet or anti-triplet (Table III in Ref. [3]). They cause transitions of anti-leptons into quarks and anti-quarks into quarks and back, transforming matter into anti-matter and back [1]. The condensate (Table 1) breaks CP symmetry. In the expanding universe, fulfilling the Sakharov request for an appropriate non thermal equilibrium, these colour triplet and anti-triplet scalars have a chance to explain the matter/anti-matter asymmetry in the universe, as well as the proton decay.
D. The theory might have a chance to explain the hierarchy of the fermion and boson masses as well.
E. The theory *predicts in the low energy regime two decoupled groups of four*

families. The fourth family to the observed three families might have masses at around 1 TeV [9]. More accurate predictions wait for better measured 3×3 submatrix, which uniquely determines the 4×4 unitary mixing matrices of quarks and leptons.
F. The theory *predicts additional scalars*, which are - as the observed higgs - the superposition of three singlets and two triplets.

The *spin-charge-family* theory seems to be a very promising candidate as a theory of the next step beyond the *standard model* - with its simple starting action (Eq. (1)) and only a few assumptions explaining all the assumptions of the *standard model*, while offering the explanation for also the existence of the dark matter and the matter/anti-matter asymmetry, promising as well answers to more open questions in the elementary particle physics and cosmology.

It is true that the assumptions in this theory, chosen in order that the theory manifests in the low energy regime the *standard model* properties, need and wait for explanation. The most needed are, of course, the experimental data confirmation of the predictions of this theory - the existence of the fourth family members, quarks and leptons, and of the additional scalar fields - making the theory trustworthy as the right next step beyond the *standard model*.

References

[1] N. S. Mankoč Borštnik, Phys. Rev. **D 91** 065004 (2015) [arxiv:1409.7791].
[2] N. S. Mankoč Borštnik, J. of Modern Phys. **6** 2244 (2015) [dx.doi.org/10.4236/jmp.2015.615230].
[3] N. S. Mankoč Borštnik, in *"What Comes Beyond the Standard Models"* (Proceedings of the 18th workshop "What comes beyond the standard models", July 11-19, 2015, Bled, Slovenia), Ed. by N. S. Mankoč Borštnik, H .B. Nielsen, D. Lukman, DMFA Založništvo, Ljubljana, December 2015, p. 87-110.
[4] A. Borštnik Bračič and N. S. Mankoč Borštnik, Phys. Rev. **D 74** 073013 (2006) [hep-ph/0301029].
[5] N. S. Mankoč Borštnik, Phys. Lett. **B 292** 25 (1992).
[6] N. S. Mankoč Borštnik, J. Math. Phys. **3** 3731 (1993).
[7] G Bregar, M. Breskvar, D. Lukman and N. S. Mankoč Borštnik, New J. of Phys. **10**, 093002 (2008).
[8] G. Bregar and N. S. Mankoč Borštnik, Phys. Rev. **D 80**, 083534 (2009).
[9] G. Bregar,N. S. Mankoč Borštnik, in *"What Comes Beyond the Standard Models"* (Proceedings of the 17th workshop "What comes

beyond the standard models", July 11-19, 2015, Bled, Slovenia), Ed. by N. S. Mankoč Borštnik, H .B. Nielsen, D. Lukman, DMFA Založništvo, Ljubljana, December 2014, p. 10-35 [arxiv:1412.5866].
[10] D. Lukman, N .S. Mankoč Borštnik and H. B. Nielsen, New J. Phys. **13**, 103027 (2011).

USE OF QUANTUM COHERENCE FOR NEUTRINO PHYSICS

M. Yoshimura
Center of Quantum Universe, Faculty of Science, Okayama University, Tsushima-naka 3-1-1 Kita-ku Okayama, 700-8530 Japan

Abstract. The conventional neutrino sources have been decay products of pion, muon and beta-nuclei. In this talk I shall present a new idea of neutrino pair beam emitted from circulating heavy ions in a pure mixed quantum state. This neutrino beam has a number of interesting features particularly suitable for CP violation measurements. Furthermore, intensive flux of the neutrino pair beam opens a possibility of using it as a tool for the earth tomography.

1 Introduction

Some time ago we proposed [1] the use of macro-coherence of a collective body of atoms to explore important unknown properties of neutrinos such as the absolute neutrino mass, distinction of mass type (either of Majorana or Dirac type), three CP violating (CPV) phase factors etc. The advantage of the process, radiative neutrino pair emission (RENP), $|e\rangle \to |g\rangle + \gamma + \nu\bar{\nu}$ where $|e\rangle$ is a metastable excited state of neutral atom, is many-fold: closeness of atomic energy levels to expected neutrino masses, high precision of laser frequencies available for a precision mass measurement, a great variety of experimental tools to manipulate their energy levels and their properties (parity, angular momentum mixing etc) etc. R and D experiments towards this goal have been initiated already for some time.

The success of atomic RENP project rests with the macro-coherence, a collective body of macroscopically many atoms made of a quantum mixture of states with phase coherence. Quantum mixture of two states, say the ground state $|g\rangle$ and the first excited state $|e\rangle$, is given by

$$|c(t)\rangle = \cos\theta_c e^{-i\epsilon_g t/\gamma}|g\rangle + \sin\theta_c e^{-i\epsilon_e t/\gamma} e^{i\varphi_c}|e\rangle \,. \tag{1}$$

An important point is that this state may be prepared by irradiation of laser (often more than two lasers of high quality) and that the pure mixed state $|c(t)\rangle$ may maintain its quantum nature for a long time after laser irradiation.

For our neutrino pair beam we use the quantum coherence of circulating ions. In particular, when lasers are irradiated after circulating ions have reached its maximum energy, its de-coherence time may be long. The coherence is defined by the off-diagonal element of the density matrix,

$$\rho_{eg}(t) = \cos\theta_c \sin\theta_c e^{-i\epsilon_{eg} t} = \frac{1}{2}\sin(2\theta_c) e^{-i\epsilon_{eg} t} \,, \tag{2}$$

with $\epsilon_{eg} = \epsilon_e - \epsilon_g$ the energy spacing of two levels. We ignore the unimportant phase φ_c.

In our para-hydrogen molecule experiments of paired superradiance (PSR) [2] the coherence of $\sim 8\%$ was realized for a macroscopic body of 15 cm long, which made it possible to observe the PSR events enhanced by more than 10^{18} orders of magnitude.

Below I ahall discuss yet another possibility of using the quantum coherence for neutrino physics. I shall concentrate on the principles and leave detailed calculations to published papers.

2 Photon emission from circulating quantum ions

What would happen when ions in this kind of quantum state are circulating in a ring of radius ρ ? The problem of photon emission may be worked out in similar fashions to the case of synchrotron radiation, however with an important difference clarified shortly. The probability amplitude for a single photon emission is

$$\mathcal{M}(t) = \int_0^t dt' \langle c(t')| \frac{e\vec{p}}{m_e} |c(t')\rangle \cdot \vec{A}(t';\vec{k},h) \tag{3}$$

$$= -i \frac{\epsilon_{eg}}{\sqrt{2\omega V}} \int_0^t dt' \langle c(t')|\vec{d}|c(t')\rangle \cdot \vec{e}_h e^{i\omega t' - i\vec{k}\cdot\vec{r}_I(t')} \,. \tag{4}$$

V is the quantization volume of emitted electromagnetic field. This is a coherent sum along the orbit trajectory (given by $\vec{r}_I(t')$) of a single ion. We assumed that the two states $|e\rangle, |g\rangle$ have quantum numbers connected by electric dipole transition.

The expectation value in eq.(4) is given in terms of the coherence we just defined:

$$\langle c(t')|\vec{d}|c(t')\rangle = \rho_{eg}(t')\vec{d}_{ge} \,. \tag{5}$$

The time dependence $\rho_{eg}(t') \propto e^{-i\epsilon_{eg}t'/\gamma}$ ($1/\gamma$ arising from the boosted motion) makes this process distinct from the synchrotron radiation in which either the ground or the excited state is considered, hence $\theta_c = 0$ or $\theta_c = \pi/2$ without the important time dependent off-diagonal part ρ_{eg}. The photon emission amplitude is, in the case of quantum circulating ions,

$$\mathcal{M}(t) = -i \frac{\epsilon_{eg}\rho_{eg}(0)}{\sqrt{2\omega V}} \vec{d}_{ge} \cdot \vec{e}_h \int_0^t dt' e^{i(\omega - \epsilon_{eg}/\gamma)t' - i\vec{k}\cdot\vec{r}_I(t')} \,. \tag{6}$$

When one computes the rate from $\partial_t |\mathcal{M}(t)|^2$, one has the phase integral of the form,

$$K(u;a,b) = \int_0^u du' \cos(bu' - a\sin u') \,, \tag{7}$$

$$u = \frac{\beta t}{\rho} \,, \qquad a = \rho\omega\cos\psi \,, \qquad b = \frac{\rho}{\beta}(\omega - \frac{\epsilon_{eg}}{\gamma}) \,. \tag{8}$$

The phase variable u that corresponds to the ion time is defined to be 2π for one turn of revolution of ions. We have introduced a coordinate system such that the emitted photon momentum
$\vec{k} = \omega(\cos\psi\cos\theta, \cos\psi\sin\theta, \sin\psi)$ and the ion orbit $\vec{r}_I(t') =$
$\rho(\sin\beta t'/\rho, \cos\beta t'/\rho - 1, 0)$. For simplicity we took the forward direction $\theta = 0$ tangential to the circulating ion in the orbit plane in the formulas for a, b.

An important point is that there are infinitely many stationary points of the phase factor, namely points of $\partial_u \cos(bu - a\sin u) = 0$ for $a > b$, which occurs only when $\epsilon_{eg} \neq 0$. This is the basic reason why this emission mechanism is so much different from the synchrotron radiation. Both a, b are very large, because this contains the product of the macroscopic length ρ in the unit of the microcscopic length $1/\epsilon_{eg}$ The product $\rho\epsilon_{eg}$ is typicaally larger than 10^{10}.

We shall make a comment on the difference from the well-known process of bremsstrahlung. The bremsstrahlung process may be described by a semi-classical formula under a sudden change of classical trajectory by a Coulomb field: its cross section is given in terms of the matrix element $\mathcal{M}_{brems}$,

$$d\sigma_B = e^2|\mathcal{M}_{brems}|^2 \frac{1}{(2\pi)^3}\frac{k^2dk}{\omega} d\Omega_k d\sigma_C\,, \tag{9}$$

$$\mathcal{M}_{brems} = \int dt \int d^3x \vec{\epsilon}_k \cdot \vec{J}(x) e^{i\omega t - i\vec{k}\cdot\vec{x}}\,, \tag{10}$$

$$\vec{J}(x) = \vec{v}\delta(\vec{x} - \vec{v}t) \quad t < 0\,, \qquad = \vec{v}\,'\delta(\vec{x} - \vec{v}\,'t) \quad t > 0\,. \tag{11}$$

This leads to the well-known factor,

$$\mathcal{M}_{brems} = -\frac{1}{\omega}\left(\frac{\vec{\epsilon}_k\cdot\vec{v}\,'}{1 - \vec{k}\cdot\vec{v}\,'/\omega} - \frac{\vec{\epsilon}_k\cdot\vec{v}}{1 - \vec{k}\cdot\vec{v}/\omega}\right)\,. \tag{12}$$

The important difference is that this is a single-shot event caused by a momentum change of charged particle. In contrast to this, the photon emission from circulating quantum ions is a continuous process from a single ion of a given trajectory.

3 Behavior of the phase integral

We show the phase integral $K(u; a, b)$ for a few choices of (a, b) values in Fig(1) Two typically different cases are clearly visible: one case in which the phase integral remains always bounded, and the other case in which an indefinite growth of integral occurs. These two cases can be understood by differences in the parameter (a, b) region belonging to the stability band and to the instability band in a Floquet system [4]. Our Floquet system is defined by three variables, $K(u)$, its integrand $\cos\Phi(u)$ and $\sin\Phi(u)$. A set of first order differential equations give a periodic function $\partial_u \Phi(u) = b - a\cos u$ as their coefficients. The periodic system gives rise to the stability and the instability band structure depending on the values (a, b). In terms of physics of photon emission these bands correspond to the non-emission and emission regions of the photon phase space of $\vec{k}$: in the large boost γ limit emitted photons are limited to the angular region within $1/\gamma$ from the tangential direction to the circulating photon, and the photon energy region of $0 \sim 2\gamma\epsilon_{eg}$. Details of angular distribution and energy spectrum are given in [4].

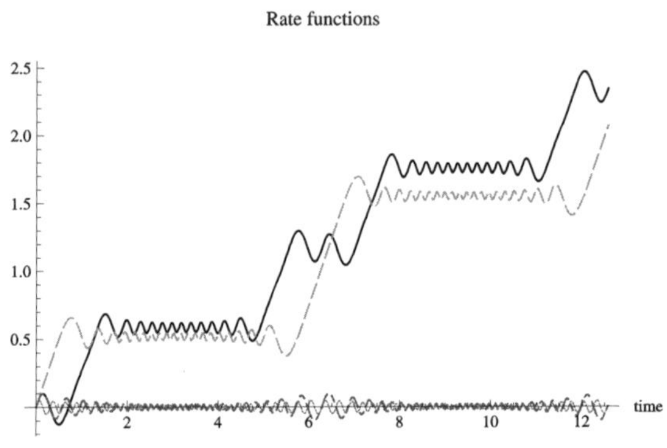

Figure 1: Phase integrals $K(u; a, b)$ in various (a, b): $(20, 20 \times 0.5)$ in solid black, $(20, 20)$ in dashed red, $(20, 20 \times 1.5)$ in dash-dotted blue, and $(20, 20 \times 2)$ in dotted black.

The photon emission from circulating quantum ions has its own merits. The process requires the coherence (quantum mixture in the pure state) at the single ion level. But if the coherence persists at the macroscopic level, namely the phase between many ions is maintained at a macroscopic level, then a coherent gamma ray beam may become feasible. This is an analgue of laser beam in the optical region. With realization of the coherent gamma ray beam it may become possible to image atoms having nuclei and electron clouds well separated. Clearly, there are many applications of this kind of coherent gamma ray beam.

4 Application to neutrino pair emission

Neutrino pairs can be produced by ionic de-excitation, but its rates are usually very small due to the high power dependence of rates to the energy spacing ϵ_{eg}. With the possibility of extended energy range of $2\gamma\epsilon_{eg}$ for quantum circulating ions the production rate of neutrino pair may become large from circulating quantum ions.

In the standard electroweak theory the neutrino pair emission is described by four Fermi interaction of neutrino with electron, and this interaction has two parts, the vector and the axial-vector contribution. In the Fierz-transformed notation, the interaction hamiltonian in the mass-eigenstate basis is

$$H_w^{(0)} = \frac{G_F}{\sqrt{2}} \int d^3x \sum_{i,j=1,2,3} (c_{ij}^V V^\beta(x) + c_{ij}^A A^\beta(x)\,)\nu_i^\dagger(x)\sigma_\beta\nu_j(x)\,, \quad (13)$$

$$C^A \equiv (c_{ij}^A)\,, \qquad c_{ij}^A = U_{ei}U_{ej}^* - \frac{1}{2}\delta_{ij}\,, \qquad C^A(C^A)^\dagger = \frac{1}{4}\,, \quad (14)$$

$$C^V = (\sum_b U_{bi}^* c_b^V)\,, \qquad (c_b^V) = (\frac{1}{2}(1 + 4\sin^2\theta_w)\,, \quad (15)$$

$$-\frac{1}{2}(1 - 4\sin^2\theta_w)\,, -\frac{1}{2}(1 - 4\sin^2\theta_w))\,,$$

with $(\sigma_\beta) = (1, -\vec{\sigma})$. We use the neutrino index convention of Roman alphabets, a, b, c, to indicate neutrino flavor states, ν_e, ν_μ, ν_τ, and Roman alphabets, i, j, k, to indicate mass eigenstates ν_1, ν_2, ν_3. For deeply bound electrons in heavy ions both axial-vector and vector parts may contribute nearly equally, since electron's motion are relativistic, while for lighter ions the axial-vector parts dominate. Depending on the spin parity relation $\Delta J^P = 1^-$ or 1^+ between two levels $|e\rangle, |g\rangle$, the production process is caused by the vector current contribution or the axial-vector contribution. One may choose the ion beam accordingly. It turns out that from oscillation experiment points the vector contribution is more important. See [6] on more of this point.

The weak mixing angle $\sin^2\theta_w$ is close to $1/4$ (more precisely ~ 0.231). In the limit of $\sin^2\theta_w = 1/4$ the Z-boson exchange contribution vanishes in the vector part, and it reduces to $(c_b^V) = (1, 0, 0)$.

Formulas of neutrino pair production rate of the vector current contribution are given in [6]. Its overall rate factor is given by

$$\frac{8}{27\sqrt{\pi}(2\pi)^4} N\sqrt{\rho\epsilon_{eg}} \frac{G_F^2\epsilon_{eg}^5}{\alpha}\gamma^{11/2} \sim 7.1 \times 10^{18}\ \text{Hz} \quad (16)$$

$$\frac{N}{10^8}\left(\frac{\rho\epsilon_{eg}}{10^{14}}\right)^{1/2}\left(\frac{\epsilon_{eg}}{10\text{keV}}\right)^5(\frac{\gamma}{10^3})^{11/2}\,.$$

Here N is the available number of ions, the coherence squared $|\rho_{eg}|^2$ times the number of circulating ions. The directional collimation to $1/\gamma$ and the energy range $0 \sim 2\gamma\epsilon_{eg}$ hold, but details of distribution are considerably different. In Fig(2) we illustrate event rates for 10 kt Fe detector (which gives the detection probability $\sim 6 \times 10^{-11}$ and $\sim 2 \times 10^{-11}$ for ν_μ and $\bar{\nu}_\mu$, respectively) placed within the forward cone of $1/\gamma$. Event rates are sufficient for precision neutrino experiments for this choice of parameters.

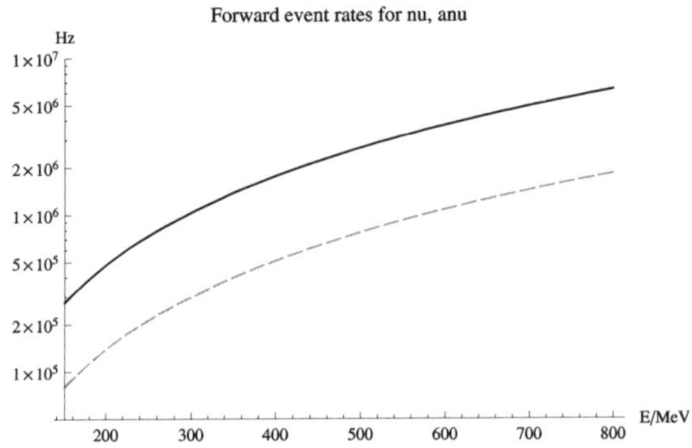

Figure 2: Forward neutrino events in the angular area of π/γ^2: ν_μ rate in solid black and $\bar{\nu}_\mu$ rate in dashed red. 10 kt Fe detector of fully covered by the pair beam is assumed, with the level spacing 50 keV and the boost factor 5×10^3.

5 Neutrino oscillation in the neutrino pair beam

As shown in [6], the vector part contribution $\propto c_b^V$ gives the propagation probability of a single neutrino (the other in the pair assumed undetected) in distance L:

$$P_a(E, L; m_i, \delta) \equiv \frac{1}{3(1-4\sin^2\theta_w)^2/4 + 4\sin^2\theta_w}$$
$$\times \left(\frac{1}{4}(1-4\sin^2\theta_w)^2 + 4\sin^2\theta_w |\sum_j U_{ej}^* U_{aj} \exp[-i\frac{m_j^2 L}{2E}]|^2 \right) . \quad (17)$$

The last formula (17) is valid when the earth matter effect is neglected. The oscillating component in eq. (17) is equivalent to the $\nu_e \to \nu_a \,, a = \mu, \tau$ appearance probability multiplied by

$$\frac{4\sin^2\theta_w}{3(1-4\sin^2\theta_w)^2/4 + 4\sin^2\theta_w} \sim 0.995 \,. \quad (18)$$

Thus, the constant off-set term $\propto (1-4\sin^2\theta_w)^2$ in eq. (17) is very small. In the limit of $\sin^2\theta_w = 1/4$ there is no contribution to the vector part from Z-boson exchange. Due to the dominance of $\nu_e \to \nu_a, a = \mu, \tau$ in the oscillating term, oscillation patterns in the pair beam have similarities to the β [11] and $\beta^\pm$ beam [12].

We refer to [6] for details of the oscillation patterns. The important point is that one can perform neutrino oscillation experiments and determine CPV parameter δ at relatively short distances such as 30 $\sim$ 100 km, very close to the ion synchrotron site. To minimize the contamination from the earth matter effect against CPV measurements, it might be a better choice to do experiments on earth. This is possible if the accelerator is placed, say $100 \sim 200$ m underground.

6 Future prospects

Since implementation the neutrino pair beam is bound to be a big project, it would be of help for the project to have multiple disciplinary objectives. We mention two interesting applications related to quantum ion circulation.

The most important application is to develop coherent gamma ray beam, which is realized when the circulating ions maintain the macro-coherence. We may expect fast imaging of atoms separating nuclei from electron cloud.

Another possibility is a direct application of the neutrino pair beam. Since the earth matter effect is not completely negligible and the intensity of the pair beam is very large, both the short range and the long range earth tomography may become feasible. We did some sample calculation treating the earth matter effect [8], [9], [10] perturbatively. The result is illustrated in Fig(3). To perform the earth tomography systematically, it should be possible to direct the neutrino pair beam into any downward direction and to have movable detector system at any point on earth or on sea.

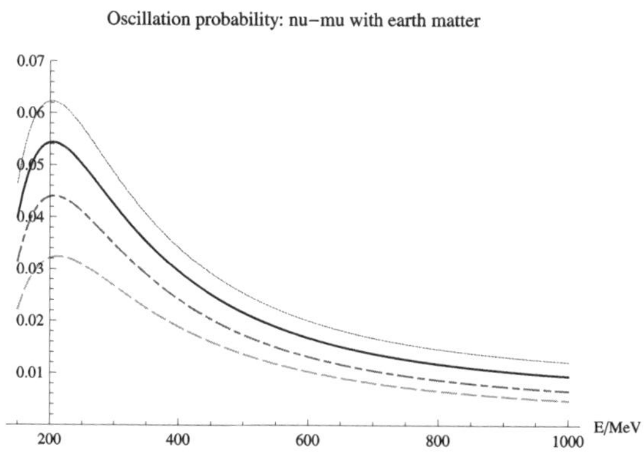

Figure 3: Effect of Fe clumps of a few sizes placed at the center of 100 km path length: without clump in sold black, 10 km clump in dashed red, 20 km clump in dash-dotted blue, and 30 km clump in dotted black. All assuming $\delta = -\pi/2$.

Clearly, both experimental R and D works of quantum coherent ion circulation and theoretical studies of candidate ions are required for further development of this new project.

References

[1] A. Fukumi et al., Prog. Theor. Exp. Phys. (2012) 04D002.

[2] Y. Miyamoto et al, Prog.Theor.Exp.Phys. 113C01(2014), and arXiv:1406.2198v2 [physics.atom-ph] (2014), and further reports.

[3] M. Yoshimura and N. Sasao, Physical Review **D92**, 073015 (2015) and arXiv: 1505.07572v2 (2015).

[4] M. Yoshimura and N. Sasao, arXiv:1512.06959 (2015).

[5] M. Yoshimura and N. Sasao, Phys. Lett. **B 753**, 465(2016), and arXiv: 1505.07663 (2015).

[6] T. Asaka, M. Tanaka, and M. Yoshimura, arXiv:1512.08076 (2015).

[7] M. Yoshimura, Phys.Rev. **D93** (2016) 1, 013016, and arXiv: 1508.027952v2 (2015).

[8] L. Wolfenstein, Phys. Rev. D17, 2369 (1978).

[9] V. Barger, K. Whisnant, S. Pakvasa, and R.J.N. Phillips, Phys. Rev. **D22**, 2718 (1980).

[10] Z.Z. Xing, Phys. Lett. B487, 327 (2000) and arXiv hep-ph/0002246 v3 (2000).
See also P.F. Harrison and W.G. Scott, Phys. Lett. B **476**, 349 (2000).

[11] P. Zucchelli: Phys. Lett. **B 532**, 166 (2002); hep-ex/0107006.

[12] A. Fukumi, I. Nakano, H. Nanjou, N. Sasao, S. Sato, and M. Yoshimura, Journal of the Physical Society of Japan **78**, 013201 (2009).

[13] We correct the sign mistake of earth matter effect in [8] and [9].

[14] K. A. Olive *et al.* [Particle Data Group Collaboration], Chin. Phys. C **38** (2014) 090001.

LIGHT STERILE NEUTRINOS IN COSMOLOGY

Stefano Gariazzo [a]

Department of Physics, University of Torino, and INFN, Sezione di Torino, Via P. Giuria 1, I–10125 Torino, Italy

Abstract. We briefly discuss the main effects of the presence of a Light Sterile Neutrino (LSν) in Cosmology and how its properties can be constrained by Cosmic Microwave Background (CMB) and other cosmological measurements.

1 Introduction

The measurements of neutrino oscillations in atmospheric, solar and long-baseline neutrino oscillation experiments firmly indicate that at least two neutrino mass eigenstates have a positive mass (see e.g. Refs. [1–3]). The mixing between the flavour neutrino eigenstates, that we indicate with ν_α ($\alpha = e, \mu, \tau$), and the mass neutrino eigenstates, indicated with ν_k ($k = 1, 2, 3$), can be written in terms of the mixing matrix $U_{\alpha k}$. The mixing relation is $\nu_\alpha = \sum U_{\alpha k}\nu_k$. The mixing matrix $U_{\alpha k}$ is described by three mixing angles, a CP violating phase and two Majorana phases, that are physical only if neutrino are Majorana particles. The description of the three neutrino mixing is completed by the existence of two squared-mass differences Δm^2_{SOL} and Δm^2_{ATM}, which are measured in neutrino oscillation experiments (see e.g. Ref. [3]).

Among the numerous neutrino oscillation experiments, it is possible to find some anomalies that cannot be explained in the context of the three neutrino mixing. These anomalies are related to the results obtained in short-baseline (SBL) neutrino oscillation experiments by the LSND experiment [4], by several experiments measuring the electron antineutrino flux from nuclear reactors [5] and in the calibration of the GALLEX and SAGE Gallium solar neutrino experiments [6] (see Ref. [7] for a full list of references). A possible explanation of these anomalies involves new neutrino oscillations, driven by a third squared-mass difference $\Delta m^2_{SBL} \simeq 1\,\mathrm{eV}$, that implies the existence of a fourth neutrino mass eigenstate. This additional *light* neutrino degree of freedom should be mainly *sterile*. We assume that the active neutrinos have a negligible mass with respect to this *light sterile neutrino* (LSν). The squared-mass difference Δm^2_{SBL} implies that the sterile neutrino has a mass $m_s \simeq m_4 \simeq \sqrt{\Delta m^2_{SBL}} \simeq 1\,\mathrm{eV}$. The opposite choice, that is an inverted sterile hierarchy with $m_4 \ll m_1, m_2, m_3$, is forbidden by the cosmological bounds on the neutrino masses [8] and by the experimental bounds on neutrinoless double-β decay, assuming that massive neutrinos are Majorana particles (see Ref. [9]).

[a]E-mail: gariazzo@to.infn.it

2 Neutrino Parameterization

To describe the contribution of the LSν in Cosmology we must separate the epoch when it was relativistic from the final part of the Universe evolution, when it is non-relativistic.

The contribution of all the relativistic species to the energy density of radiation ρ_r in the early Universe can be written in terms of the effective number of relativistic species N_{eff}, defined through:

$$\rho_r = \left[1 + \frac{7}{8}\left(\frac{4}{11}\right)^{4/3} N_{\text{eff}}\right] \rho_\gamma\,, \tag{1}$$

being ρ_γ the energy density of photons. The contribution of the three active neutrinos is $N_{\text{eff}}^{\text{sm}} = 3.046$, larger than 3 because of the non-instantaneous decoupling of the neutrinos from the electron-photon fluid and of the entropy transfer at the electron decoupling [10]. If the LSν is fully thermalized with the active neutrinos, its contribution to N_{eff} should be 1 and $N_{\text{eff}} = 4.046$. Assuming that only photons and neutrinos are relativistic in the early Universe, the contribution of the LSν can be written as [11]

$$\Delta N_{\text{eff}} = N_{\text{eff}} - N_{\text{eff}}^{\text{sm}} = \left[\frac{7}{8}\frac{\pi^2}{15} {T_\nu}^4\right]^{-1} \frac{1}{\pi^2} \int dp\, p^3 f_s(p)\,, \tag{2}$$

where T_ν is the active neutrino temperature, p is the neutrino momentum and $f_s(p)$ is the LSν momentum distribution function.

Since the neutrino temperature today is $T_\nu \propto 10^{-4}\,\text{eV}$, the LS$\nu$ is no more relativistic. The energy density of the LSν when it becomes non-relativistic can be parameterized through the effective mass m_s^{eff}, defined as [11, 12]:

$$\frac{m_s^{\text{eff}}}{94.1\,\text{eV}} = \frac{h^2}{\rho_c}\frac{m_s}{\pi^2} \int dp\, p^2 f_s(p)\,, \tag{3}$$

where h is the reduced Hubble parameter and ρ_c is the critical energy density.

Both ΔN_{eff} and m_s^{eff} depend on the LSν momentum distribution function $f_s(p)$. If the LSν is relativistic at decoupling, $f_s(p)$ does not depend on its mass m_s. Since sterile and active neutrinos are mixed, and since the LSν does not have electroweak interactions, its decoupling cannot occur later than that of the active neutrinos, that takes place at temperatures of about 1 MeV. Depending on the production mechanism of the LSν there are two main possibilities for $f_s(p)$. If the LSν is produced through some thermal (TH) process, its momentum distribution function can be written as a Fermi-Dirac function at a temperature T_s, that is $f_s^{TH}(p) = \left(e^{p/T_s} + 1\right)^{-1}$. In this case it is possible to show that $m_s^{\text{eff}} = \Delta N_{\text{eff}}^{3/4}\, m_s$. If the production is non-thermal, instead, there are several possible scenarios. A popular one is the non-resonant production or Dodelson-Widrow (DW) scenario [13]. In the DW scenario the LSν momentum distribution function can be written as $f_s^{DW}(p) = \beta\left(e^{p/T_\nu}+1\right)^{-1}$, where β is a normalization factor, and one obtains $m_s^{\rm eff} = \Delta N_{\rm eff}\, m_s$.

3 Neutrino Effects on the CMB Spectrum

The effective number of relativistic species controls the expansion rate H in the early Universe and the time of the matter-radiation equality. This influences the CMB spectrum in several ways. The redshift of matter-radiation equality is defined by $1 + z_{\rm eq} = \omega_r/\omega_m$, where $\omega_i = \rho_i h^2/\rho_c$ is the density parameter of the species i and the subscript m is for the matter component. If $N_{\rm eff}$ is increased, the equality is delayed and H at CMB decoupling is higher. The consequences are [7, 14] an increase of the first peak of the CMB, due to the early Integrated Sachs-Wolfe (ISW) effect, and a shift at higher multipoles of the angular scale of the acoustic peaks. These effects can be seen comparing the black ($\Delta N_{\rm eff} = 0$) and the red ($\Delta N_{\rm eff} = 2$) spectra in Fig. 1. If one wants to restore the matter-radiation equality to the former redshift, it is necessary to rescale the matter density. In order to maintain the redshift of the matter-Λ equality, where Λ indicates the cosmological constant, it is necessary to vary also the Λ energy density. If one varies together ω_m and $N_{\rm eff}$ (blue), or ω_m, ω_Λ and $N_{\rm eff}$ (green), the effects of an higher $N_{\rm eff}$ can be partially compensated. In the last case, the effect of an higher energy density at all the times produces a suppression of the CMB spectrum at high-ℓ, also known as Silk damping effect.

As its mass is around 1 eV, the LSν is still relativistic at the time of matter-radiation equality. As a consequence, its mass can influence only the late-time evolution. In particular, the impact of m_s on the CMB spectrum is visible mainly through the slope of the CMB spectrum at low multipoles and through the effect of neutrino free-streaming. The slope of the CMB spectrum at low multipoles is related to the late ISW effect, originated after the Universe enters the Λ-dominated phase. Neutrino free-streaming occurs since they have a large velocity and they cannot cluster at small scales. After the non-relativistic transition, however, their velocity diminishes and they start to cluster, as dark matter does. The comoving free-streaming scale reaches a maximum at the time of the non-relativistic transition, that occurs at the wavemode $k_{\rm nr} \simeq 0.0178\, \Omega_m^{1/2}\, (T_\nu/T_s)^{1/2}\, (m_s/1\,{\rm eV})^{1/2}\, h\,{\rm Mpc}^{-1}$. All the wavemodes below $k_{\rm nr}$ are not affected by the neutrino free-streaming, while all the wavemodes $k \gtrsim k_{\rm nr}$ are suppressed. The amount of the suppression depends on the neutrino masses, but it is not particularly strong in the CMB spectrum. The best way to measure the neutrino masses in Cosmology, in fact, is to study the free-streaming effect in the power spectrum of matter perturbations. Future experiments such as DESI [15] and Euclid [16] will put strong constraints on the neutrino masses, and possibly on the neutrino mass hierarchy [17, 18].

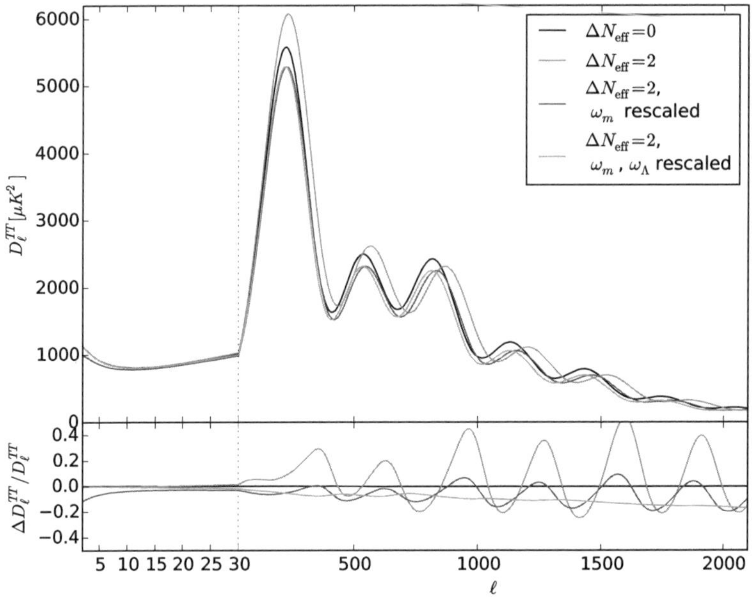

Figure 1: Comparison of different CMB temperature spectra predicted by the theoretical model. We show the effects of altering $N_{\rm eff}$ alone or in combination with ω_m and ω_Λ.

4 Current Constraints from Cosmology

Several joint analyses of cosmological and SBL neutrino oscillation data have been performed in the past. Here we present the results obtained in Ref. [19], where the joint analyses are based on the CMB temperature spectrum from the Planck 2013 release [20], on the low-ℓ polarization spectra from WMAP [21] and on the high-ℓ spectrum calculated through the ACT/SPT likelihood [22].

To derive constraints on the LSν properties we consider an extension of the well-known ΛCDM model, expanded with the LSν presented in the previous Sections. The constraints are obtained from the cosmological datasets alone or adding the results of the analysis of SBL neutrino oscillations data as a prior for the LSν mass. In the left panel of Fig. 2 we show the marginalized constraints in the $m_s^{\rm eff}$-$N_{\rm eff}$ plane obtained from the analysis of the CMB data mentioned above, both without SBL prior (black) or with a SBL prior for the mass of a DW (red) or a TH (blue) LSν. Dotted (dashed) lines indicate constant masses, with values written next to each line, for the DW (TH) model. As we can see, the SBL prior forces the physical mass to lie in the range allowed by SBL experiments, with the consequence that not all the values of $m_s^{\rm eff}$-$N_{\rm eff}$ favoured

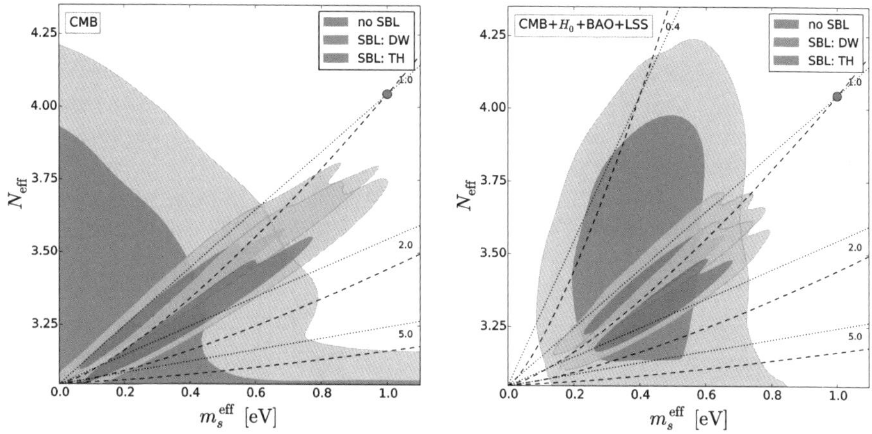

Figure 2: Constraints at 68% and 95% CL in the $m_s^{\rm eff}$-$N_{\rm eff}$ plane from the CMB data alone (left panel) and from a combination of CMB+H_0+BAO+LSS data (right panel). The green point indicates a fully thermalized LSν with $m_s = 1\,\text{eV}$. Dotted (dashed) lines indicate constant masses, for values written in the figures, for the DW (TH) model. From [7].

by the CMB data are compatible with the SBL data. A fully thermalized LSν is allowed only if it has a very small mass, but for the SBL masses an incomplete thermalization is required, being $N_{\rm eff} \lesssim 3.75$ at 95% CL. Larger LSν masses are allowed only if $N_{\rm eff}$ is close to 3.046.

The ΛCDM model explains the majority of the cosmological observations, but some tensions exist. In particular, in the context of the ΛCDM model a tension between the determinations of the matter fluctuations and of the Hubble parameter at high- and low-redshift may appear, possibly because of unaccounted systematics. The value of H_0 determined by Planck [8, 12] is in tension with the local determinations, that give higher values [23, 24]. At the same time, the amount of matter fluctuations at small scales, encoded in the parameter σ_8, is smaller if obtained from local measurements of weak lensing [25] or from cluster counts [26] than when derived from CMB constraints [8, 12].

A LSν has the potential to reconcile local and cosmological observations. This is the consequence of the correlation between $N_{\rm eff}$ and H_0, and of the LSν free-streaming, that reduces the matter fluctuations at small scales. In the right panel of Fig. 2 we show the bounds for the LSν as derived from an extended dataset, that includes the same CMB data as above, plus the measurements of the Baryon Acoustic Oscillations (BAO), a prior on H_0 and one on σ_8 from Large Scale Structures (LSS). As we anticipated, the fit prefers positive LSν masses in order to reduce the predicted value of σ_8 through the neutrino free-streaming, and $N_{\rm eff} > 3.046$ in order to have an higher H_0. Even in this case, however, the full thermalization of the LSν is strongly disfavoured.

5 Open Problems and Conclusions

Despite the fact that the LSν can explain some of the tensions between the current cosmological observations at high- and low-redshift, it seems that a full reconciliation is rather complicated. The main reason is that we do not have a valid explanation to the fact that the LSν is not fully thermalized. With the mixing parameters determined by SBL oscillations, in fact, the LSν should fully thermalize with the active neutrinos [27]. Moreover, it seems that the LSν cannot fully reconcile both the H_0 and the σ_8 tensions, since the direction of the degeneracies in the σ_8-H_0 plane is not optimal [8].

Unless a new mechanism that explains the low thermalization of the LSν is found (see e.g. Ref. [7] for some possibilities), the current status of the global analyses shows that the SBL neutrino seems to be excluded by Cosmology. Future investigations of SBL neutrino oscillations and of Cosmology are required to grow a more complete knowledge of the theories and to explain of the present tensions between the different experimental probes.

Acknowledgments

The work of S. G. is supported by the Theoretical Astroparticle Physics research Grant No. 2012CPPYP7 under the Program PRIN 2012 funded by the Ministero dell'Istruzione, Università e della Ricerca (MIUR).

References

[1] Carlo Giunti and Chung W. Kim. Fundamentals of Neutrino Physics and Astrophysics. 2007.
[2] Samoil Bilenky. *Introduction to the physics of massive and mixed neutrinos*, volume 817. Springer, 2010.
[3] K.A. Olive et al. Review of particle physics. *Chin.Phys.*, C38:090001, 2014.
[4] A. Aguilar-Arevalo et al. Evidence for neutrino oscillations from the observation of anti-neutrino(electron) appearance in a anti-neutrino(muon) beam. *Phys. Rev.*, D64:112007, 2001.
[5] G. Mention et al. The Reactor Antineutrino Anomaly. *Phys. Rev.*, D83:073006, 2011.
[6] Carlo Giunti and Marco Laveder. Short-Baseline Active-Sterile Neutrino Oscillations? *Mod. Phys. Lett.*, A22:2499–2509, 2007.
[7] S. Gariazzo et al. Light sterile neutrinos. arXiv:1507.08204 [hep-ph].
[8] P. A. R. Ade et al. Planck 2015 results. XIII. Cosmological parameters. arXiv:1502.01589 [astro-ph.CO].
[9] S. M. Bilenky and C. Giunti. Neutrinoless Double-Beta Decay: a Probe of Physics Beyond the Standard Model. *Int. J. Mod. Phys.*, A30:1530001, November 2015.

[10] Gianpiero Mangano et al. Relic neutrino decoupling including flavor oscillations. *Nucl. Phys.*, B729:221–234, 2005.
[11] Mario A. Acero and Julien Lesgourgues. Cosmological constraints on a light non-thermal sterile neutrino. *Phys. Rev.*, D79:045026, 2009.
[12] P. A. R. Ade et al. Planck 2013 results. XVI. Cosmological parameters. *Astron. Astrophys.*, 571:A16, November 2014.
[13] Scott Dodelson and Lawrence M. Widrow. Sterile-neutrinos as dark matter. *Phys. Rev. Lett.*, 72:17–20, 1994.
[14] Maria Archidiacono et al. Cosmic dark radiation and neutrinos. *Adv. High Energy Phys.*, 2013:191047, 2013.
[15] Michael Levi et al. The DESI Experiment, a whitepaper for Snowmass 2013. arxiv:1308.0847 [astro-ph.CO].
[16] R. Scaramella et al. Euclid space mission: a cosmological challenge for the next 15 years. *IAU Symp.*, 306:375–378, 2015.
[17] Andreu Font-Ribera et al. DESI and other dark energy experiments in the era of neutrino mass measurements. *JCAP*, 1405:023, 2014.
[18] Benjamin Audren et al. Neutrino masses and cosmological parameters from a Euclid-like survey: Markov Chain Monte Carlo forecasts including theoretical errors. *JCAP*, 1301:026, 2013.
[19] S. Gariazzo et al. Light Sterile Neutrinos in Cosmology and Short-Baseline Oscillation Experiments. *JHEP*, 11:211, 2013.
[20] P. A. R. Ade et al. Planck 2013 results. I. Overview of products and scientific results. *Astron. Astrophys.*, 571:A1, November 2014.
[21] C. L. Bennett et al. Nine-Year Wilkinson Microwave Anisotropy Probe (WMAP) Observations: Final Maps and Results. *Astrophys. J. Suppl.*, 208:20, 2013.
[22] J. Dunkley et al. The Atacama Cosmology Telescope: likelihood for small-scale CMB data. *JCAP*, 1307:025, 2013.
[23] Adam G. Riess et al. A 3% Solution: Determination of the Hubble Constant with the Hubble Space Telescope and Wide Field Camera 3. *Astrophys. J.*, 730:119, 2011.
[24] George Efstathiou. H0 Revisited. *Mon. Not. Roy. Astron. Soc.*, 440:1138–1152, 2014.
[25] Catherine Heymans et al. CFHTLenS: The Canada-France-Hawaii Telescope Lensing Survey. *Mon. Not. Roy. Astron. Soc.*, 427:146, 2012.
[26] P. A. R. Ade et al. Planck 2013 results. XX. Cosmology from Sunyaev–Zeldovich cluster counts. *Astron. Astrophys.*, 571:A20, 2014.
[27] Steen Hannestad et al. Thermalisation of light sterile neutrinos in the early universe. *JCAP*, 1207:025, 2012.

PROMPT NEUTRINO FLUX IN THE ATMOSPHERE REVISITED

Maria Vittoria Garzelli [a] Sven-Olaf Moch, and Günter Sigl
II Institute for Theoretical Physics, University of Hamburg, Luruper Chaussee 149, D-22761 Hamburg, Germany

Abstract. Prompt neutrino fluxes due to the interactions of high-energy cosmic rays with the Earth's atmosphere are backgrounds in the search for high-energy neutrinos of galactic or extra-galactic origin performed by Very Large Volume Neutrino Telescopes. We summarize our predictions for prompt neutrinos, showing their basic features as emerging from the calculation in a QCD framework capable of describing recent charm data from the Large Hadron Collider.

1 Motivation of this work

Very large volume neutrino telescopes (VLVνTs) provide a unique window to the high-energy universe [1], because they allow for the detection of high-energy neutrinos which can reach the Earth after having travelled from far galactic or extra-galactic sources, almost undeflected by cosmic magnetic fields and insensitive to electromagnetic and strong forces. In this respect neutrinos propagate quite differently with respect to charged cosmic rays (CRs), and are thus precious messengers capable of providing direct information on the regions of the Universe where they are produced.

A first detection of high-energy leptonic events up to an energy of a few PeV was reported by the IceCube collaboration. The initial evidence from the first two-year data analysis [2] was subsequently strenghtened by increased statistics in the three and four-year analyses [3, 4]. So far, more than 50 high-energy starting events have been recorded. The discussion on the origin of these events is quite lively, and a number of hypotheses concerning the possible sources has been formulated. Although it is believed that most of the neutrino events detected so far have an extraterrestrial origin, theoretical calculations predict the existence of a terrestrial component due to the decays of the unstable hadrons produced by the interactions of high-energy and ultra-high-energy CRs with the Earth's atmosphere [5]. This terrestrial component should be properly taken into account as a background and subtracted in any astrophysical analysis. The terrestrial component is usually separated into two contributions: the conventional neutrino flux, coming from the decay of $\pi^{\pm}$ and $K^{\pm}$ mesons, and the prompt neutrino flux, originating from the decay of heavier mesons and hadrons, in particular those including charm quarks as valence quarks in their composition. The prompt component is supposed to become dominant with respect to the conventional one only at energies high enough. Several computations of conventional neutrino fluxes exist (see e.g. Ref. [6]), together

[a]E-mail: maria.vittoria.garzelli@desy.de

with experimental evidence. On the other hand, no experimental evidence for the existence of a prompt flux has emerged so far. It is thus important to provide precise and refined theoretical predictions of prompt neutrino fluxes, to understand if the lack of detection is just due to the still low statistics of the VLVνT experiments at high energy, or if the methodologies and/or theories used so far to predict these fluxes have some flaws.

2 Charm hadroproduction in QCD and prompt neutrino fluxes

In Ref. [7] we have recently computed prompt neutrino fluxes making use of QCD, with microscopic interactions described by event generators including up-to-date information from the QCD theory community working at hadron colliders. As shown in the following, we have used the same generators adopted in our astrophysical study, to produce predictions which can be compared with experimental data recently collected at the Large Hadron Collider (LHC).

Nowadays, state-of-the-art tools for predicting many differential distributions at the LHC are represented by Shower Monte Carlo (SMC) codes, matched with hard-scattering processes evaluated in perturbative QCD including next-to-leading-order (NLO) radiative corrections at least [8]. For a few selected processes, calculations at the differential level involving the resummation of different kinds of large logarithms to all orders, or next-to-NLO (NNLO) QCD radiative corrections, start to be available as well. However, this is not yet the case for the processes involving charm quarks we are interested in in this paper. Interfacing hard scattering amplitudes including radiative corrections with the parton shower algorithms (PS) embedded in the SMCs requires a careful matching, to avoid double counting effects and to ensure a proper coverage of the whole phase space available for light parton emissions. At present, one of the most widely used NLO QCD + PS matching schemes is POWHEG [9,10], which has been automated in dedicated tools, like e.g. the `POWHEGBOX` [11]. At NNLO the situation is even more complicated, and efforts to extend matching schemes with PS at this level of accuracy are ongoing, as well as those for pushing the accuracy of PS far beyond the present one.

In this paper we consider $c\bar{c} \to c$-hadron + X hadroproduction evaluated by means of `POWHEGBOX` + `PYTHIA 6` [12]. This allows to generate differential distributions for the hadroproduction of different kinds of charmed mesons and hadrons at NLO QCD + PS accuracy. The PS ensures at least leading logarithmic (LL) accuracy in all collinear QCD emissions following the first light partonic one accompanying $c\bar{c}$, which is instead evaluated by `POWHEGBOX` with NLO accuracy.

This computation requires to fix some input. To establish the central values of the renormalization and factorization scales μ_R and μ_F, of the charm mass m_c and a set of parton distribution functions (PDFs) suitable for the calculation

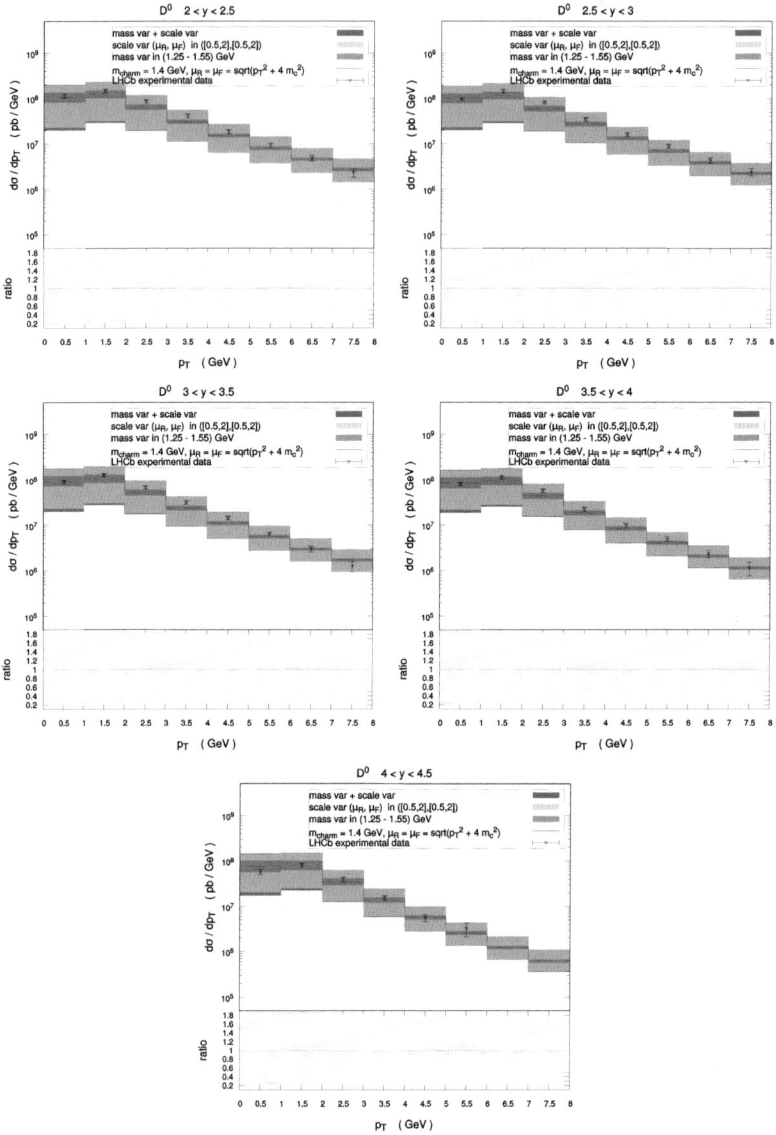

Figure 1: Theoretical predictions for D^0 hadroproduction by POWHEGBOX + PYTHIA 6 compared to LHCb data at 7 TeV. Transverse momentum distributions in different rapidity bins are shown in different panels. The green and violet bands refer to scale and mass variation, respectively, summed in quadrature in the blue band. Central values for all input parameters are fixed as explained in the text.

of prompt neutrino fluxes, taking into account that a broad range of center-of-mass energies $\sqrt{s}$ are involved, including both energies much lower and much higher than those explored so far at colliders, we have studied the behaviour of the total $c\bar{c}$ hadroproduction cross-section at different orders/degrees of accuracy. We have compared predictions including LO, NLO and NNLO QCD corrections, as well as the behaviour of the cross-sections when adopting the $\overline{MS}$ scheme as an alternative to the on-shell one in the renormalization of the charm quark mass. The procedure leading to our choice of the central values of these parameters and of their range of variation, useful to determine the associated uncertainties, is detailed in Ref. [7]. Here we summarize our choices. The central μ_R and μ_F scales were fixed to $\mu_0 = \sqrt{p_{T,c}^2 + 4m_c^2}$, and allowed to vary in the range $(\mu_R, \mu_F) \in$ (0.5, 0.5), (2,2), (0.5, 1), (1, 0.5), (1, 2), (2,1) μ_0. We did not include the variations (0.5, 2) and (2, 0.5) μ_0, as suggested in Ref. [13]. The central charm mass value was fixed to $m_c = 1.4$ GeV in the on-shell scheme and allowed to vary in the range $\pm$ 0.15 GeV. We have used the ABM11 PDFs [14] at NLO including the PDF and α_s variations. We emphasize that the choice of all these parameters was uniquely derived by theoretical considerations, taking into account in particular the convergence of the perturbative series of radiative corrections, and the behaviour of different sets of PDFs already available in the `LHAPDF 6.1.5` interface [15] in an extended Bjorken-x range, also covering the region of low x. No comparison with any experimental data was used to tune these parameters or retune other parameters entering `PYTHIA 6`, for which we have adopted one of the most recent already available Perugia tunes [16], ensuring a transverse-momentum ordered PS.

Comparison with experimental data was done a-posteriori, to test the theoretical predictions. In particular the LHCb collaboration has recently measured the hadroproduction of charmed mesons and baryons in different rapidity and transverse momentum bins, at both 7 and 13 TeV [17, 18], covering the total range $2 < y < 4.5$, corresponding to mid-peripheral collisions. A comparison between our theoretical predictions for D^0 hadroproduction and experimental data at 7 TeV is shown in Fig. 1. The theoretical predictions turn out to agree with experimental data within the theoretical uncertainty bands due to scale and mass variation in all rapidity bins. Similar agreement is found for charged D-mesons.

The large uncertainties due to μ_R and μ_F scale variation, shown in Fig. 1 for LHC predictions, are almost independent of $\sqrt{s}$, i.e. remain of the same order of magnitude even at other energies, and propagate to prompt neutrino fluxes. Therefore, they can be considered to be the dominant QCD uncertainty, as follows from the comparison of Fig. 2.a with with Fig. 2.c. In particular, the non-diagonal μ_R and μ_F variations, neglected in calculations completed before our one, like e.g. that in Ref. [19], turn out to enlarge the uncertainty band on prompt neutrino fluxes of several tens percent, as shown in Fig. 2 (compare Fig 2.a with Fig. 2.b or Fig 2.c with Fig 2.d).

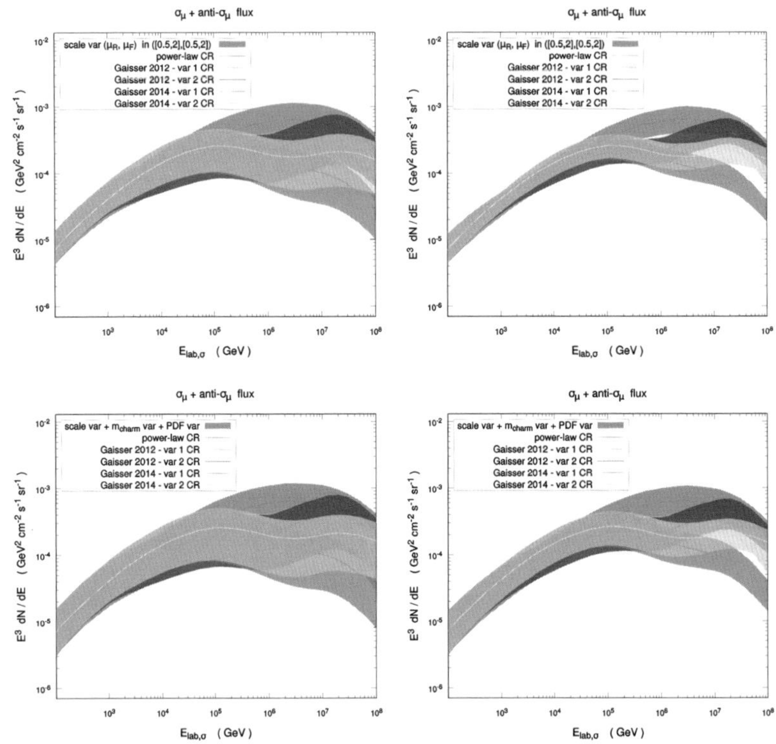

Figure 2: Our predictions for $(\nu_\mu + \bar{\nu}_\mu)$ fluxes as a function of E_ν. Uncertainties due to scale variation around the central value $\mu_R = \mu_F = \sqrt{p_T^2 + 4m_c^2}$, are shown in panel 2.a (upper-left), considering all possible (μ_R, μ_F) scale combinations (0.5, 0.5), (2, 2), (1, 2), (2, 1), (0.5, 1) and (1, 0.5) μ_0, and in panel 2.b (upper-right) by disregarding the non-diagonal contributions (0.5, 1) and (1, 0.5) μ_0. The addition in quadrature of the uncertainties in panel 2.a (2.b) with those due to m_c and PDF variation gives rise to the QCD uncertainty band in panel 2.c (2.d). In each panel, different colors refer to fluxes computed with different primary CR all-nucleon spectra, including power-law and four different variants of Gaisser CR spectra [20, 21]. See text for more detail.

At the highest energies, uncertainties related to our poor knowledge of the composition of the primary flux of CRs entering the upper layer of the Earth's atmosphere, turn out to become even larger than those of QCD origin, as shown in Fig. 2, and dedicated efforts from extended air shower (EAS) experiments are indeed needed to reduce them, together with the improvement and reduction of uncertainties in the theoretical models describing EAS formation.

Acknowledgments

We are grateful to the Conference organizers and partecipants for many lively discussions. This work has been supported by Deutsche Forschungsgemeinschaft in Sonderforschungsbereich 676.

References

[1] A. Kappes for the IceCube collaboration, arXiv:1501.07798, in Proc. of the 37th Int. Conf. on High Energy Physics (ICHEP 2014) July 2 - 9, 2014, Valencia, Spain.
[2] **IceCube** Collaboration (M. Aartsen et al.), Science **342**, 1242856 (2013).
[3] **IceCube** Collaboration (M. Aartsen et al.), Phys. Rev. Lett. **113**, 101101 (2014) [arXiv:1405.5303].
[4] T. Montaruli, arXiv:1512.07978, in Proc. of the Cosmic Ray International Seminar: The Status and the future of UHE Cosmic Ray Physics in the post LHC era (CRIS 2015), September 14 - 16, 2015, Gallipoli, Italy.
[5] P. Gondolo, G. Ingelman, and M. Thunman, Astropart. Phys. **5**, 309 (1996) [hep-ph/9505417].
[6] M. Honda, T. Kajita, K. Kasahara, S. Midorikawa, and T. Sanuki, Phys. Rev. **D 75**, 043006 (2007) [astro-ph/0611418].
[7] M. V. Garzelli, S. Moch, and G. Sigl, JHEP **10**, 115 (2015) [arXiv:1507.01570].
[8] P. Nason and B. Webber, Ann. Rev. Nucl. Part. Sci. **62**, 187 (2012) [arXiv:1202.1251].
[9] P. Nason, JHEP **11**, 040 (2004) [hep-ph/0409146].
[10] S. Frixione, P. Nason, C. Oleari, JHEP **11**, 070 (2007) [arXiv:0709.2092].
[11] S. Alioli, P. Nason, C. Oleari, and E. Re, JHEP **06**, 043 (2010) [arXiv:1002.2581].
[12] T. Sjostrand, S. Mrenna, and P. Z. Skands, JHEP **05**, 026 (2006) [hep-ph/0603175].
[13] M. Cacciari, S. Frixione, N. Houdeau, M. L. Mangano, P. Nason *et al.*, JHEP **10**, 137 (2012) [arXiv:1205.6344].
[14] S. Alekhin, J. Blümlein, and S. Moch, Phys. Rev. **D 86**, 054009 (2012).
[15] A. Buckley, J. Ferrando, S. Lloyd, K. Nordstrom, B. Page *et al.*, Eur. Phys. J. **C 75**, 132 (2015) [arXiv:1412.7420].
[16] P. Z. Skands, Phys. Rev. **D 82**, 074018 (2010) [arXiv:1005.3457].
[17] **LHCb** Collaboration (R. Aaij et al.), Nucl. Phys. **B 871**, 1 (2013) [arXiv:1302.2864].
[18] **LHCb** Collaboration (R. Aaij et al.), arXiv:1510.01707.

[19] A. Bhattacharya, R. Enberg, M. H. Reno, I. Sarcevic, and A. Stasto, JHEP **06**, 110 (2015) [arXiv:1502.01076].
[20] T. K. Gaisser, T. Stanev, and S. Tilav, Front. Phys. China **8**, 748 (2013) [arXiv:1303.3565].
[21] T. Stanev, T. K. Gaisser, S. Tilav, Nucl. Instr. Meth. **A 742**, 42 (2014).

HIGGS EFFECTIVE LAGRANGIANS AND PHYSICS BEYOND THE STANDARD MODEL

Eduard Massó [a]

Grup de Física Teòrica, Dept. Física and IFAE-BIST,
Univ.Autònoma de Barcelona, E-08193 Bellaterra (Barcelona), Spain

Abstract. Some aspects of the effective Lagrangian approach to describe effects of New Physics are briefly described.

1 Introduction

The discovery of the Higgs boson [1–3] at the LHC with $m_h \simeq 125$ GeV has given a significant boost to the Standard Model (SM) of particle physics. Indeed, the discovery completes the spectrum of the model; also, the measured Higgs properties are fairly consistent with the SM expectations.

In the SM, symmetry breaking occurs as a result of the form of the Higgs potential

$$V(H) = -\mu^2|H|^2 + \lambda|H|^4 \tag{1}$$

with μ^2 and λ positive. Because of the form of the potential, the Higgs doublet H acquires a non-zero vacuum expectation. The physical Higgs particle h arises as the excitations around this vacuum

$$H \to \frac{1}{\sqrt{2}}\begin{pmatrix} 0 \\ v+h \end{pmatrix} \tag{2}$$

This simple phenomenon gives rise to the characteristic property of the Higgs h, namely that it couples to mass. Indeed, the coupling to the electroweak bosons W, Z is

$$m_W^2\left(W^{+\mu}W_\mu^- + \frac{1}{2\cos^2\theta_w}Z^\mu Z_\mu\right)\left(1+\frac{2h}{v}+\frac{h^2}{v^2}\right)\,, \tag{3}$$

and the coupling to fermions f is

$$m_f\,\bar{f}f\left(1+\frac{h}{v}\right)\,, \tag{4}$$

The Higgs particle couples also to massless particles, i.e. to photons γ and gluons g, through loops of massive particles. The couplings $h\gamma\gamma$ and hgg turn out to be crucial from the experimental point of view, as they describe the main process of Higgs production and the main process of decay to detect the Higgs particle at the LHC. The coupling of the Higgs to mass is being tested

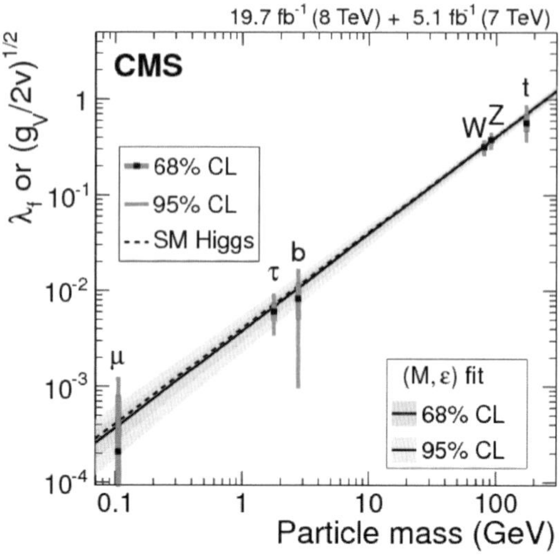

Figure 1: CMS summary of fits for the Higgs couplings as function of the particle masses [4]. ATLAS has a similar plot [5].

for W, Z and for the heavier fermions at a precision level of about 10 %; see Figure 1.

One of the main LHC objectives is to elucidate how the electroweak (EW) symmetry is broken. The data so far is compatible with the standard Englert-Brout-Higgs mechanism of EW symmetry breaking described above in equations (2) and (1) with $\mu^2 > 0$. In the context of the SM, the LHC findings have determined the last parameters of the SM to be measured

$$\mu \simeq 88 \text{ GeV} \qquad \lambda \simeq 0.13 \tag{5}$$

LHC will look for signals of Beyond the Standard Model (BSM) physics at high energies. So far no clear signal has been found. At the time of this writing some LHC data may indicate decays of high-mass resonances, but it is too soon to claim it is BSM. More data will help to clarify the situation.

If the Higgs is indeed an elementary scalar we are confronted with the well-known problem of naturalness. In the usual effective-theory framework where physics at the electroweak scale is described by cutting-off momenta of virtual particles at a high energy scale Λ, the corrections δm_h to the physical Higgs mass are

$$\frac{\delta m_h}{m_h} \sim \left(\frac{\Lambda}{500\,\text{GeV}}\right)^2 \tag{6}$$

[a]E-mail: masso@ifae.es

This suggests that in order not to need to excessively fine tune m_h some new phenomena at LHC energies should show up. In the case that it is confirmed that no new signal is found at LHC, the argument of naturalness should be revised.

In this contribution, we summarise some aspects of effective Lagrangians to describe model-independent BSM Higgs physics.

2 BSM and effective Lagrangians

A powerful method to study potential deviations from the SM is the effective Lagrangian approach. Here I will describe the so-called linearly-realized effective Lagrangian which consists in considering the Higgs field as part of an isodoublet H, and integrate the BSM heavy degrees of freedom.

When integrating out heavy particles, one obtains that the SM Lagrangian is modified by a tower of high-dimensional operators. We will focus on the lowest-dimensional operators conserving baryon and lepton number, namely on a Lagrangian formed by dimension-six operators, suppressed by a scale Λ and with arbitrary Wilson coefficients

$$\mathcal{L}_6 = \sum_a \frac{c_a}{\Lambda^2}\,\mathcal{O}_a \,. \tag{7}$$

The sum is over a complete basis set of operators. Since the number of operators in a basis is large and different bases can be used, one should proceed with care.

A way to proceed is to start with a complete basis of dimension-six operators and select the group of operators which lead to physics involving the Higgs. In this group, one can select different sectors. In one sector there are some operators which are already very constrained by precision EW measurements at the Z-pole. For example, in the effective Lagrangian one obtains terms

$$\frac{v^2}{\Lambda^2}\left(1+\frac{2\,h}{v}\right) Z_\mu J^\mu + \ldots \tag{8}$$

where

$$J^\mu = \sum c^{rs}\,\bar{f}^r \gamma^\mu f^s \tag{9}$$

and the sum is over $f = u_{L,R}, d_{L,R}, e_{L,R}, \nu_L$ and over family indexes r, s, and of course restricted to neutral combinations. We see that the new vertex $hZff'$ is already very constrained by LEP bounds on the Zff' vertex.

Operators that can be only restricted in Higgs physics have the general form $(H^\dagger H)\mathcal{O}_4$ where $\mathcal{O}_4$ is a dimension-four operator. It is clear that in vacuum, $(H^\dagger H) \to v^2/2$, the contribution is just an innocuous renormalization of an operator $\mathcal{O}_4$ present in the SM Lagrangian. This procedure shows which directions of the dimension-six operator space can be constrained using Higgs physics. It amounts to the coupling of h to gg, γZ, $\gamma\gamma$, to fermions, and to the weak bosons,

$$h\left(W^{+\mu}W^{-}_{\mu}+\frac{1}{2\cos^{2}\theta_{w}}Z^{\mu}Z_{\mu}\right) \tag{10}$$

which as we see it has the custodial preserving form. In addition there is the h^3 interaction which is difficult to measure at LHC and should wait for improved experiments. More details can be found in [6, 7].

Here I have just mentioned an aspect of effective Lagrangians applied to Higgs physics. There are many works containing other interesting findings, like bounding the Wilson coefficients using available LHC data, determining renormalisation group equations of dimension-six operators, etc.

Acknowledgments

This work is partially supported by *MINECO* grant number CICYT-FEDER-FPA2014-55613-P, by *Generalitat de Catalunya* grant 2014 SGR 1450, and by the *Centro de Excelencia Severo Ochoa* SEV-2012-0234.

References

[1] S. Chatrchyan *et al.* [CMS Collab.], Phys. Lett. B **710** 403 (2012) [hep-ex/1202.1487].

[2] G. Aad *et al.* [ATLAS Collab.], Phys. Rev. Lett. **108** 111803 (2012) [hep-ex/1202.1414].

[3] For the last data analyses see talks by E. Torrence and P. Spagnolo at *The Third Annual Large Hadron Collider Physics Conference*, St. Petersburg, Russia, August 31 September 5, 2015. Link http://lhcp2015.com/

[4] http://cms-higgs-results.web.cern.ch/cms-higgs-results/Comb/HIG-14-009-Paper/sqr_m6summary_fitmu.png

[5] http://atlas-physics-updates.web.cern.ch/atlas-physics-updates/wp-content/uploads/2015/03/fig_17.png

[6] R. S. Gupta, A. Pomarol and F. Riva, Phys. Rev. D **91**, no. 3, 035001 (2015) [arXiv:1405.0181 [hep-ph]].

[7] E. Masso, JHEP **1410**, 128 (2014) [arXiv:1406.6376 [hep-ph]].

EXOTIC HIGGS DECAYS IN U(1) EXTENSIONS OF THE MSSM

Peter Athron
ARC Centre of Excellence for Particle Physics at the Terascale, School of Physics, Monash University, Melbourne, Victoria 3800, Australia
Margarete Mühlleitner
Institute for Theoretical Physics, Karlsruhe Institute of Technology, 76128 Karlsruhe, Germany
Roman Nevzorov and Anthony G. Williams
ARC Centre of Excellence for Particle Physics at the Terascale and CSSM, School of Chemistry and Physics, University of Adelaide, Adelaide SA 5005, Australia

Abstract. We study the decays of the lightest CP–even Higgs boson into a pair of pseudoscalar Higgs states within $U(1)_N$ extensions of the MSSM.

1 Introduction

A new bosonic state with a mass around 125 GeV, which was discovered at the LHC a few years ago, can provide a window into new physics beyond the Standard Model (SM). The signal strengths associated with this particle are currently consistent with the SM Higgs boson. However further exploration of this resonance can lead to new exciting discoveries. Physics beyond the SM may change the Higgs decay rates to SM particles and give rise to new decay modes. Here we consider the decays of the SM-like Higgs boson into a pair of pseudoscalar states within $U(1)$ extensions of the Minimal Supersymmetric Standard Model (MSSM). In particular we focus on the E_6 inspired SUSY models which are based on the low–energy SM gauge group together with an extra $U(1)_N$ gauge symmetry defined by:

$$U(1)_N = \frac{1}{4}U(1)_\chi + \frac{\sqrt{15}}{4}U(1)_\psi \,. \tag{1}$$

Gauge $U(1)_\psi$ and $U(1)_\chi$ symmetries can originate from the breakings $E_6 \to SO(10) \times U(1)_\psi$, $SO(10) \to SU(5) \times U(1)_\chi$. Different aspects of the phenomenology of these E_6 inspired supersymmetric (SUSY) models have been extensively studied in [1–13].

2 The $U(1)_N$ Extensions of the MSSM

These SUSY models imply that just below the scale M_X the gauge symmetry in the E_6 Grand Unified Theory (GUT) is broken down to $SU(3)_C \times SU(2)_W \times U(1)_Y \times U(1)_N \times Z_2^M$, where $Z_2^M = (-1)^{3(B-L)}$ is a matter parity. To ensure anomaly cancellation, gauge coupling unification and the appropriate breakdown of the gauge symmetry at low energies the matter content of the SUSY models under consideration involves three 27 representations of the E_6, two $SU(2)_W$ doublets L_4 and $\overline{L}_4$ from extra $27'_1$ and $\overline{27'}_1$, two SM singlets S and $\overline{S}$ that carry non–zero $U(1)_N$ charges and stem from another pair of $27'_2$ and $\overline{27'}_2$ as well as a SM singlet ϕ that does not participate in gauge interactions. Thus these models involve extra matter beyond the MSSM. In particular, they include three pairs of $SU(2)_W$–doublets (H_i^d and H_i^u) that have the quantum numbers of Higgs doublets, and three pairs of colour triplets of exotic quarks $\overline{D}_i$ and D_i with electric charges $+1/3$ and $-1/3$ respectively. To suppress flavour changing neutral currents as well as the most dangerous baryon and lepton number violating operators one can impose a discrete $\tilde{Z}_2^H$ symmetry, under which all superfields except one pair of H_i^d and H_i^u (say $H_d \equiv H_3^d$ and $H_u \equiv H_3^u$), L_4, $\overline{L}_4$, S, $\overline{S}$ and ϕ are odd.

The conservation of Z_2^M and $\tilde{Z}_2^H$ symmetries implies that these models contain at least two states which are absolutely stable and can contribute to the dark matter density. One of these states is a lightest SUSY particle (LSP) whereas another one is the lightest ordinary neutralino. Using the method proposed in [14] it was argued that there is a theoretical upper bound on the mass of the LSP [7]. In the simplest phenomenologically viable scenarios the LSP has a mass around 1 eV or even smaller forming hot dark matter in the Universe. The presence of such very light neutral fermions in the particle spectrum may have interesting implications for neutrino physics (see, for example [15]). In these scenarios the lightest ordinary neutralino can account for all or some of the observed dark matter density.

3 Non-Standard Higgs Decays

The Higgs sector of the SUSY models under consideration contains two Higgs doublets H_d and H_u as well as S, $\overline{S}$ and ϕ. The scalar components of the superfields S and $\overline{S}$ acquire large vacuum expectation values (VEVs) that break the $U(1)_N$ gauge symmetry inducing the masses of exotic particles and the Z' boson. The interaction of S, $\overline{S}$ and ϕ is determined by the following terms in the superpotential that can be written as

$$W_S = \sigma\phi S\overline{S} + \frac{\kappa}{3}\phi^3 + \frac{\mu}{2}\phi^2 + \Lambda\phi\,. \qquad (2)$$

When the couplings κ, μ and Λ are rather small these models possess an approximate global $U(1)$ symmetry. This symmetry is spontaneously broken by the VEVs of S, $\overline{S}$ and ϕ leading to the presence of a light pseudoscalar state in the particle spectrum. Our analysis indicates that the couplings of this state to the SM particles are always small. Therefore even if this pseudoscalar is substantially lighter than 100 GeV it could escape detection at former and present collider experiments. Moreover this state can be so light that it gives rise to the decays of the SM-like Higgs boson into a pair of these pseudoscalar states.

The branching ratio of such decays depends rather strongly on the values of the couplings that explicitly violate the global $U(1)$ symmetry.

To simplify our analysis we set $\mu = 0$ as well as $\Lambda = 0$ and examine the dependence of the branching ratio of the exotic Higgs decays on κ. For $\kappa \sim 0.001$ the pseudoscalar Higgs boson with a mass of $40 - 60$ GeV can be obtained without fine-tuning but the branching ratio of the SM-like Higgs decays into a pair of such particles is smaller than 10^{-4} [13]. Thus it will be rather difficult to observe such decays at the LHC. If $\kappa > 0.01$ a fine tuning of at least 1% is required to get the lightest CP-odd Higgs state with mass of $40 - 60$ GeV. However the branching ratio of the corresponding exotic Higgs decays can be larger than 1% in this case [13]. After being produced from such decays the lightest pseudoscalars decay into a pair of b-quarks or τ-leptons resulting in four fermion final states.

Acknowledgments

This work was supported by the University of Adelaide and the Australian Research Council through the ARC Center of Excellence in Particle Physics at the Terascale.

References

[1] S. F. King, S. Moretti, R. Nevzorov, Phys. Rev. **D 73**, 035009 (2006).
[2] S. F. King, S. Moretti, R. Nevzorov, Phys. Lett. **B 634**, 278 (2006).
[3] S. F. King, S. Moretti, R. Nevzorov, Phys. Lett. **B 650**, 57 (2007).
[4] S. F. King, R. Luo, D. J. Miller, R. Nevzorov, JHEP **0812**, 042 (2008).
[5] P. Athron, S. F. King, D. J. Miller, S. Moretti, R. Nevzorov, Phys. Lett. **B 681**, 448 (2009).
[6] P. Athron, S. F. King, D. J. Miller, S. Moretti, R. Nevzorov, Phys. Rev. **D 80**, 035009 (2009).
[7] J. P. Hall, S. F. King, R. Nevzorov, S. Pakvasa, M. Sher, Phys. Rev. **D 83**, 075013 (2011).
[8] P. Athron, S. F. King, D. J. Miller, S. Moretti, R. Nevzorov, Phys. Rev. **D 84**, 055006 (2011).
[9] R. Nevzorov, Phys. Rev. **D 87**, 015029 (2013).
[10] P. Athron, S. F. King, D. J. Miller, S. Moretti, R. Nevzorov, Phys. Rev. **D 86**, 095003 (2012).
[11] R. Nevzorov, S. Pakvasa, Phys. Lett. **B 728**, 210 (2014).
[12] R. Nevzorov, Phys. Rev. **D 89**, 055010 (2014).

[13] P. Athron, M. Mhlleitner, R. Nevzorov, A. G. Williams, JHEP **1501**, 153 (2015).
[14] S. Hesselbach, D. J. Miller, G. Moortgat-Pick, R. Nevzorov, M. Trusov, Phys. Lett. **B 662**, 199 (2008).
[15] J. M. Frere, R. Nevzorov, M. I. Vysotsky, Phys. Lett. **B 394**, 127 (1997).

RENORMALIZATION OF THE COUPLING CONSTANT AND THE HD REGULARIZATION IN SUPERSYMMETRIC THEORIES

Konstantin Stepanyantz [a]

Faculty of Physics, Moscow State University, 119991 Moscow, Russia

Abstract. It seems that the loop integrals contributing to the β-function of $\mathcal{N}=1$ supersymmetric theories regularized by higher covariant derivatives are integrals of double total derivatives. This allows obtaining the NSVZ relation for the renormalization group functions defined in terms of the bare charges at least in the Abelian case. For the renormalization group functions defined in terms of the renormalized charges we discuss the prescription giving the NSVZ scheme in all orders for the Abelian theories. In the non-Abelian case we present the exact expression for the Adler D-function, which follows from the similar structure of quantum corrections.

1 Introduction

In $\mathcal{N}=1$ supersymmetric theories the β-function is related with anomalous dimension of the matter superfields by the so called "NSVZ relation" [1–4]. For a simple particular case of the $\mathcal{N}=1$ supersymmetric electrodynamics (SQED) with N_f flavors the exact NSVZ β-function has the form [5, 6]

$$\beta(\alpha)=\frac{\alpha^2 N_f}{\pi}\Big(1-\gamma(\alpha)\Big). \tag{1}$$

Obtaining this equation by direct summation of supergraphs appears to be a complicated problem. For example, using the dimensional reduction and $\overline{\text{DR}}$-scheme the NSVZ relation in the lowest loops is obtained only after a specially tuned finite renormalization [7–9]. In this paper we briefly describe how the NSVZ β-function can be obtained in the perturbation theory, and why it is convenient to use the higher derivative (HD) regularization for this purpose.

2 Quantum corrections in $\mathcal{N}=1$ SQED with the HD regularization

The HD regularization [10, 11] is a consistent regularization, which does not break supersymmetry. To regularize a theory by HD it is necessary to add a term with higher degrees of covariant derivatives. Then divergences remain only in the one-loop approximation. These remaining divergences are regularized by inserting the Pauli–Villars determinants into the generating functional [12]. The HD regularization can be generalized to the $\mathcal{N}=1$ supersymmetric case [13, 14]. Also it can be constructed for $\mathcal{N}=2$ supersymmetric theories [15–17]. In the Abelian case the NSVZ β-function can be obtained in all orders by direct summation of supergraphs using the $\mathcal{N}=1$ supersymmetric HD regularization.

[a] E-mail: `stepan@m9com.ru`

It relates the renormalization group (RG) functions defined in terms of the bare charge

$$\beta(\alpha_0) \equiv \frac{d\alpha_0(\alpha,\Lambda/\mu)}{d\ln\Lambda}\Big|_{\alpha=\text{const}}; \qquad \gamma(\alpha_0) \equiv -\frac{d\ln Z(\alpha,\Lambda/\mu)}{d\ln\Lambda}\Big|_{\alpha=\text{const}}. \qquad (2)$$

They are scheme independent for a fixed regularization [18–20], but depend on the regularization. The NSVZ relation follows from the fact that with the HD regularization the loop integrals giving the β-function defined in terms of the bare charge are integrals of (double) total derivatives [21, 22]. This statement has been confirmed by numerous explicit calculations, see, e.g., [23–25]. It allows to calculate one of the loop integrals analytically and to obtain the NSVZ relation for the RG functions defined in terms of the bare coupling constant. In the Abelian case this has been done in all loops [26, 27] due to the identity

$$\begin{aligned}\frac{\beta(\alpha_0)}{\alpha_0^2} &= \frac{d}{d\ln\Lambda}\Big(d^{-1}(\alpha_0,\Lambda/p)-\alpha_0^{-1}\Big)\Big|_{p=0}\\ &= \frac{N_f}{\pi}\Big(1-\frac{d}{d\ln\Lambda}\ln G(\alpha_0,\Lambda/q)\Big|_{q=0}\Big) = \frac{N_f}{\pi}\Big(1-\gamma(\alpha_0)\Big). \qquad (3)\end{aligned}$$

The RG functions defined in terms of the renormalized coupling constant

$$\widetilde{\beta}(\alpha) \equiv \frac{d\alpha(\alpha_0,\Lambda/\mu)}{d\ln\mu}\Big|_{\alpha_0=\text{const}}; \qquad \widetilde{\gamma}(\alpha) \equiv \frac{d\ln Z(\alpha(\alpha_0,\Lambda/\mu),\Lambda/\mu)}{d\ln\mu}\Big|_{\alpha_0=\text{const}} \qquad (4)$$

are scheme-dependent. It was shown in [19] that they coincide with the RG functions defined in terms of the bare coupling constant, if the boundary conditions $Z_3(\alpha,x_0)=1$ and $Z(\alpha,x_0)=1$ are imposed on the renormalization constants, where x_0 is an arbitrary fixed value of $\ln\Lambda/\mu$, e.g., $x_0=0$.

In the three-loop approximation all these statements can be verified by explicit calculations [18–20]. In particular, it is possible to find the three-loop RG functions in the above NSVZ scheme and to check that they satisfy the NSVZ relation. The results are also known for the MOM-scheme [20] and for the $\overline{\text{DR}}$-scheme [7]. It was shown that they do not satisfy the NSVZ relation.

3 Exact expression for the Adler D-function in $\mathcal{N}=1$ SQCD

Following Refs. [28, 29], let us consider $\mathcal{N}=1$ (massless) SQCD interacting with the Abelian gauge field, which is described by the action

$$S = \frac{1}{2g_0^2}\text{tr}\,\text{Re}\int d^4x\, d^2\theta\, W^a W_a + \frac{1}{4e_0^2}\text{Re}\int d^4x\, d^2\theta\, \boldsymbol{W}^a \boldsymbol{W}_a$$

$$+\sum_{f=1}^{N_f}\frac{1}{4}\int d^4x\,d^4\theta\Big(\Phi_f^+e^{2q_f\boldsymbol{V}+2V}\Phi_f+\widetilde{\Phi}_f^+e^{-2q_f\boldsymbol{V}-2V^t}\widetilde{\Phi}_f\Big). \tag{5}$$

The gauge group is $SU(N_c)\times U(1)$, and the matter superfields Φ_f and $\widetilde{\Phi}_f$ lie in the (anti)fundamental representation. V is non-Abelian gauge superfield, and $\boldsymbol{V}$ denotes the Abelian gauge superfield. The considered theory contains two coupling constants $\alpha_s=g^2/4\pi$ and $\alpha=e^2/4\pi$. Quantum corrections to the electromagnetic coupling constant α, which appear due to the quark loop with internal gluon and quark lines are encoded in the Adler D-function [30]. Treating $\boldsymbol{V}$ as an external superfield, in terms of the bare coupling constant the Adler D-function is defined as

$$D(\alpha_{0s})=\frac{3\pi}{2\alpha_0^2}\frac{d\alpha_0}{d\ln\Lambda}. \tag{6}$$

Thus, it depends on regularization, but is independent of a subtraction scheme. The HD regularization is introduced similarly to the Abelian case, but in this case it is necessary to use the covariant derivatives. Then it is possible to prove that the integrals for the Adler D-function are integrals of double total derivatives and to derive for it the following NSVZ-like exact expression

$$D(\alpha_{0s})=\frac{3}{2}\sum_f q_f^2\cdot N_c\Big(1-\gamma(\alpha_{0s})\Big). \tag{7}$$

This implies that in the considered non-Abelian case all features of quantum corrections are the same as in $\mathcal{N}=1$ SQED.

4 Conclusion

It seems that the NSVZ relation in $\mathcal{N}=1$ supersymmetric theories is obtained in the case of using the HD regularization for the RG functions defined in terms of the bare coupling constant (independently of the renormalization prescription). This statement is proved exactly in all loops for the Abelian case and for the Adler D-function in the case of $\mathcal{N}=1$ SQCD by using the fact that the integrals which determine the β-function defined in terms of the bare coupling constant are factorized into integrals of double total derivatives. If the RG functions are defined in terms of the renormalized coupling constant, the NSVZ β-function is obtained in a special subtraction scheme, called the NSVZ scheme. In the case of using the HD regularization this scheme can be obtained by imposing simple boundary conditions on the renormalization constants.

Acknowledgments

This work was supported by the RFBR grant No. 14-01-00695.

References

[1] V.A.Novikov, M.A.Shifman, A.I.Vainshtein, V.I.Zakharov, Nucl.Phys. **B 229**, (1983) 381 (1983).
[2] D.R.T.Jones, Phys.Lett. **B 123**, 45 (1983).
[3] V.A.Novikov, M.A.Shifman, A.I.Vainshtein, V.I.Zakharov, Phys.Lett. **B 166**, 329 (1986); Sov.J.Nucl.Phys. **43**, 294 (1986).
[4] M.A.Shifman, A.I.Vainshtein, Nucl.Phys. **B 277**, 456 (1986); Sov.Phys.JETP **64**, 428 (1986).
[5] A.I.Vainshtein, V.I.Zakharov, M.A.Shifman, JETP Lett. **42**, 224 (1985).
[6] M.A.Shifman, A.I.Vainshtein, V.I.Zakharov, Phys.Lett. **B 166**, 334 (1986).
[7] I.Jack, D.R.T.Jones, C.G.North, Phys.Lett. **B 386**, 138 (1996).
[8] I.Jack, D.R.T.Jones, C.G.North, Nucl.Phys. **B 486**, 479 (1997).
[9] I.Jack, D.R.T.Jones, A.Pickering, Phys.Lett. B **435**, 61 (1998).
[10] A.A.Slavnov, Nucl.Phys. **B 31**, 301 (1971).
[11] A.A.Slavnov, Theor.Math.Phys. **13**, 1064 (1972).
[12] A.A.Slavnov, Theor.Math.Phys. **33**, 977 (1977).
[13] V.K.Krivoshchekov, Theor.Math.Phys. **36**, 745 (1978).
[14] P.C.West, Nucl.Phys. **B 268**, 113 (1986).
[15] V.K.Krivoshchekov, Phys.Lett. **B 149**, 128 (1984).
[16] I.L.Buchbinder, K.V.Stepanyantz, Nucl.Phys. **B 883**, 20 (2014).
[17] I.L.Buchbinder, N.G.Pletnev, K.V.Stepanyantz, Phys.Lett. **B 751**, 434 (2015).
[18] A.L.Kataev, K.V.Stepanyantz, Nucl.Phys. **B 875**, 459 (2013).
[19] A.L.Kataev, K.V.Stepanyantz, Theor.Math.Phys. **181**, 1531 (2014).
[20] A.L.Kataev, K.V.Stepanyantz, Phys. Lett. **B 730**, 184 (2014).
[21] A.A.Soloshenko, K.V.Stepanyantz, Theor.Math.Phys. **140**, 1264 (2004).
[22] A.V.Smilga, A.Vainshtein, Nucl.Phys. **B 704**, 445 (2005).
[23] A.B.Pimenov, E.S.Shevtsova, K.V.Stepanyantz, Phys.Lett. **B 686**, 293 (2010).
[24] K.V.Stepanyantz, arXiv:1108.1491 [hep-th].
[25] A.E.Kazantsev, K.V.Stepanyantz, J.Exp.Theor.Phys. **120**, 618 (2015).
[26] K.V.Stepanyantz, Nucl.Phys. **B 852**, 71 (2011).
[27] K.V.Stepanyantz, JHEP **1408**, 096 (2014).
[28] M.Shifman, K.Stepanyantz, Phys.Rev.Lett. **114**, 051601 (2015).
[29] M.Shifman, K.V.Stepanyantz, Phys.Rev. **D 91**, 105008 (2015).
[30] S.L.Adler, Phys.Rev. **D 10**, 3714 (1974).

ELECTROWEAK INTERACTION OF PARTICLES WITH ACCELERATED MATTER AND ASTROPHYSICAL APPLICATIONS

Maxim Dvornikov[a]

Pushkov Institute of Terrestrial Magnetism, Ionosphere and Radiowave Propagation (IZMIRAN), 142190 Troitsk, Moscow, Russia

Physics Faculty, National Research Tomsk State University, 36 Lenin Avenue, 634050 Tomsk, Russia

Abstract. The description of physical processes in accelerated frames opens a window to numerous new phenomena. One can encounter these effects both in the subatomic world and on a macroscale. In the present work we review our recent results on the study of the electroweak interaction of particles with an accelerated background matter. In our analysis we choose the noninertial comoving frame, where matter is at rest. Our study is based on the solution of the Dirac equation, which exactly takes into account both the interaction with matter and the nonintertial effects. First, we study the interaction of ultrarelativistic neutrinos, electrons and quarks with the rotating matter. We consider the influence of the matter rotation on the resonance in neutrino oscillations and the generation of anomalous electric current of charged particles along the rotation axis. Then, we study the creation of neutrino-antineutrino pairs in a linearly accelerated matter. The applications of the obtained results for elementary particle physics and astrophysics are discussed.

Nowadays it is understood that noninertial effects are important in various areas of modern science such as elementary particles physics, general and special relativity, as well as condensed matter physics [1]. Recently in Refs. [2–4] it was realized that the electroweak interaction of particles with accelerated matter leads to interesting applications in physics and astrophysics. In those works, the treatment of the particle evolution was made in the comoving frame, where matter is at rest, with the noninertial effects being accounted for exactly. In the present work we review our recent results on the particle interaction with accelerated matter.

Our study of the fermion propagation in an accelerated matter is based on the Dirac equation in a comoving frame. In this situation one can unambiguously define the interaction with background matter. It is known that the motion in a noninertial frame is equivalent to the interaction with an effective gravitational field having the metric tensor $g_{\mu\nu}$. The Dirac equation for the particle bispinor ψ in curved space-time has the form [3],

$$[\mathrm{i}\gamma^{\mu}(x)\nabla_{\mu} - m]\,\psi = \gamma_0(x)\left\{\frac{V_{\mathrm{L}}}{2}\left[1-\gamma^5(x)\right] + \frac{V_{\mathrm{R}}}{2}\left[1+\gamma^5(x)\right]\right\}\psi, \quad (1)$$

where $\gamma^{\mu}(x)$ are the coordinate dependent Dirac matrices, $\nabla_{\mu} = \partial_{\mu} + \Gamma_{\mu}$ is the covariant derivative, Γ_{μ} is the spin connection, m is the particle mass, $V_{\mathrm{L,R}} \sim$

[a]E-mail: maxdvo@izmiran.ru

$G_F n_{eff}$ are the effective potentials of the interaction of left and right chiral projections with background matter, G_F is the Fermi constant, n_{eff} is the effective density of background particles, $\gamma^5(x) = -(i/4!)E^{\mu\nu\alpha\beta}\gamma_\mu(x)\gamma_\nu(x)\gamma_\alpha(x)\gamma_\beta(x)$, $E^{\mu\nu\alpha\beta} = \varepsilon^{\mu\nu\alpha\beta}/\sqrt{-g}$ is the covariant antisymmetric tensor in curved space-time, and $g = \det(g_{\mu\nu})$.

In Ref. [2] we found the solution of Eq. (1) for an ultrarelativistic neutrino moving in a rotating matter. Note that, in case of neutrinos, we should set $V_R = 0$. Choosing the appropriate vierbien vectors, we obtained ψ, which is expressed in terms of the Laguerre functions. Then we generalized our result to include different neutrino eigenstates and mixing between them. We obtained that the resonance condition is shifted by the matter rotation contrary to our previous claim in Ref. [5]. This effect can have the implication for the explanation of great linear velocities of pulsars since there is a correlation between the linear and angular velocities of a pulsar [6].

In Ref. [3], we obtained the solution of Eq. (1) for ultrarelaticistic electroweakly interacting electrons and quarks in the rotating matter. Using this solution we derived the nonzero electric current along the rotation axis in the form,

$$\mathbf{J} = \frac{q\boldsymbol{\omega}}{\pi}\left(V_R\mu_R - V_L\mu_L\right), \tag{2}$$

where $\boldsymbol{\omega}$ is the angular velocity, q is the electric charge (including the sign) of a test fermion, and $\mu_{R,L}$ are the chemical potentials of right and left fermions. The existence of the nonzero current in Eq. (2) is attributed in Ref. [3] to the new *galvano-rotational effect* (GRE). GRE is analogous to the chiral vortical effect [7], in which the induced current is $\mathbf{J} \sim \boldsymbol{\omega}(\mu_L^2 - \mu_R^2)$. However, in the later case the current is vanishing in the equilibrium at $\mu_L = \mu_R$, whereas $\mathbf{J}$ in Eq. (2) is nonzero in this situation.

GRE can be used for the generation of a toroidal magnetic field (TMF) in neutron and quark/hybrid stars. It is well known that, in a star, a purely poloidal magnetic field, which is observed by astronomers, is unstable. A toroidal component, which lays inside a star and can be of the same magnitude as a poloidal one, is required. In Ref. [3] we estimated the strength of TMF generated owing to GRE as $B_{tor} \sim |\mathbf{J}|R$, where $R \sim 10\,\text{km}$ is the star radius. Using the characteristics of the background matter in a compact star, one gets that $B_{tor} \sim 10^8\,\text{G}$ can be generated [3]. This TMF strength is comparable with the observed magnetic fields in old millisecond pulsars [8].

In Ref. [4] we solved Eq. (1) for an ultrarelativistic neutrino interacting with a linearly accelerated matter. In this case ψ is expressed via the Whittaker functions. The obtained solution turned out to reveal the instability of the neutrino vacuum leading to the creation of the neutrino-antineutrino ($\nu\bar{\nu}$) pairs. This phenomenon is analogous to the well known Unruh effect [9] consisting in the emission of the thermal radiation by an accelerated particle, with the effective temperature $T_{eff} = a/2\pi$, where a is the particle acceleration.

Requiring that the probability of the creation of $\nu\bar{\nu}$ pairs is not suppressed, in Ref. [4], we obtained the upper bound on the neutrino mass,

$$m \lesssim m_{\mathrm{cr}}, \quad m_{\mathrm{cr}} = 2\sqrt{\frac{|V_{\mathrm{L}}|a}{\pi}}. \tag{3}$$

If we study the creation of $\nu\bar{\nu}$ pairs in a core collapsing supernova (SN) at the bounce stage, one gets that $m_{\mathrm{cr}} \sim 10^{-7}\,\mathrm{eV}$. The obtained upper bound is comparable with the constraint on neutrino masses established earlier in Ref. [10], where we studied the $\nu\bar{\nu}$ pairs creation in SN at the neutronization.

In conclusion we mention that we have studied various phenomena happening with particles electroweakly interacting with accelerated background matter. We have considered two types of the acceleration: due to rotation and a linear acceleration. The exact solutions of the Dirac equation for a test fermion, accounting for both the matter interaction and the noninertial effects, have been found. Then we have discussed the influence of the matter rotation on the resonance in neutrino oscillations, the generation of the electric current flowing along the rotation axis, and the creation of $\nu\bar{\nu}$ pairs in a linearly accelerated matter. Finally we have considered the possibility of the implementation of our results in various astrophysical media such as neutron and quark/hybrid stars as well as SNs.

Acknowledgments

I am thankful to the Tomsk State University Competitiveness Improvement Program and to RFBR (research project No. 15-02-00293) for partial support.

References

[1] D. Alba, *et al.*, Int. J. Mod. Phys. **A 21**, 2781 (2006); H.W. Crater, *et al.*, Int. J. Geom. Meth. Mod. Phys. **11**, 1450086 (2014); D. Chowdhury, *et al.*, Annals Phys. **335**, 166 (2013).

[2] M. Dvornikov, JHEP **10** (2014) 053, arXiv:1408.2735; Mod. Phys. Lett. **A 30**, 1530017 (2015), arXiv:1503.01431.

[3] M. Dvornikov, JCAP **05** (2015) 037, arXiv:1503.00608.

[4] M. Dvornikov, JHEP **08** (2015) 151, arXiv:1507.01174.

[5] M. Dvornikov, in Proceedings of the 14th Lomonosov Conference on Elementary Particle Physics. Ed. by A.I. Studenikin, World Scientific, Singapore, 2010, pp. 183–185, arXiv:1001.2690; Azerbaij. Astron. J. **6**, 5 (2011), arXiv:1001.2516.

[6] S. Johnston, *et al.*, MNRAS **364**, 1397 (2005).

[7] A. Vilenkin, Phys. Rev. **D 20**, 1807 (1979).
[8] E.S. Phinney, *et al.*, Annu. Rev. Astron. Astrophys. **32**, 591 (1994).
[9] L.C.B. Crispino, *et al.*, Rev. Mod. Phys. **80**, 787 (2008).
[10] M. Dvornikov, *et al.*, Phys. Rev. **D 89**, 105028 (2014), arXiv:1312.2288.

PLASMON DECAY TO A NEUTRINO PAIR VIA NONSTANDARD FLAVOR VIOLATING INTERACTIONS IN A STRONGLY MAGNETIZED MEDIUM

A. V. Borisov [a]
Faculty of Physics, Moscow State University, 119991 Moscow, Russia
P. E. Sizin [b]
National University of Science and Technology MISiS, 119049 Moscow, Russia

Abstract. We calculate the neutrino luminosity of a degenerate electron gas in a strong magnetic field via plasmon decay to a neutrino pair due to nonstandard tensorial interactions with flavor violation and obtain the relative upper bounds on the corresponding effective coupling.

1. At present nonstandard neutrino interactions (NSIs) are actively investigated (see [1] and references therein). There is reason to believe that tiny neutrino masses m_ν (the recent cosmological bound on the sum of light neutrino masses is $\sum m_\nu < 0.12$ eV [2]) in comparison with charged fermion masses (leptons and quarks) m_f is due to NSIs (for the energy scale of New (nonstandard) Physics $M \gg m_f$): $m_\nu \sim m_f^2/M \ll m_f$ (the see-saw mechanism [3,4]). Tensorial NSIs are of particular interest for they directly generate (as opposed to (axial)vectorial ones) a neutrino magnetic moment (NMM) that does not depend on m_ν [1]. In the framework of the Standard Model, the NMM is proportional to m_ν, and its predicted value is much less than the present laboratory and astrophysical upper bounds (see a recent review on neutrino electromagnetic interactions [5]).

The effective Lagrangian of tensorial nonstandard interactions of neutrinos ν_α with fermions f can be written as [1]:

$$-\mathcal{L}_{\mathrm{TNSI}} = \frac{\epsilon^f_{\alpha\beta}}{M^2}\left(\bar{\nu}_\beta \sigma_{\mu\nu}\nu_\alpha\right)\left(\bar{f}\sigma^{\mu\nu} f\right) + \text{h.c.}, \tag{1}$$

where $\epsilon^f_{\alpha\beta}$ are (dimensionless) NSI couplings; $\alpha, \beta = e, \mu, \tau$ are flavor indices; $\sigma^{\mu\nu} = i[\gamma^\mu, \gamma^\nu]/2$ (we use the system of units where $\hbar = c = k_{\mathrm{B}} = 1$, $\alpha = e^2/4\pi \simeq 1/137$).

The NMM generated by tensorial NSIs (1) was calculated in Ref. [1] and, from the astrophysical bound on the NMM [6] (laboratory bounds are much weaker [5]), $\mu_\nu < 3 \times 10^{-12}\mu_{\mathrm{B}}$, where μ_{B} is the Bohr magneton, the bounds on couplings $\epsilon^f_{\alpha\beta}$ (see Eq. (1)) were derived. In particular, for electron couplings,

$$|\epsilon^e_{\alpha\beta}| < 3.9. \tag{2}$$

We note that a more stringent astrophysical bound $\mu_\nu < 1 \times 10^{-12}\mu_{\mathrm{B}}$ (see Refs. [7,8]) leads to

$$|\epsilon^e_{\alpha\beta}| < 1.3. \tag{3}$$

[a] E-mail: borisov@phys.msu.ru
[b] E-mail: mstranger@list.ru

In the present report, we investigate plasmon decay to a neutrino pair, $\gamma \to \nu_\alpha \bar{\nu}_\beta$, via the tensorial NSIs (1) in a strongly magnetized degenerate electron gas under conditions of the neutron star crust [9, 10]:

$$T \ll \mu - m, \; H > [(\mu/m)^2 - 1]H_0/2, \tag{4}$$

where T and $\mu \simeq \mu(T = 0) \equiv \varepsilon_{\rm F} = (m^2 + p_{\rm F}^2)^{1/2}$ are the temperature and chemical potential of the gas, $\varepsilon_{\rm F}$ and $p_{\rm F}$ are the Fermi energy and momentum, H is the magnetic field strength, $H_0 = m^2/e \simeq 4.41 \times 10^{13}$ G, m and $-e$ are the electron mass and charge. Under the conditions (4), electrons occupy only the lowest Landau level in the magnetic field with $p_{\rm F} = 2\pi^2 n_e/(eH)$, where n_e is the electron concentration, and the photon becomes the plasmon with nonzero mass which is equal to the plasmon frequency [14]

$$\omega_p = [(2\alpha/\pi)(p_{\rm F}/\varepsilon_{\rm F})H/H_0]^{1/2} m, \tag{5}$$

α is the fine-structure constant.

Comparing the neutrino luminosity (the rate of energy losses by a unit volume of a medium) due to the process in question, $Q_{\rm TNSI}$, with the luminosity $Q_{\rm SM}$ via the corresponding Standard Model process [11], we obtain the relative bounds on couplings $\epsilon^e_{\alpha\beta}$ that are close to the value given by Eq. (2). This (astrophysical) method has been used previously to obtain constraints on the NMM by comparative analysis of the electromagnetic and weak mechanisms of photoproduction of a neutrino pair on an electron [12] and plasmon decay to a neutrino pair [13] under the same conditions.

2. Taking into account Eqs. (1), (4), and (5), we have obtained asymptotic formulas for the luminosity $Q_{\rm TNSI}$ in three cases that are of interest for applications.

1) For $p_{\rm F} \ll m$ (the nonrelativistic case)), $\omega_p \gg T$:

$$Q_{\rm TNSI} = \frac{2}{3}\frac{\epsilon_e^2 \alpha m^5}{(2\pi)^{11/2}} \left(\frac{m}{M}\right)^4 \left(\frac{H}{H_0}\right)^2 \left(\frac{\omega_p}{m}\right)^{11/2} \left(1 + \frac{2p_F}{\omega_p}\right)^2 \left(\frac{T}{m}\right)^{3/2} e^{-\omega_p/T}. \tag{6}$$

2) For $p_F \gg m$ (the ultrarelativistic case), $\omega_p \gg T$:

$$Q_{\rm TNSI} = \frac{\epsilon_e^2 \alpha m^5}{6\,(2\pi)^{7/2}} \left(\frac{m}{M}\right)^4 \left(\frac{H}{H_0}\right)^2 \left(\frac{\omega_p}{m}\right)^{7/2} \left(\frac{T}{m}\right)^{3/2} e^{-\omega_p/T}. \tag{7}$$

3) For $p_F \gg m$, $\omega_p \ll T$:

$$Q_{\rm TNSI} = \frac{2\zeta(5)}{3\pi^4}\, \epsilon_e^2 \alpha m^5 \left(\frac{m}{M}\right)^5 \left(\frac{H}{H_0}\right)^2 \left(\frac{T}{m}\right)^5. \tag{8}$$

In Eqs. (6)–(8), the effective TNSI coupling is introduced as

$$\epsilon_e = \left[\sum_{\alpha,\beta} \left(\epsilon^e_{\alpha\beta}\right)^2\right]^{1/2}. \tag{9}$$

3. Assuming $Q_{\rm TNSI} < Q_{\rm SM}$ and $M = 1$ TeV (as in Ref. [1]), we have obtained the following relative upper bounds on the coupling (9) for the above cases (6)–(8) with the use of the corresponding asymptotic formulas for $Q_{\rm SM}$ (see Ref. [11]).

Case 1:

$$\epsilon_e < \bar{g}_V G_{\rm F} M^2/\sqrt{2} = 7.9 \tag{10}$$

for $p_{\rm F} \gg \omega_p$ with $G_{\rm F}$ being the Fermi coupling constant and the effective vector coupling of the weak current $\bar{g}_V = \big[\sum_{\ell=e,\mu,\tau} g_V^2(\ell)\big]^{1/2} \simeq 0.964$; for $p_{\rm F} \ll \omega_p$:

$$\epsilon_e < 600 {\rho_6}^{1/2} H_{13}^{-1}\,, \tag{11}$$

where $\rho_6 = \rho/(10^6\ {\rm g/cm^3})$, $H_{13} = H/(10^{13}\ {\rm G})$; for typical values $\rho_6 = 1$, $H_{13} = 150$, Eq. (11) gives

$$\epsilon_e < 4.0. \tag{12}$$

Case 2:

$$\epsilon_e < \sqrt{2}\bar{g}_V G_{\rm F} M^2/\pi \simeq 5.1. \tag{13}$$

Case 3:

$$\epsilon_e < \bar{g}_A G_{\rm F} M^2/\pi \simeq 3.2, \tag{14}$$

where the effective axial coupling $\bar{g}_A = \big[\sum_{\ell=e,\mu,\tau} g_A^2(\ell)\big]^{1/2} = \sqrt{3}/2$.

In conclusion, we have obtained the relative upper bounds (10)–(14) on the effective tensorial NSI coupling (9) which are of the order of the bounds (2) and (3) derived from the astrophysical limits on the neutrino magnetic moment.

References

[1] K. J. Healey, A. A. Petrov, and D. Zhuridov, Phys. Rev. D **87** 117301 (2013); Phys. Rev. D **89** 059904 (2014) [arXiv:1305.0584 [hep-ph]].

[2] N. Palanque-Delabrouille *et al.*, JCAP **11**, 011 (2015).

[3] M. Fukugita and T. Yanagida, *Physics of Neutrinos: And Applications to Astrophysics* (Springer, Berlin, Heidelberg, 2003).

[4] Z.-Z. Xing and S. Zhou, *Neutrinos in Particle Physics, Astronomy and Cosmology* (Zhejiang University Press, Hangzhou; Springer, Berlin, Heidelberg, 2011).

[5] C. Giunti and A. Studenikin, Rev. Mod. Phys. **87**, 531 (2015).

[6] G. G. Raffelt, Phys. Rep. **320**, 319 (1999).

[7] A. Ayala, J. C. DOlivo, and M. Torres, Nucl. Phys. B **564**, 204 (2000).

[8] A. V. Kuznetsov, N. V. Mikheev, and A. A. Okrugin, Int. J. Mod. Phys. A **24**, 5977 (2009).

[9] R. C. Duncan and C. Thompson, Astrophys. J. Lett. **392**, L9 (1992).

[10] P. Haensel, A. Y. Potekhin, and D. G. Yakovlev, *Neutron Stars 1. Equation of State and Structure* (Springer, New York, 2007).

[11] M. V. Chistyakov and D. A. Rumyantsev, JETP **107**, 533 (2008).

[12] A. V. Borisov, B. K. Kerimov, and P. E. Sizin, Phys. Atom. Nucl. **75**, 1305 (2012).
[13] A. V. Borisov and P. E. Sizin, Moscow Univ. Phys. Bull. **68**, 114 (2013).
[14] A. E. Shabad, Tr. FIAN **192**, 5 (1988).

GENERAL PROPERTIES OF GRAVITATIONALLY BOUND SYSTEMS AT DIFFERENT LEVELS OF QUANTIZATION

Michael Fil'chenkov [a]
Institute of Gravitation and Cosmology, Peoples' Friendship University of Russia, 117198 Moscow, Russia

Abstract. Levels of quantization in gravity theory have been classified. Objects and processes in gravitationally bound quantum systems and their general properties are analyzed to create a consistent theory of gravity quantization and to construct quantum models of compact astrophysical objects and the early Universe.

1 Introduction

Some approaches to quantization in gravity theory [1–3] are well developed, e. g. quantum field theory in curved space-time. To quantize gravity itself, there exist a perturbative field-theoretical approach, applicable only for weak gravitational fields, and non-perturbative quantum loop gravity, which generalizes quantum geometrodynamics at the scales comparable with the Planckian length. Other approaches concern quantum mechanics, quantum gravitational collapse and quantum cosmology dealt with gravitationally bound systems.

2 Levels of Quantization in Gravity Theory

Gravity is considered in the framework of general relativity and Newtonian theory, and quantum theory – in terms of quantum mechanics and quantum field theory. The levels of quantization in the preset space-time or Newtonian gravitational field:

- quantum mechanics in the Newtonian gravitational field with relativistic corrections,
- quantum theory of extraneous fields in curved space-time.

The levels of gravitational-field and space-time quantization:

- weak-gravitational-field quantization involving gravitons,
- quantum geometrodynamics for geometry as a whole,
- loop quantum gravity of space-time itself.

To these levels of quantization correspond the objects and processes in gravitationally bound systems as follows:

- graviatoms, which comprise miniholes capturing particles,
- the Universe's birth from de Sitter's vacuum,
- quantum gravitational collapse,
- particle creation near horizons and in the early Universe.

[a] E-mail: fmichael@mail.ru

3 General Properties of Gravitationally Bound Quantum Systems

3.1 System Energies

The graviatom energy levels

$$E = -\frac{2B^2m}{\hbar^2}\frac{1}{\left[2p+1+\sqrt{(2l+1)^2+\frac{8mA}{\hbar^2}}\right]^2}, \tag{1}$$

where $p = n - s - 1 = 0, 1, 2, ..., l \leq n = 1, 2, 3..., m$ is the mass of the particle being constituent of the graviatom, n and l the principal and orbital quantum numbers respectively, with $s(s+1) = \frac{2mA}{\hbar^2} + l(l+1)$, $A = \frac{mc^2 r_q r_g}{4}$, $B = \frac{mc^2 r_g}{2}$, where $r_g = \frac{2GM}{c^2}$ is the gravitational radius of the minihole, M its mass, $r_q = \frac{q^2}{mc^2}$ the classical radius of a particle with the mass m and the charge q.

The energy levels of the pre-de-Sitter Universe behaving as a planckeon with radiation energy (cf. Lemaître's atom) in the field of stiff matter

$$E = -\left(\frac{r_0}{l_{pl}}\right)^6 \frac{B_5^2 m_{pl} c^2}{8\left[p+\frac{1}{2}+\sqrt{\frac{1}{4}-B_6\left(\frac{r_0}{l_{pl}}\right)^4}\right]^2}, \tag{2}$$

where $p = 0, 1, 2, ...,$ r_0 is de Sitter's horizon, B_5 the contribution of perfect gas, B_6 the contribution of ultrastiff matter to the total energy density on de Sitter's horizon, m_{pl} the Planckian mass.

3.2 System Component Masses

Minihole masses M, masses of particles captured by them in the graviatom and masses of leptoquarks m born in the early Universe are related by the same ralationship:

$$Mm \sim \frac{1}{2}m_{pl}^2, \tag{3}$$

which follows from the condition of leptoquark creation being compatible with that of Jeans instability for miniholes $ct_c \sim r_g$, where t_c is the particle Compton time. The particles are created on exit from under the potential barrier as a result of de Sitter's vacuum decay.

3.3 Universe's Birth Probabilities

The probability of a universe birth as a result of gravitational collapse

$$D \sim \exp\left\{-2|B_4|^{\frac{3}{4}}\left(\frac{r_0}{l_{pl}}\right)^2\right\}, \tag{4}$$

where B_4 is the contribution of radiation to the total energy density on de Sitter's horizon, $|B_4| \gg 1$.

The probability of a new universe birth as a result of quantum fluctuation

$$D \sim \exp\left\{-\frac{2(k-B_2)^{\frac{3}{2}}}{3B_0}\left(\frac{r_0}{l_{pl}}\right)^2\right\}, \tag{5}$$

where B_0 is the contribution of de Sitter's vacuum, B_2 the contribution of strings to the total energy density on de Sitter's horizon. For $4B_0B_4 \sim (k-B_2)^2$ this formula reduces to the previous one if $B_4 \ll 1$. Thus, the universe's birth as a result of gravitational collapse is considerably more probable than that for gravitational collapse.

4 Conclusion

The graviatom and Lemaître's atom are described by Schrödinger's type equation. For Lemaître's atom this follows from Peres' equation [4] containing momentum squared.

The linkage between the relationships for graviatoms and formation of miniholes and leptoquarks in the early Universe is due to the presence of miniholes and quantum field effects in both cases. A unified formula for new universe birth as a result of gravitational collapse and quantum fluctuation is due to both processes being based on the tunnel effect.

The consideration of gravitationally bound quantum systems may play a part in the creation of a consistent theory of gravity quantization allowing loop and string cosmology to be combined [5].

Acknowledgments

I am grateful to the Organizing Committee for providing me a possibility of participating the 17th Lomonosov Conference.

References

[1] C. Rovelli, *"Quantum Gravity"* (Cambridge University Press, Cambridge, UK, 2004).
[2] C. Kiefer, *"Quantum Gravity"* (Oxford University Press, Oxford, UK, 2007).
[3] M.L. Fil'chenkov, *"Quantum Gravity"* (Peoples' Friendship University Press, Moscow, 2013).
[4] J. A. Wheeler, *"Einsteins Vision"* (Springer-Vellag, Berlin, 1968).
[5] M. Szydłowski, J. Mielczarek, J. Cosmol. Astroparticle Phys. **0808**, 014 (2008).

VACUUM EFFECTS DUE TO POSSIBLE LORENTZ VIOLATION AND ANOMALOUS MAGNETIC MOMENT

Andrey Bubnov [a] Nadezda Gubina [b] Vladimir Zhukovsky [c]
Faculty of Physics, Moscow State University, 119991 Moscow, Russia

Abstract. We calculate one-loop corrections to the effective Lagrangian of a model of a charged fermion with an anomalous magnetic moment and additional axial-vector interaction with the background field.

1 Introduction

As is well known a charged fermion, due to the Dirac equation, has an intrinsic angular momentum, i.e. spin and, as a consequence, a Dirac magnetic moment which is equal to the Bohr magneton: $\mu_0 = \frac{e\hbar}{2mc}$. In the framework of QED description an anomalous Schwinger term contribution [1] to the Dirac magnetic moment appears even in the lowest order of the fine structure constant $\alpha = \frac{e^2}{\hbar c}$: $\mu_{Sch} = \mu_0(1 + \frac{\alpha}{2\pi})$. The Schwinger anomalous magnetic moment $\Delta\mu_{Sch} = \frac{\alpha}{2\pi} \cdot \mu_0$ is denoted as an electronic vacuum magnetic moment.

Recently, much interest has been attracted to a possible violation of Lorentz symmetry. A field theoretical model, called Standard Model Extension, incorporating various possible terms with Lorentz violation (LV) was developed [2]. Subject of a special interest is investigation of one of the proposed LV terms including axial-vector fermionic interaction, which can be described by the following CPT–odd term in the Dirac Lagrangian: $\bar{\psi}\gamma^5\gamma^\mu b_\mu \psi$ (here b^μ is a constant axial 4D-vector, that can be interpreted as a certain condensate of some more fundamental higher dimensional model) (see e.g., [3]). Indeed, such terms may appear as a nontrivial vacuum expectation value of some vector field, which could lead to space anisotropy or, on the other hand, they can be a consequence of a certain torsion field induced by fermions (see, e.g., [4]). In this report, we calculate one-loop correction to the effective Lagrangian in a magnetic field due to the simultaneous presence of both anomalous magnetic moment and an axial–vector interaction terms[d].

The modified Dirac equation for a fermion with an anomalous magnetic moment[e], $\mu = \kappa\mu_0$, in the presence of an additional interaction with a constant

[a]E-mail: bfandrey@mail.ru
[b]E-mail: gubina_nadya@mail.ru
[c]E-mail: vlchzh@gmail.com

[d]The vacuum effects in the presence of each of these two terms separately was considered, e.g., in Refs. [3], [5], [6], [7].

[e]Assuming possible extension of the Standard model, some additional terms may appear in the anomalouse magnetic moment (see, e.g., [8]), besides the Schwinger term $\Delta\mu_{Sch}$, and this is included in the general coefficient κ.

axial-vector field b^μ, according to [2] can be written as follows:

$$\left(i\gamma^\alpha D_\alpha - m + \frac{\mu}{2}\sigma^{\alpha\beta}F_{\alpha\beta} - \gamma^5\gamma^\alpha b_\alpha\right)\psi = 0,\ D_\alpha = i\partial_\alpha - eA_\alpha. \qquad (1)$$

In what follows, we shall consider only a nonzero external magnetic field $\vec{H} = H\vec{e}_z$ and assume 3d-vector $\vec{b}$ is also parallel to the z-axis.

2 The Effective Lagrangian

The one-loop fermion contribution to the effective Lagrangian can be calculated with the help of the proper-time method [1]

$$\Delta\mathcal{L} = \frac{1}{4\sqrt{\pi}}\frac{eH}{(2\pi)^2}\int_{-\infty}^{\infty} dp \sum_{n,\zeta,\epsilon}\int_{\frac{1}{\Lambda^2}}^{\infty}\frac{ds}{s^{3/2}}\,\mathrm{e}^{-sE^2} \qquad (2)$$

Here E is the fermion energy, and the quantum numbers have the following meaning: p is the $z-$component of the fermion momentum, $n = 0, 1, 2\ldots$ is the Landau quantum number, $\zeta = \pm 1$ is the spin projection on the magnetic field direction, $\epsilon = \pm$ is the energy sign (Λ is the cut-off parameter). The summation over all quantum numbers should take into account that the ground state has no degeneracy, i.e. $n = 0$, $\zeta = +1$ (for $e > 0$) and $n = 0$, $\zeta = -1$ (for $e < 0$).

In what follows, we shall consider various cases with special values of the parameters.

a) anomalous moment $\mu = 0$, **axial condensate** $b_0 = b \neq 0$, $\vec{b} = 0$:
$E^2 = m^2 + (\mathrm{sgn}(p)\sqrt{p^2 + 2eHn} + \zeta b)^2$ (for this energy spectrum, cf. e.g. [3], [5], [7], [9]).

Note that in the expansion in powers of the small parameter b, the linear term will be absent due to the oddness of the expression, so up to the second order in b, one gets the regularized expression for the one loop effective Lagrangian

$$\Delta\mathcal{L} = \frac{b^2}{4\pi^2}\left[-m^2 eH\int_0^\infty ds\,\mathrm{e}^{-sm^2}\left(\coth(eHs) - \frac{1}{eHs}\right) + \Lambda^2 - m^2\log\frac{\Lambda^2}{m^2}\right].$$

Note that terms with the cut-off Λ do not actually depend on the magnetic field, and thus can be eliminated by renormalization, yielding the final result

$$\Delta\mathcal{L}_{b^2} = -b^2\frac{(eH)m^2}{4\pi^2}\int_0^\infty ds\,\mathrm{e}^{-sm^2}\left(\coth(eHs) - \frac{1}{eHs}\right). \qquad (3)$$

This result could be easily generalized to the case with the anomalous moment $\mu = 0$, axial condensate $b_0 \neq 0$, $b_3 \neq 0$. Then, by introducing a vector $\beta^\mu = \frac{1}{2}\varepsilon^{\mu\nu\alpha\beta}b_\nu F_{\alpha\beta}$, with nonzero components $\beta^0 = \varepsilon^{0312}b_3F_{12} = b_3H$

$\beta^3 = \varepsilon^{3012} b_0 F_{12} = -b_0 H$, we shall have to deal with the same intermidiate expressions as in the previous situation, and the unified result can then be presented by Eq. (3) where now b is the 4-vector $b = (b_0, 0, 0, b_3)$.

b) anomalous moment $\mu \neq 0$, **axial condensate** $b_0 = 0$, $\vec{b} = 0$ ([3], [6]):

$$E^2 = p^2 + (\sqrt{2eHn + m^2} + \zeta\mu H)^2.$$

(for this energy spectrum, cf. e.g. [3], [6]).

Assuming μH to be small let us consider terms of order $(\mu H)^2$. After substraction regularization procedure similar to the previous case one obtains

$$\Delta\mathcal{L} = \frac{(\mu H)^2}{2(2\pi)^2}\left(\Lambda^2 + m^2 \log\frac{\Lambda^2}{m^2}\right) - \frac{(\mu H)^2 eH}{2(2\pi)^2}\int_0^\infty ds \frac{\mathrm{e}^{-sm^2}}{s}\left(\coth(eHs) - \frac{1}{eHs}\right).$$

The second term converges and divergence in the first term could be eliminated by subtraction, yielding the final result

$$\Delta\mathcal{L}_{\mu^2} = -\frac{(\mu H)^2 eH}{2(2\pi)^2}\int_0^\infty ds \frac{\mathrm{e}^{-sm^2}}{s}\left(\coth(eHs) - \frac{1}{eHs}\right).$$

c) anomalous moment $\mu \neq 0$, **axial condensate,** $b_0 = b \neq 0$, $\vec{b} = 0$ (for the energy spectrum, cf. e.g. [3]).

In this more general case we shall follow [3] and introduce, for more convenience, a mixing angle Θ
$\Theta = \arctan\frac{b}{\mu H}$, then $\mu H = \widetilde{\mu}H\cos\Theta$, $b = \widetilde{\mu}H\sin\Theta$, $\widetilde{\mu}H = \sqrt{(\mu H)^2 + b^2}$.
The resulting renormalized expression for the one-loop effective Lagrangian can be written in the second order in the effective anomalous magnetic moment $\widetilde{\mu}H$ and Lorentz-violating parameter b

$$\Delta\mathcal{L} = -(\widetilde{\mu}H)^2\frac{eH}{(2\pi)^2}\int_0^{+\infty} ds e^{-m^2 s}\left(m^2\sin^2\Theta + \frac{1}{2s}\cos^2\Theta\right)\left(\coth(eHs) - \frac{1}{eHs}\right).$$

It is clear that this general result can be easily transformed into the above special cases a) and b)
$\underline{\Theta = 0}$ implements the case $b = 0$, then $\widetilde{\mu}H = \mu H$ and we have $\Delta\mathcal{L} = \Delta\mathcal{L}_{\mu^2}$
$\underline{\Theta = \frac{\pi}{2}}$ implements the case $\mu = 0$, then $\widetilde{\mu}H = b_0$ and we have $\Delta\mathcal{L} = \Delta\mathcal{L}_{b^2}$
The result obtained in this work, when both condensates are nonvanishing, generalizes the previously considered particular cases, when only one of the two possible terms, b_μ or μ was taken into account.

References

[1] J. Schwinger, Proc. Nat. Acad. Sci. USA, **40**, 132 (1954).
[2] D. Colladay, V.A. Kostelecky, Phys. Rev. **D 58** 116002 (1998).
[3] I.E. Frolov, V.Ch. Zhukovsky, J. Phys. **A 40**, 10625 (2007).
[4] I. L. Shapiro, Phys. Rept. **357**, 113 (2002).
[5] Y.A. Sitenko, K.Y. Rulik, Eur. Phys. J. **C 28** 405 (2003).
[6] V. N. Rodionov, J. of Exp. and Theor. Phys. **98** 395 (2004).
[7] K. Fukushima, D.E. Kharzeev, H.J. Warringa, Phys. Rev. **D 78** 074033 (2008).
[8] Y.V. Stadnik, B.M. Roberts, V.V. Flambaum, Phys. Rev. **D 90** 045035 (2014).
[9] A.F. Bubnov, V.Ch. Zhukovsky, Moscow Univ. Phys. Bulletin **65** 105 (2010).

MAGNETIZATION IN (2+1)-DIMENSIONAL GROSS–NEVEU MODEL WITH APPLICATION TO CARBON NANOTUBES

Dietmar Ebert [a]
Institute of Physics, Humboldt-University Berlin, 12489 Berlin, Germany
Vladimir Ch. Zhukovsky [b], Pavel B. Kolmakov [c]
Faculty of Physics, Moscow State University, 119991 Moscow, Russia
Konstantin G. Klimenko [d], Roman N. Zhokhov [e]
Institute for High Energy Physics, 142281 Protvino, Moscow Region, Russia

Abstract. (2+1)D Gross–Neveu model on the cylinder is investigated under the influence of finite temperature and magnetic field. Effective potential of the model is calculated. Phase diagram and magnetization graphs are plotted.

1 Introduction

1.1 Gross–Neveu model

Massless Gross–Neveu model [1] is one of the simplest 4-fermion interaction models with the Lagrangian written as

$$\mathcal{L} = \sum_{a=1}^{N} \overline{\psi^a} \imath \gamma^\mu \partial_\mu \psi^a + \frac{G}{2N} \left(\sum_{a=1}^{N} \overline{\psi^a} \psi^a \right)^2, \tag{1}$$

where $\hbar = c = 1$ units are used. Omitting "flavour" index a and introducing auxiliary field $\sigma(x) = \frac{G}{N}\overline{\psi}(x)\psi(x)$ (Hubbard–Stratonovich transformation), we rewrite the model Lagrangian as

$$\mathcal{L} = \overline{\psi} i \gamma^\mu \partial_\mu \psi + \overline{\psi}\psi\sigma + \frac{N}{2G}\sigma^2. \tag{2}$$

that is equivalent to the initial Lagrangian (1) if the equation of motion for the field σ is satisfied. Assuming N is large enough we can use the mean field approximation $\sigma = \frac{G}{N} < \overline{\psi}\psi >= \text{const}(x)$. If $\sigma \neq 0$ one can see that the term $\overline{\psi}\psi\sigma$ breaks the chiral symmetry of the initial model.

In 3D space-time with 4-component reducible representation for fermions motivated by graphene application (see section 1.2) we can introduce effective potential—for the model under investigation it is

$$V_{\text{eff}} = -2 \int \frac{d^3p}{(2\pi)^3} \ln(p^2 + \sigma^2) + \frac{\sigma^2}{2G} = \frac{|\sigma|^3}{3\pi} + \frac{\sigma^2}{2}\left(\frac{1}{G} - \frac{1}{G_c}\right), \tag{3}$$

[a] E-mail: debert@physik.hu-berlin.de
[b] E-mail: zhukovsk@phys.msu.ru
[c] E-mail: pavel.b.kolmakov@yandex.ru
[d] E-mail: konstantin.klimenko@ihep.ru
[e] E-mail: rzn95@mail.ru

where $G_c = \frac{\pi^2}{\Lambda}$ with Λ being the cut-off momentum. It is easy to see that for $G > G_c$ this potential has a non-trivial minimum $\sigma_0 = \pi\left(\frac{1}{G_c} - \frac{1}{G}\right)$, this phenomenon is called *dynamical symmetry breaking* and it is typical for GN and GN-like models. In the present paper it will be assumed that $G > G_c$.

1.2 Dirac equation in graphene

Nowadays it is well-known that low-energy electrons on the graphene honeycomb lattice can be described with free Dirac Lagrangian (see e.g. review [2])

$$\mathcal{L}_0 = \overline{\psi}\left[i\gamma^0\partial_0 + iv_{\mathrm{F}}\gamma^1\partial_x + iv_{\mathrm{F}}\gamma^2\partial_y\right]\psi = \overline{\psi}i\gamma^\mu\tilde{\partial}_\mu\psi, \tag{4}$$

where $\psi^t = \left(\psi_K^A, \psi_K^B, -i\psi_{K'}^B, i\psi_{K'}^A\right)$, with functions $\psi_{K(K')}^{A(B)}$ describing the states of the electrons in the $A(B)$ triangular sublattice near the "Dirac point" $K(K')$.

In addition to free Dirac Lagrangian, phonon-mediated interactions with coupling strength G in graphene can be approximated by contact 4-fermion interaction that has the same form as interaction Lagrangian in GN model [3].

2 Gross–Neveu model on a cylinder under the influence of external parameters

Now we compactify coordinate x_2 to the circle with radius of R and circumference of $L = 2\pi R$ (hence momentum p_2 becomes quantized $p_{2\ell} = \frac{2\pi\ell}{L}$) and introduce finite temperature $T = 1/k_{\mathrm{B}}\beta$ by using a technique of Matsubara frequencies $\omega_n = \frac{2\pi(n+1/2)}{\beta}$ (see [4]) and external magnetic field parallel to the cylinder axis $\mathbf{H} = H_0\mathbf{e}_z$ producing Aharonov–Bohm effect that effectively modifies periodic boundary conditions by adding a phase of $\phi = \frac{eH_0R^2}{2}$ and Zeeman effect that effectively adds the chemical potential depending on z-projection of spin $\delta\mu_s = \pm\frac{g_L}{2}\mu_B H_0$, where μ_B is Bohr magneton and g_L is Lande factor. The effective potential (3) should be modified by adding the terms $V_{(L)}$ that includes the terms independent of temperature and Zeeman effect

$$V_{(L)} = 2\mathrm{Re}\left(\tfrac{1}{\pi L^3}\mathrm{Li}_3(e^{-L\sigma+2\pi i\phi}) + \tfrac{\sigma}{\pi L^2}\mathrm{Li}_2(e^{-L\sigma+2\pi i\phi})\right), \tag{5}$$

where $\mathrm{Li}_\nu = \sum\limits_{k=1}^{\infty}\frac{x^k}{k^\nu}$ are polylogarythm special functions and $V_{(\mu T)}$ that includes influence of temperature and Zeeman effect

$$V_{(\mu T)} = -\frac{2}{\beta L}\sum_{\ell=-\infty}^{\infty}\int\frac{dp_1}{2\pi}\left[\ln\left(1+e^{-\beta\left(E^+_{\ell,p_1}\right)}\right) + \ln\left(1+e^{-\beta\left(E^-_{\ell,p_1}\right)}\right)\right], \tag{6}$$

where $E^\pm_{\ell,p_1} = \sqrt{p_1^2 + (\frac{2\pi}{L})^2(\ell+\phi)^2 + \sigma^2} \pm \delta\mu$. Further calculations and graphs drawing were performed via numerical calculations. Magnetization of the system can be calculated using the general rule (more detailed calculation and

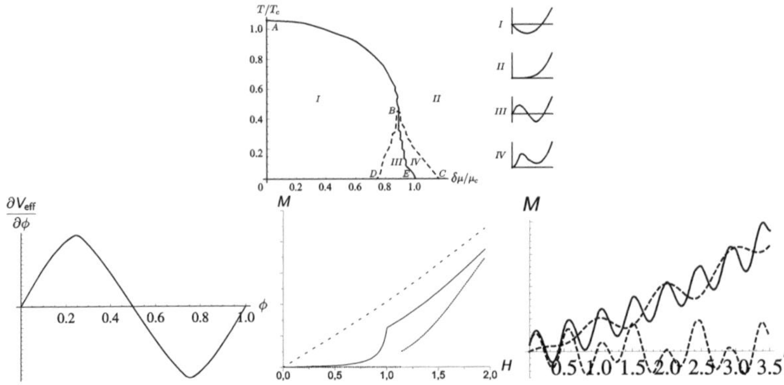

Figure 1: Top graph: Phase diagram of the model under the influence of chemical potential and temperature. Bottom left graph: system magnetization due to AB effect as a function of magnetic field H. Bottom middle graph: System magnetization M as a function of magnetic field H for $T = \frac{1}{3}T_c$ (dashed top line), $T = \frac{4}{5}T_c$ (solid line). $T = \frac{3}{2}T_c$ (dashed lower line) H is given in the units of H_μ that restores symmetry in the zero temperature model. Bottom right graph: Total system magnetization (solid line), Zeeman magnetization (dashed top line), and AB magnetization (dashed lower line) for $H_\mu = 2\,H|_{\phi=1}$.

discussion can be found in [5] and [6]) $M = -\frac{\partial V_{\text{eff}}}{\partial H}$. Figure 1 shows the phase structure of the model under the influence of Zeeman effect and the system magnetization under the influence of Zeeman and Aharonov–Bohm effects, in the diagram for Zeeman effect one can clearly observe phase transitions of first and second kind occurring in the system and shown in the upper graph.

3 Summary and conclusion

In the present paper the Gross–Neveu model on a cylinder was investigated under the influence of temperature and external magnetic field. The main result of this work is the effective potential being calculated that allowed to plot different graphs representing properties of the modeled system, such as the system magnetization due to external magnetic field and allows to investigate over properties such as for example phase structure of the model.

References

[1] D. J. Gross, A. Neveu, Phys.Rev. **D 10** 3235 (1974).
[2] A. H. C. Neto et al, Rev. Mod. Phys. **D 81**, 109 (2009).
[3] N.A. Viet, H. Ajiki, T. Ando, J. Phys. Soc. Jpn **63**, 3036 (1994)

[4] J. I. Kapusta, C. Gale, *"Finite-temperature Field Theory: Principles and Applications."* (Cambridge University Press, Cambridge, 2006).
[5] R. N. Zhokhov, V. Ch. Zhukovsky, P. B. Kolmakov, Moscow Univ. Phys. Bull. **70** 226 (2015).
[6] D. Ebert et al, arXiv:1509.08093 [cond-mat.mes-hall]

ON CONSTRUCTION METHODS OF THE FUNDAMENTAL INTERACTIONS THEORY

Valery Koryukin [a] and Artem Koryukin [b]
Institute of Physics, Kazan Federal University, 420008 Kazan, Russia

Abstract. We introduce the postulates, which are used for the construction of the interaction theory in the new physical cosmology.

We are based on the postulate, that in the real world it can determine only the most plausible physical laws. We shall not confuse the real world with the mathematical one, in which will be used sets of continuum cardinality. As a result we state a modest goal: to build the most plausible physical theory in which there will be no the divergences. Naturally, in this case it is necessary to use the principle of least action and the theory of generalized functions. In addition, we will use the principle of privilege: the description should be the most simple. We develop the Dirac hypothesis on the presence of sea electrons in the Universe for the provision of the radiation stability of electrons with the positive energy relative to the photon emission. That is, it is assumed that the observable fermions are stable, because they are found in the excited state and all levels of particles in the ground (degenerate) level are occupied. In this case, the particle charge is determined by the ratio of the emission frequency of particular virtual bosons to the frequency of the collision with background particles of the Universe. We follow Pauli, who preferred to work with closed systems, in which there are conservation laws (in contrast to open systems). He suggested to introduce an unobservable (as then it seemed to him) particle, called subsequently (Fermi) as a neutrino, for a description of a beta decay. Precisely neutrinos constitute a particle background, being a catalyst of stochastic processes and characterized by two parameters (the density ρ and the temperature T). Because the neutrino background of the Universe in our opinion exists in two degenerate states (the Fermi liquid, the Bose liquid), the density of its Fermi liquid state is defined by the Fermi energy E_F, which should significantly exceed its temperature T (we shall use the energy system of units: $h/(2\pi) = c = 1$, where h is the Planck constant and c is the velocity of light). As a result the statistical physics acquires a high weight for us and the Boltzmann equation will have the fundamental importance.

We shall use also the postulate: the fundamental interaction must be the vector type [1]. As a consequence, such standard interaction as the gravitation is not the fundamental one, but it is generated by collective oscillations in the neutrino sea of the Universe [2]. Hence the weak interaction acquires the global value, because it causes the bond of neutrinos both one with the other and with charged leptons and quarks. Further, we shell rely on the cold plasma theory,

---

[a] E-mail: vmkoryukin@gmail.com
[b] E-mail: akoryukin@gmail.com

which was developed for the first-kind superconductor and for the second-kind superconductor. By this a nuclear matter is an analog of second-kind superconductors in relation to gluons, whichs as vortices penetrate in a Bose condensate consisting of neutrino Cooper pairs, while the vacuum in relation to gluons is an analog of first-kind superconductors. We assume that the interaction energy E must depend on the number of particles and quasi-particles participating in this interaction, defining its dependence to space coordinates by means of an average number of bosons, whichs are exchanged two particles. As a result in the general case (n is a number of bosons):

$$E = -2 \int \left[\sum_{\sigma_1,\sigma_2} \left(\rho_1 \sigma_1 \rho \sigma_2 \rho_2 \sum_{n=0}^{N_{\sigma_1}} n e^{-2n\rho\sigma_1 r/T} \Big/ T \sum_{n=0}^{N_{\sigma_1}} e^{-2n\rho\sigma_1 r/T} \right) \right] dV_1 dV_2, \tag{1}$$

where ρ_1 is the energy density of first body particles; ρ_2 is the energy density of second body particles; ρ is the energy density of quasiparticles, T is theirs temperature; σ_1 and σ_2 are cross-sections with the emission or the absorption of particles or quasi-particles. Hence (assuming that $N_{\sigma_1} \to \infty$) for the description of fundamental interactions it can obtain the potential in the form

$$U(r) = -\sum_{\sigma} \left[C_\sigma / \left(e^{B_\sigma r} - 1 \right) \right] \tag{2}$$

($B_\sigma \propto \sigma_1$, $C_\sigma \propto \sigma_1, \sigma_2$) and which can be the foundation for the construction of vector fields. Because the symmetry, characterizing the physical system, is selected on the basis of experimental data, the geometrical structure is only the most plausible. As a result it follows that for the description of particles only the spaces of the affine connection with the torsion can be used. The assumption on the "sea" of quarks in the ground state allows to use the Landau theory of the Fermi liquid, considering observable particles as quasi-particles on the background of "sterile" neutrinos and "sterile" antineutrinos. The properties of the latters must define the geometrical and topological properties of the space-time, which must not be a singly connected one if only in consequence of the electric charge quantization. Note that the transition to the description of the matter background subsystem with the help of the space-time manifold is carried when the Fermi energy of "sterile" neutrinos tends to infinity.

References

[1] V. Koryukin, in Contributed Papers of 10th International Conference on General Relativity and Gravitation (Padova, Italy, 4-9 July 1983), Vol. I, Consiglio Nazionale Delle Ricerche, Roma, 406, 1983.

[2] V. Koryukin, in Proceedings of III International Symposium on Weak and Electromagnetic Interactions in Nuclei (Dubna, Moscow Region, Russia, June 16-22, 1992), Ed. by Ts. D. Vylov, World Scientific, Singapore, 456, 1992.

PHYSICAL APPROACH FOR NON-HERMITIAN FERMION MODEL WITH A MAXIMAL MASS

Vasiliy N. Rodionov [a] and Galina A. Kravtsova [b]
Plekhanov Russian University of Economics, Moscow, Russia
Faculty of Physics, Moscow State University, 119991 Moscow, Russia

Abstract. We study the connection between geometrical approach of quantum theory with maximal mass and algebraic model for the fermions with γ_5-restriction for mass.

1 Geometrical Theory with Maximal Mass

In 1956 P.A.M. Dirac in his report at the conference in Russia in the Joint Institute of Nuclear Researches of the Dubna has formulated a number of the concrete problems, which in his opinion will excite of future physicists. Seven problems were written by Dirac, part of them are relevant to our days: *1). infinities; 2). fine structure constant; 3). nucleons and mesons; 4). ratio of the masses of elementary particles; 5). fundamental length; 6). gravitation; 7). neutrino.* We concern here problems number 5, 4, and partially 7.

It should be noted that if we use the notation of Fundamental length l_f and a Maximal mass M, we can write the simple expression $l_f = 1/M$. So we come to the problem of Fundamental mass.

The idea about the existence of a maximal mass in a mass spectrum of elementary particles, which is equal to the Planck mass, was suggested by Moisey Markov in 1965. This new particle was called by the author the Maximon:

$$M_{maximon} = M_{Planck} \approx 10^{19} GeV.$$

Closely with this concept was introduced and the notion of the Fundamental length $l_f \approx 10^{-33} cm$. How we can characterize the status of the concepts of Maximal mass and Fundamental length in the development of contemporary physics? For decades we have, that Standard Model (SM) plays the best role in explaining properties and regularities observed in experiments with elementary particles. In agreement with the contemporary experimental data this structure has been disclosed for any fundamental particles of SM, up to distances of the order of $10^{-17} cm$. So the question arises: up to what mass values one may apply the concept of a local quantum field? Formally, the standard QFT remains a logically impeccable scheme even when in the elementary act of interaction of the particles there are masses which are comparable, say, with the masses of cars or a buses. Under the absurdity of the situation, one should think

[a]E-mail: rodyvn@mail.ru
[b]E-mail: gakr@chtc.ru

about how may be modified theory to eliminate these physically meaningless violations of QFT. The variant such program was suggested by Kadyshevsky and his colleagues (see, for example [1] – [3]).

The key idea of this approach is the following hypothesis: the mass spectrum of elementary particles, i.e. the objects, which are described by local fields, has to be limited by a certain value $M = M_{max}$: $m \leq M$, which was named by the Maximal mass of mass spectrum of elementary particles. This statement has to be accepted as a new fundamental principle of Nature, which similarly to the relativistic and quantum postulates should underlie modified QFT. To realize this hypothesis ([1] – [3]) the value M is considered in a pure geometric way as the radius of the five-dimensional hyperboloid whose surface is a realization of a curved momentum four-space or the space of anti-de Sitter $p_0^2 - \vec{p}^2 + p_5^2 = M^2$, where the new operator of the KleinGordon may be written in the form $KG(p, M) = 2M(p_5 - M\cos\mu)$. Here also the following notation is used $\cos\mu = \sqrt{1 - m^2/M^2}$, and p_5 is the fifth component of the momentum of particle m.

It is very important that in the fermion sector Kadyshevsky geometrical theory [1] – [3] we can see the emergence of new terms in the modified Dirac operator

$$D(p, M) = p_\nu \gamma^\nu + (p_5 - M)\gamma_5 - 2M \sin\mu/2$$

and then one can obtain corresponding γ_5-depended non-Hermitian but PT-symmetric Hamiltonian:

$$H(p, M) = \vec{\alpha}\vec{p} + \beta(m_1 + \gamma_5 m_2),$$

where $\gamma_5 = -i\gamma_0\gamma_1\gamma_2\gamma_3$, $\beta = \gamma_0$, $\vec{\alpha} = \beta\vec{\sigma}$, $\vec{\sigma}$ - Pauli matrix, $(H(p, M) \neq H^\dagger(p, M))$.

2 Non-Hermitian Hamiltonians

The problem of the spectrum reality in models with non-Hermitian Hamiltonians was solved by C.Bender [4]. It is a consequence of PT -invariance of the theory, i.e. a combination of spatial and temporary parity of the Hamiltonian: $\Big[H, \mathcal{PT}\Big]\psi = 0$. For these systems the new positive-defined scalar product is proved to be $\langle C\psi|\psi\rangle$, where the operator C is linear and defined by the following team three algebraic equations: $C^2 = 1$; $\Big[C, \mathcal{PT}\Big] = 0$; $\Big[C, H\Big] = 0$. This surprising results explain the growing interest in this problem. For the past a few years has been studied a lot of new non-Hermitian PT invariant systems.

It should be noted that the use of the non-Hermitian PT - symmetric Hamiltonians may be a very fruitful environment for the considering of new physics beyond the SM.

3 Modified Dirac Model

Let us now consider the modified Dirac equations with γ_5 - depended massive term for free particles m. In this case the approach is similar to the known reception Dirac in which we can represent the Klein-Gordon operator in the form of a product of two commuting matrix operators. However now we have deal with non-Hermitian PT - symmetric operators:

$$\partial_\mu^2 + m^2 = (i\partial_\mu\gamma^\mu - m_1 - \gamma_5 m_2)(-i\partial_\mu\gamma^\mu - m_1 + \gamma_5 m_2),$$

where the physical mass of particles $m^2 = m_1^2 - m_2^2$ is expressed through new two parameters m_1 and m_2 (see [5]). For so the function would obeyed to the equations of Klein-Gordon one can demand that it also satisfies one of equations of the first order, for example

$$(i\partial_\mu\gamma^\mu - m_1 - \gamma_5 m_2)\widetilde{\psi}(x,t) = 0$$

It is obvious that the $\mathcal{PT}$ - symmetrical Hamiltonian, associated with this equations, is non-Hermitian, because it contains γ_5-dependent mass component:

$$H(p,M) = \overrightarrow{\alpha}\overrightarrow{p} + \beta(m_1 + \gamma_5 m_2),$$

which is quite close to Kadyshevsky model, and all notations where mentioned above.

It is considered that the PT-symmetry of Hamiltonian systems is unbroken when the mass m is real, i.e. $m_1^2 \geq m_2^2$ is accomplished [4]. Note that in this model this single condition isn't enough because $m^2 = m_1^2 - m_2^2$ posses the infinity number of m_1 and m_2 values. Very important condition may be obtained from the theorem for geometrical and arithmetical average if one use m and m_2 for this theorem [5]. From this theorem we have

$$m \leq \frac{m_1^2}{2m_2} = M(m_1, m_2).$$

In particular with this parameter, can would put an upper bound on the mass spectrum of particles by analogy with geometrical theory [1] – [3]. Now we have the system of equations $m_1^2 - m_2^2 = m^2$; $m_1^2/2m_2 = M$, and the solutions are the following:

$$m_1^{\mp} = \sqrt{2}M\sqrt{1 \mp \sqrt{1 - m^2/M^2}}$$

$$m_2^{\mp} = M\left(1 \mp \sqrt{1 - m^2/M^2}\right).$$

It is easy to verify that these formulas in the case of the upper sign (-) is agreed in the limit $M \to \infty$ with the ordinary Dirac equations that is corresponded to the transition to Hermitian limit $m_1 \to m$, $m_2 \to 0$. But when the lower sign (+) is chosen this transition to the Hermitian limit is absent. Thus in the considered model appear yet unknown particles are described by new equations which has no counterpart in the SM. Should be noted that analogical particles appeared earlier in the geometrical theory [1] – [3] and where they were named as "exotic particles". For each of ordinary particle it may exist the new partner, which possesses the same mass and possibly has a number of another similar properties.

Namely, we considered also the interaction of fermions with intensive uniform magnetic fields, focusing on their $(g-2)$ gyromagnetic factor [6], [7]. Due to effective research procedures, we derive the exact solutions of the energy spectra of pseudo-Hermitian Hamiltonians, taking into account the spin effects of the fermions. It is very not trivial that the ordinary and exotic fermions may be significantly separated with the help of the intensive magnetic field already at low energies of fermions.

If the exotic particles are really exist, their detection with the help of magnetic field may be used as indirect proof of the existence of the Maximal mass. This circumstance can drastically change the basic paradigm of contemporary experimental physics.

References

[1] V. G. Kadyshevsky, Phys. Part. Nucl., **29**, 227 (1998).

[2] V. G. Kadyshevsky, M. D. Mateev, Nuovo Cimento **A 87**, 324 (1985).

[3] V. G. Kadyshevsky, M. D. Mateev, V. N. Rodionov, and A. S. Sorin, Dokl. Phys. **51**, 287 (2006); (hep-ph/0512332). *"Towards a maximal mass model"*. (CERN TH/2007-150; hep-ph/0708.4205).

[4] C. M. Bender. *"Making sense of non-Hermitian hamiltonians.* (arXiv:hep-th/0703096). C. M. Bender, H. F. Jones, and R. J. Rivers. Phys. Lett., vol. B 625 (2005) 333.

[5] V. N. Rodionov, G. A. Kravtsova, Moscow Univ. Phys. Bull., **69(3)**, 223 (2014). Theor. Math. Phys. **182(1)**, 100 (2015).

[6] V. N. Rodionov, Int. J. Theor. Phys., 04.11.2014. **DOI** 10.1007/s10773-014-2410-4 4065. Physica Scripta **90**, 045302 (2015).

[7] V. N. Rodionov, *"Non-Hermitian $\mathcal{PT}$-symmetric relativistic Quantum mechanics with a maximal mass in an external magnetic field."* (arXiv: 1404.0503). *"Exact Solutions for N on-Hermitian Dirac-Pauli Equation in an intensive magnetic field."* (arXiv: 1406.0383). *"Non-Hermitian $\mathcal{PT}$-Symmetric Dirac-Pauli Hamiltonians with Real Energy Eigenvalues in the Magnetic field."* (arXiv:1409.5412).

KINEMATICS OF VORTEX RADIATION FROM RELATIVISTIC CHARGED PARTICLES

Vladimir Bordovitsyn [a] and Anastasiya Kulikova [b]
Faculty of Physics, Tomsk State University, 634050 Tomsk, Russia

Abstract. Two definitions of angular electromagnetic momentum are compared. One of them is based on a standard description of energy-momentum density tensor that is produced by arbitrarily moving charges. The second one is related to canonical energy–momentum tensor by D. Ivanenko and A. Sokolov. Despite principal differences of these two approaches in defining of angular electromagnetic momenta, both methods produce the same results in the wave zone of radiation.

1 Introduction

The hypothesis which holds that electromagnetic waves have their proper angular momentum and field moment of force connected with it was put forward by A. I. Sadowsky [1] in 1897. It was he who proposed a method of measurement of light angular momentum, which is based on transmission of a light ray through an anisotropic crystal slab.

In 1936 B. Beth [2] in the USA and A. Holborn [3] in England proved experimentally the existence of angular momentum of circular polarized light. These precise experiments did not only provide evidence for Sadowsky hypothesis but also allowed to evaluate Plank's constant within the accuracy of 10%.

D. Ivanenko and A. Sokolov [4] as well as F. Rohrlich, C. Teitelboim et al. [5, 6] made a major contribution to the development of theory of angular momentum of electromagnetic field (AMEF). Nevertheless, the general definition of AMEF is argued over among physicists until today. Debates about the adequacy of the theory of AMEF and its radiation arise from time to time. [7–11]

2 AMEF of relativistic charge and its power

With respect to Teitelboim's assumption [6], density tensor of AMEF can be defined as follows:

$$\mathcal{M}^{[\mu\nu]\lambda} = \frac{1}{c}\left(R^{\mu}\mathcal{E}^{\nu\lambda} - R^{\nu}\mathcal{E}^{\mu\lambda}\right) = -\mathcal{M}^{[\nu\mu]\lambda}. \quad (1)$$

Here $R^{\mu} = r^{\mu} + \tilde{r}^{\mu}$ is a radius-vector of an observation point, $r^{\mu} = (ct, \mathbf{r})$ describes orbital motion of a charge, $\tilde{r}^{\mu} = \left(c\left(\tilde{t} - t\right), \tilde{\mathbf{r}}\right) = (\tilde{r}, \tilde{\mathbf{r}})$ is a light-like vector from a charge to an observation point, $\mathcal{E}^{\mu\nu}$ — symmetrical tensor of energy-momentum density.

$$\mathcal{E}^{\mu\nu} = -\frac{1}{4\pi}\left(H^{\mu\rho}H_{\rho}{}^{\nu} + \frac{1}{4}g^{\mu\nu}H^{\alpha\beta}H_{\alpha\beta}\right) = \mathcal{E}^{\nu\mu}. \quad (2)$$

[a]E-mail: vabord@sibmail.com
[b]E-mail: anasta-kulikova@yandex.ru

Generally, expression for $\mathcal{E}^{\mu\nu}$ has a complicated explicit form and depends on the position of an observation point relative to a radiating particle. It should be noted that energy-momentum density tensor $\mathcal{E}^{\mu\nu}$ naturally divides into three parts:

$$\mathcal{E}^{\mu\nu} = \bar{\mathcal{E}}^{\mu\nu} + \tilde{\bar{\mathcal{E}}}^{\mu\nu} + \tilde{\mathcal{E}}^{\mu\nu}. \tag{3}$$

$\bar{\mathcal{E}}^{\mu\nu} \sim 1/\tilde{r}^4$ and $\tilde{\bar{\mathcal{E}}}^{\mu\nu} \sim 1/\tilde{r}^3$ relate to so-called convective zone of space, where electromagnetic field,which is created by charge, is still appreciable. $\tilde{\mathcal{E}}^{\mu\nu} \sim 1/\tilde{r}^2$ corresponds to wave zone of charge radiation, where radiation in the form of electromagnetic waves prevails. It should be pointed out that symmetry of energy-momentum density tensor has been used essentially in order to obtain AMEF density tensor.

Density of radiation angular momentum can be expressed as the sum of two independent parts:

$$\mathcal{M}^{[\mu\nu]\lambda} = \Lambda^{[\mu\nu]\lambda} + \mathcal{S}^{[\mu\nu]\lambda},$$

$$\Lambda^{[\mu\nu]\lambda} = \frac{1}{c}\left(r^\mu \tilde{\mathcal{E}}^{\nu\lambda} - r^\nu \tilde{\mathcal{E}}^{\mu\lambda}\right), \quad \mathcal{S}^{[\mu\nu]\lambda} = \frac{1}{c}\left(\tilde{r}^\mu \tilde{\bar{\mathcal{E}}}^{\nu\lambda} - \tilde{r}^\nu \tilde{\bar{\mathcal{E}}}^{\mu\lambda}\right). \tag{4}$$

It is natural to associate tensor $\Lambda^{[\mu\nu]\lambda}$ with orbital radiation angular momentum, while $\mathcal{S}^{[\mu\nu]\lambda}$ can be relate to spin angular momentum of radiation.

As a result of covariant integration, total radiation power is obtained in this form:

$$\frac{d\tilde{L}^{\mu\nu}}{d\tau} = \frac{2}{3}\frac{e^2}{c^5} w_\rho w^\rho r^{[\mu} v^{\nu]} = r^{[\mu} \frac{d\tilde{P}_e^{\nu]}}{d\tau}, \quad \frac{d\tilde{\Pi}^{\mu\nu}}{d\tau} = \frac{2}{3}\frac{e^2}{c^3} v^{[\mu} w^{\nu]} = \frac{2}{3}\frac{e^2}{c}\Omega^{\mu\nu}_{Th}. \tag{5}$$

It is interesting that the total radiation power of the orbital angular momentum is directly related to the power of four-dimensional charge radiation momentum. As far as the radiation power of the spin angular momentum is concerned, it is directly related to the kinematic part of the spin precession of radiating particle, namely Thomas precession at the frequency [12]

$$\Omega^{\mu\nu}_{Th} = \frac{1}{c^2} v^{[\mu} w^{\nu]},$$

that is known to be not a source of relativistic radiation and has a pure kinematic origin. In connection with this fact, we can use more adequate term for this radiation, namely, vortex radiation.

3 The original theory proposed by Ivanenko and Sokolov

The first appearance of spin AMEF in Ivanenko–Sokolov theory was caused by the fact that they used non-symmetrical canonical energy-momentum density tensor for orbital angular momentum:

$$\mathcal{T}^{\mu\nu} = -\frac{1}{4\pi}\left\{\frac{1}{4}g^{\mu\nu}H_{\alpha\beta}H^{\alpha\beta} + \left(\mathcal{D}^{\mu}A^{\lambda}\right)H_{\lambda}{}^{\nu}\right\} \neq \mathcal{T}^{\nu\mu}. \tag{6}$$

This fact didn't permit them to obtain differential conservation law for orbital angular momentum sort of one for total angular momentum in order to find radiation power of orbital AMEF.

In connection with this fact, an idea to introduce subsidiary spin angular momentum

$$\mathcal{S}^{\mu\nu\lambda\prime} = \frac{1}{4\pi c}A^{[\mu}H^{\nu]\lambda} \tag{7}$$

has arisen. As a result, density of total angular momentum $\mathcal{M}^{\mu\nu\prime} = \Lambda^{\mu\nu\prime} + \mathcal{S}^{\mu\nu\prime}$ has become to satisfy the differential conservation law $\mathcal{D}_{\lambda}\mathcal{M}^{\mu\nu\prime} = 0$. Moreover, by using the explicit form of canonical energy-momentum density tensor we can notice that it becomes symmetrical and coincides with $\mathcal{E}^{\mu\nu}$ in the wave zone.

Thus, despite all external differences, methods that were developed for the description of AMEF by Ivanenko-Sokolov, and Teitelboim et al., at the end provide the same results in the wave zone of the radiation.

Acknowledgments

This work is supported by Tomsk State University Competitiveness Improvement Program.

References

[1] A. Sadowsky, Zhurn. Russ. Fiz.-Khim. Ob.-va. **2**, 82 (1897).
[2] B. Beth, Phys. Rev. **50**, 115 (1936).
[3] A. Holborn, Nature **137**, 31 (1936).
[4] D. Ivanenko, A. Sokolov, *"Classical Field Theory"* (GITTL, Moscow-Leningrad, 1949).
[5] F. Rohrlich, *"Classical Charged Particles"* (World Scientific, Singapore, 2007).
[6] C. Teitelboim, D. Villarroel, Ch. van Weert, Rivista del Nuovo Cimento **3**, 1 (1987).
[7] H. Ohanian, Amer. J. Phys. **54**, 500 (1986).
[8] G. Rosenberg, Usp. Fiz. Nauk **40**, 328 (1950).
[9] K. Wulfson, Usp. Fiz. Nauk **152**, 668 (1987).

[10] I. Sokolov, Usp. Fiz. Nauk **161**, 175 (1991).
[11] A. Barabanov, Usp. Fiz. Nauk **163**, 75 (1993).
[12] V. Bordovitsyn, O. Konstantinova, E. Nemchenko, Russ. Phys. J. **55**, 44 (2012).

NEUTRINO PRODUCTION OF e^-e^+ PAIRS IN A STRONG MAGNETIC FIELD

A. V. Kuznetsov [a], D. A. Rumyantsev [b]
Division of Theoretical Physics, Yaroslavl State University, 150000 Yaroslavl, Russia
V. N. Savin [c]
Yaroslavl Higher Military School of Air Defence, 150001 Yaroslavl, Russia

Abstract. The processes of neutrino production of electron positron pairs, $\nu\bar{\nu} \to e^-e^+$ and $\nu \to \nu e^-e^+$, in a magnetic field of arbitrary strength, where electrons and positrons can be created in the states corresponding to excited Landau levels, are analysed. The results can be applied for calculating the efficiency of the electron-positron plasma production by neutrinos in the conditions of the Kerr black hole accretion disc considered by experts as the most possible source of a short cosmological gamma burst.

Modelling of a short cosmological gamma-ray burst (SGRB) is an important and long-standing problem of modern astrophysics. A concept which is considered by experts as one of the most possible is that the Kerr black hole accretion disk could be the source of SGRB. In the disc surface, there are produced gigantic fluxes of neutrinos and anti-neutrinos, which partially annihilate above the disc into $e^{\mp}$ pairs forming in turn the gamma-ray burst. This process was investigated in many papers, a list of references could be found e.g. in [1, 2]. In Ref. [1], in addition to the basic process of annihilation, $\nu\bar{\nu} \to e^-e^+$, the contribution of the magnetic field-induced process $\nu \to \nu e^-e^+$ to the neutrino energy deposition rate around the black hole was also included. The authors [1] concluded in part, that the process $\nu \to \nu e^-e^+$ could dominate over the basic process $\nu\bar{\nu} \to e^-e^+$. They used the result for the energy deposition rate in the process $\nu \to \nu e^-e^+$ obtained in [3, 4] in the crossed field limit, while in those physical conditions (B to 180 B_e, E_ν to 25 MeV) the approximation of a crossed field is poorly applicable (as well as the approximation of a superstrong field when e^-e^+ are created in the ground Landau level). The next Landau levels can be also excited, as we have shown in our paper [2]. Furthermore, the authors [1] considered the process $\nu\bar{\nu} \to e^-e^+$ without taking account of the magnetic field influence.

In [2], the results are presented of our calculation of the mean neutrino energy losses caused by the process $\nu \to \nu e^-_{(n)} e^+_{(\ell)}$, when the electron and the positron are created in the nth and ℓth Landau levels, in a moderately strong magnetic field, i.e. in the conditions of the Kerr black hole accretion disk. It was shown that the crossed field limit gives the overstated result which is in orders of magnitude greater than the sum of the contributions of lower excited Landau levels.

[a]E-mail: avkuzn@univ.uniyar.ac.ru
[b]E-mail: rda@univ.uniyar.ac.ru
[c]E-mail: vs_post07@mail.ru

This would mean that the conclusion [1] that the contribution of the process $\nu \to \nu e^- e^+$ to the efficiency of the electron-positron plasma production by neutrino exceeds the contribution of the annihilation channel $\nu\bar{\nu} \to e^- e^+$, and that the first process dominates the energy deposition rate, does not have a sufficient basis. A new analysis of the efficiency of energy deposition by neutrinos through both processes, $\nu\bar{\nu} \to e^- e^+$ and $\nu \to \nu e^- e^+$, in a hyper-accretion disc around a black hole should be performed, with taking account of our results [2] for the process $\nu \to \nu e^- e^+$.

The local energy-momentum deposition rate due to the process $\nu\bar{\nu} \to e^- e^+$ is defined by the equation:

$$Q^{\alpha}_{\nu\bar{\nu}} = \int \frac{\mathrm{d}^3 p}{(2\pi)^3} f_\nu(p) \int \frac{\mathrm{d}^3 p'}{(2\pi)^3} f_{\bar{\nu}}(p') \, (p+p')^{\alpha} \, \frac{(pp')}{EE'} \, \sigma(\nu\bar{\nu} \to e^- e^+) \, , \qquad (1)$$

where σ is the cross-section of the process, p and p' are the four-momenta of the neutrino and antineutrino, $f_\nu(p)$ and $f_{\bar{\nu}}(p')$ are the local distribution functions depending on the distribution functions at the surface of the black hole accretion disc, and on the details of propagation.

In a strong magnetic field, the cross-section has to be obtained by the summation over the Landau levels n, ℓ, where the upper limit is defined by the condition $(M_n + M_\ell)^2 \le (p+p')^2_{\parallel}$, $(q^2_{\parallel} = q^2_0 - q^2_z$, if z is along $\mathbf{B}$).

Unlike the cross-section in vacuum where it depends on the Mandelstam parameter S only, the cross-section in a magnetic field depends on the set of kinematic variables, e.g.: energies E, E', two polar angles and one azimuth angle.

The analysis shows that under certain conditions, the influence of a magnetic field significantly alters the final numerical evaluation of the characteristics of this reaction. The further study of the question is required of how the external field can affect the characteristics of the processes observed in various astrophysical objects.

Acknowledgments

The study was supported in part by the Russian Foundation for Basic Research (Project No. 14-02-00233-a).

References

[1] I. Zalamea, A.M. Beloborodov, Mon. Not. R. Astron. Soc. **410**, 2302 (2011).

[2] A.V. Kuznetsov, D.A. Rumyantsev, V.N. Savin, Int. J. Mod. Phys. **A 29**, 1450136 (2014).
[3] A.V. Kuznetsov, N.V. Mikheev, Phys. Lett. **B 394**, 123 (1997).
[4] A.V. Kuznetsov, N.V. Mikheev, Phys. At. Nucl. **60**, 1865 (1997) [Yad. Fiz. **60**, 2038 (1997)].

PARTICLE QUANTUM STATES WITH INDEFINITE MASS AND NEUTRINO OSCILLATIONS

A. E. Lobanov [a]

Faculty of Physics, Moscow State University, 119991 Moscow, Russia

Abstract. Spaces of particle states are constructed in such a way that the neutrinos are combined in a multiplet with their components being considered as different quantum states of a single particle. The same is done for the charged leptons and the down- and up-type quarks. In a theory based on the Lagrangian of the fermion sector of the Standard Model modified in accordance with this construction, the phenomenon of neutrino oscillations arises. By the example of the pion decay it is shown that the states of the neutrino produced in the decay process can be described by a superposition of states with different masses and identical 3-momenta with very high accuracy.

The Standard Model of electroweak interactions based on the non-Abelian gauge symmetry of the interactions, the generation of particle masses due to the spontaneous symmetry breaking mechanism, and the ideology of particle generation mixing is universally recognized. Its predictions obtained in the framework of perturbation theory are in a very good agreement with the experimental data, and there is no serious reason at least at the energies available at present) for its main propositions to be revised.

However, in describing such an important and firmly experimentally established phenomenon as neutrino oscillations (see [1]), a phenomenological, in fact, theory based on the pioneer works by B. Pontecorvo and Z. Maki et al. is used. This theory is well developed and is consistent with the experimental data. Its statements are given in various review articles and monographs.

Later numerous attempts based on the quantum theory of fields have been made to justify the conclusions of the phenomenological model. The results obtained partially explain the oscillation mechanism, but in general the theory still has a number of paradoxical properties, and what is most important, the approaches used tend to be very distinct from the regular methods, which are efficient in Lagrangian theories with a small parameter. The reason is the difficulties arising when one works with superpositions of one-particle quantum states with different masses.

In relativistic quantum field theory a particle is usually associated with an irreducible representation of the Poincaré group [2]. The eigenvalue of the Casimir operator formed from the translation operators squared is identified with the observed mass of the particle. In the framework of quantum field theory it was proved that the existence of states which are superpositions of states with different masses contradicts the relativistic invariance of the theory, if the canonical momentum operator is identified with the translation operator.

[a]E-mail: `lobanov@phys.msu.ru`

It obviously follows from the fact that the metric is defined on the hyperboloid in the momentum space determined by the value of the particle mass.

To overcome this difficulty it seems natural to associate some set of particles (a multiplet) with an irreducible representation of a wider group. However, the theorems proved in the group theory (see, i.e., [3]) indicate that the only reasonable extension of the symmetry group of the theory is the direct product of the Poincaré group and a group of internal symmetry. As a consequence, the translation operator squared is again invariant for such a structure. It is concluded that since the irreducible unitary representations of the direct product of groups are unitary equivalent to the tensor product of the representations of these groups with the same values of the invariants, the masses of the particles in the multiplet should be equal.

However, this conclusion is not correct if there is more than one multiplet with this symmetry in the theory, and these multiplets interact with each other. As a matter of fact, in the general case the representation spaces for two multiplets cannot be simultaneously reduced to the direct sum of representation spaces of the Poincaré group. Even if the representation of the internal symmetry group is considered finite, which seems quite natural for a finite set of particle masses in the multiplet, the representations for both multiplets would be unitary equivalent, but not identical. Therefore, although the values of the invariants related to the translation operators squared will be fixed (for different multiplets they may vary), there is no need to treat them as the observed squared masses of the particles. The masses will be the eigenvalues of an operator in the enveloping algebra of the Lie algebra of the internal symmetry group which naturally commutes with the elements of the Lie algebra of the Poincaré group.

Such considerations conform to the spirit of the Standard Model, where the masses of all particles are generated due to the phenomenon of spontaneous symmetry breaking and are proportional to the vacuum expectation value of the Higgs field.

In paper [4] we put forward a modification of the electroweak interaction theory, in which the fermions with the same electroweak quantum numbers are placed in fermion multiplets and are treated as different quantum states of a single particle. In this model, the mixing and oscillations of the particles arise as a direct consequence of the general principles of quantum field theory.

The Lagrangian of our model formally coincides with the Lagrangian of the Standard Model. However, the fermion wave functions describe not the individual particles, but the multiplets in general. So there is no necessity to introduce the mixing matrices into the Lagrangian of the interaction between the fermion fields and the vector boson fields and the Higgs field in an explicit form. The action defined by the Lagrangian of free fields is explicitly invariant with respect to the direct product of the Poincaré group and $SU(3)$, so when quantizing the model, the multiplet can be considered as a single particle.

Then the one-particle states in the Fock space are defined as usual, the creation and annihilation operators satisfy the standard commutation relations. However, these operators carry an additional discrete quantum number that is associated with the mass of the state. Each element of a multiplet can be either in one of the three mass states, or in a pure quantum state that is a superposition of the states with fixed masses. In a certain sense we can say that there exist only four fundamental fermions in our model.

The developed approach enables one to calculate the probabilities of the processes taking place at long distances from the detector. Calculations of higher order processes including the computation of the contributions due to radiative corrections can be performed in the framework of perturbation theory using the regular diagram technique.

Our approach adequately describes the phenomenon of neutrino oscillations, which is, in fact, the phenomenon of the oscillation of the probability of the processes involving neutrinos. The flavor state, constructed as a superposition of the mass states with the same momentum, is distinguished by the fact that it is very close to the state which is formed in the decay process. The possibility to use such flavor states, when describing neutrinos at distances from the source that are larger than the oscillation length, directly follows from the smallness of the neutrino masses compared to their energy, which is really observed for neutrinos. Therefore, the phenomenological theory of oscillations, based on the ideas of B. Pontecorvo (see, i.e., [5]), is a very good approximation for quantum field theory.

Acknowledgments

The author is grateful to A.B. Arbuzov, A.V. Bolsinov, A.V. Borisov, A.D. Dolgov, D.V. Gal'tsov, A.V. Kartavtsev, E.M. Murchikova, I.P. Volobuev, and V.Ch. Zhukovsky for fruitful discussions.

References

[1] K. Olive *et al.* (Particle Data Group), Chin. Phys. C **38**, 090001 (2014).

[2] N.N. Bogolubov, A.A. Logunov, A.I. Oksak, and I.T. Todorov, *"General Principles of Quantum Field Theory"* (Kluwer Academic Publishers, Dordrecht, 1990).

[3] A. Barut and R. Raczka, *"Theory of Group Representations and Applications"* (Polish Scientific Publishers, Warszawa, 1977).

[4] A. E. Lobanov, arXiv:1507.01256 [hep-ph], (2015).

[5] S. M. Bilenky and B. Pontecorvo, Phys. Rept. **41**, 225 (1978).

THE RADIATIVE DECAY OF A NEUTRINO WITH A MAGNETIC MOMENT IN A STRONG MAGNETIC FIELD WITH TAKING ACCOUNT OF THE POSITRONIUM INFLUENCE ON THE PHOTON DISPERSION

Anatoly Mosichkin [a]

Faculty of Physics, Yaroslavl P.G. Demidov State University, Sovietskaya 14, 150000 Yaroslavl, Russia

Abstract. We study the process of the radiative decay of the neutrino with a magnetic moment in a strong magnetic field, with taking account of the influence of the positronium on the photon dispersion. The positronium contribution into the photon polarization operator leads to an essential modification of the photon dispersion law, and of the neutrino radiative decay amplitude. It has been shown that the probability of the neutrino radiative decay essentially increases under an influence of the positronium on the photon dispersion.

1 Introduction

Neutrino processes such as radiative decay of the neutrino $\nu \to \nu\gamma$ also called Cherenkov neutrino process, are of great interest for a long time. The process $\nu \to \nu\gamma$ in a magnetic field has been investigated in the case of a relatively weak field [1], of a strong field [2] and for the field of arbitrary intensity [3]. In all these papers, the situation was analysed of low-energy neutrinos compared to the mass of the electron. The case of more significant neutrino energies should be more interesting in the light of possible astrophysical applications. In this case, the radiative corrections become substantially enhanced, leading to a renormalization of the photon wave function and a significant deviation from the vacuum dispersion law. These factors were taken into account for the first time in Ref. [4]. In the most of the previous papers, the analyses of the neutrino radiative decay were performed with taking account only of the contribution of free electrons and positrons in the loop correction to the photon polarization operator. The influence of a bounded electron-positron pair on the photon polarization operator was taken into account in Ref. [5] for the case of the standard electromagnetic interaction of a neutrino, arising at the loop level and thus being strongly suppressed. In this paper, we study the process of the radiative decay $\nu \to \nu\gamma$ of the neutrino with a magnetic moment, with taking account of the positronium influence on the photon polarization operator.

[a] E-mail: anatoly_mosichkin@mail.ru

2 The probability of the process $\nu \to \nu\gamma$

The interaction Lagrangian corresponding to the process $\nu \to \nu\gamma$ for a neutrino with a magnetic moment has the form:

$$L = -\frac{i\mu_\nu}{2}\Big(\bar{\nu}(x)\sigma_{\alpha\beta}\nu(x)\Big)F^{\alpha\beta}, \tag{1}$$

where $\sigma_{\alpha\beta} = \frac{1}{2}(\gamma_\alpha\gamma_\beta - \gamma_\beta\gamma_\alpha)$, $F^{\alpha\beta} = \partial^\alpha A^\beta - \partial^\beta A^\alpha$, μ_ν is the magnetic moment of a neutrino.

The expression for the probability of the radiative decay of the neutrino with a magnetic moment can be written as:

$$\frac{W}{W_0} = \xi^{\frac{3}{2}} \int\limits_0^1 \frac{dx x^2 e^{-2\xi x^2}\left(\frac{1}{\pi}H(z) + \frac{\lambda z}{1-\lambda^2-z}\right)\left(\sqrt{1-x^2} - x\arccos x\right)}{1 + \frac{\alpha\beta}{2}e^{-2\xi x^2}\frac{d}{dz}\left(\frac{1}{\pi}H(z) + \frac{\lambda z}{1-\lambda^2-z}\right)}. \tag{2}$$

The variables x and z are linked by the dispersion equation:

$$z - \xi\beta x^2 = -\frac{\alpha\beta}{2}e^{-2\xi x^2}\Big(\frac{1}{\pi}H(z) + \frac{\lambda z}{1-\lambda^2-z}\Big), \tag{3}$$

where $\xi = E^2/(eB)$, $\beta = B/B_e$, $B_e = 4.41 \times 10^{13}G$, $W_0 = 2\alpha\beta^{\frac{5}{2}}\mu_\nu^2 m_e^3/\pi^2$, $H(z) = \int\limits_0^1 dx/[1 - z(1-x^2) - i0] - 1$, $z = q_{\|}^2/(4m_e^2)$, $x = |q_\perp|/(2E)$, $\lambda = \alpha/(2\nu)$.

In this paper, we have studied the process of the radiative decay of a neutrino with a magnetic moment in a strong magnetic field with taking account of the positronium influence on the photon dispersion. Taking account of the positronium influence on the photon dispersion leads to a significant change in the behaviour of the dependence $W(\xi)$ and its essential increase. The details of the analysis will be presented in an extended paper being in preparation.

Acknowledgments

The study was supported by the Russian Foundation for Basic Research (Project No. 14-02-00233-a).

References

[1] Galtsov D. V. and Nikitina N. S. 1971 *ZhETF* **62** 2008 [*JETP* **35** 1047 (1972)]
[2] Skobelev V. V. 1976 *ZhETF* **71** 1263 [*JETP* **44** 660]

[3] Ioannisian A. N. and Raffelt G G 1997 *Phys. Rev.* D **55** 7038
[4] Gvozdev A. A., Mikheev N. V. and Vasilevskaya L. A. 1977 *Phys. Lett.* B **410** 211
[5] Anikin R. A. and Mikheev N. V. 2013 *Yad. Fiz.* **76** 1610 [*Phys. At. Nucl.* **76** 1541]

SELF-FORCE ON A CHARGE OUTSIDE THE SCHWARZSCHILD-TANGHERLINI BLACK HOLE

Yuri V. Grats and Pavel Spirin[a]
Department of Theoretical Physics, Faculty of Physics,
M. V. Lomonosov Moscow State University, 119991 Moscow, Russia

Abstract. We compute the self-energy of a static classical point-like charge, coupled with vector massless field and localized at the distance $r \gg r_g$ outside the horizon of higher-dimensional Schwarzschild-Tangherlini black hole. We use the perturbation theory and obtain first orders for the electrostatic energy and corresponding self-force. Since the formal expression for self-energy diverges, for the computation we apply the dimensional-regularization technique, well-known in the quantum theory. In the lowest-dimensional cases the results are compared with the ones known in the literature.

Introduction

Among the most intriguing effects to detect the possible extra dimensions one has to pay the particular attention to those ones, which could reveal the possible presence of extra dimensions, so one needs the graviton coupled with the SM field. Among these effects of theoretical interest, there is a phenomenon of the charge's self-action in a curved background.

In four dimensions this effect was studied for Schwarzschild metric [1, 2], for Reissner-Nordström metric [3] and for Kerr metric [4]. Later this problem attracted an attention in relation with the fields of topological defects [5]. Recently the problem of the charges' self-action has become of theoretical interest [6, 7] in the framework of higher-dimensional gravity models.

Our analysis here is based on the work [8].

1 The setup

In the isotropic coordinates $x^\mu = (t, \mathbf{x})$, where the spatial part of a metric is $n-$dimensional conformally-Euclidean hyperplane at $t = \text{const}$, the metric of spherically-symmetric static black hole takes the form:

$$ds^2 = -\frac{\left[1-(r_g/r)^d\right]^2}{\left[1+(r_g/r)^d\right]^2}\,dt^2 + \left[1+\left(\frac{r_g}{r}\right)^d\right]^{4/d}\delta_{ij}\,dx^i dx^j, \tag{1}$$

where $d \equiv n-2$, $r = \left(\delta_{ij}\,x^i x^j\right)^{1/2}$, and r_g is an isotropic gravitational radius.

Making use of the static metric, the equation on a static vector-potential $A_\mu \equiv (\Phi, 0, \dots, 0)$ with source localized at $x = b$, reads:

$$-\partial_i\left(\sqrt{-g}\,g^{00}g^{ij}\partial_j\Phi\right) = e\,\delta^n(x-b)\,. \tag{2}$$

[a]E-mail: grats@phys.msu.ru, pspirin@physics.uoc.gr

The self-energy and the corresponding radial self-force equal

$$E = -\frac{1}{2}\,e^2\,G(b,b)\,, \qquad\qquad F_{\rm em} = -\frac{1}{1-(r_g/b)^{2d}}\,\frac{\delta E_{\rm em}}{\delta b}\,, \tag{3}$$

where $G(x,x')$ is the Green's function, given by the solution of $\mathcal{L}(x,\partial)\,G(x,x') = \delta^n(x-x')$, where $\mathcal{L}(x,\partial)$ is a field equation operator in the LHS of (2). Following Schwinger, we rewrite it as $\mathfrak{L}\mathcal{G} = 1$, so that $\mathcal{G} = \mathfrak{L}^{-1}$. In coordinate representation it is given by

$$\langle x|\,\mathcal{G}\,|x'\rangle = G(x-x')\,, \qquad\qquad \langle x|\,\mathfrak{L}\,|x'\rangle = \mathcal{L}(x,\partial)\,\delta^n\,(x-x')\,.$$

Representing $\mathfrak{L}$ as $\mathfrak{L} = \mathfrak{L}_0 + \delta\mathfrak{L}$ and considering $\delta\mathfrak{L}$ as perturbation, we seek the solution in the form $\mathcal{G} = \mathcal{G}_0 + \delta\mathcal{G}$, where $\mathcal{G}_0$ is unperturbed Green's function, satisfying $\mathfrak{L}_0\mathcal{G}_0 = 1$ (so that $\mathcal{G}_0 = \mathfrak{L}_0^{-1}$), we get the series

$$\mathcal{G} = \Big[\mathfrak{L}_0\,(1+\mathcal{G}_0\delta\mathfrak{L})\Big]^{-1} = \mathcal{G}_0 - \mathcal{G}_0\,\delta\mathfrak{L}\,\mathcal{G}_0 + \dots\,. \tag{4}$$

In our case $r \gg r_g$ we have small $\alpha \equiv (r_g/r)^d \ll 1$ which in fact distinguishes orders of smallness in $\delta\mathfrak{L}$: $\delta\mathfrak{L} = \sum \mathfrak{L}^{(k)}$. The same concerns the Green's function under evaluation. Therefore, the basic (unperturbed) order in $\delta\mathfrak{L}$ is the zeroth order in expansion in α: $\mathcal{L}_0(x,\partial) = \partial^2$, hence $G_0(x-x') = \partial^{-2}\,\delta^n(x-x')$ is a well-known elementary solution of the Laplace equation. Expression for the Green's function in the coincidence-point limit is formal, due to it diverges, and we have to regularize it. We use the dimensional-regularization technique, well known in the Quantum Field Theory. Within this method the integrals of the form $\int d^n p\, p_{i_1}...p_{i_k}/p^2$ are considered as vanishing, in paricular, the zeroth-order contribution entirely. This corresponds to the tadpole diagrams in quantum theory.

2 Self-energy and self-force

The perturbation operator reads $\delta\mathcal{L} = -\partial_i\left(\sqrt{-g}\,g^{00}\,g^{ij}\partial_j\right) - \partial^2$. Expanding it in α, the first- and second-order operators are

$$\mathfrak{L}^{(1)} = 4\alpha\left[\partial^2 - (n-2)\,\frac{x\cdot\partial}{r^2}\right], \qquad \mathfrak{L}^{(2)} = 7\alpha^2\left[\partial^2 - 2(n-2)\,\frac{x\cdot\partial}{r^2}\right],$$

with corresponding contributions to the Green's function

$$\mathcal{G}^{(1)} = -\mathcal{G}_0\,\mathfrak{L}^{(1)}\,\mathcal{G}_0\,, \qquad\qquad \mathcal{G}^{(2)} = -\mathcal{G}_0\,\mathfrak{L}^{(2)}\,\mathcal{G}_0 + \mathcal{G}_0\,\mathfrak{L}^{(1)}\,\mathcal{G}_0\,\mathfrak{L}^{(1)}\,\mathcal{G}_0\,.$$

Omitting the details of computation, for odd $n \geqslant 3$ we have (in the final answer we write r instead b):

$$E = \frac{e^2}{2\pi^{n/2}}\,\frac{\Gamma(n/2)}{(n-2)}\left(\frac{r_g}{r^2}\right)^{n-2}\left[1 - 4\left(\frac{r_g}{r}\right)^{n-2}\right]. \tag{5}$$

In particular, for $n = 3$ the self-force reads:

$$\mathbf{F} = \frac{e^2}{4\pi}\frac{M}{r^3}\left[1 - 3\frac{M}{r} + \mathcal{O}\left((M/r)^2\right)\right]\frac{\mathbf{r}}{r}, \tag{6}$$

that coincides to this order with the corresponding results of another authors [1, 2] based on another regularization schemes.

For even $n \geqslant 4$ we have:

$$E = \frac{e^2}{2\pi^{n/2}}\frac{\Gamma(n/2)}{(n-2)}\left(\frac{r_g}{r^2}\right)^{n-2} - \frac{9\,e^2}{4\pi^{n/2}}\frac{\Gamma(3n/2-3)}{\Gamma(n-2)}\left(\frac{r_g^2}{r^3}\right)^{n-2}\left[\ln\frac{r}{r_g} + \mathcal{O}(1)\right].$$

In particular, for the five-dimensional black-hole ($n = 4$) one has:

$$\mathbf{F} = \frac{e^2}{\pi^2}\left(\frac{r_g}{r^2}\right)^2\left[1 - 27\left(\frac{r_g}{r}\right)^2\left(\ln\frac{r}{r_g} + \mathcal{O}(1)\right)\right]\frac{\mathbf{r}}{r^2}. \tag{7}$$

This allows us to compare our result with results of [6, 7]. First, in the leading order all three coincide. Second, all three contain logarithmic factor as first subleading contribution.

3 Conclusions

Using the dimensional-regularization technique we have computed perturbatively the self-force on the static charge outside the higher-dimensional spherically-symmetric black hole. For the case $n = 3$ we obtained the well-known result, namely, the Smith – Will force. For $n = 4$ we are in the agreement with the results in the literature, while in higher dimensions our results are new.

Acknowledgments

The Authors would like to acknowledge conversations with A. V. Borisov, D. Gal'tsov and Th. Tomaras. This work was supported in part by the EU programs "Thales" (MIS 375734) and "ARISTEIA II" as well as by the RFBR grant 14-02-01092. Finally, PS acknowledges the non-commercial "Dynasty" Foundation (Russian Federation) for financial support.

References

[1] A. Vilenkin, Phys. Rev. D**20**, 373 (1979).
[2] A. G. Smith, C. M. Will, Phys. Rev. D**22**, 1276 (1980).
[3] B. Leaute, B. Linet, Phys. Lett. A. **58**, 5 (1976).
[4] B. Leaute, Ann. Inst. Henri Poincare Sect. A **27**, 167 (1977).

[5] Yu. V. Grats, A. Garcia, Class. Quantum Grav. **13**, 189 (1996).
[6] M. J. S. Beach, E. Poisson, B. G. Nickel, Phys. Rev. D **89**, 124014 (2014).
[7] P. Taylor, E. E. Flanagan, Phys. Rev. D **92**, 084032 (2015).
[8] Yu. V. Grats, P. Spirin, Zh. Exp. Teor. Fiz **149**, iss.1, 112 (2016).

Problems of Intelligentsia

THE INTELLIGENTSIA AND RELIGION

John Kuhn Bleimaier [a]
15 Witherspoon Street, Princeton, New Jersey 08542 USA

Abstract. This is the 20th anniversary of the Moscow University Conference on the Intelligentsia. It is fitting that a conference dedicated to that class of individuals who cultivate deep thought and abstract reasoning should take place in Russia. The word "intelligentsia" was first coined in the Russian language and the cadre of exceptional individuals who first constituted the Intelligentsia had their principal 19th Century home within the boundaries of the Russian Empire. This conference has examined numerous aspects of the experience of the Intelligentsia, and has produced a unique body of scholarship. This year the topic of the conference is "The Intelligentsia & Culture." This is an apt topic because, by definition, the members of the Intelligentsia are the custodians the cultural heritage of mankind.

Interestingly the number of individuals participating in the Moscow Conference has declined in recent years. This may bespeak a generalized lessening of interest in the Intelligentsia in a world of ever increasing professional specialization and ebbing concern for the intellectual "big picture." It should, in any case, be mentioned that the original Moscow Conferences on the Intelligentsia were held in both the English and the Russian languages. Many papers were delivered in Russian with spontaneous translation into English. While the Conference has become more cosmopolitan and the proceedings are now conducted exclusively in English, it may be that the largest concentration of conscious members of the Intelligentsia continue to be found in Russo-phone societies. Thus, it is respectfully suggested that Russian should, once again, be designated as an official language of our conference.

The quintessential question faced by the Intelligentsia has always been existential. The ultimate abstract concept relates to the nature of reality and the position of man in the universe. This matter has been at the root of the great debates which have defined the Intelligentsia and have divided it into differing schools of thought. First and foremost the existential debate has centered around God as a concept, or, the complete rejection of God as an alternative faith based system. Our topic, the "Intelligentsia & Religion" goes to the heart of the Intelligentsia's existence. It also addresses one of the basic components which have shaped human culture from the beginning of time. Religion represents a set of beliefs concerning the cause, nature, and purpose of the universe. These are issues about which the Intelligentsia has ruminated for centuries.

1 Introduction

"In the Beginning was the Word and the Word was with God and the Word was God." This quote from the New Testament is the first verse of the Gospel of John. It is perhaps the most poetic passage from the Bible, and the most thought provoking. Language is what makes us human. All higher mammals are capable of thinking, analyzing and planning. But in the absence of language the animal world thinks in terms of objects. In order to think in terms of concepts we require language in order to define abstractions. It is language, not our erect posture or our juxtaposed thumb which has made us human. Thus, if language is humankind's gift from the Creator

[a]B.A., Columbia College of Columbia University; Master of International Affairs, Columbia University School of International Affairs; Juris Doctor, Saint John's University School of Law; member of the New York, New Jersey & United States Supreme Court bars; formerly Fellow of Mathey College of Princeton University

it is the essence of the spirit of God which dwells within us. "In the Beginning was the Word and the Word was with God and the Word was God."

Now, abstract thinking is the essential sphere of the Intelligentsia. As we have seen, language is the tool used by the Intelligentsia in the pursuit of abstraction. Thus, God's gift of the Word is the genesis of all intellectual analysis. The Intelligentsia utilizes language to contemplate the nature of the universe and to give meaning to observed phenomena. This represents the basic starting point of the relationship between the Intelligentsia and religion.

2

When we look for the foundations of our civilization we go back to the "people of the book," the ancient Biblical figures who first set down the rules upon which society has been constructed. Before the "people of the book," in the Hobbsian state of nature, life was nasty brutish and short. The rules of societal intercourse were the first iteration of the Word in forming a society from a disorganized aggregate of human individuals.

If we are to seek an ancient predecessor of the Intelligentsia we need look no further that the prophets, priests, scribes, lawyers, Pharisees and Sadducees of the New Testament era. These religious leaders were also the custodians of history, culture and custom. They compiled, copied and preserved the texts which form the raw material from which is constructed our first conception of the universe. Interestingly, at the time of Christ, the ecclesiastic Intelligentsia was divided into two large groups, the Pharisees and Sadducees. The former, priestly sector, affirmed the existence of an afterlife, while the latter, social elite, denied that life had any significance beyond the here and now. Similar schisms continue to divide the Intelligentsia to this day.

The early followers of Christ, the saints, were also deeply thoughtful and analytical in their beliefs. The apostle Peter explicated the Old Testament scriptures so as to convince the Hebrew scholars of the day that Jesus Christ represented the personal fulfillment of a rich literature of prophesy. The apostle Paul engaged in intellectual debate with Greek Sophists on Mars Hill in Athens in order to convert the Gentile intellectual elite to the new Christian religion.

The historic grouping which we have identified as the biblical Intelligentsia was also the source of some of the earliest recorded poetry. The Psalms and Proverbs, as well as certain passages from the Old Testament prophets and the New Testament Gospels and Epistles possess great intensity of expression pertaining to feelings and ideas.

3

The intimate relation between religion and the world of ideas is not confined to the Biblical Hebrews and early Christians. The top rung of the intellectual ladder in ancient Hindu society was occupied by Brahmins, the priestly caste. It was the Brahmins who devoted themselves to contemplation, to poetry and the arts. Gurus who gave themselves up entirely to abstract contemplation might be counted as members of the Intelligentsia, as well. However, the Intelligentsia is always an interacting group. Therefore, a hermit who completely shuns human contact might fail to qualify as a member of the Intelligentsia, whereas a contemplative teacher would surely be included.

Confucianism has dominated intellectual life in China for centuries, along with influence from both Buddhism and Taoism. While these may be thought of as philosophical systems they also are recognized as religions. The scholar/administrators of ancient China may have constituted an intelligentsia. Melding intellectual activity with an organized system for viewing the universe is one of the functions of the Intelligentsia.

In primitive tribal cultures the social brain trust is concentrated in the person of the shaman or witchdoctor. He is the person who personifies tradition, preserves group history and also engages in rudimentary scientific work, attempting to master the healing arts and the secrets of nature. Thus the religious figure is also the person most directly involved in intellectual activity in the least developed human environment. At the most basic level of human civilization, society consists of peasants, warriors and priests. If there could be an Intelligentsia in subsistence based tribal community, the religious leadership would qualify for membership.

4

The poetry, prose and drama of classical Greece and Rome are concentrated almost exclusively on the relationship between man and the classical deities. From Homer to Euripides; from Virgil to Cicero; the gods are vital characters in classical belles lettres and an elaborate religious mythology precisely defines the relationship between man and his environment. The creative thinkers who produced this literature constituted the classical Intelligentsia. Even the mighty Greek philosophers accepted the classical mythology as a given.

Furthermore, the physical remnants of the classical civilizations are almost exclusively tied to the religious life of the society. The architecture of ancient Greece and Rome which has survived into the 21st Century consists largely of temples and shrines. The commanding position of the Parthenon in Athens bespeaks the dominant role played by religion in the society. The technical and aesthetic achievements of the ancients witness the hand of creative thinkers, the classical Intelligentsia, if you will.

5

As has been noted previously, the early church fathers were plainly intellectuals capable of explicating the Old Testament scriptures in order to make the case of the New Testament message before an Intelligentsia consisting of Pharisees and Sadducees. The ministry of the first evangelists was devoted largely of proselytizing the contemporary Intelligentsia. The early church synods considered esoteric questions of doctrine and reached consensus on the meaning of abstract notions. From its inception, part of Christian theology has been an intellectual response to a spiritual yearning.

After the reign of the Emperor Constantine, when Christianity became an official religion of the Roman Empire, the church structure became more rigid and its organization more formal. The priests came to monopolize the access to the books of scripture. Icons acted as pictographs to tell the Bible story to those unschooled in letters. The priesthood became the Intelligentsia, controlling access to "the Book."

Was there an Intelligentsia during the Dark Ages in Europe? The darkness to which the appellation refers is precisely the absence of enlightenment in the intellectual sense. Nevertheless, Western civilization managed to survive the sack of Rome by the barbarians and the retreat of culture to within the narrow confines of the feudal system. It was precisely behind the fortress like defenses of the monasteries that the heritage of the earlier Intelligentsias was preserved for posterity. The custodian monks not only preserved the ancient texts, they copied and illustrated them, creating a new art form.

The medieval ecclesiastical Intelligentsia did not only preserve the intellectual attainments of the past, they read them closely and undertook careful analysis. The philosophies of the ancients, of Aristotle and Plato, were accommodated to Christian doctrine. Mathematical and scientific advances of the pre Christian era were studied and found not to be inconsistent with scriptural teaching.

6

Not only did the Church preserve the works of classical antiquity, the Church absorbed them and enshrined them. When it comes to philosophy Thomas Aquinas and scholasticism created a rigorous intellectual discipline and began a sacred tradition of Catholic scholarship. The spiritual teachings of Jesus were alloyed with Aristotelian logic and Socratic insight. This phenomenon took place largely after the great schism when the Roman and Byzantine churches split. While the Roman Catholic Church absorbed the attainments of pre Christian civilization, the Eastern Rite Church retained its orthodox purity and eschewed classicism. As a result, the Intelligentsia's relationship to the Church was more intense in Western Christendom than it was in the East. Western theology evolved into an elegant intellectual construct. Where as the essence of Orthodox theology was the preservation of early Christian ritual and belief.

The Catholic Church's love of learning and erudition lead to the evolution of complex intellectual paradigms. The blending of classical philosophy and Christian scripture resulted in the development of a body of Catholic doctrine with only tenuous relation to the original teachings of Christ. The Catholic Church tantalized its Intelligentsia with novel intellectual creations.

The intellectually tantalizing reasoning of Catholic scholasticism may possess its attractions. However, the Church's forays into the world of science were less felicitous. As the repository of ancient mathematical and astronomical texts the Catholic Intelligentsia of the Middle Ages extrapolated the empirical findings of the ancients and melded them with Church doctrine. This ultimately turned the Church's friendly relationship with science into bitter enmity. For example, early and incomplete understanding of the solar system led the Church to stake its credibility on the concept that the earth was the center of the universe. This proposition seemed to have good theological justification in light of the scripture's assertion of God's love and concern for mankind at the pinnacle of His creation.

When new theories and new empirical evidence established the sun as the center of our solar system and the placement of this solar system itself in a seemingly insignificant position in an endless universe, the Catholic Church balked. The travails of Galileo, Copernicus, and Giordano Bruno evidence the rift between the Catholic Intelligentsia and the newly emerging scientific Intelligentsia. Because the Byzantine Church never dabbled in science no such friction between religious and scientific thinkers ever developed in Eastern Christendom.

7

Perhaps the embarrassment of the Catholic Church on account of its scientific teachings emboldened a group of radical northern clerics to question the infallibility of Catholic doctrine on its theological home turf. The result was the Reformation, the ultimate battle between religious Intelligentsias. The renegade monk, Martin Luther, had the temerity to compare the convoluted doctrines created by Catholic scholasticism with the straightforward teachings of Jesus as found in the Bible. Luther convincingly demonstrated that the Catholic Intelligentsia had strayed far from the basic teachings of the church. The force of Luther's reasoning was so compelling that it ultimately engendered the Counter Reformation whereby the Roman Church ultimately, largely reformed itself.

The Reformation may be viewed as an anti intellectual movement whereby simple reasoning is juxtaposed to elegant intellectualism. This is likely the way Martin Luther viewed himself. However, initially the Reformation was really a battle between Intelligentsias. The Catholic Intelligentsia had arrogated unto itself exclusive control of religion, even to the extent of forbidding the laity to read the scriptures. The Protestant Intelligentsia, on the other hand, translated the Bible into the vernacular and promulgated the revolutionary doctrine of the priesthood of all believers. The emphasis on faith and grace was indeed simple but nonetheless elegant and intellectually satisfying.

The Reformation contributed immeasurably to the expansion of the Intelligentsia. The freedom of conscience inherent in the priesthood of all believers opened the possibility to question authority and engage in free intellectual enquiry. The legitimacy of sources of power was opened to debate. The Reformation in England had a unique character. Henry VIII broke with Rome for reasons of personal and political convenience likely unrelated to the contemporary theological turmoil. Thus the emergent Church of England under the Archbishop of Canterbury remained outwardly conservative and not radically evangelical. Puritanism, John Wesley and Methodism, as well as the emergence of the Society of Friends were the manifestations of Protestant intellectualism in the British Isles.

However, the freedom of conscience expounded by Martin Luther had a radical political impact in Britain. The Puritan Revolution took the reform foment into the streets and toppled the monarchy for a time placing Oliver Cromwell at the head of a militantly protestant and politically egalitarian state. The intellectual emancipation fostered by the Reformation lead to works of political philosophy by the likes of John Locke and Thomas Hobbes who critically examined the hierarchical organization of society.

8

The Enlightenment, a uniquely secular intellectual movement, was nonetheless the product of the religious emancipation initiated by the Reformation. The new, worldly Intelligentsia did not cease to contemplate the place of God in relation to creation and human society. Voltaire's famous quote, "If God did not exist, it would be necessary to invent him," points clearly to the continuing centrality of religion in the analysis of secular polity. Descartes attempted to form a fundamental intellectual framework analyzing existence by starting at ground zero, "I think therefore I am." Further extrapolation lead the French philosopher to intuit the existence of God independently of any revealed truth.

All of this illustrates the continuing interaction between the Intelligentsia and religion near the dawning of the republican experiment. The theoretical underpinnings of modern democracy derive precisely from the Reformation and the Enlightenment, from the Intelligentsia's extrapolation of religious freedom into the sphere of civil polity.

9

The American Revolution was a further watershed in the history of the Intelligentsia in its interaction with religion. It is not without significance that the Declaration of Independence of the American colonies makes specific reference to inalienable rights bestowed upon the populous by their Creator. Thomas Jefferson further posited that it was not merely the right but the duty of the free individual to assert these rights in the face of a sovereign who violated the laws of God and nature. Thus the Protestant freedom of conscience was transformed into the supremacy of conscience.

What a mighty Intelligentsia that was, which gathered in Philadelphia on the edge of an untamed continent in order to set forth new principles for self governance, first in 1776 in declaring independence from monarchal authority and then in 1789 to adopt a Constitution. The language of the resultant documents is rich with Biblical imagery and peppered with the ecclesiastic turn of phrase. These founders of the American republic were well schooled intellectuals yet steeped in a deeply religious tradition.

Interestingly the American Constitution guaranteed freedom of religion. This is quite distinct from the "freedom from religion" that a later cadre of revolutionaries would espouse. For the 18th Century founders of the republic, freedom from religion would have been unthinkable, and if ever conceived, would be undesirable. As Voltaire had posited, in the absence of God, an authority figure must be created. It is precisely to the Creator that Thomas Jefferson refers as the source of the rights he has articulated. Religion gives the color of legitimacy to what would otherwise have been an unruly rabble.

The freedom of religion is the most cherished religious right, the right to select one's own belief system. The analysis of existence and the universe in this religious context is the mission of the American revolutionary Intelligentsia. But a certain commonality of beliefs is required in order to share a common morality. It is impossible to legislate in the absence of commonly held moral values. And morality is always the product of some religious based value system. To attempt to derive laws on the basis of mere utility is only to beg the question. For the determination of what is utile is a moral choice, determined by religious values.

10

The French Revolution was a very different creature from its American predecessor. It is tempting to view the American Revolution as a Protestant societal phenomenon, while the French Revolution was a uniquely Roman Catholic experience. In France the revolutionaries took on a sovereign who was intimately intertwined with the religious establishment, who was crowned by the Pope and claimed the authority by divine right. In France religion was not the revolutionaries' coconspirator but rather their implacable adversary. The church in pre-Revolutionary France was the bulwark of the monarchy the quintessence of the status quo.

The French revolution was essentially anticlerical. The revolutionaries seized the properties of the Roman Catholic Church, they harried the Roman Catholic clergy. The French revolutionaries even had the temerity to establish a new calendar with the storming of the Bastille rather than the birth of Christ as defining the year one. Religion was important to the French revolutionary Intelligentsia. However, it was an important adversary. However, when the dictatorship of reason transformed itself into "the terror," the Intelligentsia blanched. The brief Napoleonic period was, from the standpoint of intellectual history, a mere interregnum separating the foment of anti-Catholic revolution from the reactionary restoration of the Bourbon state.

The Congress of Vienna at the conclusion of the Napoleonic Wars restored the old values on the surface, but Pandora's box can never be successfully resealed. The French Revolution spawned a new and skeptical Intelligentsia, one that looked askance at the verities of the Roman Catholic Church. Indeed, the Masonic movement began to proliferate after 1815 as the educated elite sought a framework for an ordered society which would yet be free from the old stultified structure. The Free Masons paid lip service to the values of the old religion in the absence of its historic baggage.

11

The revolutionary movements of 1848 exhibited a matured version of the French Revolutionary fervor. With roots in the developing middle class and overtones of nationalistic fervor, the revolutionary Intelligentsia of 1848 sought to unseat the monarchies whose power structure was still derived from feudal institutions which had no relevance to the developing modern world. The unrest of the year 1848 did not result in any immediate change in the social landscape. However, it reflected the fundamental changes which were proceeding under the surface.

On the heels of 1848 the arrival on the intellectual scene of Karl Marx is of great significance. Marx openly attacked religion as the opiate of the masses and an obstacle in the path of societal progress. Marx enshrined materialism in reaction to the latent spiritualism of religion. However, by naming religion as an enemy, Marx elevated the subject of religion to a position of critical importance in the ruminations of the Intelligentsia. Furthermore, the creation of a Marxist ideology brought with it the evolution of a distinct Marxist Intelligentsia, a sort of irreligious priestly class.

During this period the impact of both Charles Darwin and Friedrich Nietzsche upon the Intelligentsia and religion was noteworthy. Darwin's theory of the origin of species challenged religious teachings pertaining to the Creation. The concept of the survival of the fittest in the evolution of the natural world created a new basis for societal morality. And it was Nietzsche who took the notion of evolution and extrapolated it into the human social context with the concept of the superman. In an age of ascendant nationalism it did not take long for members of the Intelligentsia to establish a racial pecking order in the human environment and to pose the ideal development of the superman on the basis of the survival of the fittest.

In various circles of the Intelligentsia there developed a face off between nationalism and religion. At a later stage this was particularly marked in Germany and Italy. However, In Poland an entirely different phenomenon arose where the Intelligentsia was obsessed with Polish nationalism and Catholic religiosity simultaneously. This was no doubt the result of Poland's geographic partition between three foreign sovereigns, two of which were not Catholic, Orthodox Russia and Lutheran Prussia.

12

Russia has been the cradle of the Intelligentsia in the modern era. It is from the Russian language that we derive the term "Intelligentsia." From Russia we derive the realization that the societal cadre of abstract thinkers constitutes a class onto themselves. It might be said that the Russian Intelligentsia is a political class in an apolitical state. While members of the Russian Intelligentsia have always been intensely interested in the governance of their homeland, they have traditionally lived under governments which eschewed the political arts and embraced autocratic government.

Russia experienced a cultural eruption in the wake of the reforms undertaken by Peter the Great. When he opened Russia to the West he brought new influences to a heretofore isolated society. However, Peter also unleashed an intellectual giant unto the world scene with the breath of fresh air associated with a newcomer.

The history of the Russian Intelligentsia cannot fail to note Mikhail Lomonosov, a true Renaissance man, and the university which he helped to establish in 1755. Alexander Pushkin released the latent poetry of the Russian language. Ivan Turgenev awakened social conscience. And then the literary giants, Fedor Dostoyevsky and Lev Tolstoy came thundering on the world stage with novels of surpassing significance.

In Russia the question of religion has always been a central topic of discussion for the Intelligentsia. To read Russian literature is to delve deeply into the question of God and His relation to mankind. The nihilist movement in 19th Century Russia represented a stark reaction to the traditionalism of the Russian Orthodox Church and the rigidity of the monarchal system. The Slavophil nationalist Intelligentsia, on the other hand, enshrined the traditions of old pre Petrine Russia and the doctrines of the immutable Byzantine Church. At the beginning of the 20th Century it might have been possible to overlook the Russian Marxist Intelligentsia divided into Menshevik and Bolshevik factions. But subsequent developments have writ large the significance of this group with its particularly antagonistic attitude toward religion.

13

Religion and the Intelligentsia are both central to the Russian consciousness. But religion comes first. The baptism of Rus in the Dnieper River in 988 under Prince Vladimir is the genesis of the Russian national self awareness. The sense of Russian nationhood was forged on an anvil hammered by Tartar Islam, Shamanist Mongols and Roman Catholic Poles. The suppression of the Old Believers by the Orthodox patriarchate in the 17th Century brought home a consciousness of religious divisions across the nation. The Petrine social reforms took place against a background of subordinating the Russian Orthodox Church to the sovereign tsar.

The socially inferior position of the clergy in tsarist Russia in relation to the nobility resulted in a clerical Intelligentsia in Russia which was distinctly politically liberal. Indeed, when the Russian Revolution broke out in 1917 the Russian Orthodox Church did not unequivocally support the monarchy. The Church synod immediately took advantage of the collapse of the tsarist government in order to elect a patriarch for the first time since the reign of Tsar Peter the Great. Ultimately the communist state decimated and subordinated a puppet Orthodox Church, but a lively Orthodox Intelligentsia thrived outside the Soviet Union during the entire communist era. A new Russian Orthodox Church, resurrected after the fall of the Soviet Union has, nevertheless, apparently fallen pray to the tantalizing favors showered upon it by the regime of Vladimir Putin.

14

The Jews have played an important role in the life of the Intelligentsia since the French Revolution and their general emancipation from a ghetto existence. As a people possessed of a vibrant theological literature and a tradition of intellectual dispute, it is not surprising that Jews have aspired to prominence in the ranks of the Intelligentsia. Their attainments in law, the professions and commerce have given the Jews the financial wherewithal to obtain education and follow intellectual pursuits. Show business and the media, which represent the commercialization of drama and journalism, have proven fertile fields for blending business and intellectual activity. The role of the Jews in these spheres can hardly be overestimated. The contributions of Sigmund Freud to the literature about human consciousness have introduced Jewish people to a key area of interest to the Intelligentsia, psychology.

The position of the Jews as traditional outsiders, yet economic leaders has resulted in fascinating social developments. The emancipation of the Jews in Western Europe following the Napoleonic wars resulted in a massive infusion of money and manpower into the ranks of the liberal, left-leaning, Intelligentsia. Because of their historic pariah status in European monarchal society the Jews had little stake in the old establishment and quite naturally joined the ranks of radical reformers. The historic animosity of the Roman Catholic Church in relation to the Jews naturally resulted in the majority of the Jewish Intelligentsia adopting an anticlerical stance. Thus, into the first half of the 20th Century the ranks of revolutionaries, anarchists, socialists and communists were rife with Jewish intellectuals.

Interestingly the second half of the 20th Century and the beginning of the 21st Century have witnessed a major shift in the ideological position of the Jews. The phenomenon of populist, anti Semitic fascism gave the Jewish intellectuals pause in contemplating the heretofore assumed virtue of the masses. Furthermore, the enormous economic dislocations associated with two World Wars and the Great Depression resulted in the inordinate concentration of economic power in the hands of Jews in Europe and the Americas. Suddenly the outsiders were transformed into insiders. Not surprisingly the Jewish Intelligentsia began to shift from left liberalism to right conservatism. Self interest has an amazing capacity to clarify political philosophy. Thus we have experienced the development of neo-conservatism. The so called "neo-con" movement has been lead by members of the Jewish Intelligentsia.

Judaism is arguable a religion, a nationality and an ideology, all at the same time. The history of the Jews in Europe since the Crusades provides fertile ground for further study of the relationship between religion and the Intelligentsia.

15

In the 20th Century ideology and religion have contended for the attention of the Intelligentsia. Communism and materialism; fascism and neo-paganism; all have captivated certain elements of the Intelligentsia. Ideology can be seen as a substitute for religion. Ideology claims to be revealed truth. Ideology provides a uniquely colored world view for its adherents. Is atheism a religion or an ideology? Is belief in the primacy of the Big Bang an article of faith? Is the impossibility of "intelligent design" implacable dogma?

At various points, particularly in the twentieth century, it has seemed that religion was going into decline. Militantly anti religious movements have swept onto the scene with mass support and have engendered seemingly little resistance. Attendance in churches and identification with particular religious denominations has decreased, The all-pervasiveness of mass media has supplanted a niche in the life of the common man which had been occupied by religion. Nevertheless there is something in the human condition which craves a belief system and a systematic explanation for existence. The anti religious doctrine of Marxism did not prevent Marxist societies from marketing their ideology with all the trappings of religiosity, from revealed texts to iconic figures. The mass ideologies of fascism and National Socialism enlisted pagan imagery and mystical belief systems in order to fill a religious craving in the populous.

A dispassionate historical perspective reveals that, while certain manifestations of religion can change radically, there is an inescapable religious component in the human psyche. The Intelligentsia both reflects and shapes mankind. Thus, the topic of the Intelligentsia and religion is not merely a historical study. It continues to be a vital subject for ongoing analysis.

16

In our time liberalism has a significant role to play in the interaction of religion and the Intelligentsia. A doctrine of universal relativism is attractive in a time when so many absolutes have been discredited. The carnage of the First World War and its aftermath have put the fundamental verities of Western culture, and the faith in the inevitability of progress, in doubt. Conservatism has been tarnished or totally discredited in light of the excesses of fascists and rightwing extremists in the mid 20th Century. The role of the organized church in the life of everyday people has been in decline in Europe since the Second World War. Television and mass media have created a fluid but pervasive conventional wisdom which successfully questions all absolute doctrine.

However, there is something fundamentally bipolar in human social organization. This, of course, is not meant in the psychiatric sense of the term, although a play on words is tempting. Society naturally divides into pros and antis. Thus the advances of liberal agnosticism naturally engender a reaction of conservative religiosity. The action-reaction social phenomenon guarantees that neither religion nor atheism will pass from the earth until the second coming.

17

In the face of all the foregoing there appears to be an insurgent evangelical Intelligentsia reexamining some of the old concepts which have interested intellectuals for centuries. Theological inquiry appears to enjoy a new-found attractiveness for some members of the Intelligentsia. There is an undeniable, general appeal felt for unchanging, fundamental verities. There appears to be a contemporary analysis of environmental issues from the perspective of the existence of good and evil. The preservation of a healthy natural environment represents a fundamental good. Whereas, the pollution or outright destruction of the planet is an absolute evil. Not everything is colorless, relative or unknowable. One view of religion might conclude that the pursuit of scientific inquiry is a moral imperative.

18

The resurgence of Islamic fundamentalism is an undeniable force in the 21st Century global community. This mass movement has many causes. The threat posed to the holiest shrines of Islam by the occupation of the old city of Jerusalem by a militantly anti Islamic state has engendered a sense of urgency across the Muslim world. The concentration of enormous petroleum based wealth in the hands of militant, proselytizing, Persian Gulf principalities has fueled expansion of Islamic religious education internationally. The presence of a large Muslim minority within often-hostile European societies has strengthened Muslim self awareness. International, instantaneous mass communication has united diverse and dispersed communities. The confluence of these factors has created a militant movement, at least in part, dedicated to the establishment of a global caliphate.

As is often the case with religious movements, radical Islam has exercised its strongest attraction at the extremes of the educational spectrum. The appeal of fundamentalist Islam is greatest among the highly educated and among the unschooled masses. The Muslim Intelligentsia seeks to rediscover its old scholastic preeminence and establish itself as a counter weight as against the secular West. The average Muslim on the street, who has always been closer to the strict religious traditions of the village, also seeks out a mass movement which promises liberation and escape from poverty.

Islam, like Judaism, has a strong intellectual history of interpretation of texts and scholarly disputation. There has long been an Islamic Intelligentsia. The modern twist is that the Muslim clerics and intellectuals now seek to inject themselves into the administration of the state structure and participate in the universal implementation of sharia, Islamic law. Contemplative disquisition has been replaced by the powerful urge to impose a religiously based system of governance.

Historically Islam knew an early period of intense growth and expansion in its first centuries of existence. Mohammed's message held strong attraction for multiple peoples emerging from paganism who had already had fleeting exposure to the charismatic teachings of early Christianity. Islam provided an alternative monotheism which addressed fundamental religious cravings but did not impose the heavy burdens of conscience, pacifism and anarchism associated with early Christianity. Early Islam satisfied widely held yearnings in the Middle East and elsewhere.

The political decline of Islam was inextricably intertwined with the decadence of the Ottoman Turkish Empire. The central political unit of the Muslim world degenerated into a corrupt, mismanaged and restive political unit torn by centrifugal forces. After Ottoman legions were expelled from the Iberian Peninsula and turned back from the headwaters of the Danube, Islam gradually lost its political militancy and withdrew into the purely spiritual realm. The presence of vast Muslim populations under European colonial domination further served to emasculate fundamentalist Islam.

The 20th Century decline of the West in the wake of the European civil wars, which we know as World Wars One & Two, provided Islam with an irresistible opportunity. In the face of Christendom's massive experience of self doubt an opening developed for an unselfconscious religion to fill the gap. The Islamic Intelligentsia provided answers to questions from which Christian societies retreated. The successes of radical Islam in Europe and in the former provinces of the Ottoman Empire should have been easy to anticipate.

The rise of radical Islam has led some members of the Western Intelligentsia to look to their own religious roots in search of motivation. Careful analysis of the contemporary intellectual scene leads to the conclusion that the reemergence of a militant Christian Intelligentsia is to be anticipated. The example set by the successes of a radical Islam will engender emulation. Young Western intellectuals cannot fail to recognize the powerful unifying force of a universal spiritual awakening. Were it not for the overpowering anti-Christian efforts of mass media in the West this resurgence of Christianity would already be well under way.

In any event, the current situation of the Islamic Intelligentsia in the global political mix further demonstrates the relevance of our topic, "the Intelligentsia and Religion."

19 The Intelligentsia & Contemporary Moral Turmoil — A Personal Reflection

As we have seen, the interaction between the Intelligentsia and religion has been ongoing since the dawn of civilization. The priestly class was the original custodian of knowledge and repository of cultural tradition. Law and morality had their genesis in theology and it is the shared sense of moral purpose which gave birth to the institution of the Intelligentsia. The spiritual liberation embodied in the teachings of Jesus Christ rendered intellectual inquiry an egalitarian exercise. The Protestant Reformation established freedom of conscience which, in turn, led to political emancipation guided by a vanguard from the Intelligentsia. However, in the 19th and 20th Centuries elements of the Intelligentsia rebelled against the concept of God as the autocratic Ruler of the universe. Extrapolating some concepts from humanism, Marxists and exponents of other atheistic movements sought to establish a universal temporal regime fashioned exclusively with human hands. Forgetting that their moral underpinnings derived from religious doctrine, the rebel intellectuals sought to create a utopia out of a disembodied abstraction. The Marxist, atheist edifice ultimately collapsed of its own accord. However, the quaint notion that man is exclusively his own master and compass in the infinite expanse of the universe continues to retain adherents among some elements of the Intelligentsia.

Those who have not learned the lesson taught by the disastrous fate of aggressively secular intellectual constructs continue to seek out a future devoid of an omnipotent, intelligent Source of universal design. In the 21st Century we continue to note the presence of movements within the Intelligentsia which reject God and perceive the revealed truth of religion as an enemy of human nature. Presently, among the torchbearers of this attitude are the adherents of the Lesbian, Gay, Bisexual, Transgender & Queer movement (hereinafter LGBTQ). The LGBTQ movement rejects the authority of established religious doctrine when it comes to sexual morality and the family and would establish a self-willed freedom from any inconvenient moral restraint.

First let us establish that the behavior advocated by the LGBTQ movement is violative of established religious doctrine. Throughout this essay the author has consciously taken a Euro-centric approach to intellectual history. While giving the nod to the existence and validity of other intellectual traditions it is respectfully urged that the institution of the Intelligentsia has its original and firmly established roots in Christendom. Furthermore, your author recognizes that his personal qualifications derive from this cultural tradition. In order to presume to speak with authority I will continue to address the subject at hand from this perspective which is familiar to me. The LGBTQ movement advocates a lifestyle and world view which is unequivocally at variance with Christian doctrine and with the religious tradition from which Christianity derives. Leviticus 18:22 & Romans 1:18-27 &1 Corinthians 6:9-10 & 1 Timothy 1:9-10 & Jude 17.

Let us examine a representative passage from the New Testament. "For the wrath of God is revealed from heaven against all ungodliness and wickedness of those who by their wickedness suppress the truth. For what can be known about God is plain to them, because God has shown it to them. Ever since the creation of the world his eternal power and divine nature, invisible though they are, have been understood and seen through the things he has made. So they are without excuse; for though they knew God, they did not honor him as God or give thanks to him, but they became futile in their thinking, and their senseless minds were darkened. Claiming to be wise, they became fools; and they exchanged the glory of the immortal God for images resembling a mortal human being or birds or four-footed animals or reptiles. Therefore God gave them up in the lusts of their hearts to impurity, to the degrading of their bodies among themselves, because they exchanged the truth about God for a lie and worshipped and served the creature rather than the Creator, who is blessed for ever! Amen. For this reason God gave them up to degrading passions. Their women exchanged natural intercourse for unnatural, and in the same way also the men, giving up natural intercourse with women, were consumed with passion for one another. Men committed shameless acts with men and received in their own persons the due penalty for their error." Romans 1:18-27 Is this not an apt description of our time? Is not contemporary humanism, as espoused by some members of the Intelligentsia, the exchange of the glory of the immortal God for a falsely idealized mortal image? Does our society not engage in futile thinking? Do we not give ourselves up to vain discussions seeking controversy where meanings are plain? Are not those who claim to be wise always fools? Have we not been instructed that the foolishness of God surpasses the wisdom of men? (1 Corinthians 1:25) Are we not fools by God's definition when we attempt to explain away the plain language of the scripture or unilaterally disregard it? Has not 21st Century man exchanged the truth about God for a lie? From communism to fascism to consumerism is not contemporary man exchanging God's definition of goodness for a golden calf?

From an objective, historical point of view the most striking social phenomenon at the turn of the 21st Century has been the propagation of the concept that sodomy is an acceptable, alternative lifestyle. This unprecedented assault on the institutions and moral fabric of civilization represents an abomination. There is no point in mincing words. The concerted effort of mass media over several decades to recast homoeroticism as acceptable activity constitutes an affront to civilization. There can be no question that this aberration will presently come to an end. Nature deplores the unnatural. Historic evils have inevitably passed away.

The propagandists of homoeroticism within the Intelligentsia have attempted to subvert language to sanitize their miscreant activity. First, mass media virtually banished the words sodomy and sodomite from the popular vocabulary. Initially they substituted the clinical sounding word, homosexuality. In recent times "homosexual" has been replaced by the inapt word, "gay." This represents an attempt to associate a happy face with a dirty business. Such a misuse of language is itself an abomination.

Will we next call pederasts and pedophiles, "the friendlies?"

The advocates of homosexuality would argue that this aberration is just another choice of lifestyle or that it is an innate, inborn characteristic of certain individuals. Neither proposition is persuasive. Engaging in activity contrary to nature is not an appropriate choice. If a person wishes to insert food into his ear rather than into his mouth such behavior is unacceptable. Just so the homosexual act is unacceptable. If anyone is born with a predilection to do that which is unnatural or antisocial they must be actively discouraged. People born with violent tempers learn to curb their emotions. The same applies to kleptomaniacs. Similarly, if some individuals are born with a homosexual predilection it should not be encouraged. All organized society is based on rules of conduct.

Human society and its emergence from the Hobbsian state of nature has been premised on the institution of the family and the existence of sexual morality. The family is an economic and social unit, which allows for procreation and cooperation, for the rearing of a new generation, for the attainment of happiness and security. Sexual morality allows for societal organization and joint effort between people. When men and women observe moral strictures they do not universally look upon one another as sex objects or sexual competitors. They can work together efficiently. Thus the family and morality create the framework for the development of civilization. Upholding the family and upholding morality is not merely the way or religious righteousness, it is expedient for secular society.

Historically civilized society has been committed to upholding the family and abhorring sodomy as a vice. Thousands of years of accepted wisdom ought never to be lightly discarded. Interestingly, until December 15, 1973 the American Psychiatric Association designated homosexuality as a mental disorder. The removal of homoeroticism from the list of mental illnesses was the result of a well financed political struggle within the professional organization rather than a scientific discovery or theological revelation. Whether a vice or an illness, the homosexual lifestyle violates the norms which have been the underpinning of human civilization from the earliest times.

Why has this abomination befallen our society in our time? From a Christian perspective, those who believe in a personification of good, whom we call God, and a personification of evil, whom we refer to as Satan, the answer is simple. The force of evil is always tempting man to break Gods rules. For one generation the force of evil encourages war and genocide. For another generation the force of evil propagates sodomy and makes a frontal attack on the institution of the family. The study of history is the study of the conflict between good and evil in various guises.

A secular explanation of this abomination would attribute it to an economic phenomenon. At the tail end of the 20th Century a chance concentration of wealth in the hands of a small number of homosexuals allowed them to financially back a media blitz aimed at conveying a false and alluring image for homoeroticism. It is certainly no secret that the movie industry and show business in general are rife with advocates of homosexuality. In the 21st Century, television commentators and movie characters are the peer group for modern man who is estranged from roots, family and community. If the electronic media determine to push any particular activity it is sure to soon have adherents. Even if a lifestyle is contrary to nature, the power of visual media is such that many wayward individuals will find it irresistible.

The fact that some religious leaders and health care professionals within the Intelligentsia have become apologists for homosexuality is hardly significant. There have always been bizarre theological heresies which have lead some folks astray. Similarly, the history of fleeting and false medical and scientific theories could fill many textbooks.

Obviously Christians must refrain from hating any individual or group. It is the social malaise which we oppose, not the unfortunate victims who have been subverted and infected by this abomination. However, if we remain silent and do not confront this societal crisis we do a disservice to today's poor sufferers from homosexuality and the members of the future generation who could be at risk. Christian members of the Intelligentsia have a tradition of courageously taking on the powers that be, when those powers have fallen into grievous error.

Indeed, God is love. (1 John 4:8) But His love is a parent's love, which entails instruction and reproof. In the first chapter of the New Testament book of Romans we see what provokes the wrath of a loving God. And it is in this context that we see the unambiguous description of same sex relationships. The balance of the book of Romans shows us how we can become the beneficiaries of God's love by faith through grace, on the basis of the sacrifice which has been made on our behalf by Jesus Christ.

If we prayerfully read the scriptures it forms our conscience. It gives us the basis to determine what is right and what is wrong. When we, nevertheless, do that which is wrong we must repent. When we repent, confess our wrongdoing and ask for forgiveness, God in His infinite love and mercy forgives us. This is the essence of the Christian religion. We are all imperfect. God established rules relating to sex and the family among a myriad of other rules. From time to time we all chafe at the various constraints which God established. We are tempted and at times we transgress. But in order for us to be forgiven we must repent and request God's mercy. In order to repent we must understand that we have done wrong. Thus we require a well formed conscience, a conscience formed by the scriptures. It is one of the functions of the church to teach the scriptures and to form the conscience. The church is supposed to remind us what is right and what is wrong. If the church blesses conduct which the scriptures state is wrongful, the church is thwarting God's love. The church is discouraging repentance and preventing forgiveness. For this reason the church may not bless same sex marriages.

We must love all our neighbors, including those who fail to comprehend the difference between right and wrong in some contexts. However, we must always bear true witness. We must joyfully give and receive brotherly and sisterly counsel. We must never become futile in our thinking and never exchange the truth about God for a lie. This ought to be the hallmark of the Christian Intelligentsia.

It is impossible to disregard the fact that among the advocates of the LGBTQ movement there are those who claim to be adherents of the Christian religion. While this is a relatively insignificant group their existence is hardly surprising. The established freedom of conscience includes the undeniable freedom to err. Indeed the history of the Christian Church is replete with heresy. The New Testament admonishes that there will be false prophets and those who disingenuously claim to come in the name of Christ. The existence of both good and evil is axiomatic. The genius of the Protestant doctrine of the priesthood of all believers is that it entrusts the individual well-formed conscience with the task of separating the wheat from the chaff. We can be certain that the doctrine advanced by advocates of the LGBTQ movement who claim to be Christian, nevertheless represents intellectual chaff. They do not claim to have received a new revelation but merely arrogate unto themselves the authority to relegate all teaching which is contrary to their chosen lifestyle to the litter basket of irrelevant literature and historic anomaly. The apologists of the LGBTQ movement contend that the only core teaching of Christianity is "love" and that all reproof, or discipline, is in conflict with that core value. What the advocates of the LGBTQ movement fail to realize is that the love of God is a parent's affection, which is always alloyed with instruction and the imposition of authority.

The sale of indulgences by the Roman Church in the 15th Century, whereby a sinner might seek remission of punishment for a transgression by the payment of a gift or the performance of a good work, was at the heart of the Protestant Reformation. The reformers upheld the Biblical principle that sins are forgiven by the Grace of God for all who repent and seek reconciliation with God through Jesus Christ. Not by payment of money, nor by performance of good works, but by faith alone can man obtain salvation. This is axiomatic and was recognized by the Roman Catholic Church in the Counterreformation. .

In the 21st Century we find ourselves in the anomalous position that some theologians in the Intelligentsia are granting a blanket indulgence for a certain sin on the basis of the financial strength and societal organization of a group which advocates a particular transgression. I have in mind the churchmen who would have us disregard the unequivocal Biblical admonition against sodomy. God's directive not to engage in sodomy is no less positive than His directive that we avoid adultery, incest, polygamy, fornication or pedophilia. Yet in the face of a long standing and highly organized public relations campaign undertaken by a well financed and organized group of homosexuals, some church leaders have decided to declare a de facto indulgence for this particular transgression

The history of man is the chronicle of the continuing struggle between good and evil. The Christian Intelligentsia recognizes that humankind is by nature inclined to be tempted by evil. Through Jesus Christ we can resist this evil and by accepting His sacrifice on our behalf, we can be forgiven for our transgressions when we repent. But our human part of this Divine equation is that we recognize our wrongdoing, ask for forgiveness and accept God's gift. According to Christian doctrine, if we fail to repent we cannot be forgiven.

Herein lays the grievous error of those who would grant indulgences. The indulgence misleads the transgressor into believing that he or she need not repent and seek God's forgiveness. Those 21st Century theologians who would grant a blanket indulgence for homosexual activity do the greatest injustice to that group whose interests they ostensibly seek to advance. They mislead homosexuals into believing that their activity is sanctioned by God. All Biblical authority is to the contrary.

In our society there is little doubt that homosexuals as a group have attained substantial economic success and have achieved a societal position commensurate with their great wealth. They are able to pay for media acceptance, even adulation of their life style. But it is the function of the Christian Intelligentsia to articulate God's directives and to speak His truth. Men and women who believe in God and accept Christ as their Savior must uphold His directives. We must not call that which is wrong, right. We must not arrogate unto ourselves the power to grant a false indulgence.

The past, present and future of the interaction of the Intelligentsia and religion is a chronicle of lively debate. That does not signify that there is no right and wrong. Rather it represents proof that the human condition is dialectic and characterized by free will. There is good and there is evil and it is the function of the members of the Intelligentsia to lead the inquiry and to seek out the truth.

20 Conclusion

From my perspective I take solace in the inescapable conclusion that, The foolishness of God surpasses the wisdom of men.

SEVENTEENTH LOMONOSOV CONFERENCE ON ELEMENTARY PARTICLE PHYSICS

ELEVENTH INTERNATIONAL MEETING ON PROBLEMS OF INTELLIGENTSIA:
"The Intelligentsia and Culture"

Moscow State University, Moscow, 20 – 26 August, 2015

Under the Patronage of the Rector of Moscow State University Victor Sadovnichy

Dedicated to the 60th Anniversary of Joint Institute for Nuclear Research (1956-2016)

International Advisory Committee

E.Akhmedov (Max Planck, Heidelberg), S.Belyaev (Kurchatov Inst., Moscow), V.Belokurov (MSU), V.Berezinsky (GSSI/LNGS, INFN), S.Bilenky (JINR, Dubna), J.Bleimaier (Princeton), M.Danilov (ITEP, Moscow), A.Dolgov (Novosibirsk State Univ., INFN-Ferrara & ITEP, Moscow), N.Fornengo (University of Torino & INFN), C.Giunti (University of Torino & INFN), M.Itkis (JINR, Dubna), A.Masiero (INFN, Padua), V.Matveev (JINR, Dubna), L.Okun (ITEP, Moscow), M.Panasyuk (MSU), V.Rubakov (MSU & INR, Moscow), D.Shirkov (JINR, Dubna), J.Silk (Univ. of Oxford), A.Skrinsky (INP, Novosibirsk), A.Slavnov (MSU & Steklov Math. Inst.), A.Smirnov (Max Planck, Heidelberg), P.Spillantini (INFN, Florence), A.Starobinsky (Landau Inst., Moscow), N.Sysoev (MSU), Z.-Z.Xing (IHEP, Beijing)

Organizing Committee

V.Bagrov (Tomsk Univ.), I.Balantsev (MSU), V.Bednyakov (JINR, Dubna), V.Braginsky (MSU), A.Egorov (ICAS), R.Fabbricatore (MSU), D.Galtsov (MSU), A.Grigoriev (MSU), A.Kataev (INR, Moscow), A.Kazantsev (MSU), O.Kharlanov (MSU), Yu.Kudenko (INR, Moscow), A.Lokhov (INR, Moscow) – Scientific Secretary, N.Nikiforova (MSU), A.Nikishov (Lebedev Phys. Inst., Moscow), V.Ritus (Lebedev Phys. Inst., Moscow), M.Polyakova (ICAS), Yu.Popov (MSU), V.Savrin (MSU), A.Studenikin (MSU & ICAS) – Chairman

Organizers and Sponsors

Faculty of Physics, Moscow State University
Joint Institute for Nuclear Research, Dubna
Ministry of Education and Science of Russia
Russian Foundation for Basic Research
Institute for Nuclear Research of Russian Academy of Sciences
Bruno Pontecorvo Neutrino and Astrophysics Laboratory (MSU)
Interregional Center for Advanced Studies

Programme of the 17th Lomonosov Conference on Elementary Particle Physics
/duration of 25 (20/15) min talks includes 5 (3) minutes for discussion/

20 August, Thursday

08.00 – 09.00 **Registration** (Hall in front of Central Physics Auditorium, Faculty of Physics, Moscow State University)

09.00 – 09.30 **Opening** (Conference Hall)

Alexander Studenikin, Chairman of Organizing Committee, MSU & JINR
Alexey Khohlov, Vice Rector of MSU
Andrey Slavnov, Head of Department of Theoretical Physics of MSU

09.30 – 13.20 **MORNING SESSION** (Conference Hall)

Chairman: **Michael Koratzinos**

09.30 A. Slavnov (Steklov Math.Inst. & MSU) *60 Years of Yang-Mills gauge field theories* (25 min)
09.55 M. Dubinin (SINP MSU) *Investigation of the Higgs boson in the Standard Model and its extensions* (25 min)
10.20 K. Mazumdar (Tata Inst. of Fund. Res.) *Overview of CMS experiment at CERN-LHC* (25 min)
10.45 P. Glaysher (Univ. of Edinburgh) *Search for New Phenomena at the High-Luminosity LHC with ATLAS* (25 min)

11.10 – 11.40 Tea break

Chairman: **Robert Wilson**

11.40 C. Giunti (INFN & Univ. of Turin) *A review on present status of neutrino mixings and oscillations* (25 min)
12.05 M. Smy (Univ. of California, Irvine) *Matter effects on neutrino oscillations* (25 min)
12.30 G. Raffelt (Max Planck Inst. for Physics, Munich) *Supernova neutrinos* (25 min)
12.55 A. Ereditato (Univ. of Bern) *The Fermilab international neutrino program* (25 min)

13.20 – 15.00 Lunch

15.00 – 18.30 **AFTERNOON SESSION** (Conference Hall)

Chairman: **Jonathan Granot**

15.00 V. Berezinsky (GSSI & LNGS) *UHECR: Acceleration and Propagation* (25 min)
15.25 N. Mandolesi (Univ. of Ferrara) *PLANCK experiment* (25 min)
15.50 T. Suomijarvi (Univ. Paris-Sud) *Pierre Auger Observatory* (25 min)

16.15 – 16.45 Tea break

Chairman: **Ivanka Bozovic Jelisavcic**

16.45 O. Rogachevsky (JINR) *Status of NICA project at JINR* (25 min)
17.10 A. Vasiliev (National Research Centre "Kurchatov Institute") *FAIR facility* (25 min)

17.35 B. Barish (CalTech) *The International Linear Collider and the future of Accelerator Based Particle Physics* (25 min)
18.00 N. Gehrels (NASA-GSFC) *The Transient High Energy Sky* (25 min)

18.30 – 23.30 SPECIAL SESSION (40 ^{0})
Reception banquet will be held on board of a ship that will stream along the river across the central part of Moscow; the conference buses to the ship will depart from the entrance of the Faculty of Physics at **18.30**

21 August, Friday

09.00 – 13.30 MORNING SESSION (Conference Hall)

Chairman: **Masayuki Nakahata**

09.00 J. Shirai (Tohoku Univ.) *KamLAND-Zen* (25 min)
09.25 S.-B. Kim (Seoul National Univ.) *Results from RENO* (25 min)
09.50 X. Li (IHEP, Beijing) *Recent results from Daya Bay Reactor Neutrino Experiment* (25 min)
10.15 T. Bezerra (Tohoku Univ.) *The Double Chooz experiment: June 2015 results and prospects to the near detector era* (25 min)
10.40 F. Villante (Univ. of L'Aquila & INFN-LNGS) *Recent developments in solar neutrino theory* (25 min)

11.05 – 11.30 Tea break

Chairman: **Soo-Bong Kim**

11.30 M. Nakahata (Univ. of Tokyo) *Super-Kamiokande results* (25 min)
11.55 Yu. Kudenko (INR RAS) *Hyper-Kamiokande project* (25 min)
12.20 L. Oberauer (Tech. Univ. of Munich) *LSc development for solar and supernova neutrinos* (20 min)
12.40 A. Kappes (ECAP / FAU Erlangen-Nuremberg) *Neutrino mass-hierarchy from atmospheric neutrinos* (25 min)
13.05 M. Frank (Univ. of Virginia) *First Results from NOvA* (25 min)

13.30 – 15.00 Lunch

15.00 – 19.15 AFTERNOON SESSION (Conference Hall)

Chairman: **Yuri Kudenko**

15.00 K. Helbing (Univ. of Wuppertal) *IceCube's neutrino signal and searches for BSM physics* (25 min)
15.25 D. Besson (Univ. of Kansas) *Ultra-high energy neutrino and cosmic ray detection via the Askaryan effect* (20 min)
15.45 Y. Li (IHEP, Beijing) *JUNO project* (25 min)
16.10 S. Davini (Gran Sasso Science Inst.) *Recent results of the Borexino experiment* (20 min)
16.30 M. Yoshimura (Okayama Univ.) *Neutrino mass spectroscopy: theoretical aspects* (25 min)
16.55 N. Sasao (Okayama Univ.) *Toward neutrino mass spectroscopy with atoms: experimental status* (20 min)

17.15 – 17.40 Tea break

SESSION 21.08. A (Neutrinos)

Chairman: **Matthew Malek**

17.40 M. Tenti (INFN-Bologna) *Search for sterile neutrinos in the νμ disappearance mode at FNAL* (20 min)
18.00 A. Barabash (ITEP) *The Majorana Demonstrator ^{76}Ge double beta decay experiment: present status* (20 min)
18.20 V. Belov (ITEP) *EXO-200 and nEXO* (20 min)
18.40 M. Garzelli (Univ. of Hamburg) *Prompt neutrino flux in the atmosphere revisited* (20 min)

SESSION 21.08. B (Physics at Colliders)

17.40 Y. Kolomensky (LBL) *Review on BaBar* (25 min)
18.05 V. Pantuev (INR) *Recent results from PHENIX* (20 min)
18.25 D. Druzhkin (CERN) *Technology of the TOTEM RomanPots and Future Detectors Upgrade* (20 min)

SESSION 21.08. C (Gravitation, cosmology, astrophysics)

Chairman: **Viktor Semikoz**

17.40 A. Zakharov (ITEP) *Is an ordinary supermassive black hole at the galactic center?* (20 min)
18.00 Y. Srivastava (Perugia) *Nature of the cosmic ray power law exponents* (20 min)
18.20 J. Ng (Triumf) *Majoron and dark matter and dark energy* (20 min)
18.40 S. Popov (SAI MSU) *Magnetic field evolution of neutron stars: linking magnetars and antimagnetars* (20 min)
19.00 V. Dokuchaev (INR RAS) *Weighing of dark matter in the Galactic center* (20 min)

19.30 – 22.30 Sight-seeing bus excursion in Moscow

22 August, Saturday

09.00 – 13.20 MORNING SESSION (Conference Hall)

Chairman: **Alexander Zakharov**

09.00 P. Salati (LAPTH, CNRS) *Indirect dark matter searches in the light of the recent AMS-02 observations* (25 min)
09.25 R. Sparvoli (Univ. of Rome Tor Vergata) *Direct measurements of cosmic rays* (25 min)
09.50 G. Prodi (Univ. of Trento and INFN) *Status of advanced Virgo* (20 min)
10.10 D. Martynov (CalTech) *Status of advanced LIGO* (20 min)
10.30 G. Stratta (Univ. degli Studi di Urbino "Carlo Bo") *The most promising astrophysical sources of gravitational waves* (20 min)

10.50 – 11.30 Tea break

11.30 P. Belli (INFN Roma Tor Vergata) *Recent results from DAMA/LIBRA and perspective* (25 min)
11.55 C. Burigana (INAF-IASF Bologna, Italy) *Astrophysical components from Planck maps* (20 min)
12.15 P. Natoli (Univ. of Ferrara) *Large angular scales from Planck in temperature and polarization* (20 min)
12.35 A. Taketa (Univ. of Tokyo) *Spectrometry of the Earth using neutrino oscillations* (20 min)
12.55 R. Jacobsson (CERN) *SHiP - A facility at the CERN SPS for hidden sector searches and tau neutrino physics* (25 min)

13.20 – 14.50 Lunch

14.50 – 18.25 **AFTERNOON SESSION (Conference Hall)**

Chairman: **Richard Jacobsson**

14.50 M. Yamashita (Kamioka Observatory, ICRR, Univ. of Tokyo) *Direct Dark Matter Search* (25 min)
15.15 D. Gorbunov (INR RAS) *Antibaryonic dark matter* (25 min)
15.40 J. Granot (Open Univ. of Israel) *GRB and their importance for Lorentz Invariance Violation* (25 min)
16.05 R. Desiante (INFN Torino & Udine University) *Fermi LAT observations of high-energy emission from solar flares* (25 min)
16.30 A. Morselli (INFN Roma Tor Vergata) *Astrophysics in the MeV range* (25 min)

16.55 – 17.20 Tea break

Chairman: **Norma Mankoc Borstnik**

17.20 K. Fujii (KEK) *ILC physics and recent status of ILC project in Japan* (25 min)
17.45 M. Yamaguchi (Tohoku Univ.) *Supersymmetric models with 125 GeV Higgs* (25 min)
18.10 S. K. Liu (Tsinghua Univ.) *The China Jinping underground laboratory and the CDEX dark matter program* (20 min)

23 August, Sunday

9.00 – 19.00 Bus excursion to Sergiev Posad

24 August, Monday

09.00 – 13.35 **MORNING SESSION** (Conference Hall)

Chairman: **Georg Raffelt**

09.00 F. Fraenkle (Karlsruhe Inst. of Technology) *KATRIN experiment* (25 min)
09.25 Yu. Kudenko (INR RAS) *Search for heavy neutrinos in fixed-target experiments* (20 min)
09.45 M. Batkiewicz (Inst. of Nuclear Physics PAN, Krakow) *Latest results from T2K* (25 min)
10.10 T. Ariga (Univ. of Bern) *New results from the OPERA experiment at the CNGS beam* (25 min)
10.35 A.Guglielmi (INFN-Padova) *ICARUS: present status and perspective for Sterile Neutrinos search at FermiLab* (20 min)
10.55 R. Wilson (Colorado State Univ.) *The Deep Underground Neutrino Experiment at the International Long-Baseline Neutrino Facility hosted by Fermilab* (25 min)

11.20 – 11.50 Tea break

SESSION 24.08. A (Neutrinos)

Chairman: **Yogendra Srivastava**

11.50 A. Minamino (Kyoto Univ.) *Neutrino cross section for the CP violation measurement* (25 min)
12.15 M. Krivoruchenko (ITEP) *Longitudinal component of the weak vector current of spin-1/2 nuclear isotopic doublets* (20 min)

12.35 M. Rebelo (Inst. Superior Tecnico, U. Lisboa) *On quasidegeneracy of Majorana neutrinos and the origin of large leptonic mixing* (25 min)
13.00 S. Gariazzo (INFN & Univ. of Turin) *Sterile neutrinos in cosmology* (25 min)

SESSION 24.08. B (BSM, Theory)

Chairman: **Anatoly Borisov**

11.50 M. Vysotsky (ITEP) *Double Higgs production in extensions of the Standard Model* (25 min)
12.15 X.-G. He (SJTU/NTU) *Fermion EDMs with Minimal Flavor Violation* (25 min)
12.40 E. Masso (Univ. Autonoma Barcelona) *Higgs effective Lagrangians and physics beyond the Standard Model* (25 min)
13.05 A. Nesterenko (JINR) *Hadronic contributions to electroweak observables in the framework of dispersive approach to QCD* (20 min)

SESSION 24.08. C (Physics at Colliders)

Chairman: **Mikhail Dubinin**

11.50 Ch. Royon (CERN) *TOTEM Physics Searches and Forward Physics at LHC Run 2* (25 min)
12.15 F. Nemes (CERN) *LHC Optics and Elastic Scattering Measured by TOTEM Experiment* (20 min)
12.35 V. Petrov (IHEP, Protvino) *Diffractive Processes at High Energies: Conceptual Survey* (25 min)
13.00 S. Poslavsky (IHEP, Protvino) *Precise study of XYZ charmonium states with PANDA* (20 min)
13.20 I. Spassovsky (Fusion Dept, ENEA-Frascati, Italy) *New rf drivers for future colliders* (15 min)

SESSION 24.08. D (Dark matter searches)

Chairman: **Pierre Salati**

11.50 N. Rossi (LNGS - INFN) *The DarkSide-50 Experiment: a liquid Ar target for dark matter particles* (20 min)
12.10 G. Kessler (Univ. of Zurich) *XENON dark matter project* (20 min)
12.30 V. Kozlov (KIT-IKP) *Direct dark matter search with EDELWEISS-III: recent results and future prospects* (20 min)
12.50 X. Defay (Excellence Universe Cluster, Munich) *Status of the CRESST Dark Matter Search* (20 min)
13.10 V. Solovov (Laboratory for Instrumentation and Particle Physics (LIP-Coimbra)) *The LUX direct dark matter search experiment* (20 min)

13.30 – 15.00 Lunch

15.00 – 18.35 **AFTERNOON SESSION** (Conference Hall)

SESSION 24.08. E (Neutrinos)

Chairman: **Margarida Rebelo**

15.00 S. Nasri (United Arab Emirates Univ.) *Neutrino mass models and application to cosmology* (20 min)
15.20 M. Gromov (SINP MSU) *Antineutrino physics in Borexino* (20 min)
15.40 G. Koroteev (INR RAS) *Status of "Troitsk nu-mass" experiment* (15 min)
15.55 M. Tenti (INFN-Bologna) *Search for sterile neutrino mixing in the $\nu_\mu \to \nu_\tau$ appearance channel with the OPERA detector* (15 min)
16.10 S. Zemskova (JINR) *Search for $\nu_\mu \to \nu_{\blacksquare}$ oscillations in the OPERA experiment* (15 min)

16.25 O. Suvorova (INR RAS) *Status of the Baikal-GVD project: first cluster Dubna* (20 min)
16.45 A. Chudaykin (INR RAS) *Cosmology based on f(R) gravity with light sterile neutrino* (15 min)

SESSION 24.08. F (Theory)

Chairman: **Xiao-Gang He**

15.00 V. Belokurov (MSU) *Singular substitutions in QFT functional integrals* (20 min)
15.20 R. Nevzorov (Univ. of Adelaide & ITEP) *Exotic Higgs decays in U(1) extensions of the MSSM* (15 min)
15.35 M. Tokarev (JINR) *Fractal Structure of Proton Spin* (20 min)
15.55 K. Stepanyantz (MSU) *Renormalization of the coupling constant and the HD regularization in supersymmetric theories* (20 min)
16.15 K. Zhukovsky (MSU) *High harmonic generation in the undulators for Free Electron Lasers* (20 min)

16.35-17.05 Tea break

Chairman: **Norma Mankoc Borstnik**

17.05 N. Gubina (MSU) *Fermionic vacuum in the electromagnetic field with consideration for possible Lorentz violation and anomalous magnetic moment* (15 min)
17.20 A. Kazantsev (MSU) *Relation between two-point Green functions of N=1 SQED: three-loop verification* (15 min)
17.35 G. Kravtsova (MSU) *Physical approach for a non-Hermitian fermion model with a maximal mass* (20 min)
17.55 A. Lobanov (MSU) *Particle quantum states with indefinite mass and neutrino oscillations* (15 min)
18.10 M. Fil'chenkov (Peoples' Friendship Univ. of Russia) *General properties of gravitationally bound systems at different levels of quantization* (15 min)
18.25 A. Kulikova (National Research Tomsk State Univ.) *Kinematics of vortex radiation from relativistic charged particles* (10 min)

SESSION 24.08. G (LHC Physics)

Chairman: **Ivan Spassovsky**

15.00 F. Scutti (Univ. of Bonn) *Beyond-the-Standard Model Higgs physics using the ATLAS experiment* (15 min)
15.15 Yu. Smirnov (MEPhI) *Search for exotic physics beyond the Standard Model with the ATLAS detector* (20 min)
15.35 K. Androsov (Univ. of Siena & INFN Pisa) *Higgs to tau-tau channels with CMS* (20 min)
15.55 R. Naranjo (DESY) *Top Physics on ATLAS* (20 min)
16.15 M. Filipuzzi (DESY/UHH) *Measurements of the properties of the Higgs boson using the ATLAS detector* (15 min)
16.30 N. Kondrashova (DESY) *QCD measurements of vector-boson production with the ATLAS detector* (15 min)

16.45 – 17.15 Tea break

Chairman: **Kajari Mazumdar**

17.15 D. Krasnopevtsev (MEPhI) *Electroweak measurements with the ATLAS detector* (15 min)
17.30 A. Mazurov (Univ. of Birmingham) *Heavy-quark hadron spectroscopy at LHCb* (15 min)
17.45 D. Savrina (SINP MSU & ITEP) *Rare decays in the LHCb* (15 min)
18.00 A. Artamonov (IHEP) *Production studies in the forward acceptance at the LHC* (15 min)

SESSION 24.08. H (Rare processes, QCD)

Chairman: **Alexander Nesterenko**

15.00 S. Kononov (Budker Inst. of Nuclear Physics) *Project of the Super Charm-Tau Factory at Novosibirsk* (20 min)
15.20 D. Epifanov (Univ. of Tokyo) *Study of Michel parameters in tau decays at Belle* (20 min)
15.40 H. Atmacan (Middle East Technical Univ., Ankara) *Recent Results on Charmless B Decays at Belle* (20 min)
16.00 K. Flood (CalTech) *Measurement of Angular Observables in the rare decay* $B \to K^{*}l^{+}l^{-}$ *at BaBar* (20 min)
16.20 M. Misheva (JINR) *First observation and study of* $K^{\pm} \to \pi^{\pm}\pi^{0}e^{+}e^{-}$ *decay by NA48/2 experiment* (15 min)

16.35-17.05 Tea break

17.05 H. Pirumov (DESY) *Precision QCD measurements at HERA* (20 min)
17.25 I. Korzhavina (SINP MSU) *Diffraction at HERA* (20 min)

25 August, Tuesday

09.00 – 13.30 **MORNING SESSION** (Conference Hall)

Chairman: **Phillip Guitierrez**

09.00 I. Bozovic Jelisavcic (Vinca Inst. of Nuclear Sciences) *CLIC Physics Overview* (25 min)
09.25 K. Lipka (DESY) *Proton structure in the LHC era* (25 min)
09.50 D. d'Enterria (CERN) *Physics programme of the FCC-ee* (25 min)
10.15 M. Koratzinos (CERN) *FCC-ee machine* (25 min)
10.40 H. Zhu (IHEP, China) *The CEPC project* (20 min)

11.00 – 11.30 Tea break

11.30 O. Adriani (Univ. of Florence & INFN Firenze) *LHCf physics results* (25 min)
11.55 D. Epifanov (Univ. of Tokyo) *Recent results from Belle* (25 min)
12.20 J. Milosevic (Vinca Inst. of Nucl. Sci.) *Anisotropy in pPb and PbPb collisions from CMS* (25 min)
12.45 I. Lakomov (CERN) *Overview of recent ALICE results* (25 min)
13.10 D. Khaneft (Helmholtz-Inst. Mainz) *Measurement of the two-photon exchange in the elastic* $e^{\pm}p$ *scattering at the OLYMPUS experiment at DESY* (20 min)

13.30 – 15.00 Lunch

15.00 – 19.00 **AFTERNOON SESSION** (Conference Hall)

Chairman: **Vladimir Petrov**

15.00 P. Gutierrez (Univ. of Oklahoma) *Review of recent Tevatron results* (25 min)
15.25 D. Troendle (Univ. of Hamburg) *Higgs boson at CMS* (25 min)
15.50 P. Pakhlov (ITEP) *Belle II overview* (25 min)
16.15 I. Polyakov (ITEP) *LHCb overview* (25 min)
16.40 I. Selyuzhenkov (EMMI/GSI) *The CBM experiment at FAIR* (20 min)

17.00 – 17.30 Tea break

SESSION 25.08. A

Chairman: **Giuseppe Dattoli**

17.30 E. Romero Adam (IFIC - CSIC) *Supersymmetry searches in ATLAS* (25 min)
17.55 A. Dewhurst (STFC Rutherford Appleton Laboratory) *Heavy flavour physics results from the ATLAS experiment* (25 min)
18.20 D. Di Filippo (Univ. of Napoli) *NA62 Status and Prospects* (20 min)
18.40 A. Rostovtsev (IITP RAS) *Hard QCD and heavy flavour production at HERA* (25 min)

SESSION 25.08. B

Chairman: **Georg Raffelt**

17.30 V. Semikoz (IZMIRAN) *The electroweak electron-nucleon interactions in a neutron star and magnetic fields in magnetars* (20 min)
17.50 M. Dvornikov (IZMIRAN) *The new galvano-rotational effect and its astrophysical applications* (10 min)
18.00 V. Novikov (ITEP) *A cosmological bound on e+ e− mass difference* (15 min)
18.15 A. Kuznetsov (Yaroslavl State Univ.) *Neutrino processes* $\bar{\nu}\nu \to e^- e^+$ *and* $\nu \to \nu e^- e^+$ *in a strong magnetic field* (10 min)
18.25 P. Sizin (Moscow State Mining Univ.) *Plasmon decay to a neutrino pair via nonstandard flavor violating interactions in a strongly magnetized medium* (15 min)
18.40 P. Kolmakov (MSU) *Magnetization in (2+1)-dimensional Gross-Neveu model with application to carbon nanotubes* (15 min)

SESSION 25.08. C

Chairman: **Igor Lakomov**

17.30 S. Morozov (INR RAS) *Recent results from NA61/SHINE* (20 min)
17.50 A. Sadovsky (INR RAS) *Study of azimuthal flow anisotropy of protons, deuterons and tritons produced in Au+Au collisions at* $\sqrt{s_{NN}} = 2.4 GeV$ (20 min)
18.10 A. Aparin (JINR) *On possibility to study hard cumulative processes in a fixed target experiment at STAR* (20 min)

SESSION 25.08. D

Chairman: **Alexander Zakharov**

17.30 E. Arbuzova (Dubna International Univ. & Novosibirsk St. Univ.) *Gravitational instability in non-stationary external fields* (20 min)
17.50 Yu. Grats (MSU) *Self-force on a charge outside a Schwarzschild-Tangherlini black hole (15 min)*
18.05 Y. Zhezher (INR RAS) *Spatial structure of WMAP haze* (10 min)

26 August, Wednesday

09.00 – 12.55 **MORNING SESSION** (Conference Hall)

Chairman: **Oscar Adriani**

09.00 P. Lukin (Budker Inst. of Nuclear Physics & Novosibirsk State Univ.) *Recent results from the CMD-3 detector at VEPP-2000 collider* (20 min)
09.20 A. Drutskoy (ITEP) *Status of the COMET experiment at JPARC* (20 min)
09.40 D. Kaneko (Univ. of Tokyo) *Today and tomorrow of the MEG experiment* (25 min)
10.05 V. Tishchenko (BNL) *Fermilab muon G-2 experiment* (25 min)

10.30 – 11.00 Tea break

Chairman: **Jumpei Shirai**

11.00 N. Mankoc Borstnik (Univ. of Ljubljana) *The spin-charge-family theory offers the explanation for the assumptions of the standard model, for the dark matter, for the matter-antimatter asymmetry, predicting the fourth family and several scalar fields* (25 min)
11.25 D. Mazin (Univ. of Tokyo) *Extragalactic background light studies with Fermi/LAT and Imaging Cherenkov Telescope data* (20 min)
11.45 I. Tkachev (INR RAS) *Telescope array* (25 min)
12.10 L. Simard (Univ. Paris-Sud) *The present and future of double beta decay experiments* (25 min)
12.35 M. Malek (Univ. of Sheffield) *Are Neutrinos responsible for our matter-dominated Universe?* (20 min)

12.55 – 14.30 Lunch

ROUND TABLE DISCUSSION
"Particle Physics in the Year of Light: from Maxwell's Equations to Physics beyond the Standard Model"
(Conference Hall)

Chairman: **Alexander Studenikin**

14.30 A. Starobinsky (Landau Inst.) *Inflationary models after PLANCK* (25 min)
14.55 O. Adriani (Univ. of Florence & INFN Firenze) *Overview of the PAMELA physics results* (20 min)
15.15 B. Bertucci (Univ. of Perugia) *The AMS-02 experiment on the ISS* (25 min)
15.40 A. Burtovoi (Univ. of Padova & INAF OAPd) *The Cherenkov Telescope Array observatory and the ASTRI mini-array precursor* (20 min)
16.00 G. Rubtsov (INR RAS) *Spectra of distant blazars and transparency of the Universe* (20 min)

16.20 – 16.40 Tea break

16.40 G. Dattoli (ENEA Frascati, Fusion Dept.) *Free Electron Laser program and challenges* (25 min)
17.05 F. Deppisch (Univ. College London) *Neutrinos and collider physics* (25 min)
17.30 V. Petrov (IHEP, Protvino) *Relativity and Light* (25 min)

ELEVENTH INTERNATIONAL MEETING ON PROBLEMS OF INTELLIGENTSIA:
"The Intelligentsia and Culture"
(Conference Hall)

Chairman: **Alexander Studenikin**

18.30 J. Bleimaier (Princeton) *The Intelligentsia & Religion* (25 min)

Closing of the 17th Lomonosov Conference on Elementary Particle Physics and the 11th International Meeting on Problems of Intelligentsia

SPECIAL SESSION (40 ^{0})

POSTER SESSION

N. Agafonova (INR RAS) *Measurement of the muon-induced neutron yield in Fe and Pb using Large Volume Detector at LNGS*
V. Ashikhmin (INR RAS) *Joint analysis of experimental data on the search for neutrinos from stellar collapses using the LVD and BUST detectors*
E. Bubelev (JINR) *The aim, urgent necessity and peculiarities for creation of the Artificial Intelligence in High energy physics of XXI century*
A. Chernitskii (St.-Petersbutg State Univ. of Economics) *Soliton-particle solutions of nonlinear field models*
V. Koryukin (Mari State Univ.), **A. Koryukin** (Kazan Federal Univ.) *On construction methods of the fundamental interactions theory*
A. Mosichkin (Yaroslavl State Univ.) *The radiative decay of a neutrino with a magnetic moment in a strong magnetic field with taking account of the positronium influence on the photon dispersion*
A. Nozik (INR RAS) *The DataForge software framework for scientific data processing automation*
A. Yudin (ITEP) *Generation of Neutrons by Cosmic-Ray Muons in Different Materials*

List of participants of the 17th Lomonosov Conference on Elementary Particle Physics and the 10th International Meeting on Problems of Intelligentsia

Adriani Oscar	Univ. of Florence & INFN Firenze	adriani@fi.infn.it
Agafonova Natalia	INR RAS	natagafonova@gmail.com
Akhmedov Evgeny	Max Planck Inst., Heidelberg	akhmedov@mpi-hd.mpg.de
Alexander Vasiliev	Kurchatov Inst.	Alexander.Vasiliev@ihep.ru
Androsov Konstantin	Univ. of Siena & INFN Pisa	konstantin.androsov@cern.ch
Aparin Alexey	JINR	aparin@jinr.ru
Arbuzova Elena	Dubna International Univ. & Novosibirsk St. Univ.	al.arbuzova@gmail.com
Ariga Tomoko	Univ. of Bern	tomoko.ariga@lhep.unibe.ch
Artamonov Alexander	Protvino-IHEP	alexander.artamonov@ihep.ru
Ashikhmin Vsevolod	INR RAS	vole4ka86@mail.ru
Atmacan Hulya	Middle East Technical Univ., Ankara	hatmacan@metu.edu.tr
Barabash Alexander	ITEP	barabash@itep.ru
Barish Barry	CalTech	barish@ligo.caltech.edu
Batkiewicz Marcela	Institute of Nuclear Physics PAN, Krakow (IFJ PAN, Krakow)	Marcela.Batkiewicz@ifj.edu.pl
Belli Pierluigi	INFN Roma Tor Vergata	pierluigi.belli@roma2.infn.it
Belokurov Vladimir	MSU	vvbelokurov@yandex.ru
Belov Vladimir	ITEP	belov@itep.ru
Berezinsky Veniamin	GSSI & LNGS, Italy	venya.berezinsky@lngs.infn.it
Bertucci Bruna	Univ. of Perugia	bertucci.bruna@gmail.com
Besson Dave	Univ. of Kansas	zedlam@ku.edu
Bezerra Thiago	Tohoku Univ.	thiago@awa.tohoku.ac.jp
Bleimaier John	Princeton	bleimaier@aol.com
Bordovitsyn Vladimir	National Research Tomsk State Univ.	vabord@sibmail.com
Bozovic Jelisavcic Ivanka	Vinca Institute of Nuclear Sciences	ibozovic@vinca.rs
Burigana Carlo	INAF-IASF Bologna, Italy	burigana@iasfbo.inaf.it
Burtovoi Aleksandr	Univ. of Padova/INAF OAPd	aleksandr.burtovoi@studenti.unipd.it
Chernitskii Alexander	St.-Petersbutg State Univ. of Econ.	AAChernitskii@mail.ru
Chudaikin Anton	INR RAS	chudy@ms2.inr.ac.ru
Dattoli Giuseppe	ENEA Frascati Fusion Department	giuseppe.dattoli@enea.it
Davini Stefano	Gran Sasso Science Institute	stefano.davini@gssi.infn.it
Defay Xavier	Excellence Universe Cluster, Munich	xavier.defay@tum.de
d'Enterria David	CERN	dde@cern.ch
Deppisch Frank	Univ. College London	f.deppisch@ucl.ac.uk
Desiante Rachele	INFN Torino & Udine Univ.	racheledesiante@gmail.com
Dewhurst Alastair	STFC Rutherford Appleton Lab.	alastair.dewhurst@cern.ch
Di Filippo Domenico	Univ. of Napoli	domenico.di.filippo@cern.ch
Dokuchaev Vyacheslav	INR RAS	dokuchaev@inr.ac.ru
Drutskoy Alexey	MEPhI, ITEP	Drutskoy@itep.ru
Druzhkin Dmitry	CERN	Dmitry.Druzhkin@cern.ch
Dubinin Michail	SINP MSU	dubinin@theory.sinp.msu.ru
Dvornikov Maxim	IZMIRAN	maxdvo@izmiran.ru
Epifanov Denis	Univ. of Tokyo	epifanov@bpost.kek.jp
Ereditato Antonio	Univ. of Bern	antonio.ereditato@cern.ch
Fil'chenkov Michael	Peoples' Friendship Univ. of Russia	fmichael@mail.ru
Filipuzzi Marco	DESY/UHH	marco.filipuzzi@cern.ch

Flood Kevin	CalTech	kflood@slac.stanford.edu
Fraenkle Florian	Karlsruhe Institute of Technology	florian.fraenkle@kit.edu
Frank Martin	Univ. of Virginia	mjf8u@virginia.edu
Fujii Keisuke	KEK	keisuke.fujii@kek.jp
Gariazzo Stefano	INFN & Univ. of Turin	gariazzo@to.infn.it
Garzelli Maria Vittoria	Univ. of Hamburg	garzelli@to.infn.it
Gehrels Neil	NASA-GSFC	gehrels@milkyway.gsfc.nasa.gov
Giunti Carlo	INFN & Univ. of Turin	carlo.giunti@to.infn.it
Glaysher Paul	Univ. of Edinburgh	paul.glaysher@cern.ch
Gorbunov Dmitry	INR RAS	gorby@ms2.inr.ac.ru
Granot Jonathan	Open Univ. of Israel	granot.j@gmail.com
Grats Yuri	MSU	grats@phys.msu.ru
Gromov Maxim	SINP-MSU	gromov@physics.msu.ru
Gubina Nadezhda	MSU	gubina_nadya@mail.ru
Guglielmi Alberto	INFN-Padova	alberto.guglielmi@pd.infn.it
Gutierrez Phillip	Univ. of Oklahoma	pgutierrez@ou.edu
He Xiao-Gang	SJTU/NTU	hexg@sjtu.edu.cn
Helbing Klaus	Univ. of Wuppertal	helbing@uni-wuppertal.de
Ishkhanov Boris	MSU	bsi@depni.sinp.msu.ru
Jacobsson Richard	CERN	richard.jacobsson@cern.ch
Kaneko Daisuke	Tokyo Univ.	kaneko@icepp.s.u-tokyo.ac.jp
Kappes Alexander	ECAP/FAU Erlangen-Nuremberg	kappes@physik.uni-erlangen.de
Katsanevas Stavros	Univeristy Paris Diderot	katsanevas@apc.univ-paris7.fr
Kazantsev Aleksandr	MSU	kazancev@physics.msu.ru
Kessler Gaudenz	Univ. of Zurich	gkessler@physik.uzh.ch
Khaneft Dmitry	Helmholtz-Institut Mainz	khaneftd@kph.uni-mainz.de
Kim Soo-Bong	Seoul National Univ.	sbk@snu.ac.kr
Kolmakov Pavel	MSU	pavel.b.kolmakov@yandex.ru
Kolomensky Yury	UC Berkeley	yury@physics.berkeley.edu
Kondrashova Nataliia	DESY/Universität Hamburg	nataliia.kondrashova@desy.de
Kononov Sergey	Budker Institute of Nuclear Physics	s.a.kononov@inp.nsk.su
Koratzinos Michael	CERN	michael.koratzinos@cern.ch
Koroteev Grigoriy	INR RAS	koroteev@phystech.edu
Koryukin Valery	Mari State Univ.	vmkoryukin@gmail.com
Korzhavina Irina	SINP MSU	irina@mail.desy.de
Kozlov Valentin	KIT-IKP	valentin.kozlov@kit.edu
Krasnopevtsev Dimitrii	MEPhI	Dimitriy.Krasnopevtsev@cern.ch
Kravtsova Galina	MSU	gakr@chtc.ru
Krivoruchenko Mikhail	ITEP	mikhail.krivoruchenko@itep.ru
Kudenko Yury	INR RAS	kudenko@inr.ru
Kulikova Anastasiya	Tomsk State Univ.	anasta-kulikova@yandex.ru
Kuznetsov Alexander	Yaroslavl State Univ.	avkuzn@univ.uniyar.ac.ru
Lakomov Igor	CERN	Igor.Lakomov@cern.ch
Li Xiaonan	IHEP, China	lixn@ihep.ac.cn
Li Yufeng	IHEP, China	liyufeng@ihep.ac.cn
Lipka Katerina	DESY	katerina.lipka@desy.de
Liu Shukui	Tsinghua Univ.	abc6081@163.com
Liu Wei	IHEP, China	
Lobanov Andrey	MSU	lobanov@phys.msu.ru
Lukin Peter	BINP&NSU,Novosibirsk,Russia	P.A.Lukin@inp.nsk.su
Malek Matthew	Univ. of Sheffield	m.malek@sheffield.ac.uk
Mandolesi Nazzareno	Univ. of Ferrara	mandolesi@iasfbo.inaf.it
Mankoc Norma	Univ. of Ljubljana	norma.mankoc@fmf.uni-lj.si
Martynov Denis	CalTech	martynov_d@ligo.caltech.edu

Masso Eduard	Univ. Autonoma Barcelona	masso@ifae.es
Mazin Daniel	Univ. of Tokyo	mazin@icrr.u-tokyo.ac.jp
Mazumdar Kajari	Tata Inst. of Fund. Res.	kajari.mazumdar@gmail.com
Mazurov Alexander	Univ. of Birmingham	alexander.mazurov@cern.ch
Milosevic Jovan	Vinca Inst. of Nucl. Sci.	jovan.milosevic@fys.uio.no
Minamino Akihiro	Kyoto Univ.	minamino@scphys.kyoto-u.ac.jp
Misheva Milena	JINR	milena.hristova.misheva@cern.ch
Morozov Sergey	INR RAS	sergey.morozov@cern.ch
Morselli Aldo	INFN Roma Tor Vergata	aldo.morselli@roma2.infn.it
Mosichkin Anatoly	Yaroslavl State Univ.	anatoly_mosichkin@mail.ru
Nakahata Masayuki	Univ. of Tokyo	nakahata@suketto.icrr.u-tokyo.ac.jp
Naranjo Roger	DESY	roger.naranjo@desy.de
Nasri Salah	United Arab Emirates Univ.	snasri@uaeu.ac.ae
Natoli Paolo	Univ. of Ferrara	ntlpla@unife.it
Nemes Frigyes	CERN	frigyes.janos.nemes@cern.ch
Nesterenko Alexander	JINR, Dubna, Russian Federation	nesterav@gmail.com
Nevzorov Roman	Univ. of Adelaide & ITEP	nevzorov@itep.ru
Ng John	Triumf, Canada	misery@triumf.ca
Nigmatkulov Grigory	MEPhI	nigmatkulov@gmail.com
Novikov Viktor	ITEP	novikov@itep.ru
Nozik Alexander	INR RAS	nozik@inr.ru
Oberauer Lothar	Tech.Univ. of Munich	lothar.oberauer@ph.tum.de
Pakhlov Pavel	ITEP	pakhlov@itep.ru
Pantuev Vladislav	INR RAS	pantuev@inr.ru
Petrov Vladimir	IHEP	Vladimir.Petrov@ihep.ru
Pirumov Hayk	DESY	hayk.pirumov@desy.de
Polyakov Ivan	ITEP	polyakov@itep.ru
Popov Sergei	Sternberg Astronomical Institute, MSU	sergepolar@gmail.com
Poslavsky Stanislav	IHEP-Protvino	stvlpos@mail.ru
Prodi Giovanni	Univ. of Trento & INFN	prodi@science.unitn.it
Raffelt Georg	Max Planck Inst. for Physics, Munich	raffelt@mppmu.mpg.de
Rebelo Margarida N.	Instituto Superior Tecnico, U. Lisboa	rebelo@ist.utl.pt
Rogachevsky Oleg	JINR	rogachevsky@jinr.ru
Romero Adam Elena	IFIC - CSIC	elena.romero.adam@cern.ch
Rossi Nicola	LNGS - INFN	nicola.rossi@lngs.infn.it
Rostovtsev Andrei	IITP RAS	rostov@itep.ru
Royon Christophe	CERN	royon@mail.cern.ch
Rubtsov Grigory	INR RAS	grisha@ms2.inr.ac.ru
Sadovsky Alexander	INR RAS	sadovsky@inr.ru
Saito Naohito	KEK / J-PARC	naohito.saito@kek.jp
Salati Pierre	LAPTH, CNRS	salati@lapp.in2p3.fr
Sasao Noboru	Okayama Univ.	sasao@okayama-u.ac.jp
Savchenko Andrey	MEPhI	andrew.a.savchenko@gmail.com
Savrina Daria	SINP MSU & ITEP	Daria.Savrina@cern.ch
Scutti Federico	Univ. of Bonn	Federico.Scutti@cern.ch
Selyuzhenkov Ilya	EMMI/GSI	ilya.selyuzhenkov@gmail.com
Semikoz Victor	IZMIRAN	semikoz@yandex.ru
Shirai Jumpei	Tohoku Univ.	shirai@awa.tohoku.ac.jp
Simard Laurent	Univ. Paris-Sud	simard@lal.in2p3.fr
Sizin Pavel	Moscow State Mining Univ.	
Borisov Anatoly	MSU	borisov@phys.msu.ru
Slavnov Andrei	Steklov Math.Inst. & MSU	slavnov@mi.ras.ru
Smirnov Yury	MEPhI	Yury.Smirnov@cern.ch

Smy Michael	Univ. of California, Irvine	msmy@uci.edu
Solovov Vladimir	Laboratory for Instrumentation and Particle Physics	solovov@coimbra.lip.pt
Sparvoli Roberta	Univ. of Rome Tor Vergata	roberta.sparvoli@roma2.infn.it
Spassovsky Ivan	Fusion Department, ENEA-Frascati	ivan.spassovsky@enea.it
Srivastava Yogendra	Perugia	srivastava@pg.infn.it
Starobinsky Alexei	Landau Inst.	alstar@landau.ac.ru
Stepanyantz Konstantin	MSU	stepan@m9com.ru
Stratta Giulia	Universita degli Studi di Urbino	giulia.stratta@uniurb.it
Suomijarvi Tiina	Univ. Paris-Sud	tiina.suomijarvi@u-psud.fr
Suvorova Olga	Institute for Nuclear Research	suvorova@cpc.inr.ac.ru
Taketa Akimichi	Univ. of Tokyo	akimichi.taketa@gmail.com
Tenti Matteo	INFN- Bologna	Matteo.Tenti@bo.infn.it
Tishchenko Vladimir	Brookhaven National Laboratory	vladimir.tishenko@gmail.com
Tokarev Mikhail	JINR	tokarev@jinr.ru
Troendle Daniel	Univ. of Hamburg	troendle@cern.ch
Villante Francesco	Univ. of L'Aquila & INFN-LNGS	villante@lngs.infn.it
Visotsky Mikhail	ITEP	vysotsky@itep.ru
Wilson Robert	Colorado State Univ.	wilson@colostate.edu
Yamaguchi Masahiro	Tohoku Univ.	yama@tuhep.phys.tohoku.ac.jp
Yamashita Masaki	Kamioka Observatory, ICRR, Univ. of Tokyo	yamashita@icrr.u-tokyo.ac.jp
Yoshimura Motohiko	Okayama Univ.	yoshim@okayama-u.ac.jp
Yudin Andrey	ITEP	yudin@itep.ru
Zakharov Alexander	ITEP	alex_f_Zakharov5@mail.ru
Zemskova Svetlana	JINR	sgzemskova@gmail.com